THE BOOK OF BAMBOO

"The most useful service
we can render a culture
is to add a new plant
to its agriculture."

—THOMAS JEFFERSON

DAVID FARRELLY

Sierra Club Books · San Francisco

The Sierra Club, founded in 1892 by John Muir, has devoted itself to the study and protection of the earth's scenic and ecological resources—mountains, wetlands, woodlands, wild shores and rivers, deserts and plains. The publishing program of the Sierra Club offers books to the public as a nonprofit educational service in the hope that they may enlarge the public's understanding of the Club's basic concerns. The point of view expressed in each book, however, does not necessarily represent that of the Club. The Sierra Club has some fifty chapters coast to coast, in Canada, Hawaii, and Alaska. For information about how you may participate in its programs to preserve wilderness and the quality of life, please address inquiries to Sierra Club, 530 Bush Street, San Francisco, CA 94108.

Library of Congress Cataloging in Publication Data

Farrelly, David, 1938–
The book of bamboo.

Bibliography: p. 321.
Includes index.
1. Bamboo. 2. Bamboo—History. 3. Bamboo work.
I. Title.
SB317.B2F37 1984 584'.93 84-5366
ISBN 0-87156-824-1
ISBN 0-87156-825-X (pbk.)

Jacket and text design by Catherine Flanders
Printed in the United States of America
10 9 8 7 6 5 4 3 2 1

Barefoot Bamboo

To the immense millions of the most penniless,
who would most benefit from bamboo abundance,
the people, the forever forgotten ones who
can't afford a book on bamboo or anything else,
whose bare feet never wander stores in cities
where such information is sold,
who don't have the data and dollars
to search for the species most suited
to their weather and needs,
nor land to plant nor time to wait
until a grove becomes established,
who lack traditions of bamboo know-how
to guide in care, harvest, and use;
who often have never seen and can't afford a tool
more costly or specialized than a machete,
whose daily battle for barest subsistence leaves
little curiosity or energy for the unknown;
whose long historical experience has taught
a salutary distrust of anybody better dressed
or fatter fed coming toward them
with some bright message to better their lot . . .

to those who will never see or touch or hear of it,
who couldn't read it, in any case,
this book is affectionately dedicated.

CONTENTS

1. DAWN IN THE WEST/2

Bamboo morning in the evening land/Designing designers/Culture doctor, climate doctor/Plant therapy: the leaf/people ratio/*Beginner's bamboo: especially for generalists*/Bamboophilia: lovers and vendors/The king of China and the clever child.

2. ONE THOUSAND THINGS/12

Bamboo appropriate technology/Planetary norm: the modern village/Appropriate bamboo catalog: Acupuncture needles, aphrodisiacs, arrows, baskets, beehives, beer, bikes, boats, bridges, cables, cages, crutches, dams, diesel fuel, flutes, gutters, hats, irrigation, junks, kites, ladders, looms, medicines, musical instruments, packaging, paper, rafts, rakes, rings, ropes, scaffolds, string, teahouses, tents, towers, toys, umbrellas, violins, water pistols, waxes, whistles, windmills, wine, xylophones, yurts, and zithers.

3. NUDE MODEL TO THE EAST/72

The artist's ally, the emperor's ornament, the people's friend/Galloping through rules/*The brush dances, the ink sings*/*The yellow bell*/Flutes came before fingers/*Bamboodhism.* Something about nothing/The zero zone/*Haiku housing.* Tea's easy way/The ceremony of friendship/The tao of housing.

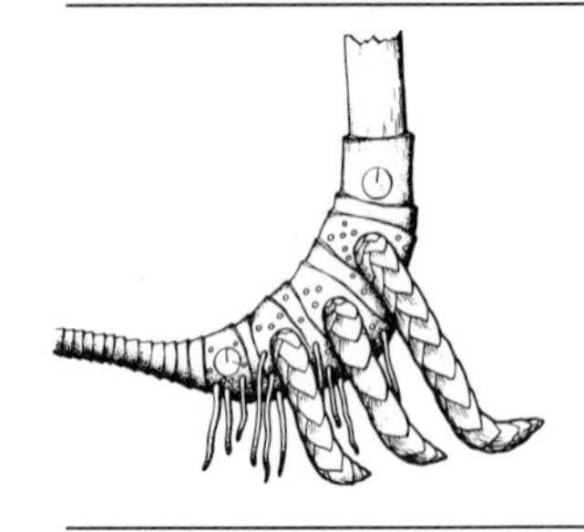

4. SHELTER: IN THE BEGINNING WAS BAMBOO/94

Third skin. Architectural activists/*Lamination*/Plybamboo/*Reinforced concrete*/*Quaking without breaking*/Disaster designs/Bamboo polyhedra/*The psychology of shelter*/Bamboo Walden/Escaping the teahouse/Uncle Sam's cabin/Thatch therapy/The cult of life.

5. POPULAR ARCHITECTURE/112

Roofing the race. The comforts of inconvenience/*Architecture with heart*/*Guadua angustifolia:* the world's most durable, the west's largest bamboo/Flexible shelter/The Death of Bamboo/Anti-extinction/*The man who loved trees*/Minimal eco-ethic.

6.

THE PLANT: BAMBOO BEHAVIORS/136
Tropical (sympodial) and temperate (monopodial) species/Rhizome/Grove form/Roots/Culm/Branch complement/Sheaths/Leaves/Flowering/Seedling/Distribution/Environment/Bamboo Adam.

7.

SPECIES: TEMPERATE AND TROPICAL/156
Crackjaw names/**Temperate Bamboos**/Arundinaria/Chimonobambusa/Phyllostachys/Pseudosasa/Sasa/Shibataea/**Tropical Bamboos**/Bambusa/Cephalostachyum/Chusquea/Dendrocalamus/Gigantochloa/Guadua/Melocalamus/Melocanna/Nastus/Ochlandra/Oxytenanthera/Pseudostachyum/Schizostachyum/Sinarundinaria/Teinostachyum/Thamnocalamus/Thyrsostachys/Pseudo Bamboos/Taxonomy/Collecting specimens for identification/Primary genera of bamboo.

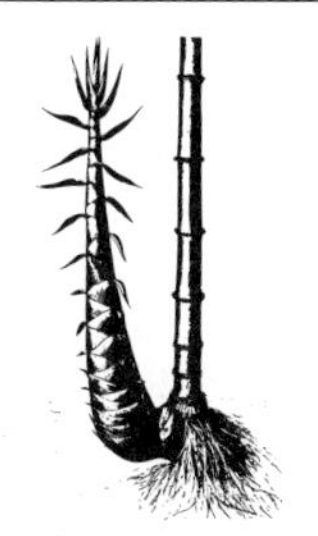

8.

CULTIVATION, HARVEST, CURING/206
Dixie bamboo/*Basic garden guidelines*/Transport/*Expanding grove*/Bonsai/Hedges/*Bamboo farm*/Harvest methods/Bugs love bamboo: problems and solutions/*Growing a graceful giant:* Moso cultivation in Japan/*Yields.*

9.

USES OF BAMBOO/234
Crafts/*Creating creation*: Bamboo community workshops/Tools/Techniques/Cutting, splitting, bending/Joinery/Knots and lashing/*Kids love bamboo*/The effort/effect ratio/Broke is beautiful/Miniature magic/Kid-buildable playgrounds/Flute-making is easy/*Constructions*/Foundations/Frames/Floors/Walls/Roof/Gutters/Thatch/Fences/Pipes/Hen house/Bamboo camp/*Paper*/Paper history/Handmade in China/Modern methods/Paper makes schools/Papermaking in your kitchen sink/*Food*/Eat your lawn/Cookbooklette/Hay and forage/*Ecoprotection*/Revegetation: Heal thy planet/Once upon a tree/Hamburger hunters/Monster makers/Agroforestry/Erosion control/The cloudy bough/Green China/*Epilogue:* The great disorder/The crop of permaculture is complete people.

10.

SOURCES AND DIRECTIONS/294
Botanical activists/Johnny's appleseed/Bamboo messenger: McClure/Bamboo classics/*Current Asian research*/Groves and centers. Whoo's whoo/*Message design/Messenger design*/Tactical attitudes/The Eden of effort and the Eden of ease/*Reknowable energy*/Cultural meristem/Kindereden: optimal orphanage/Come without a book or banner.

Glossary/318
Bibliography. Books by subject/321
Index/335

ACKNOWLEDGMENTS

A glance at the ample footnotes, bibliography and masses of quoted matter in this book will indicate to the most casual reader its collective nature. A summing up of the relevant past scholarship on bamboo in a work intended for a broad public must inevitably rely on the many long years of the work of others, often under very difficult circumstances, usually with no public recognition beyond a small circle of professional friends. Impossible to acknowledge all that sufficiently.

I have been more of an editor than an author, a crow strutting in feathers of peacocks. This exercise has been fed, sheltered, and endured by friends longer than the best of them often felt possible, and unfailingly encouraged by the living bamboo lovers whose paths crossed its composition—among them, Thomas Soderstrom and Cleofe Calderon at the Smithsonian Institution in Washington, D.C., Oscar Hidalgo and Dickens Castro in Bogotá, Colombia, and Ruth McClure, the widow of Floyd A. McClure, whose life work both inspired this book and furnished much of its matter.

The staff at Sierra Club Books in San Francisco has patiently drawn order from the confusion of thousands of scraps of manuscript, illustrations, captions, boxes, tables, and whatnot, after waiting for them some three years beyond our contracted deadline. Gratitude to Eileen Max, David Spinner, L. Jay Stewart, Cathy Flanders, and Danny Moses. Among hundreds who helped, special thanks to Stuart Chapman, who drew or retouched or redrew from poor xeroxes most of the many illustrations; to Kevin Quigley, who shared this work in many weird ways for five years and provided photographs; and to Julia Foster for drawings, as well as prolonged active support of this obsession, toiling and traveling in the shadow of bamboo through some dozen countries, while giving birth to our children, Sasa and Jesse, en route.

PREFACE

A PLANT PROFILE

Bamboo grows more rapidly than any other plant on the planet. It has been clocked surging skyward as fast as 47.6 inches in a 24-hour period. Astonishing vitality, great versatility, lightweight strength, ease in working with simple tools, striking beauty in both its natural and finished state—these qualities have given bamboo a longer and more varied role in human cultural evolution than any other plant. It has been most widely used for shelter, food, paper and countless articles of daily life like chopsticks, mats, and baskets. But in addition to its more common tasks, bamboo has run a hundred hidden errands in human history, huge and minute, crude and fine, dramatic and humdrum, from skyscraper scaffolding in Tokyo and phonograph needles in America to slide rules, skins of airplanes, and diesel fuels. Medicines for asthma, hair and skin salves, eyewashes, potions for lovers and poisons for rivals have all been extracted from different portions of the plant; and even the ashes of bamboo are used—to polish jewels and manufacture electrical batteries. The range of its uses is perhaps unequaled by any other resource: it has been employed in bikes, dirigibles, windmills, scales accurate enough to weigh crickets, and retaining walls strong enough to resist flood and tide. Split and twisted into 21-inch diameter cables, bamboo built bridges up to 750 feet long in China; and bamboo fibers provided Edison, after fruitless experiment with other materials, with a successful filament in the first light bulb, a bridge of light.

Ancient resident of earth, among the most primitive of grasses, here before people by some 100 to 200 million years, bamboo is also the first born, the Adam of the Atomic Age, through its survival of Hiroshima closer to ground zero than any other living thing. Its root-like "rhizome" and "culm"—as the stalk is called—compose perhaps the hardiest natural structure to evolve in millions of years of restless experiment in cellular life on earth, and the most efficient laboratory that has yet appeared for braiding sunlight, water, and soil into forms which have for centuries proved super-useful for human needs.

Bamboo is cosmopolitan in habitat, a plant with a thousand faces, a miracle mutant of adaptive forms: It tolerates extremes of drought and drowning from 30 to 250 inches of annual rainfall. It thrives in some twelve to fifteen hundred species that are native to every continent but Europe and the poles, from sea level to 12,000 feet in elevation, ranging in form from scrubby bushes used principally for cattle fodder to towering culms 120 feet high that provide sturdy beams with a diameter of nearly a foot and walls an inch or more thick. Abnormal culms of 150–180 feet are also recorded.

Although there are species that flower annually, bamboo normally reproduces asexually, without flowering and without seeds. New sprouts shoot up from the rhizomes, the underground growth that is squat and creeping in clumping tropical "sympodial" species, but in temperate "monopodial" free-standing bamboo resembles a subterranean spaghetti of long, tangled runners that are jointed like the culms above. The ordinary method of bamboo propagation is transplanting rhizomes, not planting seeds. However, at intervals depending on the species, sometimes as seldom as once a century, in obedience to some clock ticking in their cells all the bamboos of a given kind, even groves from rhizomes transported generations before to distant countries, flower together around the world—and die. Seeds of bamboo are produced at this time only, but in sufficient abundance to have provided famine relief in India more than once in the history of that hungry country.

The importance of bamboo at the earliest, most formative periods of human culture is suggested by the fact that it was deified in some primitive tribes, such as the Piyoma of Formosa. In others, man and woman were said to have issued from different joints of a single culm. The earliest mention of bamboo in planet letters is in perhaps the oldest book alive and breathing, the I Ching, the Book of Changes. *Disguised as an oracle, it contained the political mysticism of ancient China, a dense and handy guide to Change Design. Noting that change was the only constant, the sages of old playfully created in the book an intriguing game to help apprentice sages roll strategically with transformation rather than fruitlessly resisting it or trusting to blind luck.*

For one of China's most admired virtues, limitation, the sages chose the most distinctive feature of her favorite plant—the joints of bamboo. Observing that bamboo attained its astonishing and useful altitude by throwing a horizontal truss across at distances carefully determined by stress levels in the ascending culm, they used the node as an image of the adaptive aim that tacks with diplomacy through the crosswinds of social existence; of the thrift and measure that gives duration to all economies from the poorest peasant to the Sun of Heaven and his state; and of the central modesty that best befits the spirit of a tiny creature briefly wandering mountains and rivers without end. What constitutes proper "limits of growth" is a theme of present-day economic debate. It is interesting that bamboo, which has extended its life limits further than most organisms in duration, distribution, and rambunctious vitality, is first cited in world literature as a symbol of a sane restraint.

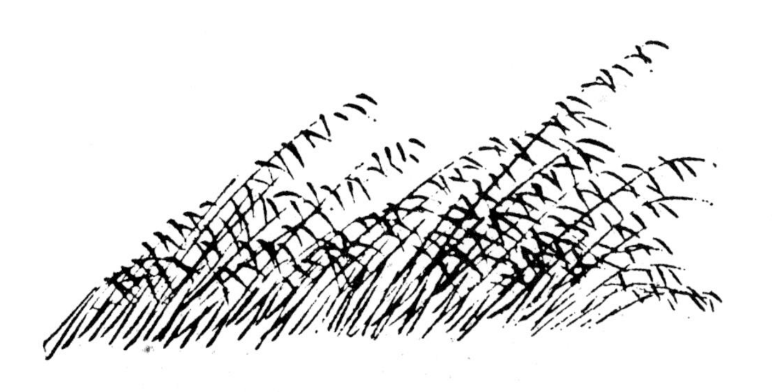

THE BOOK OF BAMBOO

1. DAWN IN THE WEST

Begin with your conclusion.

—MONTAIGNE

HERE COMES EVERYBODY

Bamboo is a plant of ancient and increasing importance for humanity. Known as "the wood of the poor" in India, "the friend of the people" in China, "the brother" in Vietnam, bamboo is much less known in the West than it soon will be. Increased population increases importance of the Great Yielders, such as bamboo, leucaena, eucalyptus . . . rabbits. Superproductive plants and animals will be increasingly investigated as our own species becomes ever more productive. And bamboo, under optimum conditions, can provide two to six times as much cellulose per acre as pine. Forests in general increase 2 to 5 percent yearly in total bulk or "biomass"; groves of bamboo increase 10 to 30 percent.

Through a tube, darkly: a brief fit of prophecy.

World shelter and paper shortages will inspire looking closer at bamboo's traditional oriental roles and modern variants. Lamination, plybamboos, reinforced bamboo-cement, and veneers will increase use of bamboo in housing as much as modern papermaking machinery in India has increased volume of bamboo used for pulp. Some 35 to 40 paper factories in India used 2.2 million tons of bamboo in 1980 to make 600,000 tons of paper, 70 percent of their total paper production. Developing countries generally will become more paper conscious, alert to paper's economic, political, and cultural implications. Schools will begin to study and make paper, the unexamined basis of modern education, curiously exiled from the curriculum it makes possible.

Bamboo ball, Burma. This form is common in many oriental countries.

Learning revolution.

The quite recent idea of universal education, now shared worldwide, is occurring at a time of unprecedented global population growth. Raw materials to make schools and to teach children to make and build and read and write in them are increasingly expensive. Cost-per-student is a critical aspect of all education design. Bamboo's abundant growth, lightweight workability, and paper history

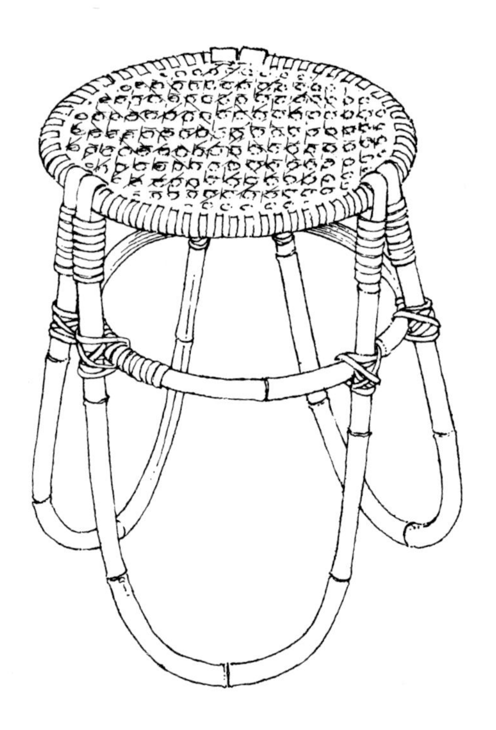

I never saw so fine a chair
as one which mostly
wasn't there,
fashioned of the vacant air
and elegantly bent bamboo.

If we did everything we do
in a style so spare and
sightly,
building could keep pace
with birth;
our lives, less littered with
our mess,
would sit more lightly
on a wilder earth
and trash it less.

suggest a possible future in schools of a dimension as yet undreamt by Western educators.

Planetary norm.
"Everybody"—this new idea in education—is the basic idea of revolution also. "Culture" for the few, purchased with ignorance of the many, is no longer a suggestible model for people who have heard other historical alternatives. Bamboo fits the everybody idea. There could never be enough silver flutes to give one to everybody in the world. There could, easily, be enough bamboo for all 50 billion fingers on the earth to make and play their own. Bamboo is a possible planetary norm in the sense that its vitality, wed to human will, could easily produce enough for everyone, for any of a thousand things.

Bamboo renaissance design.
The industrialization of oriental cultures has introduced many changes in bamboo use there, yet the total volume of bamboo consumption has not diminished but increased. Many more people are turning to its ample capacity to serve human need, and they have devised new, mechanized extensions of ancient uses. For a number of reasons, bamboo diffusion in the West has not kept pace with the expectations of agronomists and plant explorers who have witnessed its wonders in the East—but increasing human populations make it apparent that this prolific old ally of human purposes will reappear among us.

It will be more renaissance than revolution when bamboo returns because the plant was once so intimate a part of some cultures in regions of the Americas that it was mythologized as the womb of the race. Some species were so useful that wide employment without replanting led to the virtual extinction of the best bamboos over vast areas of their original terrains. A bamboo culture in Central and South America would not be an alien introduction but a resurrection of native tradition, fortified by species and know-how from the East and adjusted to the realities of present times.

To any student of change design, a few main rocks in the road of Western bamboo development are readily apparent. Too few bamboo messengers are versed deeply enough in the cultivation and use to pass on the word to others. No bamboo farm and study center exists to train those with an interest in the plant. A number of important species of oriental bamboos have been planted in this hemisphere by the USDA and a hardy handful of botanists in government experimental stations and private groves. But there have been no skilled gardeners of history, "comprehensive, anticipatory culture design scien-

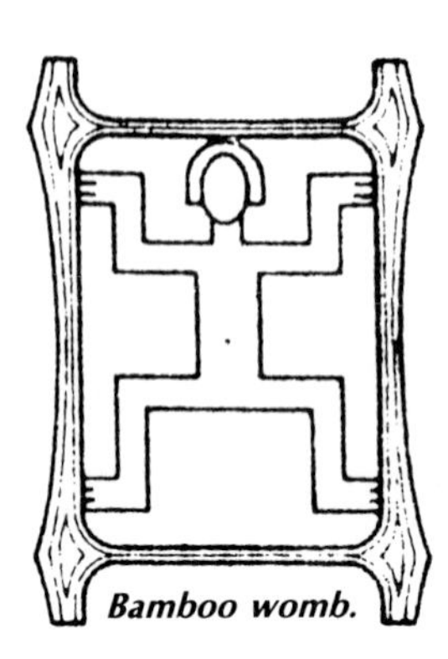
Bamboo womb.

tists," coaxing bamboo roots deep into the daily consciousness of Western peoples. So when Sierra Club Books offered to publish a book on the plant, some friends of bamboo asked themselves: What voice would whisper closest in a Western ear? What was the method, where was a wind to shove back the clouds that smothered up the glory of this bamboo morning in the evening land?

BAMBOO MORNING IN THE EVENING LAND

A century ago, in 1882, golden bamboo *(Phyllostachys aurea),* a hardy Chinese species now known on many an American lawn, was first successfully introduced into the continental United States in Alabama. That same year, Thomas Edison was beginning a light-bulb factory—the world's first—using filaments of bamboo. In 1827, black bamboo *(Phyllostachys nigra),* the first hardy oriental bamboo to find its way west, stepped off a lengthy ocean voyage from Japan into British soil. Two centuries back, in 1789, the first bamboo genus, *Bambusa,* was described properly in the finicky Latin of European botanists—who had no native bamboos to examine —on the basis of a species now called *Bambusa arundinacea,* a thorny staple of India's massive paper trade.

Whatever date we pick as a significant moment of the bamboo bridge gradually arching over oceans from the Orient to us, by comparison with the millennia of highly developed bamboo technology there, widespread growth and use of this grass giant is only dawning in the Euro-American culture that still precariously dominates the West.

But morning becomes noon. The methods for using bamboo within the format of modern industry are multiplying. And researchers are multiplying who recognize that bamboo is equipped to replace in a number of uses—or indirectly decrease consumption of—three resources our century is finding critically scarce as it staggers towards its close with an increasingly crowded globe on its shoulders: wood, metal, and oil. Bamboo is a superrenewable resource, equipped in its genes to multiply at a pace that rivals the proven reproductive capacities of our race. Bamboo is extraordinary in many ways, but the basis for its other wonders is its unparalleled physical vitality, which has made bamboo so plentiful in so many places of the planet for so many centuries that people have found a thousand ways to relate to its other physical properties of strength, weaving ease, lightness, hollowness . . . and visual charm. No other "crop" is so beautiful, with so many well-mannered and multihandy ways. None other has a past so romantic, complex, and curious, so stuffed with anecdote and legend.

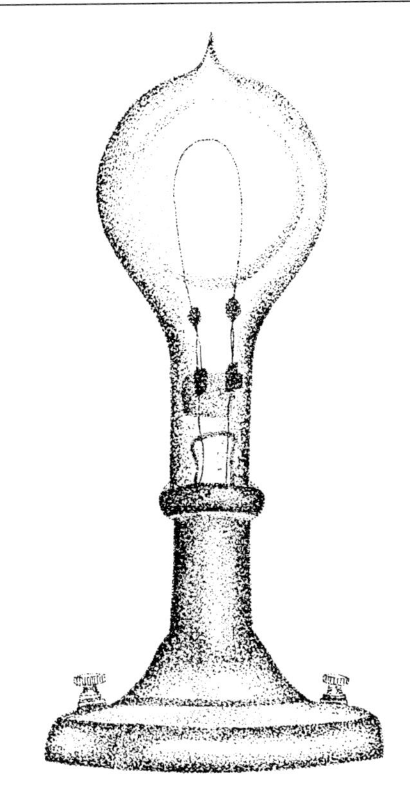

Bamboo filament light bulb of Thomas Edison (1882).

Research gap.

Yet no government in any country in the world is at present funding an adequate program of research and development in bamboo. Even the most bamboo-conscious countries, such as India or Japan, are criticized by their own experts for lagging woefully behind in the possible exploitation of their resources, while in Western countries bamboo study centers are virtually nonexistent.[1] Individual botanists and agronomists in many countries are aware of the immense potential of bamboo, but lack of government funding keeps them from acting on that knowledge with appropriate amplitude.

Some friends of bamboo traveled from the United States to Peru to survey the status of bamboo studies in the West. We found friends of the plant everywhere, lovers of bamboo struggling to make its virtues more widely known: Taiwan experts visiting El Salvador to introduce new species and demonstrate bamboo waterwheels irrigating 129 times faster than a small farmer could by hand;

corporations building tidy and attractive bamboo low-cost housing alternatives, crawling up the mountainsides in Manizales, western Colombia; large bamboo suburbs striding on *mangle* stilts out into the tides in southern Ecuador, near Guayaquil; private planters north of Lima in Peru with huge groves established first as windbreaks on citrus plantations and then used later in a dozen discovered, unexpected ways; and we found one country ready to formally explore a bamboo alliance. From July 1981 to March 1983 we lived in a small basket-making village in Nicaragua, working with various government ministries to evolve a design for comprehensive use of their bamboo resources with particular focus on schools, the small farmer, and the village economy. Our work there is ongoing.

DESIGNING DESIGNERS

Plato, in *The Republic,* raises a question never answered. Any moderately large group requires some apparatus to enact its will—a state. Every state requires guardians who will act as watchdogs against stupidity and corruption. But who will guard the guardians? Victor Papanek, in *Design for the Real World,* rephrases Plato's question in relation to industrial designers. Who shall design them, and how?

Of all the scientists who have ever lived, 90 percent are alive today, exchanging technical information through a hundred thousand journals in over sixty languages. The volume of this information is presently doubling every fifteen years. To package it, papermaking machines now produce in one 24-hour day as much paper as a skilled vat man could make in ten years of 14-hour days working by hand. Newsprint machines 30-feet wide roll out paper at a half mile per minute. The people and technologies are available for communicating viable cultural objectives and enacting appropriate industrial designs—for the many rather than the money.

But the focus of the industrial design profession, according to Papanek, is as if all doctors left general practice and surgery to concentrate on dermatology and cosmetics. Progress—if it is progress —is not keeping pace with people. There are more now without electricity than Before Edison, more without cars than Before Ford. The worldwide shortages of food, fuel, and housing are anxiously discussed on the world's shortage of paper. "There was an old woman who lived in a shoe with too many children to know what to do . . ." The art of annihilation is the only human science that has managed to keep pace with our race: Bombs are the only artifact we have in sufficient, even excessive, abundance for all. "Overkill"—a very modern word, not needed till now—describes the number of times all the existing bombs on the planet could kill all the existing people.

Reentry.

In this crunch of resources and common sense, it's inevitable that increasingly students of the art of serving life rather than snuffing it should be turning to one of life's most successful designs for living. For all the shortages described above, many looking toward the human future find themselves contemplating hints from our human past—and find bamboo standing by all the centuries on half the globe. The comfy corner supertech has constructed in part of one century of human history is tiny, brief—as selfish as it has been vulgar—and fortunately drawing to a close.

What can we do about it? Rejoin the race, reenter the real world of hunger, scale down our own needs to a feasible planetary norm, and begin to design an educational system for industrial designers in which world need is the prime determinant rather than fadmind. Shortly after this happens —as soon it must—these more humanely designed designers will begin to notice bamboo.

CULTURE DOCTOR, CLIMATE DOCTOR

When Oswald Spengler, the German philosopher of history (1880–1936), decided in *The Decline of the West* ("Evening in the Eveningland" is closer to the German title[2]) that the sun was setting on Western culture, he was obviously talking about his own cultural corner and economic group. The West wasn't dying. What was dying was the violent unification of the rest by the West. For a while, a large part of the world was united—as a jail is united by guards—by European and then U.S. dominance, through the instruments of war and communications technology, which are now serving to rip the empire apart.

But sundown for the few is dawn to many millions. By the year 2000, for example, some 500 million people will be speaking Spanish in twenty countries with a birthrate unrivaled in the West. Only 50 million or so, roughly 10 percent, will be living in the European country that gave birth, linguistically, to all the rest. By then, the United States and its uneasy alliances will be a shrinking fraction —15 percent—of the countries they confront. These excolonies are now in touch through a communications system quite different from the monopoly prevailing when they were filled with "natives" who'd never heard of one another.

Privilege surrender and class peace.
In these pages we are speaking both to those living through the sundown of their class—at least to that sane minority we imagine among them, groping for graceful ways to surrender their privileges—and to those classes experiencing a political and cultural dawn. It is interesting that "the wood of the poor," as bamboo is known in India, is also the most perennial pet plant of the rich man's garden in the Orient. Classes at odds over other issues are united in their admiration of its practicality and charm. Perhaps a similar metapolitical alliance will emerge among radical conservationists of the West, as more people realize that bamboo is a green piece of planet bulging with more wonders than Aladdin's lamp, equipped to provide people a multitude of things, while acting on the wounded landscape as a mending salve. It is one of the largest and oldest devices of the earth to place more rapid order in that thin green film, one millionth its total bulk, that wherever not gashed by human intrusion is the healthy, breathing skin of a sensitive planet, the soil.

PLANT THERAPY

It is trumpeted as a triumph of contemporary agriculture in the United States that 3 percent of the population produces the food for all the rest. Many hidden hours of hidden hands elude these statistics —deckhands and truckers coaxing oil from Arabia to Omaha tractors—but even if true, is it a desirable condition? If 3 percent of the people raise all the food, 97 percent are alienated from the soil and water and sun and subtle planet processes that hold us here. More and more people have been raised since childhood in a historically unprecedented dimension of exile from the earth itself, an exile celebrated as a blessed exodus from the bad land of toil. Contemporary with this exile, unprecedented destruction of forests, loss of topsoil, extinction of species, and elimination of native peoples have occurred. In the midst of this official mechanical madness—in which each bushel of corn is costing five bushels of eroded topsoil, each bushel of wheat, twenty—here and there is a quiet voice, a dirtied hand raised for an intimate agriculture, regional self-sufficiency, each of us once more a skillful and intimate *amigo* of the ways of plants and animals.

Hug your local oak.
Plants heal people. Living with them is a most beneficial therapy, and their unprecedented absence from our present urban designs may have more to do with the neurosis of our times than we imagine. Bismarck's physician urged him to hug an oak a half hour each afternoon as antidote to all the strains of his iron office. In order to practice this therapy, many harried Americans would have to board a bus to reach the nearest tree and wait in line to embrace it. In this cultural context, bamboo is suggested as one of the easiest of all plants to cultivate and love and among the most rapid to regreen the globe.

Leaf/people ratio: the bamboo scab.
Revegetation is critical for earth health as well as human sanity. An insane dimension of devegetation as prologue to a ruinous agriculture has lost, according to some estimates, one-third of U.S. topsoil in fifty years of industrialized and chemical agribusiness. This reckless loss of limited topsoil has complex consequences that are coming at us quickly from our increasingly fragile future. In modern cities, the leaf/people ratio is the lowest in history.

Leaves are therapeutic principally because they pump oxygen, the reduction of which in brain cells has been clinically found to increase anxiety and the whole spectrum of negative emotions that compose the familiar attitude configuration of hurried urbanites. Urban bamboo can not only pro-

duce building material on location to grow your loft or trellis or fence in your yard, it also spreads more square meters of leaves more rapidly, perhaps, than any plant alive. This airy green filter and a dense underground life renowned for erosion control combine to make bamboo a healing force for soil as well as soul. As scabs on skin, bamboo acts on earth to regreen spaces gouged by humans. It has traditionally repopulated forested areas cleared of their main timber crop or fields no longer actively farmed and is presently refoliating Vietnam in the wake of U.S. defoliation efforts during the war.[3]

Revegetation.
The per-person reserves of wood in the world are dipping drastically. Only in China and South Korea are reforestation efforts outstripping use. Since China has doubled her trees since 1949, we must conclude it isn't mere human numbers denuding our forests so much as uninstructed numbers not committed to a sane forestry policy—and a busy handful firmly committed to destructive greed. In the context of an urgent need for revegetation in many areas of the world, bamboo can take the weight off wood in a number of human uses, while playing the same roles as trees in modifying weather, controlling floods, diminishing winds, and guarding the soil.

BEGINNER'S BAMBOO: ESPECIALLY FOR GENERALISTS

> In the beginner's mind, there are many possibilities, but in the expert's there are few.
> —Suzuki Roshi, *ZEN MIND, BEGINNER'S MIND*

Specialization, and a consequent fragmentation of culture, plagues industrial societies. Monocrop mentality in modern agriculture is mirrored in modern lives, shrunk generally to the sad dimensions of a "career," a two-syllable job. In *Education Automation,* Buckminster Fuller analyzes the history of the shift in U.S. higher education from a wholistic schooling of generalists to the training of experts who knew more about less and were well conditioned to trade their deep but narrow wisdom for a wage.

> They were given the choice: king or messenger? Everyone chose to be messenger, running about with messages become meaningless, because there were no kings.
> —Franz Kafka, *PARABLES*

The tycoons of nineteenth century U.S. capitalism, the industrial pirates Fuller calls them, treasured the global overview that was the base of their power. They needed more and more well-trained, technological lackeys, and they did not need competition from generalists with a comprehensive grasp of how the empire was put together; hence the design of higher education in the United States, which they funded. The U.S. Naval Academy, before the introduction of radio aboard ships, was the only place deliberately training generalists, needed to assume intelligent direction of ships isolated at sea for long periods, able to shift decisions without consulting headquarters. But radio permitted the captain of each ship to be integrated in a command system directed from far away and far above. The U.S. Naval Academy immediately redesigned itself to train specialists.

The global schoolhouse and the global grove.
Whatever the cause of hyperspecialization in our

schools, bamboo is an interesting antidote, a prime piece of unassigned homework in Western education that could unite many directions of research, provide raw material for many forms of teaching aids and crafts, and landscape the campus. Hints for introducing bamboo at all levels of our schools as a subject to talk about and a source of projects to do are scattered through this text. We are particularly anxious to address teachers, help them reflect on the possibilities of bamboo use in their particular climate and cultural context. Schools too northern for outdoor cultivation might find constructing a greenhouse with a bamboo polyhedra frame a useful project to create garden space.

Locke, the British philosopher, in his "Essay Concerning Human Understanding," confesses himself among those rambling, unbloody hunters who prefer to start game to shooting it. In like spirit, we content ourselves here with disclosing a dozen directions rather than carefully mapping one. Inspirations rather than ready-made curriculums are intended. Occasionally, detailed descriptions for building a bamboo hen house or laying bamboo pipe are included to suggest the content of a more practical and explicit how-to bamboo.

BAMBOOPHILIA

The bamboo pioneers in the United States—David Fairchild, I. A. McIlhenny, F. A. McClure—were characteristically generalists, interested in all things human in all parts of the world, and finding in bamboo a specialization for generalists. The diversity of bamboo use, its global distribution, and the antiquity of its wedding with human will are features of the plant that make students of it residents of a rippling circle whose center is everywhere and circumference nowhere, as some Western theologian once defined God. So studying bamboo makes one a beginner for life, which may account for the longevity of many of its most famous lovers.

Lovers and venders.

This perennial freshness is more amplified for us because bamboo itself is at a point of new beginning in the West, with directions of development as yet unknown. That is part of its excitement. A field as yet unfenced by experts, a virgin grove, is all the more open to amateurs—a word whose root meaning lies in the Latin word for "love." Amateurs are always useful reminders to the sober specialist that affection is the sharpest lens, and that the most nimble scientist shares the passion of the true philosopher, who is not the husband, warden, vender, barker, or press secretary of wisdom, but merely her lover—a mobile and wonderful uncertainty rather than a frozen role.

Planet poultice.

A culture doctor, prescribing healing motions for the rheumatic West, could do worse than bamboo as a totem plant and poultice for our social pains—not least of which is a marked decline in love of life and of our kind, our human kin. Bamboo enjoys a widespread and persistent history as an aphrodisiac, reputedly giving more bounce per ounce than anything but a rhinocerous horn, which the rhizome of one species in India resembles so closely that it has stimulated a fraudulent black market of love in which only an expert (pardon the word) can distinguish the real rhino from the rhizome.

Handsome history now.

Maybe more intimate acquaintance with bamboo traditions can foster keener lovers of history, the total spoor of our species, as well as lovers of the poor planet beneath history where the trash of error lands. The truest history, for us, is the one we are making. Bamboo's vitality may inspire us to make more handsome history now, with a little help from our mutual friends.

THE KING OF CHINA AND THE CLEVER CHILD

Once upon a Zen fairy tale, the King of China called his three sons out back to the bamboo groves and said, "Before I die, I want to leave my throne to my wisest child. Tell me, please, what is the best thing in the kingdom?" The first son praised the country's natural wonders and abundance, its noble mountains and its blooming plains. The second spoke eloquently and at length of what the human hand had heaped on the land—the roads and bridges, the temples and glistening palaces. "The best thing in the kingdom," said the third, "is my own mind—for with this I judge the value of the rest."

Questing roots.

The best resource of nature in the plastic and polluted kingdom of our days is that naughty human intelligence clever enough to make a rapid mess of our planet, using—according to ancient masters and contemporary psychologists—only some 2 percent of the computers tucked between our ears. That leaves 98 percent as yet untapped to clean up the trash. It obviously isn't intelligence but clear will and agreements to learn that are lacking, a generous

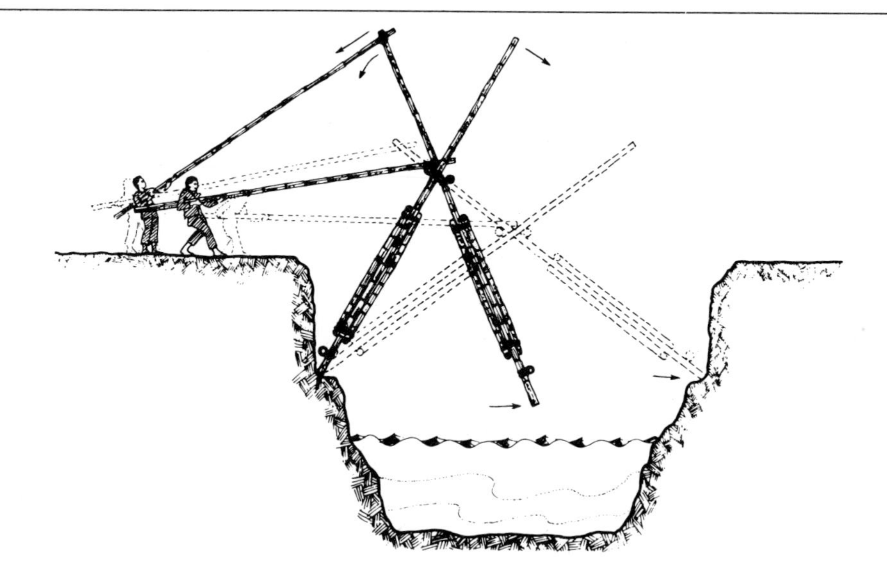

leap into a zone unfrequented by the specialized mind: thinking with the heart, learning to live and let live with our whole equipment.

> The world's two greatest underdeveloped resources are the human capacities for creative fulfillment, which are thwarted by hunger, poverty, disease, violence, and lack of education; and the mineral rich subsoils which can most efficiently be utilized by the powerful, questing roots of trees and other perennial plants.[4]

Bamboo bridges: appropriate travel and learning to learn.

The more rapid the changes around us, the more we learn that life is about learning. The Way is an ease in process, not a fixed goal. Western people, especially, have to replace dominion with dance—in schools, in agriculture, in family and foreign relations. Bamboo, ancient and adroit changer, has been riding climates roughly 200 million years now, about four hundred times the life of our species. Perhaps it can help us really wake to our limited mornings, to flourish as durably. Bamboo's diverse use is what most makes of it a maximal learning zone, a point where many interests overlap. Admiration for its virtues unites scientists and musicians, farmers and architects, craftspeople and artists, rich and poor, country and city, East and West. Ancient builder of many bridges, bamboo can also serve as a thin union between peoples of North and Central–South America, where we need, at this dangerous and fragmented moment of a dwindling century, all the bridges we can find or build.

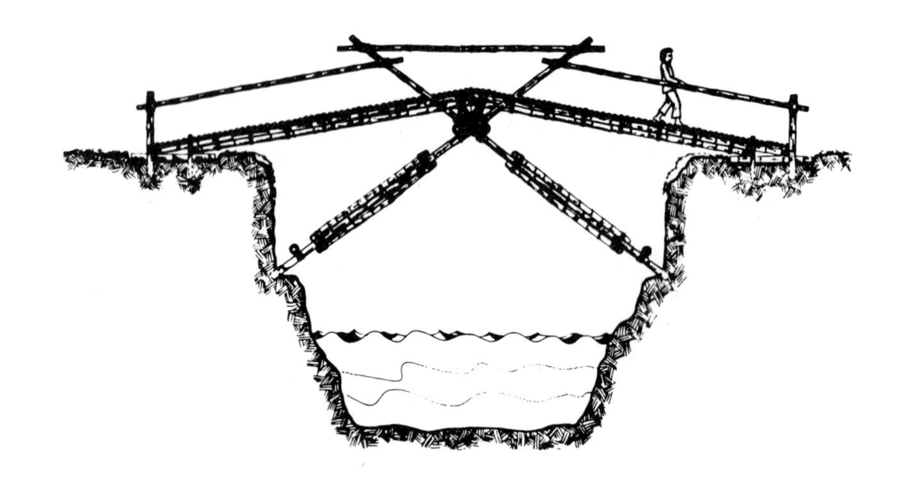

Map of the Tao, northern Sung dynasty.

Beware of bamboo.

In no way feel obligated to plod forward in these pages chapter by chapter to the end. Dip, skim, skip —enter this grove at any point, proceed in any direction. But be careful: Beware of bamboo. It is emotionally invasive. This book may begin a muddy adventure dragging you far from the safe shop where you purchase it and the home where you peruse it.

> Go . . . not knowing where. Bring . . . not knowing what. The path is long, the way unknown. None know how to arrive by themselves alone. We must seek help and guidance from more complete forces.
> —Russian fairy tale

CHAPTER 1.

1. Ueda 1960; Varmah 1980 notes a sudden but still inadequate upswing in oriental bamboo research around 1975.
2. *Abend im abendland.*
3. Drew 1974.
4. Douglas 1976, via deMoll 1978:96.

2. ONE THOUSAND THINGS

"One of the most thorough investigations of the uses of bamboo, by Hans Sporry [1903], lists 1,048 uses, from Japan only, of articles in his own collection. Another 498 ornamental uses brings the total to 1,546. These uses, with variations, would stand almost as they are for China as well. One cannot live long in a country where bamboos grow and are used by the people without feeling that bamboo has contributed a great deal to their progress, and that the mastery of its uses marks a cultural stage in the development of their civilizations. Archeologists would indeed be justified to incorporate, in their historical outlines for tropical and subtropical Asia, a definite Bamboo Age *comparable with that of Stone or Bronze.*

"It is no wonder that native craftsmen soon found such a workable material a broad field for cultivating their genius. Because of its great tensile strength, its capacity for splitting straight, its hardness, its peculiar cross-section, and the ease with which it can be grown—a combination of useful traits found together in no other plant—bamboo is one of those providential developments in nature which, like the horse, the cow, wheat and cotton, have been indirectly responsible for man's own evolution." [1]

HANDY MAGIC: BAMBOO APPROPRIATE TECHNOLOGY

Among the most useless uses of bamboo recorded, the Everest of irrelevance is that embodied in an observation of one Osbeck who published an account of his voyage to China in 1771: "Laborers are obliged to pare their nails, but people of quality let them grow as long as they will, keep them very clean and transparent, and at night put little cases of bamboo on them. Very long nails are a token of elegance, and shew that the wearers are arrived at a thorough pitch of genteel helplessness."

This tiny monument to the antihandy can stand as a perfect counterpart to the sane resourcefulness that has characterized bamboo use for millennia wherever its abundance has provided people a prolonged chance to explore alliance with the plant. Technologically, bamboo is "rampantly invasive" —as some of the species are called in nursery catalogs. If bamboo is broadly available in a culture for an extended time, its cheerful readiness to service human need creeps into people's lives at every crack and cranny . . . and seeps deeply into their affections as well. There is almost nothing that the Chinese, in particular, have not fashioned from bamboo in the course of the lengthy evolution of their technology, which Joseph Needham docu-

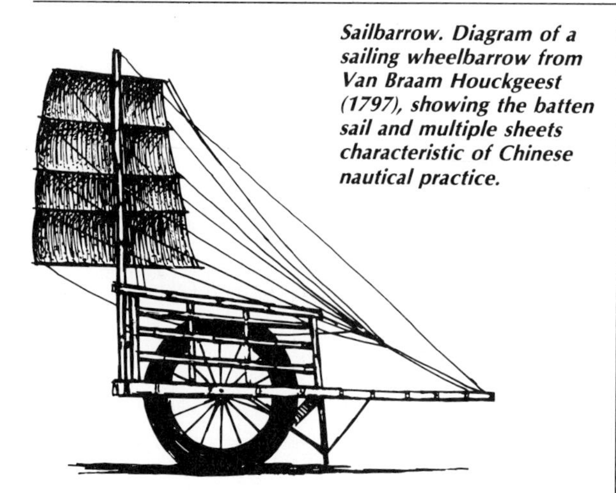

***Sailbarrow.** Diagram of a sailing wheelbarrow from Van Braam Houckgeest (1797), showing the batten sail and multiple sheets characteristic of Chinese nautical practice.*

ments in his huge witness to Chinese ingenuity and British scholarship, *Science and Civilization in China.*

Springs for mechanical toys and automata were made from bamboo laths, which served also for crossbows, door closers, and animal traps. Bamboo brine buckets 75 feet long with a 3-inch diameter lifted 28 gallons each as much as 2,000 feet up from the Szechuan salt fields to bamboo pipelines sealed with a mixture of tung oil and lime.[2] Bamboo water guns were used for fire fighting in the eleventh century,[3] at which time plybamboo was also employed for the vanes of a rotary-fan winnowing machine.[4] Bamboo and oak served as teeth in baked clay grain mills around 200 B.C.[5] Bamboo wheelbarrows tripled weights workmen could carry, even on mountain paths, "winding as the bowels of sheep."[6] Sometimes the wheelbarrows were equipped with bamboo sails, with a loading capacity of 365 pounds.[7]

A bamboo well sweep with a counterweight at one end and a bamboo pole at the other was an early and simple method of lifting water.[8] Waterwheels with bamboo cups were also common, the largest wheels being 60–75 feet in diameter.[9] Windmills with bamboo mat-and-batten sails were used for irrigation.[10] In rural Taiwan, bamboo-piped wells up to a depth of 150 meters (492 feet) are still common.[11]

A list of bamboo things, such as the very incomplete one that follows, suggests specific uses we can imitate directly or adapt—an idea bank for bamboo crafts and constructions. But the gradually amplified and modified uses of bamboo in different places and new times also indicates that its uses are virtually endless and still to be explored. Modern methods of chemical analysis, for example, have only recently pried into some secrets of the culm.

New uses for bamboo products have been developed, and new significance found in old uses. Tabasheer, found within the culm internodes of many tropical bamboos, consists almost entirely of amorphous silica in a microscopically fine-grained state; it has excellent properties as a catalyst for certain chemical reactions. From the white powder abundantly produced on the outer surface of young culms of a Chinese bamboo, many substances have been isolated, among these being a crystalline compound related in chemical composition to the female sex hormone. Liquid diesel fuel has been prepared from bamboo culms by distillation.[12]

PLANETARY NORM: THE MODERN VILLAGE

Wanting more than your share is not good ecological logic. Immanuel Kant made up a long boring name for a simple but ample idea. He called it the "categorical imperative," which meant you were to test the ethical value of any questionable act by asking: If everyone always everywhere acted as I am now acting here, would the world work? Is my style of life, my total routine of giving and taking, using and making, a possible planetary norm?

It's worth reflecting on bamboo in the context of an awakening world interest in appropriate technology[13]—a recent phrase for the old reality of creatively making do with the locally available and cheap instead of importing distant and dear solutions made elsewhere and alien to the context at hand. A sad irony in many developing countries in the process of decolonization is a schizophrenic attitude toward the dominant U.S. or European culture. Although rejecting a destiny in the dark shadow of a superpower, they are often ready to rush headlong after the disastrous lead of hyperindustrialization with its attendant pollution, its erosion of land, consciousness, and social values—all the double-ugly urban wasteland that lies in wait at the end of the freeway, still hidden behind the high horizons of their hopes. Many Third World nations, while rejecting the role of colony to the United States or to other foreign purposes, echo the tragic wish to treat the earth as though it were a colony of human will.

Current metal consumption can be cited, as one example among many, to demonstrate that present levels of use are not a feasible future for the United States or anyone else wishing to imitate this gaudy style. In the past sixty years, people have taken more metal from the earth than had been mined in the previous 6 million years.[14] But, "to raise all the 3.6 billion people of the world to the 'American standard of living' would require:

Wellsweeps from the Thien Kung Khai Wu *(1637).*

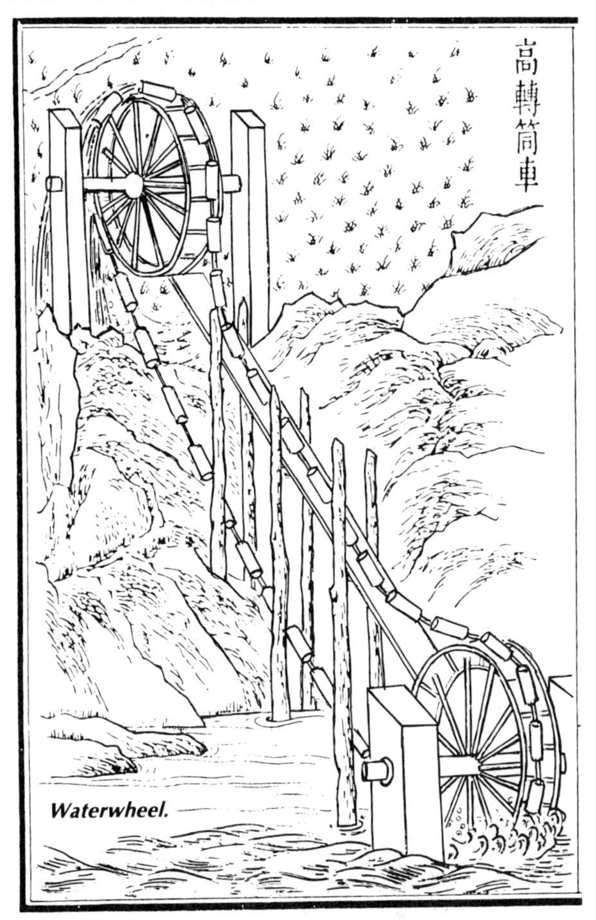

Waterwheel.

75 times as much iron, 100 times as much copper, 200 times as much lead, 75 times as much zinc, 250 times as much tin as is now extracted annually."[15]

The greater the imbalance of the metal and oil resource distribution, the greater the constant expense of more metal and oil to defend the embattled castle. The United States is a pseudo-Eden under seige. Nearly 85 percent of humanity lives outside the shiny garden. We are acting out an unreal fairy tale. A possible norm must replace the glittering exception as the objective of the great cultural quest. There isn't enough glass, there aren't enough coaches, for everyone to get the shoes and ride of Cinderella. There could be, easily, enough bamboo. For a thousand things, for 6 billion people; it is a feasible planetary norm.

Migration to margins: a sane response to urban sprawl.

A number of recent books help demonstrate that the increases in energy consumption of industrial societies cannot continue. *Soft Energy Paths* (Amory Lovins 1977) and *Rays of Hope: Transition to a Post-Petroleum World* (Denis Hayes 1977) are two that deserve mention. A cofunction of "hard energy paths" is the ever greater centralization and "implosion" that is a chief feature of our urban times. Ironically, greater physical densities seem to increase social fragmentation and a drastic decline in human intimacy. The closer our bodies are jammed together, the more our spirits and families fall apart. The hope of appropriate technologists is that small technologies can help halt the great migration to towns by making the modern village viable as a way of life, enriched, not destroyed, by a development scaled to local need and resource. Bamboo's ample possible role in this decentralization is hinted at by the brief examination in this chapter of some pieces of its past.

APPROPRIATE BAMBOO CATALOG

A A-frame houses,
activated charcoal,
acupuncture needles,
airplane wing members and stress skin for fuselage,
alarms,
alcohol,
anchors,
angklung,
antenna supports (TV and radio),
aphrodisiac,
arbors,
arrows and arrow tips,
ashtrays,
awnings.

Bamboo ubiquity

European travelers in the late nineteenth century compiled astonished lists of the daily uses of bamboo in China and Japan before the advent of extensive Western influence. A. B. Freeman-Mitford's *The Bamboo Garden* is a fair sample of Western amazement in the face of so varied a use of a single resource, a remarkable cultural phenomenon without counterpart in European traditions. "To the Chinaman, as to the Japanese, the bamboo is of supreme value; indeed, it may be said that there is not a necessity, a luxury, or a pleasure of his daily life to which it does not minister. It furnishes the framework of his house, and thatches the roof over his head, while it supplies paper for his windows, awnings for his sheds, and blinds for his veranda. His beds, tables, his chairs, his cupboards, his thousand and one small articles of furniture are made of it. Shavings and shreds of bamboo are used to stuff his pillows and mattresses. The retail dealer's measures, the carpenter's rule, the farmer's waterwheel and irrigation pipes, cages for birds, crickets, and other pets, vessels of all kinds, from the richly lacquered flower stands of the well-to-do gentleman down to the humblest utensils of the very poor all come from the same source.

"The boatman's raft and the pole with which he punts it along, his ropes, his mat-sails, and the ribs to which they are fastened, the palanquin in which the stately mandarin is borne to his office, the bride to her wedding, the coffin to the grave; the cruel instruments of the executioner, the lazy painted beauty's fan and parasol, the soldier's spear, quiver and arrows, the scribe's pen, the student's book, the artist's brush and the favorite study for his sketch; the musician's flute, mouth-organ, plectrum, and a dozen various instruments of strange shapes and still stranger sounds—in the making of all these, the bamboo is a first necessity. Plaiting and wicker-work of all kinds, from the coarsest baskets and matting down to the delicate filigree with which porcelain cups are encased, are a common and obvious use of the fibre. The same material made into great hats like inverted baskets protects the coolie from the sun, while the laborers in the rice fields go about looking like animated haycocks in waterproof coats made of the dried leaves of bamboo sewn together. See at the corner of the street a fortune teller attracting a crowd around him as he tells the future by the aid of slips of bamboo graven with mysterious characters and shaken up in a bamboo cup, and every man around him smoking a bamboo pipe."

ACUPUNCTURE NEEDLES. These were supposedly originally made from bamboo. The role of bamboo in cultural acupuncture, the art of producing maximal response through minimal force strategically applied at key points, is the meditation of these pages.

AIRPLANE SKINS. "The remarkable tensile strength of bamboo was not fully realized until recent times, when experiments were made by the Chinese Airforce Research Organization at Chengtu; it was found then that plybamboo of formidable qualities could be made by uniting layers of woven laths with aeroplane glue . . . But all through the centuries this property had been empirically utilized to the full in bamboo cables and ropes for many purposes."[16]

Woven bamboo panels were used in the Philippines during the 1950s as an experimental low-cost covering for wings and fuselage of light airplanes by the Institute of Science and Technology in Manila. *Bambusa spinosa* and *Bambusa vulgaris,* with thin walls and long internodes, were cut and seasoned in the dry months (December–May), then split into strips 1 cm wide by .6–1.2 cm thick. The inner and outermost portions of the culm wall were discarded: the inside as too weak, the skin as too slippery and hard to glue. Submerged in salt water 48–60 hours as a preservative against insects, washed in clear water, and dried, the strips were woven—at 45-degree angles to one another—in panels 2 m by .5 m. *Sawale,* a type of weave used for walls in the Philippines, was chosen. Panels were given a coat of glue liquid enough to soak well into the weave and another coat of glue mixed with fine soft-wood sawdust to even the surface, which was then sanded. Performance was good, and three years after construction the skin showed no sign of deterioration in spite of severe weather conditions.

Low in cost, requiring no complex machinery or construction techniques, bamboo was highly recommended by the experimenters as also useful for waterplanes and the landing apparatus of light planes because of its ability to absorb shock. A detailed report and tables comparing bamboo quite favorably with other materials used for conventional airplane skins is available in Hidalgo.[17]

ANGKLUNG. "The angklung of the Malays is a very agreeable instrument. It consists of a number of hollow bamboo joints of various but selected lengths and thickness which are cut out below and hang down from a bamboo frame. These give various swinging tones and strength according to their

Woven bamboo airplane skin, Philippines, 1947.

size on being beaten with a bamboo staff.''[18] (See *musical instruments.*)

APHRODISIAC. Bamboo makes the heart beat fonder—at least an old tradition has it that lovers can be more loving with treatments of tabasheer (see under *T*). But the plant probably earns its most handsome price per pound in a fraudulent traffic described by Varmah: ''The rhizome of *Dendrocalamus hamiltonii* with slight trimming and dressing, is an exact replica of a rhinoceros horn which fetches a fabulous price as an aphrodisiac. Only an expert perhaps can identify the imitation rhino horn from the real.''[19]

ARROWS. In Japan, *Pseudosasa japonica* is known as *metake* or ''arrow bamboo'' for the shafts made from its branches. A very common ornamental, it was perhaps the earliest hardy oriental bamboo introduced into the United States around 1860,[20] and groves are widely available for those who have a strung but empty bow. Above all, arrows must be straight and light, so the wood of many bamboos can be made to shape them. In the Zen tradition archery is a mirror of the soul, its practice a physical discipline that hones the mind: ''You are the target . . . if the archer misses the mark, he looks for the error *inside.'' Zen in the Art of Archery* is a well-known account of a German, Eugen Herrigel, studying under a Japanese master:

> ''The right art,'' cried the master, ''is purposeless, aimless! The more you try to learn how to shoot for the sake of hitting the goal, the less you will succeed . . . You have a much too willful will. You think that what you do not do yourself does not happen . . . We master archers say: with the upper end of the bow the archer pierces the sky; on the lower end, as though attached to a thread, hangs the earth. If the shot is loosed with a jerk there is danger of the thread snapping. For the purposeful and violent people, the rift becomes final, and they are left in the awful center between heaven and earth.''
>
> ''What must I do then?'' I asked thoughtfully.
> ''You must learn to wait properly.''
> ''And how does one do that?''
> ''By letting go of yourself, leaving yourself and everything yours behind you so decisively that nothing more is left to you but a purposeless tension.''

Austin (136–43) provides twenty-three photographs on Japanese manufacture of bamboo bows and arrows. The lamination of wood and bamboo in the bow may be the oldest surviving use of this important modern industrial practice.

B Baby carriages, bagpipes, barrels, baskets, beads, beanpoles, beds, beehives, beer, bikes, bilge pumps, blinds, blowguns and darts, boards, boat hoods, boats, bolts, bookcases, books, booms, bottles, bowls, bows (archery), boxes, bracelets, bridges, brooms, brushes, brush pots (washers, and rests), buckets, buttons.

Baskets: a bamboo boot camp.

Any design for development of bamboo as a resource in Western cultures will quickly bump up against baskets and mats and their makers. Basketmaking and weaving skills are the basis of ancient 10 by 30 foot rolls of walls delivered by oxcart to house sites in a number of oriental countries, a main component of Eastern shelter.

Basketry is also basic in lacquerware (q.v.), and in the modern manufacture of plybamboo in all forms, from building panels to laminated salad bowls and plates.

Basketmakers represent the main bamboo work force in many areas of bamboo use. Future development will start in their villages and should

be aware of the sociology of basketmakers. They tend to work in family groups, with children eased early (around 8) into the errand. Basketmakers are usually badly paid and poor. For some, drinking seems an occupational hazard. They harvest their raw material and often market their own goods, so they experience a spectrum of natural and social realities which many professions fail to roam. Working at home, at their own chosen pace and hours, at tasks possible for children to join, there is a wonderful overlap of family and workspace possible among those practising the art. Men and women, young and old join in the work, though heavier aspects of it—harvest transport, splitting large culms or weaving big cumbersome market baskets—are usually reserved for men and older boys. Basketmaking families are geared to sprints of production at crisis times, seasons of the year which require for some harvest or some fiesta a sudden rush of work. They also have that independent spirit of being their own boss, a life mood shared with all artisans; circus families; cab drivers in the modern urban scene—all those who earn their living unwatched by employer's eye.

Basket- and mat-makers as a work class represent the largest body of teachers available in any culture to introduce people new to bamboo to its basic handling from grove to market. A basket village with a large number of families involved in the craft is a ready-made school, a bamboo boot camp for those wishing to combine what you can find out from the books with what you can find out from the baskets and from the fields, from the fiber and the people splitting it.

BASKETS. China, Philippines, Japan. "Since earliest times, baskets have occupied a prominent position in Chinese civilization. In the *Book of Songs* and the ancient *Rituals,* many terms for various kinds of baskets occur. We read of round baskets of bamboo, of square shallow baskets of bamboo or straw .... The young bride offered fruit in a basket to her father-in-law, and since men and women should not touch hands, the woman should receive gifts in a basket. . . . Baskets in funeral ceremonies were placed near the coffin filled with cereals for the departed soul, a custom still of farmers around Peking, who bury their dead with an oval basket of willow twigs, which serves as a grain measure in ordinary life.

"We also hear of industrial baskets . . . fish traps set at openings of dams; and in the silk industry, the main occupation of women, the tender

Cock cage from a Hakka village in Kowloon, Hong Kong, 20th century.

leaves of mulberry on which the silkworm feeds were gathered in deep baskets, and a square basket served for depositing the cocoons. Basket trays still play an important part in the rearing of silkworms.

"In northern China, baskets are part and parcel of the rural population. Plain, practical, strong, durable, they are used chiefly for agriculture: collecting and carrying earth and manure, winnowing, storing grain, transportation. . . . The home of the artistic basket is in the Yangtse Valley and south. Here we meet in full development the flower basket with a great variety of shapes and graceful handles, the picnic basket with padlock, the neat traveling basket in which women carry their articles of toilet, and the 'examination basket' in which candidates visiting the provincial capital for the civil service examinations enclosed their books and writing materials, as well as the cozy for tea-pots, more practical and efficient than our thermos-bottles, and the curious pillow of basketry weave. The basket boxes with raised and gilded relief ornaments are also characteristic of the south.

"Chinese genius developed baskets unknown in other countries. Basketry was combined with other materials like wood, metal, and lacquer. Its appearance was enlivened and embellished through processes originally foreign to the industry. Many basket covers display a finely polished, black lacquer surface on which landscapes or genre pictures are painted in red or gold. Others are decorated with metal fittings (of brass and white metal) finely chased or treated in open work. Delicate basketry is applied to the exterior of wooden boxes and chests, even to silver bowls and cups. In this association of techniques, Chinese basketry has taken a unique development which should be seriously studied by our own industrial art-workers."[21]

Porterfield lists thirty-two types of baskets on sale in Shanghai in 1925. Among them were water-

Boat-shaped flower basket, Wen-chou, Chekiang, 19th century; 8½″ high, 10½″ wide, 14″ long.

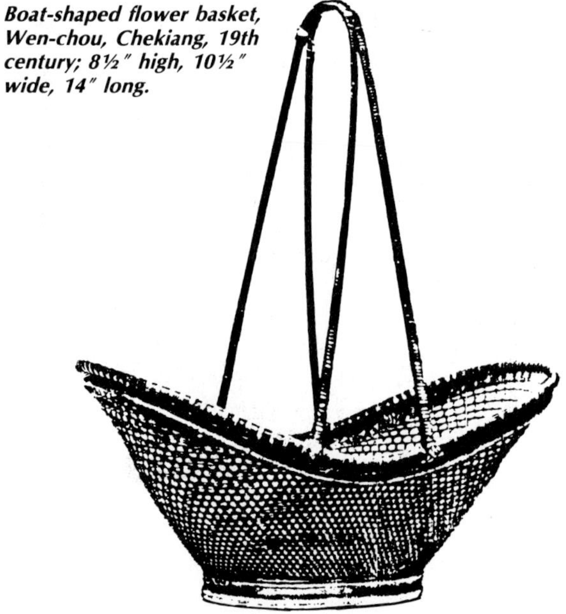

bottle baskets, sandwich, wastepaper, fruit, gardener, cake, flower, travel, clothes, picnic, tea, food, market, egg, sewing, suitcase, washing baskets. Others were named for their shape: openwork square, closed-work square, boat shaped, basin, partitioned, round, oval, and cylindrical.[22]

BEEHIVES. "Joints of the larger bamboos are also used for beehives. A closed bamboo joint or one cut in two and again tied together with strings is suspended horizontally under the roof of the house. A small hole made at one end enables the bee to enter."[23] *Dendrocalamus asper,* with an 8-inch basal diameter, is similarly used in Nicaragua. Any bamboo of ample diameter will do.

BEER. Bamboo beer is made from the dark, long-grained, ricelike seed of flowering culms of *Arundinaria hookeriana* in its native Sikkim. The species was named by Munro for its discoverer, J. D. Hooker, who reports this practice in his *Himalayan Journals.*[24]

BIKES. Bike frames are basically a triangle of joined tubes with appendages used to meet and drive wheels. They were manufactured of bamboo in Europe in the late nineteenth century.[25] "It is not at all fanciful to point out that furniture, scaffoldings, and other erections of tubular steel make use at a more neotechnic level of exactly the same principles of structural strength as the ancient applications of the bamboo tube."[26]

"The bicycle is a major mover of goods in the Third World. Loads can be tied directly to bicycles or placed in special frames, baskets, and trailers . . ."[27] "Pedal power" considers bikes from a perspective unfamiliar to most of us because the energy soup in which we swim in the United States rarely makes it necessary to generate our own. But as a power source as well as a vehicle, the bike is "in many respects the most efficient machine ever developed" on the earth.[28] The possible relation of bamboo bike frames, sidecars, trailers, and carrying baskets to Third World transport and pedal power is a research field as rich as it is fallow. China provides ample precedent for the wisdom of bike production as a way of both solving many problems of domestic transport while tooling up for heavier in-

Bamboo bike frames, manufactured in late 19th century Europe, make sense for mass transport in many developing countries today. China, with more people to move about than any other nation, has found the bike to be the most energy-efficient method of doing it.

dustrial development. Pedal power makes even more ancient sense than black, red, yellow, and so on power because physics is older than species and race. With pedal power you can run pumps, mills, light bulbs, saws, and many other small-scale power tools. In many ways, the pedal could be whirling at the fertile center of the Modern Village, everyone riding seventeen minutes a day (kids thirty) to generate independence or reduced reliance on centralized energy sources, a fat spider with skinny legs. The U.N.'s Food and Agricultural Organization (FAO) says energy from thermal and hydroelectric stations is sliced like most modern pies: 80 percent for urban industry, 10 percent for urban people, and 10 percent left for rural areas—88 percent "powerless" in poorer countries. This dark light bulb helps keep the down down in many ways. (See *windmills* for reflections on windbikes for decentralized irrigation.)

> Forty years I've been using a kerosene lamp. The second of December last year I saw an electric light, which we never had before. When we here in Galilao saw light bulbs that don't blow out in the wind, we all hugged one another, and my brother started to cry.
> —*La Barricada,* MANAGUA, NICARAGUA, 12 February 1983.

BLOWGUNS. Bamboo blowpipes or blowguns have existed a long time in many scattered locations on the earth. Known to the Iroquois, Muskhogeans, Cherokee, and other North and South American tribes, their use is perhaps most developed in Asia.

A Malayan species, *Bambusa wrayi,* has the greatest recorded internodal length among bamboos: from node to node measured 85.5 inches, making *B. wrayi* a natural choice for blowpipes in the area of its growth and helping to establish the reputation of Malayan blowguns as the best in the world.[29] "The Jacoons, or Treemen of the Malay Archipelago, shoot out poisoned arrows with such deadly aim that they even kill tigers with them."[30] Blowpipes are a handy hunting device because they require no sophisticated tool bag to produce or repair. They are light to carry, made of a material locally abundant, shoot small and equally light darts often fashioned from the same plant, and are carried easily in ample numbers in quivers cut also from bamboo. Two hundred darts are packed for hunting trips by one South American tribe. Blowpipes are accurate up to 40 yards and since their silent flight does not disturb animals, a poor shot gets a second chance. With the proper eye and lungs behind them, the weapon displays amazing accuracy, penetration, and range: A British police officer in Malaya reports seeing native hunters bury darts 3 inches deep in deer at 100 yards. The Jivaro of Ecuador can hit a hummingbird at 50 yards—with the ultimate howitzer of blowpipes, a piece of bamboo 17 feet long.

Blowpipe hunting skills were called upon as needed by many tribes in the Orient to defend themselves in wartime against invasion by hostile neighbors. The Subanon from the mountains of Mindanao in the Philippines are a peaceful and gentle people who, without their poisoned darts and blowguns, might have been exterminated centuries ago by neighboring tribes addicted to the arts of war. The Dyaks of Malaya in World War II used 8-foot drilled wooden blowpipes against Japanese sentries with fatal efficiency in the Brunei area near the Palawan Islands. Curare, the poison used by South American Indians for darts, is supposedly a painless but infallible killer that affects the end plates between muscles and nerves, disconnecting them and consequently stopping the action of the heart and lungs. In a blowgun battle, there are no wounded—only living and dead. If you're hit, you get 15 minutes to regret you didn't know more about blowguns.[31]

Boat cabins of bamboo are lightweight, easily lashed to the vessel and easily removed.

BOATS. "Although bamboo generally is not fit for the construction of boats or canoes, Chr. Costa tells us of a sort of bamboo in the Moluccos (most probably *Gigantochloa maxima*), which produces such thick halms (culms), that the single joints split in halves are used for little canoes, in which two men are said to find place! For masts and spars of small native vessels bamboo is in general use. The outriggers of canoes peculiar to the Philippines and Ceylon are all of bamboo. . . . The other parts of a boat, such as cabins, etc., are usually constructed of bamboo on the same principle as houses."[32] (Cf. also *caulking, fishing, junks, rafts,* and *sails.*)

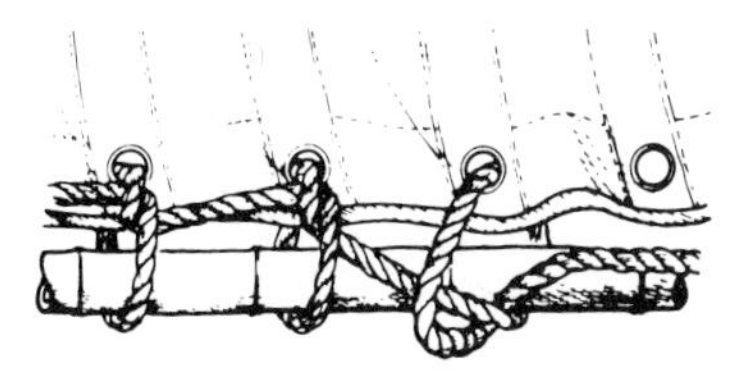

Booms are one of dozens of uses bamboo has aboard ship. Notice the method of tying the sail. Because of their light weight, bamboo frames and booms are also useful for scenery in shoestring theater, especially for ambulant street players.

BRIDGES. Bamboo bridges of many designs abound in China and throughout the Orient. In western Yunnan, primitive bridges made of bamboo cables droop across gorges with one end fixed lower than the other. Bamboo tubes greased with yak butter slide down the cables, with a cradle hanging from them in which people and animals ride. The catenary suspension bridges of west China are also constructed of bamboo cables:

> . . . made in the same way as those used for towing ships against the current of rivers, but of larger dimensions. Bamboo strips from the inner culm form a core in the center of the rope, and round them is woven a thick plaiting of bamboo strips taken from the outer silica-containing layers. The plaiting is so done that the outer portion grips the core the more tightly the higher the tension. Such ropes are generally about 2 inches thick, and three or more twisted together form one of the bridge cables. When placed in a testing machine, the straight inner strands break first, while the plaited material shows very great strength, not rupturing until a stress of *26,000 pounds per square inch* is reached, though an ordinary 2-inch hemp rope can carry a stress of only about 8,000 pounds per square inch. Moreover, the silica-containing outer surface is very resistant to wear, e.g., against rock surfaces, which is naturally important both in towing and bridge cables.[33]

Bamboo provided the fiber for the first suspension bridges, whose various designs combined whole culms, woven strips as warp to a woof of branches or small diameter bamboos, and cables braided of workable species such as **Gigantochloa apus,** ***"string bamboo," a favorite for bridges in Java.***

(See *cables,* below.) A bamboo suspension bridge over the River Min, in Szechuan, the largest of its kind in the world, is described in E. H. Wilson's *A Naturalist in Western China:*

> This remarkable structure is about 250 yards long, 9 feet wide, built entirely of bamboo cables resting on seven supports fixed equidistant in the bed of the stream, the central one only being of stone. The floor of the bridge rests across 10 bamboo cables, each 21 inches in circumference, made of bamboo culms, split and twisted together: five similar cables on each side form the rails. The cables are all fastened to huge capstans, embedded in masonry, which are revolved by means of spars and keep the cables taut. The floor of the bridge is of planking held down by a bamboo rope on either side. Lateral strands of bamboo keep the various cables in place, and wooden pegs driven through poles of hard wood assist in keeping the floor of the bridge in position. Not a single nail or piece of iron is used in the whole structure. Every year the cables supporting the floor are replaced by new ones, they themselves replacing the rails. This bridge is very picturesque in appearance, and a most ingenious engineering feat.[34]

Bamboo cables were the earliest structural element in the history of engineering to be used for suspension bridges, which originated in western China and the Himalayas. Their antiquity is not precisely known: They are mentioned as early as A.D. 399 in Chinese literature by a Buddhist monk en route to India. Of great strength—a bamboo cable

A Chinese postage stamp honors the River Min bamboo bridge, design forerunner of many Western bridges.

of 2-inch diameter can support 4 tons—these ancient bridges spanned distances up to 76 meters without central supports. A number of different kinds of suspension bridges simpler than the complex style spanning the River Min—whose cables could be "tuned" like violins—are amply illustrated in Hidalgo (1974:151–71), who also includes some eight different designs for rigid bamboo bridges, derived partly from Indonesian designs, partly from the British military bridges in Asia. Hidalgo's native Colombia and many other mountainous Latin countries that cannot afford the expense of conventional modern bridges would do well to reflect on traditional oriental designs. Kurz (223–4), in addition to describing briefly "bamboo bridges in general use all over India and Eastern Asia," mentions the use, in Java, of floating bamboo bridges as well: "true pontoon-bridges are constructed on the same island, where the pontoons are substituted by strong bamboo rafts, which rise and fall with rise and fall of the river or of the tides."

BRUSHES. The prevailing communications technology is a main shaper of all cultures. At the dawn of the East, from Jordan to China, we find the bamboo

***Chu,** the modern Chinese ideogram for bamboo. Calligraphy is intimately bound to painting, in which the elimination of the inessential is a primary rule. Here, six strokes—two culms and their attendant foliage—sum up the grove, in which each stalk can hold 80,000 leaves.*

brush and the bamboo pen (q.v.). Bamboo exerted enormous impact on the duration of cultural memory as the earliest widespread writing technique in China, and the word for it roams widely through the Chinese language. (See *fishing.*)

Ideograms were supposedly invented in the time of the *Ku Wen* or *Ancient Learning*—around 2600 B.C.—by Ts'ang Chi, a minister with four eyes who was presumably doubly clever in consequence. Among the simple radicals, the 214 roots or bones used as an etymological alphabet to classify some 49,000 ideograms, a common one is *chu,* "bamboo," a stylized but recognizable picture of two canes side by side, with strokes suggesting leaves and branches. As a radical, *chu* in turn forms part of a large number of ideograms. How deep and far *chu* has run through Chinese, how "invasive" the radical has been in the language, how many words share its strokes, would be interesting to know, and fairly easy for a native speaker to determine with a good dictionary.

The Chinese people value their past, and a large measure of their early affection for bamboo

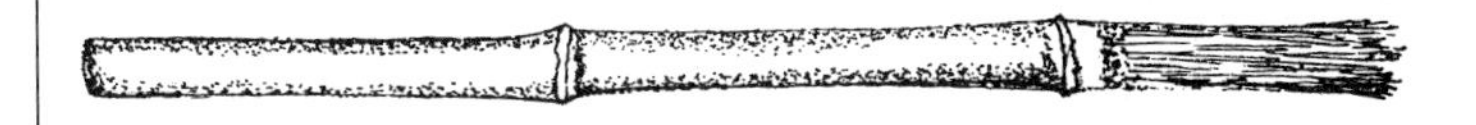

Brushes of bamboo, holding the fur of every imaginable animal, have been favored by Chinese artists for centuries. In a concession to ceremony, heavier, more ornate, and more expensive brush holders have been ritually presented and used, but the simple bamboo remains the preferred alternative. Sometimes, bamboo branches or thin culms are buried for several months, while microorganisms devour the starchy parenchyma tissue that holds the fiber in place; the brush and handle remain a seamless whole, as pictured here.

must derive from its help in preserving their traditions intact. Much of the history of Chinese custom and technology lies embedded in Chinese. Long after the solid things returned to their composing dust, the trace of them remained encased in ideograms through the fleeting and varied pressure of a bamboo brush flowing quick with ink and consciousness above thin strips of bamboo strung together like a fan. Some dug up in A.D. 281 had been buried 600 years or more—around 320 B.C.—and preserved details of Chinese history back to 2250 B.C.

> For white-washing, Chinese masons use brushes made of thin bamboo slips fastened together and secured in a handle of bamboo. The Malay has similar ones, but beats with a mallet the whole end of a bamboo joint until dissolved into fibres

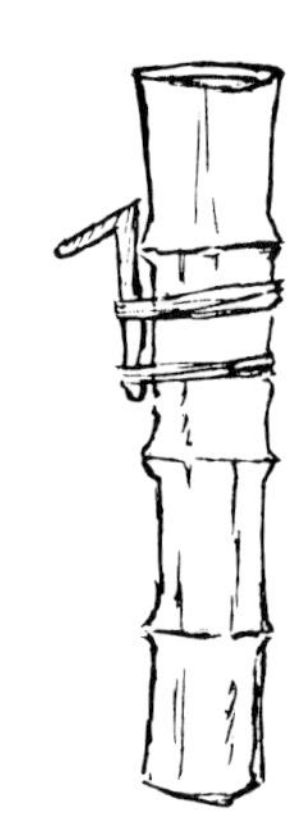

Bamboo water carrier. The four nodes are easily knocked out for a cheap, lightweight, virtually unbreakable jug. Its surface per volume is small, which means less spillage on the trail.

> . . . The small brushes, used in China for coloring pictures, are also made of fine bamboo shavings introduced into a small holder of bamboo.[35]

BUCKETS. "We do not ordinarily think of bamboo as a wood appropriate to the joiner's art. However, the making of the bamboo buckets and tubs used as containers for cooked rice is a trade in itself. Some thirty-odd tools, each with a special function, compose the kit of the maker of these bamboo vessels."[36]

"Single joints of larger bamboos serve well for small water buckets. Thinner joints are cut just below the nodes, and the Indian obtains in this way little tubes, solid below, in which he keeps fluids, honey, sugar, salted fish, or fruit just as we do in bottles and jars. Many a Javanese can be seen on market days carrying home in this tube, suspended from bamboo string, the oil, etc., that he wants in his little household."[37]

C Cables, cages (bird, cricket, tiger), candlesticks, canes, canteens, carts, castanets, catalyst (tabasheer), caulking, chairs, charcoal, chisels, chopsticks, churches, cigarette holders, clothes racks (pins, and poles), clubs, colanders, combs for hair and hand looms, cooking vessels, coops for chickens and ducks, couches, cow bells, cradles, crates, cribs, crosses, crutches and equipment for handicapped, cultures (for bacteria in lab tests), cups for drinking and waterwheels, curtains.

CABLES. "The historic travels of Marco Polo were among the first to reveal to the Western world the domestic value of bamboo. He tells how the Chinese manufactured cables for towing ships by first splitting canes their whole length into thin pieces and then by twisting these together into ropes three hundred paces long. It is interesting to note in passing that engineers experimenting for the Whangpoo Conservancy Board found twisted and plaited 4-inch ropes made with material taken from the outer eighth of an inch of bamboos being used to tow junks up against the current of the rapids in the gorges of the Yangtze River and estimated that the working stress was about 10,000 pounds per square inch of the material, this tension every now and then being doubled."[38] "On a journey up the Yangtze in 1908, Esterer made some measurements on the bamboo cables used by the junk haulers. He reckoned a tension of 7,362 pounds per square inch, of the same order as that normally taken by steel wires, yet the breakages were very few . . . Moreover, while hempen ropes lose some 25 percent of their strength when wet, the tensile strength of plaited bamboo cables increases about 20 percent when they are fully saturated with water."[39] Bamboo cables are available in lengths up to a quarter mile. As many as three hundred men may be tugging on them in the Yangtze rapids.

CANDLESTICKS. "Candlesticks are made of bamboo, superior to those empty bottles that European travelers often use. A node is left in the middle of a section of thin bamboo wide enough to receive a candle, with the portion below the node divided

into three. A stone wedged between these three furnishes a tripod candlestick."[40]

CARRYING POLES. "Big bamboo poles are used for carrying heavy loads in China. In the cities one often hears the familiar antiphonal 'heigh-ho'ing, indicating that a heavy load is being moved somewhere. The heavier the load, the louder and more agonized the chant. The load is suspended by ropes from the middle, and the ends of the pole rest on the shoulders of two men. In the case of heavier loads, the ends of the primary pole may constitute the center loads of two secondary poles, the weight then being distributed between four men instead of two. The chanting helps the men keep time, a very important factor in transporting the load easily. They get into the swing and can take advantage of the recoil of the pole to make their steps forwards. In this way the load is always heaviest when the two men, taking the first case, have both feet on the ground and lightest when they are taking a step. On the same principle, towed boats always have the towline fastened to the top of a bamboo mast because of its springiness. It gives with the step of the pullers, yet at the same time exerts an almost constant pull on the boat. A split pole about five feet long, tapered except for a small knob at both ends, serves as a carrying apparatus for one man alone, a small load suspended from each end and the whole balanced on the shoulder."[41]

"The use of bamboo for pikolan (carrying poles) is general amongst Malays, and even children are fond of appending their load (and were it only a few plantains) to a bamboo stick for the purpose of "pikol" as this mode of carrying is generally called. The bamboo halms are very strong, and can resist loads of 100 to 200 and even more pounds, but if exposed too much to the sun are apt to crack on account of heating the air enclosed in their joints. Smaller pikolans are made of a shape somewhat like bows, flattened and the edges rounded, often more or less ornamentally carved towards their ends. Loads of equal weight are fastened at both ends so as to keep the balance. When a Javanese has on one side only a load which he cannot divide, he appends as much weight (and were it only a stone) to the opposite end; so innate is custom in man. The carrying of such loads has its peculiarities, inasmuch as the carrier hastens in consonance with the elastic swingings of the bamboo, taking at the same time advantage of every swing that may lessen his burden. In this way he carries with less exertion a larger load than do the monotonously singing palkee bearers of Bengal, whose poles consist of unelastic wood."[42]

CARTS. Many kinds are made wholly or partly of bamboo. "Bamboo is also fitted for yokes of cattle, axles and even springs of the smaller carts. In Java, etc., these carts have a sort of little bamboo house built upon them with a vestibulum in front, wherein the driver comfortably sits, and often falls asleep without knowing it."[43]

The present planet importance of cart design can be brought home to a citizen of Americar only through prolonged residence in a rural area where oxen and other animals still carry or pull much of the load: "In India, it has been calculated that the total national investment in bullock carts exceeds the investment in either the national railroads or the national road network. The number of ton-miles of material moved is also comparable. (15–18 billion ton/km per year.)"[44]

> For short hauls, small loads, versatile movement over any available surface and low freight charges, the cart has no peer either in the rural areas, or, for that matter, in the towns and cities. It is still cheap, readily available, and safe. . . . [But the Indian] traditional cart is defective in design. The draught power of the animal is wasted due to friction resulting from rough bearings and crude and inefficient harnessing, etc. The wobbling rim cuts into the road surface and damages it . . . Weights run high. Traditional carts can be easily improved by: smooth bearings, lower weight, the introduction of a log-brake, better harnessing, the use of pneumatic tires on paved roads [hard rubber tires in rural areas].[45]

This analysis applies as well to Latin America. In the Nicaraguan village where we sit agreeing with Ramaswamy's analysis (above), we have been watching nearly two years now the six ox carts of the pueblo do most of the heavy hauling. Ramaswamy notes that animals provide 66 percent of the energy input in Indian agriculture; people, 23 percent; electricity and fossil fuels, 10 percent. This percentage reflects the reality in many countries, East and West.

CAULKING. Material for caulking is commonly made of shredded bamboo, prepared by scraping the culms, embedded in a putty of lime and tung oil.[46] So, bamboo is the crowning glory of junks (q.v.) in the battens of their sails—and the invisible, sine quo non caulking needful in the keel.

"The work of the Chinese shipwright, although ingeniously conceived and skillfully carried out, is of the crudest. This necessarily makes the caulkers' task a formidable one. Yawning apertures between the planks, deficiencies in the wood, careless clinching of nails and other minor errors of omission and commission demand a lavish use of putty, known as chunam, and bamboo shavings, or other material, often graving pieces of considerable size have to be inserted to fill up the larger gaps.

"In this connexion it is interesting to note that Marco Polo wrote of caulking as follows: 'The Chinese take some lime and hemp, and these they knead together with a certain wood oil; and, when the three are thoroughly amalgamated, they hold like any glue.' The mixture alluded to was, of course, what is known today in China *as chunam,* which is compound of lime and wood oil, a product of the T'ung nut. In some districts, in order to prevent the caulking from splaying out on the reverse side and necessitating frequent trimming, a plank is placed against the inner side of the seam as a basis on which to caulk. The chunam sets hard and white in about forty-eight hours with a good watertight join. In fishing craft today, caulking of large seams is carried out with a mixture of oakum and discarded fishing nets. The net is beaten soft, cut into strips, smeared with chunam, and hammered into the seam."

From the highly recommended *Sail and Sweep in China* by Worcester (1966: 8–9): Nine plates of models from the junk collection of London's Science Museum illustrate the history and development of this amazing vessel and provide hints for building bamboo model ships and rafts in schools. Worcester's book deserves reprinting, for its prolonged, firsthand, first-rate research; its lively intelligence and wit, sharp observing eye, and feeling heart. Those with no previous knowledge of junks who couldn't even imagine an interest in the subject are in danger of being infected by the author's own rambling fascination with his theme. Those concerned with appropriate village technology in developing countries now will find much to ponder. (For related matter, see *fishing, junks, rafts, sails.*)

CHARCOAL. Charcoal from bamboos is generally used in goldsmithery; it has properties that make it superior to conventional sources for use in electrical batteries.[47]

CHISELS. Chinese sculptors "use small chisels cut from the hardest part of the bamboo halms, and they are very expert in the use of them for carving plaster and such like soft material."[48]

CRUTCHES AND OTHER EQUIPMENT. A 1976 symposium at Oxford, England, on appropriate technology for the disabled in developing countries included a display of bamboo and rattan aids for handicapped children designed by J. K. Hutt, an English physiotherapist working with spastic children in Johore, West Malaysia. Her choice of material was dictated by the high cost of imported equipment, hard to repair and unsuited to the customs of the local people, in contrast with the local availability of rattan and bamboo, together with craftspeo-

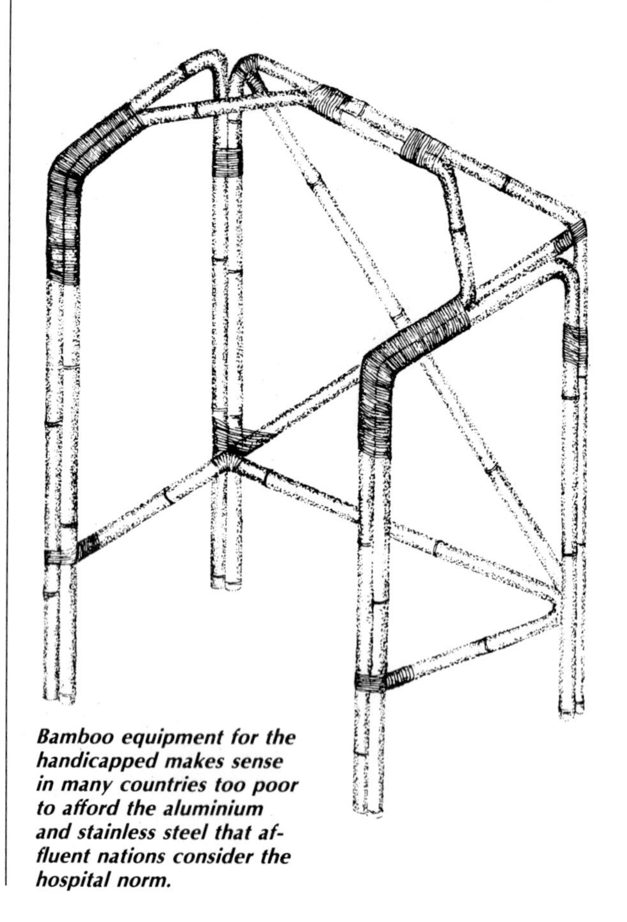

Bamboo equipment for the handicapped makes sense in many countries too poor to afford the aluminium and stainless steel that affluent nations consider the hospital norm.

ple in any village skilled in construction and repair with these native resources.

> Good quality cane also withstands the changes of a tropical climate, is cheaper than wood or metal, is more pliable and when heated can be moulded into almost any shape which it then retains. The woven network used for seats and backs of chairs, etc., allows the air to circulate and thus reduces perspiration and subsequent heat rash.[49]

The working drawings for this equipment, published in a brief volume, are being requested by countries such as Jamaica and Mexico and should prove useful in any developing country, particularly for wounded in the wake of the revolutions sweeping the planet. Drawings for a number of chairs and walkers are included, some of which would prove useful also for infants just beginning to walk. (Address in bibliography under Hutt.)

D Dams,
defensive fortifications (palisades, revetments),
deodorizers (dried leaves for deodorizing fish oils),
desks,
diesel fuel,
dikes,
dirigible,
dolls,
domes,
dowel pins,
dredge (fishing),
drogues,
dustpans.

Defense department.
"For defensive works *Bambusa blumeana,* a species very similar to *B. arundinacea,* serves well. It forms an impenetrable fence on account of its numerous dependent branchlets armed with copious sharp thorns, and such fences are generally planted around and in the trenches of the Malay fortifications and redoutes. These fences form serious obstacles to advancing troops in war and have been recognized as such by the Dutch military men who employ at present the same instead of palisades; for they prove more durable, really quite impenetrable, and against them even European artillery can do little. The same sort of bamboo is also extensively employed for fences around villages in tracts where tigers are uncomfortably numerous."[50] World War II demonstrated the sturdy capacity of bamboo to withstand A-bombs and bullets: "A flash over the city of Hiroshima on August 6, 1945, announced the arrival of the most horrible havoc imaginable. In the face of the world's first atomic bomb, within a matter of seconds streets and houses collapsed, trees and grasses were charred to bits and 200,000 souls—one half of the city's total population—perished. Some people who managed to survive the near-annihilation developed what is called the 'atomic disease' and died one after another. Today, twenty-seven years after the fateful—and to many Japanese the most unforgettable—day, a few patients succumb to the dreadful disease every month . . . In the wake of the relentless destruction, however, one living thing held out. In the very epicenter, a thicket of bamboo stood through the blast, suffering only one side to be scorched.

"The sight was an immeasurable encouragement to the war-shattered citizens. But the plants were not allowed to stay there long; they were dug out to build the Memorial Museum for Peace, and a portion of the plants is now housed in the Museum.

"During World War II, bamboo thickets often provided shelter to the Japanese soldiers under bombardment. They knew that the hard, enamel-like bamboo stalks could repel bullets and protect them. As an ex-soldier thinks back on the War, he recalls the strange sound of bullets ricocheting in the wood."[51]

While Japanese soldiers were listening to the weird *zing* of bullets searching for them in the South Pacific groves, halfway round the globe the Emperor's enemy was investigating improvements in bamboo revetment construction at the USDA Experiment Station in Puerto Rico. The U.S. Army had found the conventional revetment construction with sand-filled burlap bags unsatisfactory for two reasons: Often the bags were locally unavailable, and their life under tropical weather conditions was extremely short. A month after brand new, burlap in the tropics can be very old. A revetment of living bamboo culms offered the advantage of being a material readily available in much of the war zone providing defensive fortifications that, like wine, improved with age.

Bambusa vulgaris was selected as a readily available species quite easy to propagate. Though not a very strong bamboo, tests of water-cured *B. vulgaris* culms showed a maximum fiber stress of 14,960 pounds per square inch. It should be noted that soaking bamboo culms in water leaches out starch, which renders them less attractive to beetles, and therefore more durable, but also makes them more brittle and weak. Even so, the weakened culms of a relatively softwood bamboo compared favorably with the wood of hickory, which has a rupture modulus of roughly 20,000 pounds. The quality of fracture in bamboo is also useful for military defenses because its tenacious fiber splits but does not fragment easily: Hit by gunfire, pieces still hang on stubbornly by long stringy strands to a splintered culm.

Young bamboos whose side buds have not yet developed branches are recommended. If these are placed in contact with moist earth, a mass of roots proliferates around their base as well as from the root primordia that circle the basal nodes of *B. vulgaris* and a number of other tropical species. In a living bamboo revetment, roots at both these locations serve to nourish and anchor the wall. In the experimental revetment described, a 2-foot trench 14 inches wide was dug, care being taken to provide good drainage since bamboos love water but abhor bogs. Culm sections 18 feet long were placed in the trench, base down. Only basal sections were used, planted as close as possible to one another and then covered with clumps of grass for shading. Grass clumps were held in place by split culms tied to verticals. Horizontal hardwood poles were wired across the inner wall, at 3-foot intervals, thus uniting the entire wall. Fertile soil was used for back fill to encourage good development of basal buds, and posts were buried in it, from which guy wires were fastened to the horizontal hardwood poles in the revetment.

Large openings between the bamboos of the wall were chinked with grass clumps, their roots in the back fill helping to web and stabilize the soil. Guys were also attached to living culms planted horizontally in the back fill, their tips protruding through the palisade and wired to the horizontal hardwood members. Above the base of these horizontally planted bamboos, vertical bamboo vents were placed to assure they would not be completely smothered by the back fill. Three feet behind the palisade at 2-foot intervals, vertical bamboos with the nodes removed were buried in the back fill to facilitate irrigation and fertilizing: the culm tops protruded from the ground, and into these water and fertilizer solutions were poured.[52]

Thankfully, these building principles can also be applied to peaceable ends, such as contour barriers for erosion control on steep hillsides, windbreaks, terracing, stablizing small pond dams, live fencing, and general landscaping.

DIESEL FUEL. Soaring oil and gas prices have prompted interest worldwide in alcohol fuel. Although crop residues and tree crops could serve this purpose, attention has focused more on grain, cassava, and sugar cane. Bumper harvests in 1982 in the United States mean many more millions of dollars spent storing wheat, soybeans, sorghum, oats, barley, corn, and rye. Using emergency storage in boxcars and other improvised silos cannot fully confront the abundance, and there is talk of simply letting corn and sorghum rot unharvested in the fields to help cope with the plenty problem. In this context, fuel from grains seems to make a lot of sense—but if exporting nations begin to use up their surplus in alcohol fuel production, this amounts to an increased monopoly by the haves of the world's croplands and diminished food for the poor, particularly the city poor in developing countries.

Food or Fuel: New Competition for the World's Croplands (Brown 1980) examines the trend. New Zealand, for example, to fully meet fuel needs by the year 2000, would require an area equal to her present entire crop acreage. In Australia, 15–20 percent reliance on fuel from wheat would consume 100 percent of her crop of that grain. In the United States, 100 percent of the grain harvest would cover 30 percent of current automobile consumption. At current average use levels (10,000 miles per year), one motorist would require 8 acres annually—enough land to feed nine people in the United States or thirty-nine in underdeveloped countries where fewer grains are lost in inefficient conversion to pork and beef.

These figures don't count the liquid fuel used producing the grain crop to begin with. In Brazil, crop use by the 20 percent at the top of the economic ladder will triple—from 1 acre to 3—with the present shift to grain gas. All this in a context of rapidly increasing population and rapidly diminishing land: One-third of earth's presently arable land may be lost by the turn of the century. Third World food production must double before the year 2000 to maintain present deficient standards. Estimated available arable land by the year 2000—940 million hectares—roughly equals what abusive use has already turned to deserts: an area the size of the continental United States, too little to feed the world on, far too much to have thrown away.[53]

Granted this sobering planetary resource inventory, Brown concludes: "A carefully designed alcohol fuel program based on forest products and cellulosic materials of agricultural origin could become an important source of fuel, one that would not compete with food production."[54] In this careful design, what place bamboo? First, we could remark that as the valleys fill up, humanity is taking more and more to the hills—then shortly after our arrival, the hills leave for the valleys. China is a notable exception to our disastrous agricultural practices on sloping lands. Or, more accurately, her errors are more remote in time, and she is now setting about mending them with greater will and urgency than the rest of us. In any case, her old favorite, bamboo, provides a hillside crop whose harvest does not disturb the soil: "I know of no other type of plant growth that can yield an annual crop and at the same time serve as an effective year-round long term protector of watersheds."[55]

A second characteristic of bamboo's growth that makes it attractive for fuel production is its abundance: yearly increment in tree species is 2–5 percent. In bamboos, there is a 10–30 percent annual increase in the bulk of the grove.[56] We won't enter the complex details of fuel production from bamboo.[57] Its feasibility depends on many factors, but its productive capacity and ability to thrive on and protect marginal land recommend bamboo for study in any program investigating fuel production from vegetable sources.

DIRIGIBLE. Alberto Santos Dumont, Brazilian forefiyer of modern aviation (1873–1932), in 1901 constructed, of bamboo, the dirigible *Brasil,* flying in it from the park Saint Cloud to the Eiffel Tower and back (11 km) in 30 minutes.

DOLLS AND APPROPRIATE TOYS. Just as models serve as a useful tool-up for child architecture, handmade, homemade dolls can function as a miniature training zone for a sense of character and costume that can later be amplified into the art of puppetry and theater. Since we forget process in product, children are more and more conditioned to believe that the only real dolls are those that come in a box. They are not born from their own fingers, but adopted from a machine trained to make them but not care for them. Rilke considered dolls as they existed in his time and place a sad initiation into the received, ready-made nature of modern life. He never experienced the vital role of dolls in other cultures such as the Hopi, whose magnificent katchinas are a cottonwood Chartres. They function in the Hopi culture as stained glass did in medieval Christendom. Art can be distinguished from argument or ad or mere theology or party line because many non-Christians can "believe" in the breathless beauty of the rose window who do *not* believe that Jesus is the only log floating in our shipwrecked seas, and you don't have to be a Hopi to appreciate the complex celestial anatomy embodied in katchinas.

Appropriate toys, made by local people of local materials, are not built to confuse the child with plastic longings, but to make life clear and initiate the participation and relationship that is the essence of our social existense and social universe. *Tools* were recommended by Plato, Queen Victoria, and many others between them and since then as the most appropriate toy. With tools, the child himself becomes the doll, living out and imitating work rhythms vital for survival. Toys are not marginal to schooling in traditional cultures, but aids at the blooming center of learning to pass the culture on from mind to mind and hand to hand. We imagine the modern urban child's life as more "rich" in educational options than traditional children's, but examine the number of materials that modern children can explore and get to know in their toys and how much real texture is left in them, unerased by machine production and computer design. Compare this with the generous range of fiber and color, feather and leaf provided everywhere by

This bamboo pig with movable head can be made in many sizes, from very small—an inch or two—to riding size for a child, with a drum stretched on the rear end. The multiple use of space characteristic of the Japanese house is imitated in multiple-function toys that serve as friend, furniture, and another instrument for the family band.

earth to any village child. Coached early to look for them, our children could find as many dyes as George Washington Carver, the "Wizard of Tuskeegee," coaxed from the clays and plants of Alabama, including a dazzling hue in which Egyptologists were amazed to find the lost radiance of Tutankhamen's blue (Tompkins 1974).

Bamboo dolls and little animals have been the standard do-it-ourselves playtool of oriental kids for centuries. Twig bones, leaf body, sheath hat . . . thousands of variations have been demonstrated. Find them in the twigs and trash of your neighboring grove, along with others never found before your fingers gave birth to them in the watching eyes of your child.

Better a laughing and learning child on the planet, planting a carrot for his homemade doll with his real tiny spade than a man on the moon planting the flag of whatever imaginary nation . . . whose name, pride, and fall the moon will not remember even if you pave her surface with memorial plaques. An appropriate history is the luxury of appropriate cultures, impossible without the appropriate child. Toys are the magical *other* to the child; the environment is the prince that kisses our capacities to awaken creation. We *are* only in relationship. The early qualities offered by our environment for relationship determine largely our later capacities for creative exploration of the other, the new; toys are supercritical in learning to learn.

The boldest stroke for survival of any nation now would be to replace its war budget with a toy budget of equal magnitude. The deluge of creative human energy that would unleash could float all human navies in the palm of its mind like chips in the Pacific. War will end when we find more contagiously engaging and seductive beckonings to our energies. By invading a nation of complete children, soldiers of more sober, industrious nations never provided in their dull infancy with appropriate toys would become so enchanted with the world and people they had been sent to destroy that they would simply forget the war.

For want of a toy, that world is lost. The chief interest of our history now is that it teeters at that point called in bullfighting "the moment of truth": Will we select the Man and his Bomb, sooty as the devil in a medieval puppet play, or the Child and Toy?

DOMES. See "Bamboo polyhedra" in Chapter 4, pp. 106–107.

DOWEL PINS. "Dowel pins of bamboo are commonly employed by carpenters for joining boards edge to edge in the making of certain articles of furniture such as beds and wardrobes. The best dowel pins are made from the rind wood of *Arundinaria amabilis. Phyllostachys pubescens* is also used."[58]

DRYING AND STORING CROPS. Life after harvest is probably the central problem confronting small farmers and large agribusiness around the world. Loss following harvest in Latin America runs 25 to 50 percent; in Africa, around 30 percent; in Southeast Asia, roughly 50 percent. "This means that simple, low-cost, small-scale systems for storage and preservation are critically important, with the potential for more effect on the supply of food available to humans than dramatic gains in agricultural productivity."[59]

Ironically, superproducing new hybrids are often associated with the most dramatic losses: with monocrop hybrids, synthetic fertilizers, and pesticides, greater yields are realized—with the ecological cost in soil fertility, health cost in a less diversified, more packaged diet in the shift from subsistence to market farming, and social cost in increased dependence of country on town. Without wholistic designs that anticipate increased yields with increased capacity to process and store the abundance, the productivity becomes meaning-

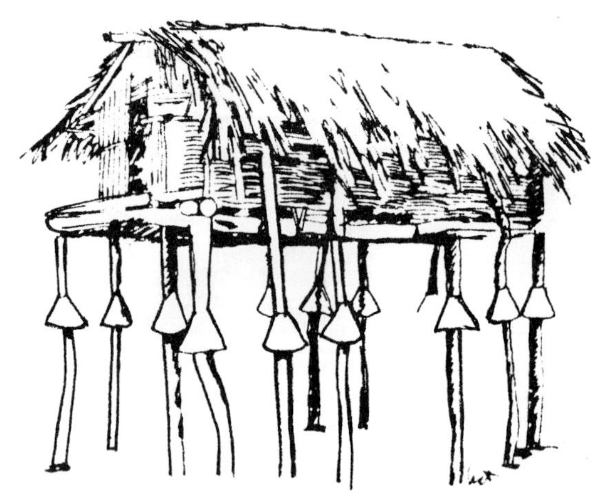

"An improved 2-ton storage bin with rat-guards" (Darrow 1981: 484). The theory and practice of rural technologies, such as the crop storage techniques shown here, are being studied in the village schools of developing countries. Studies and experimental works need to be related to the basic shelter, food, and health needs of the community, and develop rather than destroy the renewable resources presented by the local landscape.

less because it never reaches the poor person's plate; rodents, insects, and molds grow fatter than people do by most modern farming methods.

In areas of its growth, bamboo mats, racks, baskets, sunning floors, and storage cribs are all important in traditional harvest, curing, and keeping of crops . . . from rice in the Philippines to coffee in western Colombia—where ingenious small bamboo buildings have roll-off roofs that can be rapidly returned to cover the building when rain threatens and opened again when the clouds pass by.

> The hill tribes of Hainan, the Philippine Islands, and the adjacent mainland of Asia harvest their rice in short "hands" made up of the heads plus a 6 or 8-inch portion of the stalk. These "hands" are cured on a long narrow rack consisting of a row of posts set firmly in the ground with slender bamboo culms bound to them in a horizontal position at close intervals, and to a height of about 6 feet. The "hands" of rice are thrust between these bars in close order and allowed to remain there until they are thoroughly cured before being removed to the granaries. A narrow thatched roof protects them from rain. In the threshing, winnowing, and transportation of the grain, bamboo baskets, trays, and scoops are all important.[60]

Fences for animals, scarecrows for birds, spray guns for insects—all are fashioned of bamboo in the Orient to protect crops both before and after harvest. Other uses are more subtle, local, and specialized: "In the citrus groves of southern China where a certain species of predacious red ant is colonized on the trees to keep down parasitic scales and other insects, bamboo poles serve as a means of intertree transit for the ants."[61]

Bamboo matting is an omniuseful artifact in the curing of much farm produce: "Fruits and other products which would be spoiled by contact with the soil are spread out to dry on squares or rectangular pieces of coarse bamboo matting. Similar mats are used as overnight covers or during showers to protect farm produce being cured or dried in the sun."[62] Drying well is *the* critical factor in proper storage. Open-air drying is at one end of the scale in complexity and capital investment, with mechanical driers operating on some sort of fuel at the other. In between are solar driers of varying complexity, from a simple roof and woven walls through which air can pass over a stored grain to devices constructed of high-tech materials like glass, aluminum foil, and sheet metal. Fairly low-cost hybrids of industrialized and local materials are under experimental use: bamboo or other framing with clear or black plastic skin that raises the heat and diminishes drying time.

Crop drying and storage depends on many factors: the crop, local weather, materials, and traditions. Bamboo has proven a very useful material where available, and techniques have evolved in some areas—as in the Colombian coffee-drying houses with their easily removable roofs—which might find relevant application to other crops in other locations. Two constants of alternative technology are to be remembered: *adapting* distant alternatives to local conditions rather than blindly *adopting* alien solutions unadjusted to local contexts; and keeping community involvement in the foreground of all efforts. In reforestation, water supply, cook-stove innovation—whatever change design—workers worldwide have found that an absolute condition for success is *participation of the people.* Central to alternative-technology crop keeping is "understanding the principles of good grain storage and basing improvements on traditional techniques rather than the transfer of an alien grain storage technology."[63]

A successful grain storage effort in Tanzania evolved the following method: (1) They pooled local knowledge; none of the villagers knew how much they knew collectively. (2) A team of outsiders—Tanzanians and foreigners—provided alternatives from elsewhere in an open way, not as "solutions" but as options to be selectively embodied in local design at the villagers' discretion. (3) The eight-week project was integrated with the primary school; grain storage experiments were performed by the children. (4) Basis for tighter future relationship with the agriculture faculty of the local university was established through concluding the project with a seminar, taught by the villagers to fifty crop husbandry and rural economy students, on village grain storage. "An excellent example of proper, humble technical assistance in the context of real community participation."[64]

In light of the seriousness of the problem of crop storage; considering the extent to which bamboo has traditionally been used with an enormous range of crops in widely varying climates; acknowledging the usefulness of pooling global experience to inspire new hybrid technologies—it would seem relevant for some international organization such as FAO to research the subtle details of bamboo for crop curing and storage in Asian, African, and American agricultural areas. Findings should be published as *graphically* as possible for farmers, not bureaucrats. Complementary uses of bamboo for farm use should be indicated, as well as its possibilities as an environment doctor for erosion control, windbreaks, or waterway stabilization. Relevant species must be indicated. Where they don't grow presently, they have to be planted in government gene banks at experimental stations. Where they do, the species have to be made more effectively available to the people. As in the Tanzanian project, the effort must be connected with the educational system at all levels.

E

EGGCUPS. One of the most surprising things about bamboo is the poverty of objects made from it beginning with *e*. The enzymes and extracts mentioned above by McClure seemed too abstract.[65] Erosion control is not a thing but a function, though we considered using it to fill out the *E* section when an obscure article by Freeman-Mitford saved us with a brief treatise that includes the economics of eggcups in England, made of bamboo and imported from France in the nineteenth century.[66] Bamboo fits well with chickens in many ways: Groves provide a natural roost, protection from predators and weather, and a dense leaf fall sheltering lots of bugs to scratch for. Leaves are also chopped into chicken food as a vitamin A supplement (Squibb 1953, 1957). Some say bamboo leaves also help firm the eggshell. Harvested culms make excellent hen houses. Chickens, in return for these services, provide excellent dung to fertilize the groves. The bamboo eggcup provides a final tiny sample of this interdestiny.

(*U, v, x, y,* and *z* are other letters almost equally empty of bamboo artifacts. We have sought refuge in various ethnic instruments—dressed in Western names—and exotic architecture to inhabit these waste spaces. In this battle for a bamboo alphabet, we console ourselves with the reflection that, among the orders of existence, though alphabetical order may be among the most useful, it is the least significant. However, in defense of a catalog or dictionary as an art device, it should be remarked that the bursts of unrelated data yield interesting neighbors. Synchronicity imposes randomness on newspaper layout. The alphabet achieves the same arbitrary juxtaposition of news, hinting new associations. Randomness is an important feature of oracles: it bursts the bounds of reason and taste, those confining columns you could tie a donkey to for a thousand years. Inspiration comes in disorderly bits and flashes too hot for the hyperorderly left brain to handle. The alphabet becomes a useful antiorder.)

F

Fans,
farming uses,
fences,
fenders (for ships or docks),
fertilizer,
fiesta assistant,
fifes,
firearms,
fire starters,
firewood,
fireworks,
fishnets,
fish poles (and traps),
flagpoles,
flails (for nut harvesting and so on),
floats (for fishing),
flooring,
flowerpots,
flutes,
flying art,
food (shoots and animal fodder),
forage,
forms (for bean meal in oil press, for reinforced concrete),
frames (for pictures, silkworm culture, and so on),
fruit pickers,
fuel,
furniture.

FANS. These are as associated with the Orient as bamboo—from which cheap and sometimes quite elegant varieties are made. Folding fans, introduced into China in the eleventh century from Japan, were more used by men; round, fixed fans—the more ancient style—by women. Fans were so common a part of life in China that the condemned even carried them to their executions. An abandoned wife was known as an "autumn fan"—discarded after a hot spell.[67] Austin's *Bamboo* provides an extensive series of photos documenting Japanese fan-making in Kyoto:

> Traditionally, the operations have long been divided for the sake of economy in manufacture. There are over twenty-five steps in making a sophisticated fan, and even today they are

> carried out in not less than seven different places, each of which has its own specialty. These are family groups, working deftly in small rooms, whose network of cooperation spreads over the city like the roots of bamboo itself . . . Of late, however, fan-makers have been complaining that air conditioning is threatening their livelihood.[68]

FARM FRIEND. *The Samaka Guide to Homesite Farming* is a plan for Philippine families who wish to produce their own food on a small plot. First published in 1953, the *Samaka Guide* has gone through a number of editions and proved to be a quite popular "how-to" for small farmers in a land where bamboo is readily available and widely used. As such, the guide is an index of the omnihandy nature of bamboo on a subsistence farm where the plant is presumed to be available.

Bamboo use is as everywhere as wood was on an early American farm. The house, to begin with, is framed in wood or bamboo and covered with *sawale,* a bamboo matting. The tripod for drilling a well, the walls to line it, and its water pipes are all made of bamboo—and the water will run to the kitchen and shower from the sun-warmed drum on the bamboo tower. Seeds in bamboo seedboxes on a bamboo stand are sheltered by bamboo sunshades till transplanted to a garden (protected by a bamboo fence) to be hoed and raked, trellised and staked with bamboo, harvested in a bamboo basket and eaten, with bamboo chopsticks, in a bamboo bowl. "Homemade fertilizer"—rice straw, leaves, manure, garbage, and weeds—is gathered in bamboo baskets and "cooked" in large bamboo composting compartments.

The guide focuses equally on gardening and livestock, and bamboo seems equally employed in both. A goat shed and holder to still a goat while milking, a bamboo chicken cage with feeders and water troughs of bamboo, duck pens and pig sheds —animal shelter of all sorts and sizes—and a

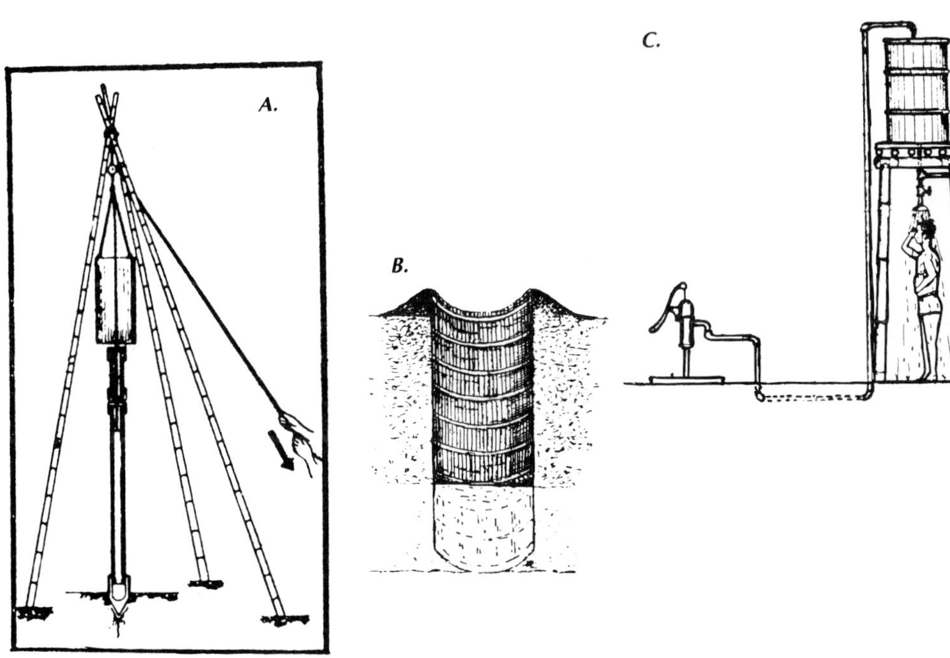

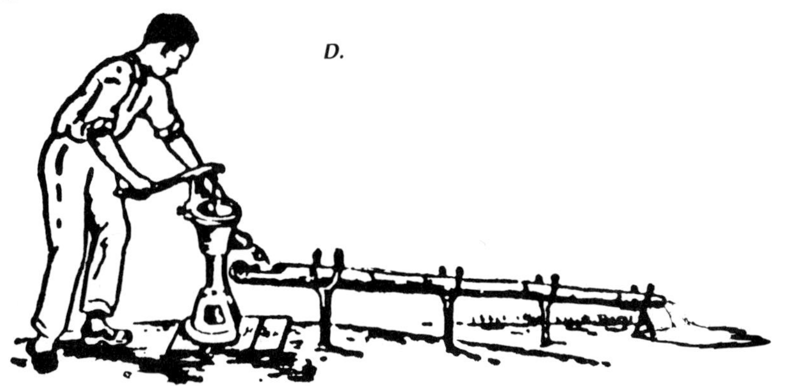

Water systems. From the tripod for hanging the drill (a) to the slats for walls that line the finished well (b); for plumbing in the house (c) or irrigation in the garden (d), bamboo provides homesteaders in the Philippines with a raw material they can grow on location as windbreak and living fence.

slaughtering table of split bamboo are pictured, together with a bamboo fish trap and dip nets for "our family fishpond."

Although bamboo is the material in thirty-three of the ninety-three illustrations in the guide, no mention is made of its propagation, cultivation, or harvest. Fences, houses, sheds, and gates are all constructed of it in the back cover map of the ideal 600–1,000 square meter yard, but no bamboo is listed among the recommended plants. The presumption that because bamboo is plentiful in a given area, it will continue to be so has proved

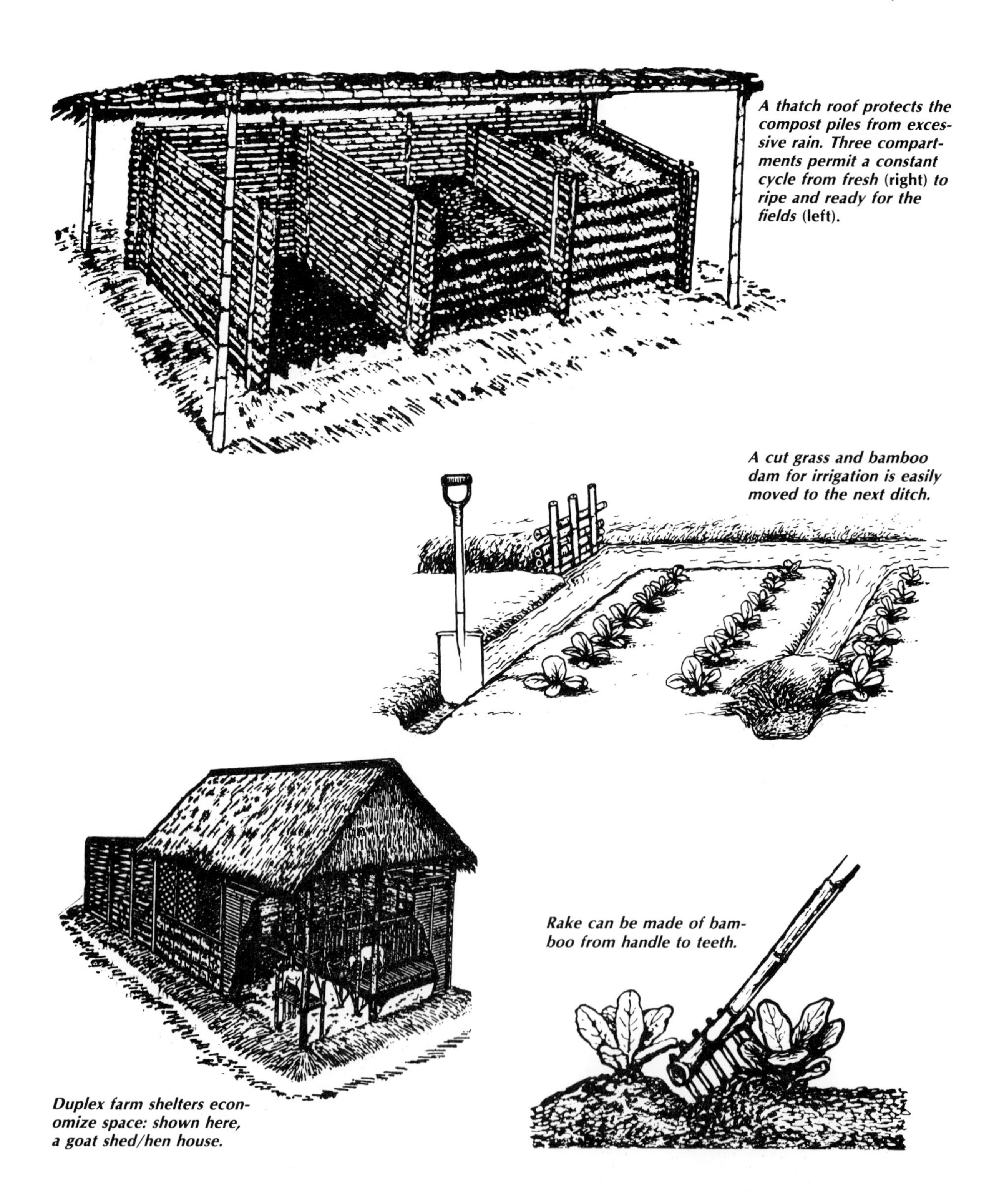

A thatch roof protects the compost piles from excessive rain. Three compartments permit a constant cycle from fresh (right) *to ripe and ready for the fields* (left).

A cut grass and bamboo dam for irrigation is easily moved to the next ditch.

Duplex farm shelters economize space: shown here, a goat shed/hen house.

Rake can be made of bamboo from handle to teeth.

incorrect in a number of countries. For the guide to neglect teaching people how to grow their own bamboo is a serious omission. The future of bamboo is actually in the hands of those who could profit most from its use—the poorest people living directly from the earth. Their governments, unfortunately, have not been able or willing to subsidize the spread of bamboo rhizomes and information, so the ordinary people in most countries where bamboo thrives are very unsophisticated in its cultivation and have no book knowledge of how bamboo is used in other countries. Relevant information and preferred species should be available to the people at a cost within their means.

In Colombia, for example, although bamboo is actually a very important part of the economy and is much used even in urban construction, it is completely unmentioned in school books about the economy and agricultural produce of the country for grade and high school students. This neglect was characteristic of all Latin American countries we visited from Mexico to Peru.

FENCES. Those made of bamboo are endless in their extent and variation in the Orient, greater, collectively, than the Great Wall—of which bamboo supposedly formed an original component.

In upright, horizontal, or diagonally lashed whole culms, in bunches of branches, in lathes woven in an infinite manner of ways, in functional walls around properties or merely decorative divisions within a garden, bamboo has perhaps made as many miles of fencing as any single construction material in the world. Tough, light, relatively inexpensive, bamboo fences are most durably constructed sustained by hardwood posts, elevated a few inches above ground level, and set vertically to drain more readily. A large culm split in two and placed as a watershed above poles—or a small thatched roof—gives a more finished look and increases fence life. In unroofed fences of whole culms, the top is ideally cut flush with a node to prevent collection of rainwater.

> The chief value of (woven) bamboo fences lies in the fact that they are all-inclusive. Anything, even small chickens, put in an enclosure surrounded by bamboo is really enclosed. Furthermore, it is very difficult to scale them—they are so shakey and springy. They seem so unstable, but in reality a bamboo fence in good condition is superior to anything except the woven wire stock and poultry fences used in the U.S. It may not be heavy enough to withstand a violent attack by a water buffalo—the posts might give way or the wire fastenings might break—but the fibres would never tear.[69]

Unless you have a water buffalo living next door, you might consider bamboo for alternative fencing.

In sheer bulk, there are few things people use —other than houses—that require more raw material to make than fences. To look backwards down them a bit into their brief history on this continent: Although Ben Franklin experimented with wire for cattle pasture and an "Account of Wire Fencing" was read at the Philadelphia Agricultural Society in 1816, it is only roughly a century since barbed wire (1873) and woven wire (1883) came to loop from post to post across America . . . and then the world. Before this, farms in southern states had even been abandoned for lack of adequate fencing materials when hardwoods such as locust, cedar, chestnut, walnut, or white oak—preferred in that order— were no longer massively available. After the Civil War, the U.S. Army's inventory of 7 million miles of wood fences added up to over $2 billion invested —at $300 a mile. Does anyone know the United States or world investment in fences now? As there are more of us and more fences, as the cost of metal and wood soars, the use of bamboo for fencing, proven for centuries in the Orient, makes ever more immense sense. Landowners should be encouraged to grow their fences on location rather than wasting precious energy transporting them, prefabricated, from somewhere else.[70]

Many living plants can provide uprights for a bamboo fence. Pruning at the top creates a denser, bushier barrier. Cassava-bamboo, a favorite Philippine combination, is shown here.

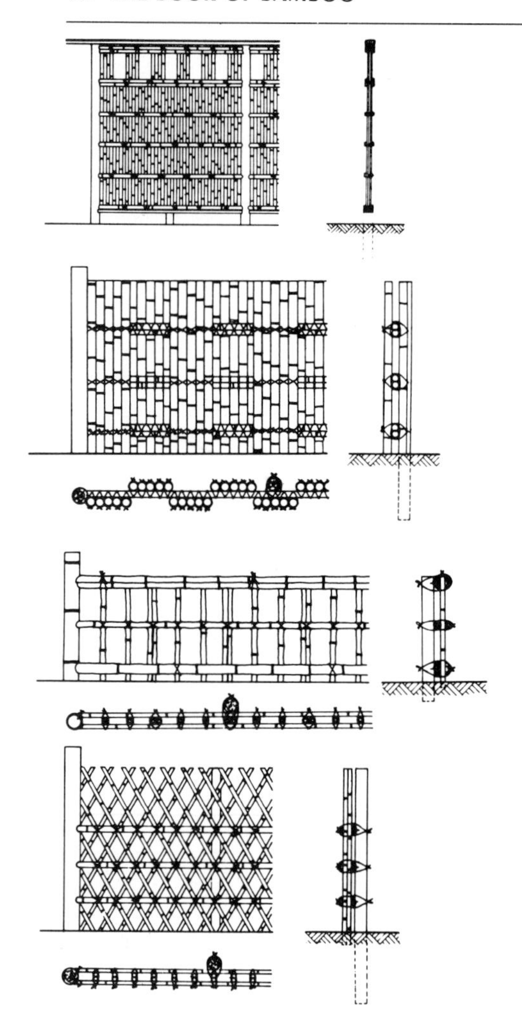

FENDERS. These are used on oriental docks and boats like old tires in the West: "Twenty-five feet of split, tapering bamboo, swiftly twisted into a springy circle and hammered over a fishtub . . . is able to withstand for years the punishment it gets on trawler and quayside."[71] Its durability in this use further testifies to the amazing toughness of bamboo fiber.

FERTILIZER. "The leaves of bamboos that grow wild in the mountains are extensively used as green manure by farmers in Japan. When culms and branches are harvested, it's wise to let them lie in the grove till the leaves have all fallen from them, for the leaves serve as fertilizer for the bamboo. Much used for paddy fields, the leaves of wild sasa are best gathered during the blooming of the wild cherry trees. Sasa leaves are also an effective fertilizer in growing ginger, helping to brighten its red color and increase the crop."[72] Contrary to popular belief in some areas that bamboo depletes soil fertility, studies by Ferrar-Delgado (1951) comparing various crop yields in lands cleared of bamboo with regional averages suggest that bamboo maintains or increases the vitality of the soil.[73]

FIESTA ASSISTANT. Bamboo improvisations embellish ceremonies and celebrations in all countries where the plant is native. Immense towers for fireworks, shelters for patron saints, poles for banners, frames for giants, floats of infinite descriptions in parades, portable altars in processions—bamboo's lightweight and rapid assemblage make it a favorite fiesta assistant that plays dozens of roles in many cultures. Related to these flashy functions is its extensive use in circuses, puppet theaters, and mobile shows of many sorts.

Triumphal arches are another standard structure in bamboo's repertoire. "In the highly ornamented gateways erected over the road to be traveled by an honored guest or conquering hero, the versatility of bamboo as a building material and decorating medium is exhibited to fine advantage," remarks McClure of his experience in China.[74] Kurz speaks of the same art in Java: "No one who has not seen them can fairly appreciate the skill and taste displayed by the Javanese in this sort of work. Yesterday you saw nothing but a heap of fresh-cut bamboo halms, and today these rude bamboo poles gradually become converted into arches, gates, and structures of the most exquisite patterns, filled out with skillfully wreathed trellis work. Broad and thin strips of bamboo and the soft yellow sheaths of the plantain leaves, taken from the interior layers of the trunk, are folded into artificial stars or flowers and ornamentally arranged, garlands of bamboo material intermingled with natural flowers gracefully hanging in tasteful designs, a few gorgeous bouquets added as a finish."[75]

FIREARMS. Fire starters, firewood, and fireworks: bamboo is as intimately linked with fire as it is with water—appropriately, perhaps, since it is a thirsty plant that flourishes best on sun-drenched tropical riverbanks. Bamboo fiber makes excellent candlewicks; old cables are used for torches; tinder and fire starters of different designs are used in various locations; and some theorists even attribute the origins of human use of fire to the observation of spontaneous combustion in bamboo groves arising from the friction of old, dried culms rubbing against each other.[76]

According to the Greek myth, Prometheus

Bamboo was used to make guns in 12th century China. Metal barrels soon replaced bamboo, but the metalworkers continued long afterwards to include the bamboo nodes in their designs.

stole fire from heaven—perhaps a metaphor attributing the first human hearth to fires started by lightning. Such fires could have been used and preserved by primitive peoples; but not having lightning at their disposal, it was not a technique for kindling fire that they could imitate. Friction of culm on culm, however, provided an example within reach of human hands. The technique persists in many locations where bamboo is available and matches aren't.

> In various out of the way places the people easily make fire by friction from the stems of such bamboos as *B. polymorpha,* and from other bamboos with large cavities and relatively thin walls. The internode is split in two lengthwise, a narrow slit cut through the round back, the hollow beneath this filled with bamboo-shaving tinder, a sharp-edged bamboo strip rubbed across this at right angles like a saw, making hot powdered bamboo drop down through the slit onto the tinder below. When a spark develops in the hot powder, it is blown to a flame in the tinder. Fire may be secured in 30 seconds by this method.[77]

Marco Polo reported the use of green bamboo tied near campfires to explode as the air in the internodes heated, providing an audial wall against tigers and other savage beasts. The effect on human witnesses was sufficiently impressive to fix a name on the plant supposedly deriving from this quite peripheral use: BAMBOOOM! Whether or not bamboo taught people to make fire, it's suggested that these explosions were the first fireworks, predating by several millennia the first mention of gunpowder—or "firedrug" as it was called—for theatrical performances around A.D. 600.[78] Gunpowder was in military use in land (mines) and air (arrows and catapulted fire bombs) for some two centuries before bamboo "fulfilled its most fateful destiny," as Needham remarks, as the grandfather of all guns.

> In 1233, the use of a "fire gun" by the Chin Tartars is recorded, but this instrument seems to have been a sort of flame thrower. By 1259 an actual gun barrel is recorded, though made of bamboo reinforced with fiber wrappings; the first metal gun barrel is recorded in 1275. For some time after their introduction, however, metal gun and cannon barrels were made ribbed, as though in imitation of the bamboo they replaced.[79]

FIREWOOD. Asked to describe their least favorite disaster—the one of many ugly candidates that most endangers the earth—*deforestation* would lead the list of many most-informed students of how the world's vulnerable balance is being kicked asunder by human mismanagement. Blundering lumbering and clearing forest land for cows, crops, and human settlements are all taking their toll. But a larger chunk of woods than most realize goes up in smoke for fuel. How can bamboo relate to this critical problem? Its magnitude justifies at least a brief sketch of its complexity.

> The humid forests of the tropics once occupied at least 1,600 million hectares [4,000 million acres] and have not only been the main centers

Quiz

The underdeveloped countries' share of total world wood use is:
5% 10% 30% 90%

Worldwide, the greatest single consumption of wood occurs in:
forest fires construction pulp manufacture firewood

Most of the world's people cook on:
gas kerosene woodstoves open fires

In the national energy budgets of the poorest countries, firewood needs take up:
15% 45% 75% 95%

Energy use in open wood fires, compared with gas stoves, is:
80% less 20% less 20% more 400% more

In Guatemala, Nepal, and Upper Volta, monthly cooking fuel costs are the equivalent of a U.S. family paying:
$0.40 $4.00 $40.00 $400

Every year, a village doubles the wood it cuts. After 19 years of cutting, half the forest still remains. The other half will disappear in:
38 19 9½ 1 year(s).

(In each case, the correct answer is the last.)

for living species on earth, but have held the lands together, moderated and modified world climates, and helped to maintain a desirable balance of atmospheric gases. Now they are vanishing at an incredible rate. There are reported to be 935 million hectares in actual humid tropical forest, a 40 percent reduction in total area. They are disappearing at a rate of 16 million hectares per year.[80]

Some 630 million people live on the dry lip of the expanding deserts of the world, which have already swallowed one-third of the planet's land. The Sahara alone claims 250,000 more acres annually. Modern agricultural practices, overgrazing, exploding populations are all important factors, but perhaps the most critical contributor to the global dust bowl is deforestation—the trunkless stumps left in the parched wake of the 1 billion, 500 million people gathering firewood to burn beneath their dwindling pot of food. From 50–70 percent of *all* wood cut in the world is used for cooking. The average wood-burning family uses 4 tons per year, double the wood consumption of richer countries for construction, paper, furniture, and firewood combined. Firewood in the developed countries represents 0.4 percent of energy used; in the developing world, 25 percent. In some countries, the figure is as high as 87 percent (Nepal) to 94 percent (Upper Volta). In richer countries, firewood claims 10 percent of total wood use; in poorer countries, 90 percent.

Forests influence the wind, temperature, humidity, soil, and water in ways often discovered only after the trees are cut, and these functions—usually beneficial to the people—are sabotaged. Forests assist in the essential global recycling of water, oxygen, carbon, and nitrogen —and without any expenditure of irreplaceable

Forty to fifty pounds (twenty kilos) of firewood makes a good load.

fossil fuels. Rainwater falling on tree-covered land tends to soak into the ground rather than rush off; erosion and flooding are thus reduced, and more water is likely to seep into underground pools and springs.[81]

As the world's forests disappear, and petroleum products soar ever farther out of reach of the planet's poor, dung is increasingly used in rural areas as fuel, which accelerates soil deterioration still more. Land becomes more scarce with greater population, less fertile with ever greater abuse: More people move to nearby cities. There, charcoal replaces firewood, using up to three times as much wood for creation of equivalent energy. It took five hundred thousand years, at least, after the emergence of the human race as a distinct species for our population, in 1830, to reach 1 billion. This figure was doubled in a century (1930). By 1960, thirty years later, 3 billion people were quarreling over the planet's resources; by 1975, 4 billion. At the present rates, 1988 will find 5 billion, 1998, 6 billion people increasing and multiplying on the face of the earth. Trees cannot keep pace with human use—without the careful collaboration of human consciousness and sweat.[82]

The rate of deforestation in an area is not a steady curve. For centuries, people get their firewood from a local forest. As population grows,

Four loads supply a week's wood; they stack about as tall as the little old woman often gathering them, and are about twice her weight.

wood consumption increases about 2 percent per year. At a certain moment, consumption and natural production are equal. From this point on, overcutting begins to reduce the forest, at first unnoticeably. After nine years, 10 percent of the forest has disappeared. Concern begins to be felt. After twelve years, 20 percent of the forest is gone. Only ten years later, the forest has been completely felled. Beyond a certain point, it is almost impossible to stop deforestation.[83]

Once use exceeds natural growth, the dangerous and descending spiral increases momentum at a shocking pace. Reforestations in the Sahel, for example, cover 2 percent of the annual loss. ". . . to supply 30 million Sahelians in the year 2000 . . . 3 to 6 million hectares will have to be planted . . . 150,000 to 300,000 hectares of forest annually up to the end of the century. . . . Nepal will require about 1.3 million hectares."[84] Forested lands in China have doubled since 1949, however. Green China indicates that reforestation *is* possible through a national policy in which "tree planting is

A year's supply—more than two hundred loads—dwarfs the supplier.

everybody's business."[85] South Korea is another nation where reforestation is actually happening.[86]

> Each person in towns in the semiarid areas verging on the Sahara needs one cubic meter of stacked firewood a year for cooking and heating. . . . It takes two hectares of natural forest to supply one townperson's needs and, as cities grow, that means a wider and wider search for fuel. It is estimated that by 1990 people will be hauling firewood into Ouagadougou from a radius of 150 km around the Upper Volta capital. But a single hectare of Australian eucalyptus in a plantation irrigated from a river can supply the needs of 50 people.[87]

The one bright spot in this rather bleak hearth horizon is—stove design. "Simple stove models already in use can halve the use of firewood. A concerted effort to develop more efficient models might reduce this figure to one-third or one-quarter, saving more forests than all the replanting efforts planned for the rest of the century . . . Use of improved cookstoves has an *immediate and lasting* effect on wood consumption with a 25:1 estimated cost advantage over tree planting, providing benefits without a long waiting period . . . directly to the households which participate."[88]

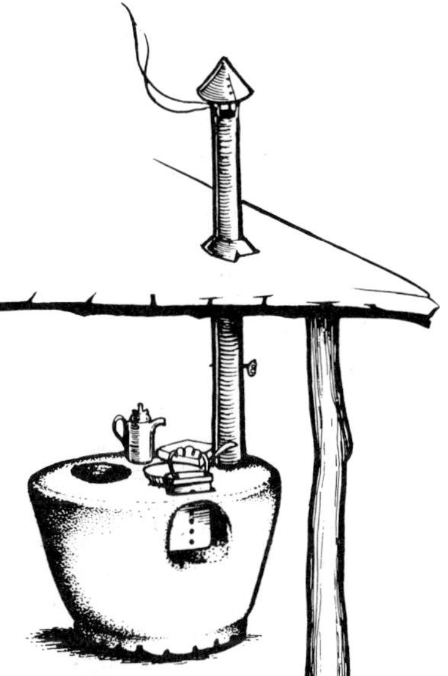

Lorena stove: (lodo-arena in Spanish) owner-built stoves of mud and sand developed in Guatemala by Aprovecho, an appropriate technology collective in Eugene, Oregon, feature a tight door to control drafts, pots tailor-fitted to pot holes, and a chimney with draft control. These simple design improvements cut wood consumption by 25% or more and eliminate indoor smoke.

The book providing this analysis—*Helping People in Poor Countries Develop Fuel-Saving Cookstoves* (Aprovecho 1981) faces a problem that directly or indirectly touches every person living now or in the burning future. Any action that serves to reduce the acuteness of the world's wood shortage should begin quickly, which returns us to one of the quickest life forms swarming the planet

Dip-net fishing raft.

—bamboo. Bamboo is used as firewood already wherever it abounds and hardwoods don't. But nowhere in print are there reports of any serious government or private efforts to establish groves of the most suited species to take the weight off wood for fuel consumption. *Dendrocalamus strictus,* famous for resistance to drought, might be one useful species.

Bamboo's role as a wood saver in paper pulp production, construction, and manufactured articles is already significant. To these uses, world governments could profitably add fire-bamboo so that the "wood of the poor" could not only shelter them but help cook as much as possible of the world's soup. Providing rhizomes and awareness of the how and why of bamboo cultivation would give people a means for personally, directly kindling solutions to a chilly problem confronting the race. One rhizome of *Bambusa vulgaris* produced 200 culms in five years in one experiment in Colombia. If this plenty could be fed to cookstoves designed for bamboo, it could contribute much to the peace and contentment of our crowded planet: *panza llena, corazón contento,* "full belly, happy heart."[89]

FISHING. If you ask most people in the United States about their associations with bamboo, the most typical response usually includes the springy feel of a bamboo fishing pole from a childhood morning. Bamboo, which grows so well and so conveniently for people who fish by streams, ponds, lakes, and rivers, has many and deep connections with fishing and other aqua-industries of the Orient as well, such as clam digging, and oyster and seaweed cultivation. Its lightness, its float-, flex-, and weaveability are among the chief reasons for its long associations with island cultures like the Philippines or Japan, where as many as 1,000 culms of *moso* bamboo may be used for the floats of a single giant net. Oshima (1931) reports one thousand boxcars, carrying 900 culms of moso each—nearly a million culms—used annually for Japanese seaweed culture. The antiquity of fishermen's alliance with bamboo is embedded in the Chinese language itself.

> This fact may be verified by anyone, even though he may not be privileged to see the varied bamboo gear that is an essential part of the Oriental fisherman's paraphernalia. It is sufficient to look up the names of these objects in a Chinese dictionary, for a great many of these complex terms (ideographs and pictographs) contain the symbol for bamboo. This fact signifies that even before their names were first reduced to writing, bamboo was employed in the making of the devices themselves. It is perhaps sufficient for our purposes to mention a few of them: traps, weirs, sluices, barriers, poles for hook-and-line fishing, spears, sea anchors, floats, trays and poles for drying fish and baskets for supporting them, netting needles, poles for drying nets, punting poles, and scaff or dip nets, including karojals and salambas. The dredges, punting poles, sieves, and sea anchors of Oriental clam-dredging equipment are all made of bamboo.[90]

Just as agriculture evolved from hunting and gathering, "water farming" is rapidly evolving in many countries now—from ancient roots. With increasing population, there is ever less land per capita, and that less is ever more eroded and debilitated. The world is two-thirds water to begin with, so more and more people are turning to aquaculture and "ocean ranching" as a partial reply to our race's hunger pains. Turtles, frogs, clams, shrimp, lobsters, eels, and various seaweeds are among the flora and fauna cultivated in addition to many species of fish. Water creatures don't waste energy supporting their weight or regulating body temperature, and the three-dimensional medium provides cubic acres rather than merely acres to support life, so the yields of water cultures can be considerably greater than terracultures: In Thailand, a fish farmer earns six times as much as a land farmer . . . who in turn is earning more than the average coastal fisherman in that country.[91]

Aquaculture is receiving increased interest recently, but its practices are traditional lore of many countries, where it is not an alternative but a complement to land farming. This balance has in some cases been upset by modern agricultural methods, as in Indonesia where the family rice paddy was a complex ecosystem providing a life opportunity for shrimp, frog, eel, and fish cultures along with vegetables, legumes, and tree crops grown with the rice, on embankments, and between crops. A hybrid monoculture rice, using lots of expensive fertilizers and pesticides, reduced or wiped out this delicate balance of food sources, diminished the water creatures and increased the role of bankers in the farmers' lives. Bamboo, in the Orient, has always formed an element in the traditional diversity of agriculture. In aquaculture, it can help construct and preserve the banks of a small pond to begin with. Rafts, nets, docks, floats, rope, and line—much of the bamboo gear of traditional fishing is used in modified ways in aquaculture, and bamboo should be considered a relevant crop on the bank of any aquaculture system. Cage culture, one format of fish farming, presents obvious possibilities for bamboo use. Oyster cultivation in Taiwan, seaweed farming in Japan, are examples of bamboo's alliance with aquaculture we will see elsewhere. In China, 4 million tons of fish per year are produced;

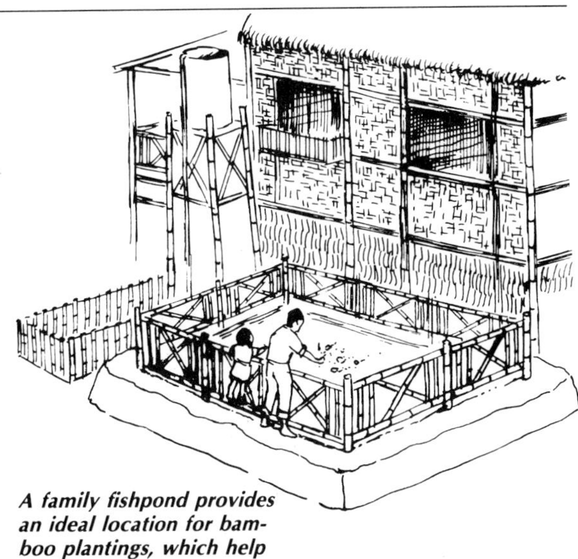

A family fishpond provides an ideal location for bamboo plantings, which help stabilize the banks they drink from. Small ponds can be built above ground in high water zones. A bamboo fishtrap facilitates selection of mature fish.

and their breeding jars, where this abundance begins, use bamboo "showers."[92]

> In the time of the Chin dynasty the people of Ch'ien Tang made a bamboo dam in which they caught a million fish a year. In consequence, it was called a "million worker dam." In the Ming period, a fence of plaited bamboo was built in ponds used for rearing fish. The most ingenious fishing baskets in this period were made of small plaited bamboos. The cover was of woven bamboo splints to which hairy or bristling bamboos were fixed. The basket gradually decreased in size from the mouth to allow the entrance of the fish, but not their exit.[93]

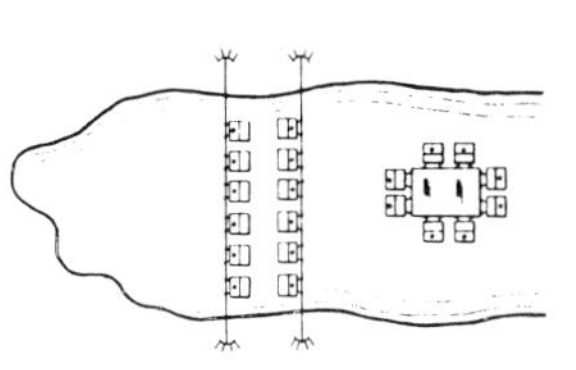

Cage culture: fish farmers in Thailand earn roughly six times as much as fishermen. Cages can be connected to the dock, to parallel cables anchored at both ends to shores, or fastened to a floating raft.

Tall tale

Fishermen are notorious liars, but here's one story they tell in Japan: Once upon a time in a small mountain village, a small boy went to cut bamboo shoots. The day was warm and he hung his jacket on the tip of a "bamboo child" *(takenoko)* as the new shoots are called in Japan. Then he walked slowly through the grove between the towering culms, barefoot, carefully listening with his toes to the forest floor for the slight bulge, which you could feel before you even saw it, a sure sign that a sweet white shoot was waiting in the rich black ground beneath, just ready to break through the surface of the soil. There you would scrape together a mound of loose soil so that the shoot could grow a bit more before reaching sunlight, which quickly made the tender flesh of the new culm tough

> and bitter, excellent in time for a thousand things, but no longer possible to eat.
>
> A bright but bamboo-muted sun filtered softly down through deep masses of tiny foliage far overhead, leaves alive always—even when the breeze died—with a slight tremor that made the very air tremble with vitality. And when the wind lifted again, culm leaning and scraping on tall culm hidden somewhere high above in those gangs of green loveliness, groaned like grandfathers in a troubled sleep . . . sounds eerie and ancient, full of a huge hollowness, murmuring resonant mysteries, moaning and creaking like a door in dreams, sounds weird and wonderful, which took all time away.
>
> At that moment, something in the boy prompted him to glance back to the shoot where he had left his jacket. The culm was growing so quickly that his coat was already out of reach! He ran back and began clambering up to retrieve it—but the culm was stretching up so fast that he soon found himself above the tallest bamboos in the grove . . . above the forest . . . higher than the surrounding mountains . . . taller than the clouds. Far, far below he heard, a little later, his parents calling him. "Up here!" he yelled back down to them. "Hold on tight," they shouted. "We'll cut you down." When the culm started falling, he held on tight for a day and a night, and the next morning—*thump*—the giant culm landed, its tip just reaching the ocean beach. The boy stood up and gazed in amazement out to sea. Nobody in his mountain village had ever seen it before.
>
> Meanwhile, back in the mountains, his parents and friends had climbed up on the immense culm and had begun walking down it to find him. Each node was so enormous that crossing was like climbing wall after wall . . . but finally they reached the tip and found not only their son, but the sea as well. And they found fish and lobsters, oysters and pearls and seaweed—all the abundance of the deep ocean —led by bamboo.

> No single gloss exhausts this tale, but among its meanings lies a magical shorthand for the ancient link between bamboo and the fishing industries.

The culm reaching from the mountains to the sea suggests as well the bamboo rafts floated from inland groves to ocean ports, in some times and places in the Orient up to ten thousand poles in a raft (Piatti 1947). Used for transport of freight and passengers as well, these giant rafts were made of a multitude of culm-length rafts stacked end overlapping end, like roof tiles—or like sections of the bamboo culm itself.

Bamboo not only led oriental peoples *to* the sea, but onto it as well. Just as Hidalgo (1974: 206–10) traces the formative influence of bamboo on classical expressions of Indian architecture such as the dome of the Taj Mahal, Joseph Needham, in *Science and Civilization in China,* returns again and again to the ghost of bamboo hovering within oriental ship design. He notes that the "ancient pictogram for a boat shows ends which are square and not pointed" (Needham iv.3:396). The reason is simple:

> The basic principle of Chinese ship construction was derived from the example of the bamboo stem with its septa, and indeed . . . the earliest vessels of East Asia were rafts of bamboo. This led directly to the rectangular horizontal plan. The following corollaries resulted: *i)* The absence of stem-post, stern-post, and keel; *ii)* the presence of bulkheads, giving a hull very resistant to deformation and leading naturally to *iii)* the system of water-tight compartments, with its many advantages. These were almost surely in use by the +2nd century, but were not adopted in the West until the end of the +18th. Provenance was then recognized. *iv)* The possibility of free-flooding compartments, found useful both on river rapids and at sea. This was not adopted to any extent in Europe. *v)* The existence of a vertical member to which the axial rudder could be attached, in "line closure" rather than "point closure." (Needham iv.3.sec.29 Shipping: 395)

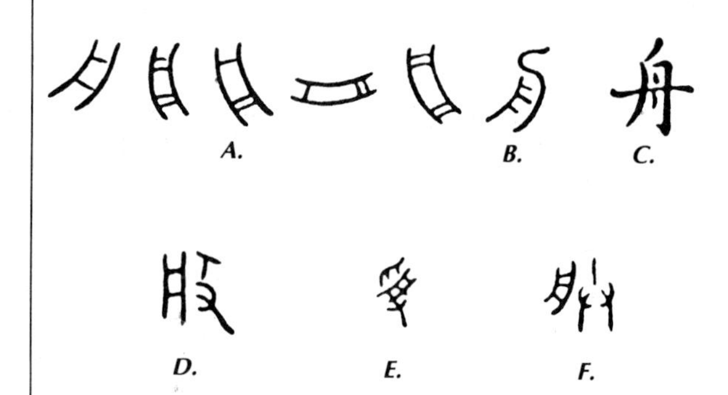

"Boat" ideogram. The Chinese word for boat has blunt ends in the early pictograms, and cross lines suggesting both the nodes of the living plant and the bulkheads of Chinese boats. The curved, dancing motion of most of the ideograms suggests both the rocking of water life and the curved line of early rafts.
Chou: *a boat. (A) Oracle bones, Shang dynasty, 1766–1122 B.C. (B)* Shou wen *(small seal), A.D. 100 (C) Modern form. Notice the node.*
Bronze inscriptions, Chou dynasty, 1122–255 B.C. (D) To row (a boat, oar, and hand). (E) To receive (a boat between hands loading and unloading it). (F) To caulk (a boat with two hands stuffing a seam).

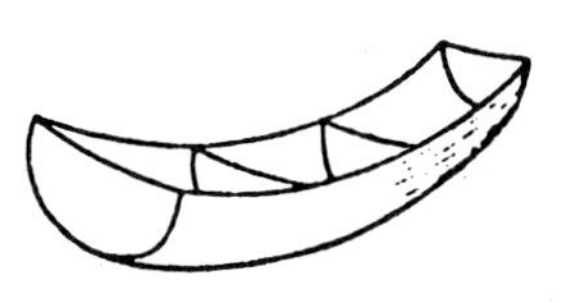

The nodes of a bamboo culm inspired watertight bulkheads in Chinese boats that evolved from bamboo rafts. Western shipbuilders began imitating this method of keeping a damaged vessel afloat about 250 years ago.

Bamboo was a model, therefore, not only to Eastern painters and moral philosophers, but to shipbuilders as well. Earlier in his study, as enormous as the civilization it describes, Needham briefly mentions a seminal influence of the plant on the form of Chinese architecture.

> Here we need only stop to remark the striking similarity between the bulkhead structure of the Chinese ship and the prominence of the transverse partitions or frameworks so fundamental in Chinese architecture. If the latter prevented a longitudinal vista and permitted the classical curve of the roof, the former provided distinct holds, rendered the vessel extremely strong, and gave it the typical bluff bow and stern of large Chinese craft. One cannot but feel that both systems were inspired by the bamboo, that plant so familiar to every Chinese from a thousand uses, with its transverse nodal septa. (Ibid.:391)

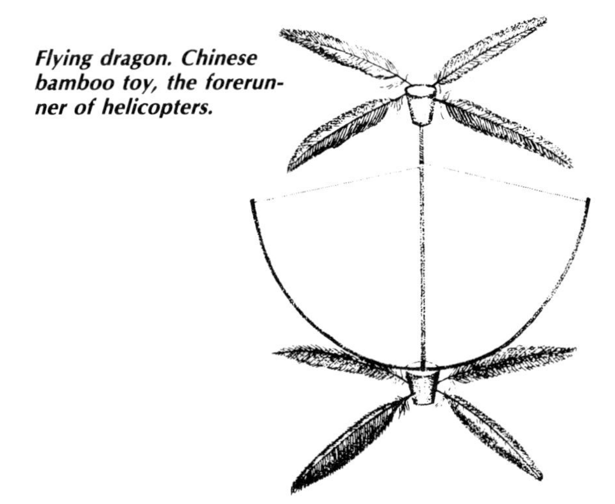

Flying dragon. Chinese bamboo toy, the forerunner of helicopters.

FLYING ART. Bamboo's use in flying art preceded by some two millennia the art of flying. A fourth century B.C. bamboo bird flew three days[94]; "wind zithers" and "hawk lutes" were early forms of bamboo musical kites; and bamboo whistles were attached to the tail feathers of pigeons.[95] The "flying dragon," a famous fourth century (A.D.) wind toy, has been called by Needham not only the "direct ancestor of the helicopter," but also "godfather of the aeroplane propeller." In 1792 "it stimulated Sir George Cayley, who may truly be called the father of modern aeronautics, to his first experiments on what he afterwards called 'rotary wafts' or 'elevating fliers.' " For more on how bamboo helped father human flight, see *kites*, below.

FURNITURE. Bamboo furniture has enjoyed waves of periodic popularity in the West. Freeman-Mitford mentions some thirty bamboo furniture manufacturers in London in the late nineteenth century.[96] Bamboo furniture seems increasingly visible in West Coast stores in the United States, so perhaps such a wave is about to crest among us again. It has been the mainstay of the oriental home for millennia. Building bamboo furniture was introduced into Philippine schools as a good way for training students in manual arts while providing needed furnishings to the poorer homes (see Cocannouer 1913).

G Gabions, games, garments, gates, gate springs (and bars), grain, grain storage, graters, greenhouses, guns, gutters, gypsy vans.

GABIONS. Bamboo "sausages," 10–20 feet long, woven of bamboo strips and then filled with rocks, are a characteristic feature of Chinese hydraulic engineering in sea walls and river embankments. They seem to have been introduced as early as 28 B.C. "The relative lightness of the gabions permitted their use on alluvial subsoils without deep foundations, and their porosity gave them a most valuable

Gabions, an early and surviving mainstay of Chinese civil engineering, are essentially huge, loosely woven bamboo baskets full of rocks used to stabilize riverbanks and waterfronts. An empty gabion is pictured.

shock-absorbing function, so that surges and sudden pressures did no damage to the defenses."[97] Footbridges were constructed across streams in the same way, composed of fieldstones held in place by strips of bamboo, forming a series of steppingstones of what was essentially a very open basket of rocks.

GAMES AND GAMBLING. The Chinese are such avid gamblers they even bet on the number of seeds in an orange. The word for gambling in Chinese contains the radical for bamboo, supposedly because of its early use for tallies and slips to record scores. In many games, also, bamboo serves as playing pieces. Mahjong, for one, uses bamboo tiles. In some dance games, long bamboo canes are used, as in the Philippine *tinakling* poles, clapped together by two children as a third dances in and out. "Legend says that these giant rhythm sticks and the dancers that move around them are artistic representations of the Philippine lake shore where the long-legged crane wades through windblown reeds."[98]

GRAIN. Bamboo was originally classified as one of the seven thousand varieties of rice, to which, as a grass, it is related. Most species flower so infrequently that we don't ordinarily think of it as a grain; but in China and India there is a saying that bamboo flowering foretells famine, and the plant generously spills its seed at this time to help the hungry people. According to oral tradition, this has occurred often enough that some think flowering is caused by prolonged drought. In any event, many records from many locations tell of bamboo grain providing famine relief. "In 1864 there was a general flowering of the bamboo in the Soopa jungles in western India, and a very large number of people, estimated at 50,000, came to collect the seed. Each party remained about ten to fourteen days, taking away enough for their own consumption during the monsoon months, as well as some for sale. Mr. Stewart adds that 'the flowering was a most providential benefit during the prevalent scarcity.' Six years ago (1921) there was a famine in Hunan, but the bamboo flowered and saved the people. Enough was gathered for food and some in addition for sale in the markets. However, calamity as well may follow flowering of bamboo. In Brazil, as well as in India, sudden production of great masses of rich grain in widespread localities increases food supply for rats and mice to multiply in extraordinary numbers. After consuming the fruit of the bamboo, they overflow into the neighboring fields and devour the crops. The German colonies in Rio Grande do Sul and Santa Catarina were visited by this plague, called 'ratadas' in Brazil, at intervals of about thirteen years, which apparently represents the periodicity of the species covering that region."[99]

It should be remembered that natural stands of bamboo are sometimes immense: Groves of *Melocanna baccifera* can cover up to 700 square miles. The tropical groves especially can be quite dense, and among species that fruit heavily you can literally scoop the seed up in your hands. In places, it seems to have fallen not off a tree, but off a truck. The seed generally resembles other grasses. Of *Arundinaria hookeriana,* a species from Sikkim later named after him, Hooker writes: "The fruit is a dark, long grain, like rice; it is boiled and made into cakes, or into beer, like Murwa."[100] An analysis of the seed of *Bambusa tulda* showed: water, 13.5 percent; protein, 10.8 percent; starch, 71.6 percent; oil, 0.6 percent; fiber, 2.1 percent; ash, 1.4 percent.[101]

GREENHOUSES. Bamboo and plastic greenhouses are a rapid way of getting more garden space in northern climates. Their construction is particularly recommended for schools. Polyhedra framing (see pp. 106–107) or conventional construction can be used. Plans for a renter's greenhouse in wood, easily built and unbuilt, are available from Aprovecho.[102] Other greenhouse designs are widely available in gardening books and can be adapted to bamboo framing.

H Hairpins, hampers, handles (rake, hoe, umbrella, and so on), hats, hawsers, hay and forage, hedges, helmets, hen houses, hinges, hoops, hookahs, houseplants, houses, humidors (tobacco), hummers.

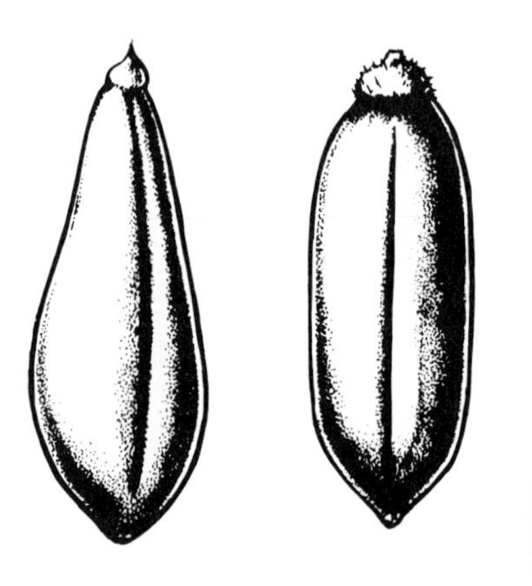

Bamboo seed.

Headrest, Szechuan, early 20th century. 4″ high, 5″ wide, 13½″ long. Portable, with legs folding inward. A good bench design, on a slightly larger scale, for mobile furniture.

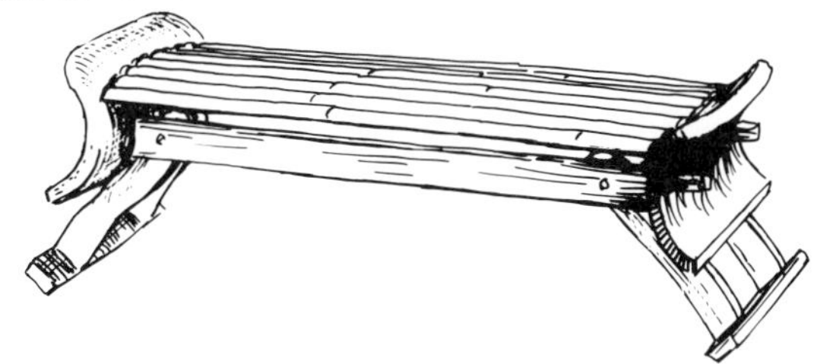

HATS. "In most parts of Burma, the farmers' sun hats are made from the culm sheaths of giant bamboos, bound around the edges with split bamboo."[103] Oriental bamboo hats are sometimes up to a yard in diameter, more a head umbrella than a hat. Helmets of bamboo were formerly worn by police and soldiers, cone-shaped to deflect blows.

I Iceless cooler, incense sticks, insect cages, irrigation waterwheels and pipes.

Jacket, 18th–19th century. Tiny tubes of bamboo branches, about ¾″ long, were strung on thread in a lozenge pattern then bound in linen with linen bottons and loops. As a summer undershirt beneath a robe, it insulated the body, helped pores to breathe, and kept the robe from sticking to skin. Detail.

Headrest, unknown province, 18th century. 6″ high, 4¼″ wide, 16½″ long. Expanded, an elegant bench.

ICELESS COOLER. In a cool corner of the kitchen, away from the stove, set a bamboo basket with a loose cover on three or four bricks or stones resting in a water container some 12 inches deep, such as part of an old oil drum. Sew burlap loosely over lid and basket so that it hangs below the basket and acts as a wick in the water. The basket itself should be above the water line. Dampen lid occasionally. This iceless ice basket cools food effectively.[104]

J Jackets, jars, jewelry, joss sticks, junks.

JUNKS. Junks! Spreading their sails . . . cleaving the waves a thousand *li.* Bamboo begat rafts, rafts begat junks, junks begat the rest of Chinese vessels, and these in turn begat critical design modifications in Western ships.

> Junk design, exemplified in the oldest and least modified types, has a carvel-built hull wanting in all the three components which elsewhere were regarded as essential—keel, stempost, and sternpost. The bottom may be flat or slightly rounded, and the planking does not close in towards the stern, but ends abruptly, giving a space which would remain open if it were not filled in by a solid transom of straight planks. . . . The hull may be compared to the half of a hollow cylinder . . . bent upwards towards each end, and there terminated by final partitions—like nothing so much as a longitudinally split bamboo. . . . Alan Villiers has written: "I think the Chinese were the greatest of all Asian seamen, and their junk the most wonderful ship. Hundreds of years ago, the seagoing junk embodied improvements only relatively recently thought out in European ships—watertight bulkheads to isolate hull damage and so keep the ship afloat, the balanced rudder which makes steering easier, and sails extended with battens."[105]

China has been the most nautical of nations, wetter longer than any other. With several thou-

War junks. Woodcuts from Ming dynasty, c. 1522.

Junk

The word comes from Portuguese *junco,* cordage or rush, from Latin *juncus,* a bulrush. Rivermen are father to seamen. The rushes on the banks of their road worldwide were a natural caulking choice. The word then came to refer to worn out, unravelled cordage aboard that could be used for caulking in emergency, then to any old and discarded bit of trash, then to the vessels caulked. "The stone the builders rejected became the cornerstone." *Caulking* (q.v.), the least noticeable and smallest part of a ship was in fact its essense, what actually holds it afloat, the bones of the boat without which the most able seaman is fish food. As we have ever less per capita resource and more per capita waste, the judo of junk, creative garbage disposal, becomes an art of ever greater urgency. For that art, the junk and its seams provide a useful emblem. Junk life, and the central importance of bamboo in its traditions, is more relevant than may appear at first bounce of the eye. The boat people of China form a sizable portion of her population. Water villages were one solution to densities imposed on the Chinese earlier than we began to feel them in the West. Crowded conditions will impose a more careful scrutiny of water living in the West, just as increased transport costs will inspire an increased use of waterways. One of the most ancient and complex water living modes is the junk and its related rafts. Children particularly should be introduced to this water world from the infancy of nautical technology that still flourishes in the age of nuclear subs.

Bucky Fuller imagined tetrahedron archipelagos of floating towns stretched across the sea routes as ocean inns and sea settlements; nodes of ocean routes as cities are the nodes of roads on land. A more immediately useful version of his vision would be water villages along navigable rivers of developing countries that use their banks and fluid acres as ingeniously and economically as the Chinese junk people have for centuries used theirs. With a handful of fuel, they cook a meal for a dozen people. Their survival designs deserve more attention in our crowded futures.

The shore and shore waters represent immense global acreage, considering all inlets and world rivers. Living on the edge of land becomes more common as land gets scarce. Chinese boat people and junk culture represent the oldest inhabited cultural design for shore survival. It should be included in the curriculum of all the planet's children from their earliest years, beginning with models in park lakes, the family tub, and the pond that deserves a place in any campus. Water should join reading, writing, and arithmetic in basic curriculum, survival on its edge a natural prelude to free roaming the oceans. Complete survivors are the issue of complete schools.

Worcester's *Sail and Sweep in China* is among those authentic books distilled over a lifetime from the blood of the author, well and amply illustrated with plans and drawings from his own hand, chiefly distinguished by the broad and compassionate sanity—as present on each page as his obvious, graceful command of the material. Ten thousand joss sticks should be lit in honor of this work of primary relevance for appropriate planetary design.

sand miles of seacoast, navigable rivers, and lakes, the Chinese became not only expert sailors engaging in complex naval maneuvers of more than seven hundred ships as early as the twelfth century, they also became a nation of "boat people," permanently living in floating dwellings. In Hong Kong alone there are over 250,000, and nearly every large city on the edge of water has overflowed onto it with houseboats:

"The Canton water roads are well known. Here boats are tied up to the banks of the river and to each other in regular rows, 'packed like the scales of a fish,' as the junkmen say, and constitute nothing less than a floating city. Lanes are left at intervals of twenty or thirty boats to facilitate communication. This is very necessary, for the nautical 'commuter' returning from work may, as likely as not, find that his floating home has moved to another 'street.' No phase of life is unrepresented among this population. Kitchen boats supply hot food at a low rate. The barber calls in a small sampan which he rows himself, calling attention by ringing a bell. The river doctor also gives notice of his approach by beating a drum; and, when his medicines prove of no avail, there are floating mortuaries.

"When he handles bamboo, the junkman's ingenuity finds its widest scope. He eats it in the shape of bamboo shoots, drinks out of it when it is made into a cup, sleeps on it when it is cut into bamboo shavings, which make excellent stuffings for mattresses. He uses it as a medicine and is finally carried to his grave by means of bamboo poles. Among a thousand and one nautical products for which it can be employed may be mentioned rope, thole-pins, masts, sails, net-floats, basket fishtraps, awnings, food baskets, beds, blinds, bottles, bridges, brooms, foot rules, food, lanterns, umbrellas, fans, brushes, buckets, chairs, chopsticks, combs, cooking gear, cups, drogues, dust pans, pens, nails, pillows, tobacco pipes, boat hooks, anchors, fishing nets, fishing rods, flagpoles, hats, ladders, ladles, lamps, musical instruments, mats, tubs, caulking materials, scoops, shoes, stools, tables, tallies, tokens, torches, rat traps, flea traps, backscratchers, walking-sticks, paper, joss sticks, and rafts."[106]

Up to 200 feet in length, junks are distinguished by off-center masts, as many as seven in number, which are oddly tilted fore and aft, giving a junk in full sail more the look of a slightly asymetrical object of nature than the product of a rigid human geometry. The squared-off stem—like the

Floating kitchen. Centuries of survival by Chinese boat people provide models for the increasingly crowded West.

node of the bamboo culm that provided the ancient pattern for junk design—permitted a style of rudder steering of unparalleled efficiency that, unknown in Europe until the fourteenth century, was later adopted throughout the world. It permitted the Chinese to launch and manage larger ships earlier than any other people. War fleets are recorded in 486 B.C., and there was an expedition of two thousand ships in the first century A.D. (See Aero 1980.)

K

KIOSK. The kiosk is a round, thatched structure that takes fuller advantage of the physical properties of bamboo than any other shelter in the rural regions of western Colombia where a particularly durable species, *Guadua angustifolia* (q.v.), is used. The kiosk was originally a family dwelling, commonly 20 feet in diameter, its conical roof supported by a king post in the center. The walls were of *esterilla,* large bamboo board made by cracking culms at all nodes with ax blows an inch or so apart around the circumference of each node. A single long continuous crack is then made to open the culm, and the nodal diaphragms are removed. Bamboos 6–8 inches in diameter yield 18–24 inch boards cut to any desired length, which function as

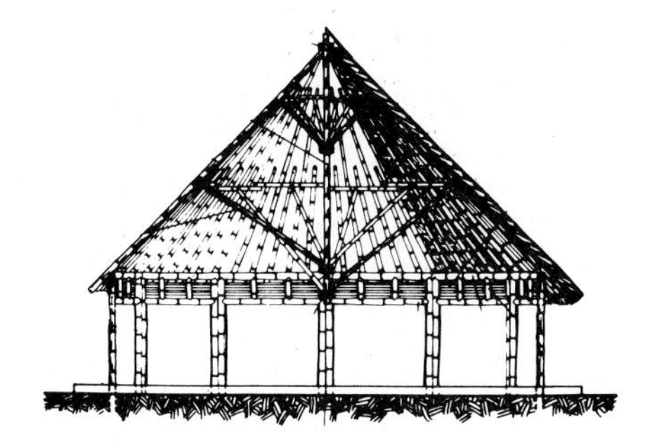

Kiosk.

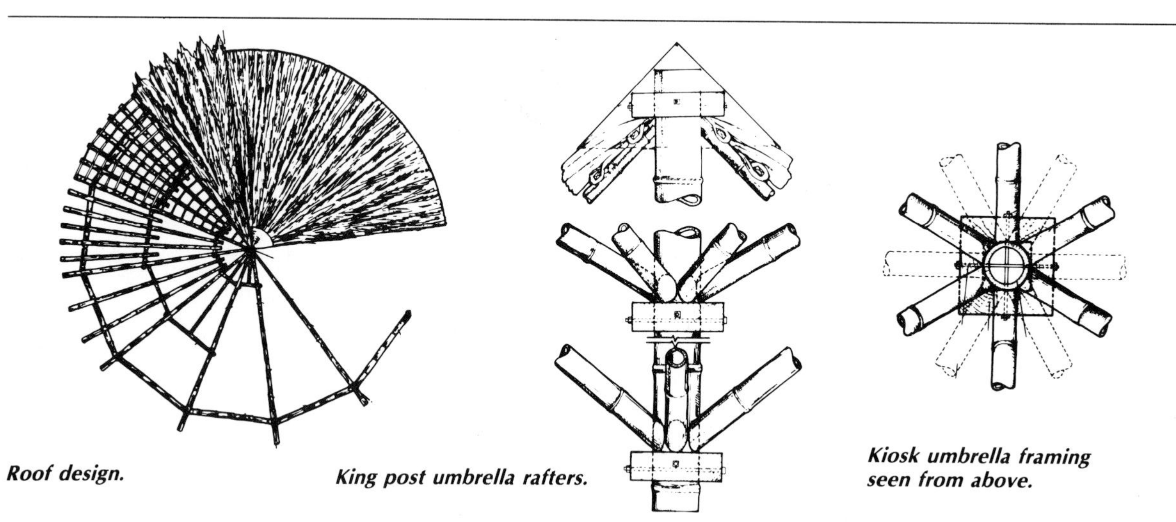

Roof design.

King post umbrella rafters.

Kiosk umbrella framing seen from above.

hoops in a barrel to keep the columns around the circumference of the kiosk from being shoved outward at the top by the thrust of the rafter and roof weight. When the early kiosks were adapted, in time, for use as communal meetings and recreation, the walls were eliminated, for more light and easier access, and their structural function was assumed by peripheral strips of bamboo ringing the kiosk at the top of the columns, which are usually no more than 10 feet apart. Since the king post blocked vision and movement at ground level, its base was eliminated and an umbrella-like framing system was devised for the roof. Building a small model is advisable before your first full-scale kiosk. Hidalgo offers fairly complete details.[107] The kiosk is an excellent intro to architecture for young builders. Try a 6-foot diameter, 6-foot roof, hexagonal plan with columns 3 feet apart. For more on *guadua* architecture, see Chapter 5.

KITE. Kite tales in the Orient are plentiful and old since this art was flying in the clouds of China 2,200 years or so before the art of flying landed in Ohio with the Wright brothers in the present century. Kite use for military signaling was recorded in China around 200 B.C., and many early mentions of kites involve their use in war. One general, Han Sin, measured the distance from his troops to a besieged palace by flying a kite over it. From the length of the string, he figured his distance to the walls by simple geometry, dug a tunnel, and attacked successfully from underground. Another Chinese general of the Han dynasty (206 B.C.–A.D. 220) routed an invading army by flying kites with bamboo hummers over their camp at night. The weird sounds were taken for gods threatening the invaders, and they fled.

Around A.D. 1500, a large falling star was interpreted as bad luck by the soldiers of a Korean general, who sneaked out the next evening and suspended a lantern from a kite. The kite light was taken as a positive omen by his soldiers who, full of confidence, routed their enemy the following day.

Warriors of the wind

A kite-fighting festival has been an annual event in Hamamatsu, Japan, since the 1500s. A sublimation for fists imposed by a ruler, according to one legend, these air battles now attract 2 million people each year. The kites honor first-born sons, who either touch the line as the kite rises or are kept strictly at home to avoid seeing their alter ego's dizzy altitudes, in the firm conviction that their presense would bring bad luck in the fight. A seventy-four-year-old participant: "I am still fascinated by the fresh greenness of bamboo and feel challenged to make a gallant kite." (Eliot 1977:555)

The Japanese often sent spies aloft by kites to scan enemy encampments, and Japanese thieves have taken the hint to break and enter from the air. One famous bandit stole the gold fins off dolphins atop a castle in Nagoya. Captured later, he was boiled in oil, and large kites were long forbidden in Japan. One Japanese architect lifted construction materials with large kites.

Fishing kites are used in Malaysia, where fighting kites—a highly developed form in many cultures—have been outlawed because the passionate contests led to such bitter disputes. Fighting kites have the string near the kite soaked in starch or glue and dipped in powdered glass. The trick is to jerk this stiff, sharp blade of string down quickly on the string of your opponent.

Although Marco Polo had reported seeing kites on his travels, their use reached Europe in the eighteenth century, not long before Ben Franklin's famous experiment with electricity in 1752. Flying a kite into a thundercloud with a wire replacing the string and a large house key fixed at his end, he got a sizable shock . . . and the idea for his subsequent invention of the lightning rod. As a child, Ben Franklin had managed to pull himself across a pond with a kite—a method of transport he suggested would work also in the English Channel. In 1903 box kites in fact did tug a boat the 20 miles across the channel from England to France.

Box kites were a notable breakthrough in kites designed by Lawrence Hargrave in 1893 (New South Wales). George Pocock had already applied the same principle to land movement in England, traveling as fast as 20 miles per hour in a two-kite carriage. Pocock was also among the first in the West to experiment with people-lifting kites. He flew his daughter up 100 feet in 1825; his son, 200 feet somewhat later.

Kites became sophisticated laboratories for some scientists, like Alexander Graham Bell, the inventor of the telephone, who built multicellular tetrahedron kites at the turn of the century to explore the aerodynamics of flight and apply these to airplane design. Orville and Wilbur Wright built a two-winged kite in 1899 with strings fastened to the tips to control flight, and they experimented with this as a steering device. In 1903 they incorporated their kite findings in the wing-bending controls of the world's first airplane. Their distinguished contemporary, Marconi, the inventor of the radio (1894), used kites to publicize his invention to an indifferent public. In 1901, he flew a kite from Newfoundland to pick up the first transatlantic message—transmitted from Cornwall, England, to his kite-elevated aerial 400 feet up.

The largest kites have been flown in Japan; they measure 50–60 feet across and require up to 200 people to launch.

Meanwhile, across another ocean in Japan, kite fanatics were competing to get the largest possible monsters aloft. In 1906, a 5,000-pound giant built of 4-inch wide bamboo pieces, a round kite 60 feet in diameter and trailing a 480-foot tail, was flown by some 200 people. That, apparently, is the record for size.

The greatest altitude was reached in Germany, 1919, where a train of eight kites flew some 6 miles high (31,955 feet). The weight of the string limited altitudes greatly until 1749 when the Scotch doctor, Alexander Wilson, invented a train of six kites to fly thermometers cloud high in weather experiments. After this, kite trains were used by meteorologists throughout the world: seventeen U.S. weather stations used kites till 1933, after which kites were not used again for serious research until 1967 when the new airfoil or parafoil kites (invented in the United States by Domina Jalbert in 1964) were used to study weather inside a cap cloud above Chalk Mountain in Colorado.

In World War II, ancient oriental military uses of kites were resurrected by Germany. Triblade, motorless helicopter kites flew sailors above subs to scan the horizon for convoys to attack.

The tiniest kite on record is a 3-inch bow kite of balsa wood and tissue paper with a silk thread string, flown about 20 feet high by Clive Hart in New South Wales.[108]

L Lacquerware (plates, cups, boxes, trays, bowls), ladders, ladles, lamps, lampshades, landing docks, landscaping, lanterns, lathing, laundry poles, levees, light-bulb filaments, lofts, looms.

LACQUERWARE. Many products of bamboo lacquerware are intended more to be seen than used. They are curios not sufficiently durable for daily use. However, in Kyaukka, Burma, a fairly elastic bamboo lacquerware is produced that is tough enough to withstand eight to ten years of rough daily use without dents or paint wearing off. It's unlikely that aluminumware would be without dents and scratches after a year or two of use in the average Burmese home, and furthermore the cups, bowls, trays, baskets, small boxes, and tiffin carriers of Kyaukka are far easier to wash than aluminum.

The manufacture of lacquerware begins with the bamboo skeleton. In Kyaukka, *Cephalostachyum pergracile,* preferred because of its great flexibility, is cut into convenient lengths, excluding nodes, and split radially into small strips.[109] (Radially split pieces are much stronger than those split tangentially.) A strip is bent into a circle, its ends tied together with thread, leaving an open portion where a second strip can be slipped in and bent into a circle. Its free end is similarly tied with thread. Strips are added in this way, with great skill and speed, in a spiral form until the desired shape is obtained. Next this bamboo base is coated with a paste made of a black oleoresin from the *thitsi* tree *(Melanerrhaca asitala)* mixed with bone ash. This forms a tough concrete when dry, which is vigorously rubbed with sandstone, a critical phase of the production that, properly done, greatly increases the lacquerware's durability. Finally, the inside only is painted with lacquer, the outside generally remaining black, with the resin and bone ash being well rubbed in with the hands. The pieces are then stored eight to ten days in a dust-free space until thoroughly dry.[110]

Kurz mentions lacquerware from Palembang so elastic it can be turned inside out without cracking or damage.[111] Plastic articles imply an economic leak from country to town, from less to more developed countries, from poor to rich. Lacquerware provides rural employment, uses local materials, decentralizes production, requires no complex machinery or heavy capital investment, and is a cottage industry that can reduce imports and increase exports.

LADDERS. These are a natural for bamboo, granted its aptitude for lightweight altitude. Hypereasy to handle, even up to 30 feet long, the two main pieces are usually wired together to keep the rungs in place. Cracks should be wired upon first appearance to prolong ladder life. Culms of larger species are also used singly as ladders, by making a triangular cut at the node. Kurz reports an ingenious bamboo ladder for climbing trees:

"Ladders for climbing lofty trees, especially for gathering fruit or obtaining beeswax, are constructed by means of bamboo pegs driven into the trunk. These pegs are made of old thick bamboo, split to about 2 inches wide. Each is cut above a joint, which forms a solid head to bear the blows of the mallet, and the point is flat and broad, cut away carefully to the siliceous outer coating. To the head of each is strongly tied a strip of the rough rind of a water plant. The climber carries forty or fifty of these pegs in a basket by his side and has a wooden mallet suspended round his neck; he has also prepared a number of strong, but slender bamboos, each 20 to 30 feet long. One of these he sticks firmly in the ground at the foot of the tree, and close to it; he then drives a peg as high as he can reach, and ties it firmly by the head to the bamboo. Climbing up on this, he drives in and ties successive pegs, each about 3 feet apart. He soon reaches the top of his pole, when another is handed up to him, and being bound to the one below, he ascends in the same way another 20 feet. When his pegs are exhausted, a boy brings a fresh full basket up to him, and a long cord enables him to pull up the bamboos as he requires them. This mode of ascent looks perilous, but is in reality perfectly secure. Each peg holds as tightly as a spike nail, besides which the weight is always distributed over a great number of them by means of the vertical bamboos. The same mode prevails in Java, amongst the Nagas of Eastern Bengal, and the Karens of Burma."[112] Tree-house architects and apprentice monkeys, ponder well.

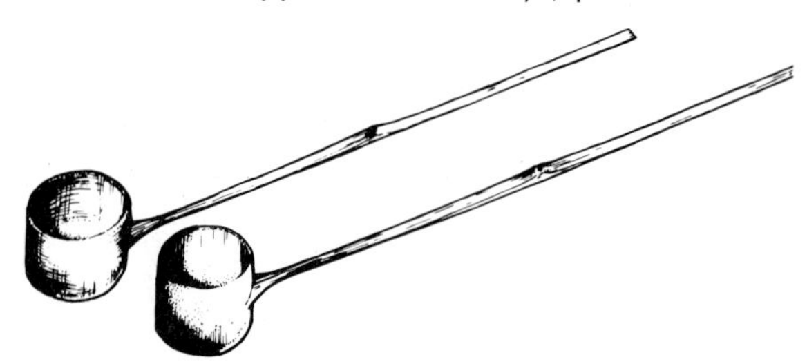

LADLES. Many shapes and sorts are made of bamboo. For a water ladle, the cup is cut just above a node, to the desired size, a hole carved into it, and a handle fitted. Since the ladle lives in water, no glue is needed to bind the pieces that compose it. The bamboo swells to seal the joint: Its glue is use.

LAUNDRY POLES. Bamboo poles are run through the sleeves of shirts or dresses and stuck out the window as drying racks. "This is especially convenient for boat people as well as villagers who live over their shops."[113] Leaving the stub of a branch on at one node gives a convenient support for a clothesline.

LIGHT BULB. The world's first light bulb, over a

century old, still burns with bamboo in the Smithsonian Institution in Washington, D.C. "In the process of inventing the incandescent lamp, Thomas Edison confronted a difficult problem—finding the appropriate material for a filament. In 1880, he learned that finely shredded bamboo serves as a firm support for round hemp palm leaf fans. Taking his clue from this hint, he collected all available varieties of bamboo from throughout the world, including Southeast Asia and Japan. After repeated tests, he ascertained that a variety from Kyoto would best serve his purpose. In 1882, he set up a company to produce incandescent lamps from the filaments of Japanese bamboo and illuminated the nights of New York."[114]

M Marimbas, markers, masts, mattangs, matting, mattresses, medicines, mills, mobiles, mushroom culture houses, musical instruments.

MARKERS. "The U.S. Department of Defense even found bamboo ideal for use as markers on the Greenland icecap. It happens that metal poles placed in the ice are warmed by the sun and sink out of sight. Bamboo does not conduct as much heat and so stayed where it was put."[115]

MATTANGS. Here is a seaschool visual aid. Throw a bean into your bathtub, or a pebble into a pond, and watch the ripple pattern produced. Objects that break the water surface—a brick set up in your bathroom experiment, an island in real sea—alter this pattern in learnable ways. Pond or Pacific, wave motions obey the same principles of liquid mechanics. The study of these patterns was developed to a highly refined science-art by the Polynesian peoples some three thousand years ago in the course of colonizing migrations that in time settled almost every island in the Pacific, and they remained in touch as a cultural entity across vast seascapes so effectively that even today a Maori of New Zealand can stretch his native tongue to the Hawaiian islands and remain understood. In the nineteenth century, captains of Western ships were still losing themselves in brief 150-mile voyages between islands in the South Pacific, their vessels equipped with all the instruments of the most up-to-date navigation and charts.

For millennia before them, the Polynesians in pairs of 60- to 80-foot dugout canoes fashioned with tools of shell and bone, lashed together with a deckhouse between of bamboo or other materials, had found their way by reading the map the sea

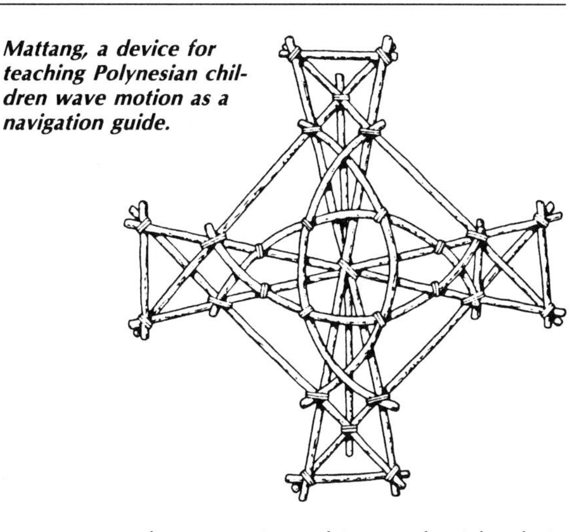

Mattang, a device for teaching Polynesian children wave motion as a navigation guide.

was constantly generating of its nearby islands in the motions of its waves. It was a braille of the whole body: they would not merely *look* at the waves but crouch in the hull of the bow as close as possible to the most precise device for navigation they had discovered, the ocean's motions, and *feel* wave patterns bouncing off islands 100 miles away. Map and medium, sailor and tool of his trade, were one. The visual aid to teach this teletactic art to child sailors of the crew was the *mattang,* a simple web of bamboo strips or branches providing a codified summary of possible wave patterns. The gradual mastery of this code of tribal memory graven in bamboo permitted them in time to navigate by touch and teach the art in turn to their children, bound for islands none of them had ever before felt quivering their locations to their skins as they listened in the bottom of their boats.

"Miniaturization" is the term Bucky Fuller used to describe the gradual decrease in matter and increase in organizing mind that marks technological evolution. The dramatic Alice-in-Wonderland shrinkage from early crystal sets to transistor radios was, for Fuller, a useful example of this process in his own time. Following this trajectory, we can imagine a technique of such density—that it disappears. Some "primitive" technologies appear less "advanced" to civilized peoples bristling with instruments to lug about precisely because the human body-mind had become such a subtle master of its own capacities that no other equipment was required. This is certainly the most appropriate technology we can evolve towards, however, because it's available to all without investment in additional accessories to burden the equipment possessed already by each owner-in-residence, the local human, floating in his bag of skin, full of blood perception, and inventive energy.

When we can feel history approaching as far off as seabums in the primitive Pacific could feel the distant beach, we will find kinder harbors as well and sail, with less costly equipment, more significant seas.

MEDICINES. "The Chinese used bamboo medicinally as a tonic for the stomach, as a cure against dysentery, and as a remedy against toothache. When medical opinion proved in doubt, they would explode it in fire to drive any demons away."[116] Coughs, asthma, cancer, eye ailments, and paralytic complaints were also treated with various preparations from the plant, as was a sluggish love life and hair and skin problems. (See *tabasheer.*)

MUSHROOM CULTURE HOUSES. "Some 20 million poles of Makino bamboo *(Phyllostachys makioni)* were used in the construction of 6.8 million square meters of mushroom culture houses. The poles were treated with PCP–sodium salt, since creosote-treated bamboo is unsuitable for this purpose. Poles so treated last around five years, two to three times as long as untreated ones in this industry."[117]

MUSICAL INSTRUMENTS. "The two most characteristic acoustic features of the music of the Chinese culture area (extending from Korea to Indonesia) are the high proportion of chime-idiophones and the prominence of the bamboo plant (and the pitch pipes derived from it)" (Needham iv.3:142). From North Korea to northern Australia is a sweep of latitudes equaling that from Vancouver, Canada, to Lima, Peru, in our hemisphere. Roughly half the people in the world live there. Music is a central part of their life, and bamboo is a central part of that music. Its eminence is ancient, recorded in the oldest surviving classification of instruments we have in the world. Over four thousand years ago in China, under the Emperor Shun (2225–2206 B.C.), instruments were divided into eight materials, each with a corresponding cardinal point of the compass, a season of the year, and an element of nature represented by one of the eight trigrams of the *I Ching.* Bamboo was associated with pipes and the east, the direction of the rising sun, which linked it naturally with spring, when culms are rising with all the thrusting energy of the sun itself. Among the elements of nature, bamboo was associated with the mountain, where temperate, monopodial species have especially been at home because of their preference for a moist but well-drained soil.

The mountain was the "youngest son" in the family of hexagrams, associated with the end of movement, tranquillity, keeping still. Home of her

sages and hermits—the most respected members of Chinese culture—subject of her favorite paintings, forming an element in her most basic concepts such as *yin* and *yang,* the mountain for the Chinese was not simply a hunk of rock but a revered and animate being. Bamboo's identification with it was an honored one, no less significant than the plant's link with spring and dawn, the most vigorous season and hour.

This vigor has given bamboo a long history as a martial instrument or club whose rhythms have stirred many a land and bruised many opponents. The *shakuhachi* is sometimes described as the only musical instrument in the world that doubled as a weapon. But on the island of Trinidad in the Caribbean, bamboo also evolved as a weapon-instrument to replace prohibited drums. When the bamboo stick bands were in turn outlawed in the late 1930s, the people—whose invention is often a cofunction of oppression—evolved the steel bands whose engaging rhythms have since spread throughout the world.

Briefly, the pre-Lent Carnival ("farewell, flesh," from Latin) was introduced to Trinidad by French planters in an upper-class format in the late 1700s. The slaves out back of the Big House had their version, too, which came out of the closet with liberation in 1837. "Masks allowed action without fear of reprisal. Bands grew larger and louder. Songs were purposely lewd and pointedly anti-upper class. The planters were horrified. Police tried to squelch Carnival with little success . . ." Church propped up State to hold back the blacks: "The original slave celebrations had moved to the beat of the congo drums—the same drums used in religious rituals to summon believers to worship and gods to receive sacrifices. The official church, as part of its drive to eliminate 'foreign' religions, outlawed the drums. But no celebration in Trinidad works right without rhythm. 'Bamboo tamboo' stick bands were devised to fill the musical gap." The bands were hundreds of young people, a roaming rhythm section whose music proved so potent it was outlawed by the police. The people, musically stimulated, would take no sass from the brass, and had several hundred heavy-duty clubs in their orchestra to back their will.[118]

The Dutch had a similar problem with popular music in West Java, which is a bamboo musical center with over twenty different percussion, wind, and string instruments as ancient as they are ample. The *calung* and *gambang* are carved in stone in the eighth century temple in Borobudur, and bamboo instruments predate Hinduism in Java. Their development is supposedly closely related to the pre-Christian Polynesian migration. "The Sundanese used bamboo instruments in honor of Dewi Sri, goddess of rice and agriculture in Javanese mythology. The melodious sounds of the slender, gracefully constructed instruments express well the cheerful character of the Sundanese people."[119] The *angklung* is a bamboo percussion instrument that originated in West Java and is now known round the world. Two, three, or four bamboo pieces, usually tuned to as many octaves, closed at the bottom by a node, have three-quarters of their circumference removed to about half their lengths to form a sort of tongue by which they are hung vertically, their lower ends sliding horizontally in a slot in the bottom of the bamboo frame. When shaken, "the sound is curiously bright. The air column in the tubular part of the mobile bamboo segments produces, when blown, the same tone-pitch as shaking the entire segments."[120] The Baduy of South Banten still shake three or four angklungs when ending work on the sacred arable land. Anciently an instrument of farming festivals, it was also used to rouse soldiers and was suppressed as such by the Dutch for fear it would inspire revolt against colonial power. By the 1920s it had become merely a child's toy, used by beggars in the 1930s to attract a coin. A musician from Bandung then began an angklung renaissance, incorporating Western-style music and modern arrangements with local traditions: Hybrid vigor.

Java and the surrounding islands of Indonesia is an area rich in its variety of bamboo instruments. A few among them: The *calung* of West Java is made of twelve, fourteen, or sixteen bamboo tubes closed by a node at one end, with a diagonal cut at the other, tied together with cords like a rope ladder. From top to bottom, tubes get bigger and lower in pitch. The top is hung from house or tree; the lower end tied to the left knee of a sitting player or to the waist of one who is standing. It is "beaten with two *penakols,* sickle-shaped wooden sticks of slightly different sizes."

The *gambang bambu* evolved recently from a wooden version, "the Sudanese counterpart of the Western piano." Bamboo tubes are tied horizontally together by cords on a bamboo platform and played like the *calung,* which the *gambang* resembles but with more tubes. A similar instrument—with different music—is played in North and South Celebes.

The *rengkong* is one long bamboo pole, carried on the shoulder of a dancer, with bushels of newly harvested rice hanging from each end by split bamboo strings. It is played in a ceremony to Dewi Sri, the Javanese goddess of rice and agriculture, after a poor harvest. "The rhythmic movement of the dancer's steps causes the suspended rice to swing, and the friction of the string against the bamboo produces the characteristic musical sound."

The *kohkohl* is a complete internode with a slit along one side beaten with a soft wooden stick. The sound is determined by the length and diameter of the bamboo tube and the width of the slit.

The *hatong,* a panpipe, is usually played in pairs—a large one with two or three tubes and a small one with ten to fourteen small tubes. Three different sorts are used in stag hunts to communicate among the hunters, with one, two, and three pipes, respectively.

The *kolecer* is a weathercock, with propellers attached to various size tubes. A number of different sizes are set out to produce as many notes.

The *celempung* is the Sudanese bamboo string instrument, "a complete internode closed at ends by nodes, provided with a long slit and two bamboo strings cleverly split from the slit. This instrument is widely distributed and can be found from Madagascar, Vietnam, Malaya, and all over Indonesia to the

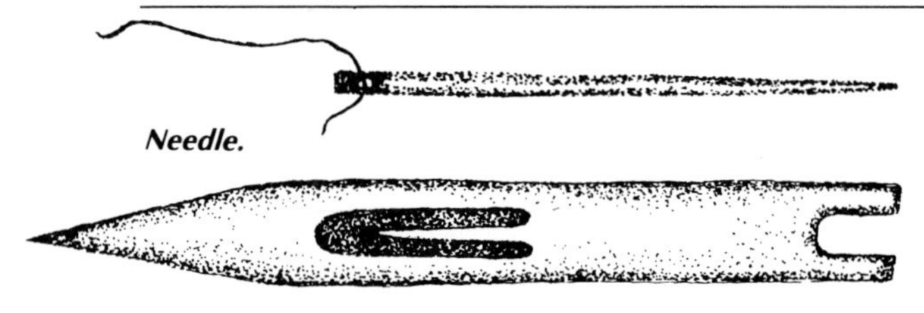

Needle.

Netting needle.

Tasadays in the Mindanao of the Philippines. Each area normally has a characteristic number of strings, varying from one to four."[121]

Musics of Many Cultures and *Music of the Whole Earth,* both one-volume efforts to sum up the globe of music which we are, contain many references to bamboo.[122]

N Nails, napkin rings, net floats, nets, netting and sewing needles, *netsuke*.

NAILS. "Limaye [1943] found that bamboo nails [more strictly, pegs] hold as well as iron nails, although the bamboo nails require predrilling of the wood."[123]

NET FLOATS. Bamboo's natural bouyancy, which inspired sealed compartments in Chinese ship design, also makes it a natural for net floats. Oshima (1931) mentions huge nets requiring a thousand culms each of moso *(P. pubescens),* the giant of temperate bamboos (70 feet by 7-inch diameter), with culms noted for their gradual taper.

NETSUKE. "Netsuke were the decorated objects men wore on one end of the cord by which a small box for writing materials or medicine called an *inro* was suspended from the formal Japanese belt."[124] Netsuke and brush pots were among the objects most favored by oriental bamboo carvers to demonstrate their amazing virtuosity.[125]

O Organs, ornaments, outriggers, ox cart beds (sides, yokes), ox goads, oyster cultivation.

OYSTER CULTIVATION. "In shallow coastal waters in some areas around Taiwan, extensive bamboo racks have been built for oyster cultivation. The yield of oysters raised on rayon lines attached to these racks can be ten to twelve times the yield by the conventional method of driving bamboo splits directly into the ocean floor."[126]

P Packaging, paper cutters, paper pulp, pegs, pen and pencil holders, pens (ink and animal), pillows, pins, pipes (drain, gas, irrigation, opium, organ, tobacco, water, and so forth), plant labels (shades and stakes), plates, plybamboo, poison, poles (boat, carrying, clothes, flag, fishing, fruit picking, garden, punting, telephone, tent, TV, and vaulting), polo balls, polo mallets, posts, pots, printing pads, propellers, props (for banana and various trees), punishments.

PACKAGE TACTICS. Japan, as a small island, entered the experience of population densities and consequent resource crunch earlier than most of the cultures on the planet, which discovered much later that all earth is an island. Increased precision and efficiency of design were the necessary cultural adaptations to these denser conditions, and bamboo provided the raw material for centuries of cultural experiment. The Japanese traditions of craftsmanship and patient addiction to detail ripened slowly around bamboo, and the delicate skills fostered in the production of fans, flutes, tea whisks, and dozens of delicate bamboo artifacts found expression in the twentieth century design of electronic circuit boards. More omnipresent, so generally less visible and considered, are the packaging traditions of Japanese culture.

How one person makes one package is a small matter, but the package tactics of an entire culture can determine the use of sizable resources and generate enormous trash. Industrial cultures apparently prefer dirty lots and national parks full of old plastic, bent aluminum, and broken soda bottles to the work of considering more frugal and tidy alternatives of package design. America created her habits growing up as a spendthrift young culture with plenty of wilderness, a colossal backyard where you could get more or dump the ugly. But Japan evolved as an island culture. This means limited resources, so don't waste. Make it well, make it to last, make it small and neat. Design multiple use of one space or one object, and remember that unvarnished simplicity, the grain of natural objects, is often the grandest ornament.

Japan's traditional packages were therefore carefully designed of local materials, renewable and biodegradable resources that sank back into the earth they were thrown upon. Bamboo was

Possible planet pen

The cheapest, oldest, and most widespread pen in the planet village is also the most possible as a planetary norm in the global schoolhouse of modern universal education. Bamboo or reed writing pens have been used in Jordan since 3000 B.C.

One end of a piece of dry bamboo roughly 6 inches by 3/8 inch by 3/16 inch is whittled to desired width for fine writing or bold block lettering, then shaved down to a flexible point formed by the durable outside skin. The end is cut straight across, then shaped to your personal writing angle by writing gently on sandpaper—without ink. Drill a hole roughly 3/32 inch in diameter about 1/8 inch from the point to serve as a retaining hole for ink. A plate can be added for less frequent refilling.[127]

Pen.

more widely used than any other material. The sheath wrapped countless pastries, candies, fish, and soybean lunches; 6-inch culm sections served as jars for soy sauce or delicacies like pickled mountain burdock roots. Dozens of liquid or semi-liquid dishes and sauces were transported home in bamboo containers—which then were used around the house as jars and small buckets after whatever came in them was gone.

Miniature bamboo objects became the trademark package of different areas. Small boxes of bamboo leaf over a frame of split bamboo; model boats, houses, and hats of sheath or leaf and bamboo strips; bamboo snowshoes filled with vegetables pickled in soybean paste; tiny mats and baskets in the style characteristic of the region producing the treat inside—the package evolved from a functional wrapper to an object in its own right.

Photographs of traditional Japanese packaging in bamboo and other materials are beautifully packaged in *How to Wrap 5 Eggs* (Oka 1975).

PLYBAMBOO. Japanese researchers found that bamboo fibers have a lamellar structure in which seven to nine fiber layers are alternately parallel and perpendicular to the axis.[128] Bamboo-fiber anatomy, therefore, anticipated the basic structure of plybamboo, whose early use was explored in eleventh century China as vanes for rotary winnowing machines. Modern processing of plybamboo is well developed in a number of oriental countries. Here, from India, is how to make your own:

> Bamboo mats are woven from strips, dried, treated with phenol formaldehyde resin (alcohol soluble), conditioned and pressed in a hot-press at temperature of 150°C and pressure of 28–35 kg/cm². Duration of pressing time depends on product thickness: board can be prepared from single or multi mats. Veneers, veneer mats, sawdust, shavings, etc., can be used as core material or faces in board production. Boards can be made also in corrugated form. Stringer type boards can be made using a mould and mandrels. Bamboo boards can be used as light partitions or ceilings, for production of moulded furniture, attache cases, suitcases, table tops, chair seats, windmill blades, etc.[129]

Investigations were carried out in the Composite Wood Branch at Dehra Dun (India) Forest Research Institute, and the process has since been patented (Indian patent No. 42228). Plybamboo boards were used in all parts of test houses built at Dehra Dun. Panels buried seven years were dug up still sound.[130] Plybamboos offer the plant's abundance in a format convenient for industrialized mass shelter. Their use will probably claim a progressively large share of the world's bamboo harvest. Experimental use should be high priority research in all ministries of housing in countries with significant bamboo reserves. (See Chapter 4, pp. 96–98, for more on laminated bamboo.)

POLO BALLS. "Polo balls are made in India from the hard, even-grained rhizome of a common species of Dendrocalamus."[131]

POTS. "People in the jungle can use the internodes of any large bamboo as a pot in which to boil water or cook food, as the cooking has finished before the green stems burn through."[132]

PROPS. A large volume and common use of bamboo in many parts of the world is for cheap, light, easily made and transported props. *Bambusa vulgaris* was used to prop bananas for many years by United Fruit in Central America, and this use is one of a number of rapidly expanding bamboo industries in Taiwan, where islanders with limited space are multiplying shrewd uses of bamboo's rapid altitudes. "Four bamboo treating plants preserve banana props with creosote, with a volume growing from 1.5 million poles treated in 1971 to about 3 million in 1973. Cultivation hasn't kept pace with use, so green poles have risen around 30 percent in cost. Treatment began around 1966 by the open tank method, later by the Bethell process (90°C, 14kg/cm^2)."[133]

PUNISHMENT. "In a fearful punishment, formerly used in Bali, the criminal was strained horizontally over the young growing shoots of a bamboo stock, of which the longer halms have been removed. As they grow quite rapidly, the very hard silica-rich shoots pierce through the unfortunate sufferer."[134] A cruel and unusual punishment involving bamboo is told of the tyrant Kao Yang (A.D. 550–559) who forced prisoners to attempt flight from a 100-foot tower: "On one occasion the emperor visited the Tower of the Golden Phoenix to receive Buddhist ordination. He caused many prisoners condemned to death to be brought forward, had them harnessed with great bamboo mats as wings, and ordered them to fly down to the ground. This was called a 'liberation of living creatures.' All the prisoners died, but the emperor contemplated the spectacle with enjoyment and much laughter."[135]

"Under the Chou and Han dynasties, convicted criminals could look forward to either: being branded on the forehead, having their noses cut off, maiming, castration, or death. . . . later times these pleasantries were amended to: bambooing, bastinadoing, banishment, exile, and the ever-popular death (choice of strangulation or decapitation)."[136]

A Victorian traveler, Colonel Barrington de Fonblanque questions, "What would a poor Chinaman do without bamboo?" The thatch of his house, mat where he sleeps, cup for his drink, chopsticks to eat sprouts of the same plant, pipe to water his crops, rake, sieve, and basket to harvest, clean, and carry them, the mast of the junk for boat people, pole of the cart on land—all of bamboo . . . and if he strays briefly from this path of civil virtue: "He is flogged with a bamboo cane, tortured with bamboo stakes, and finally strangled with a bamboo rope."[137] The colonel neglects to mention the funeral procession that follows, customarily led with —a spring of bamboo.[138]

The colonel is mercifully silent as well about the public and private use of bamboo poison. In a special part of a special bamboo, "there is a poisonous secretion extremely irritating to the throat and nose, causing itching which brings on a bad skin infection. This has been known to the Chinese for a long time, for the poison was formerly used in criminal cases. A drink was prepared for the condemned. Death followed, but not before much agony had been suffered. This bamboo is not planted near wells."[139]

***Punting poles. A preferred species is* Bambusa tuldoides.**

R Racks (for curing and drying food), rafts, raincoats, rainspouts (and guttering), rakes, rattles, rayon, record needles, reeds (for woodwind instruments), reinforcement (for concrete and adobe), rings, ritual objects, roofing, ropes (towing, drilling salt wells, hoisting brine), rug poles, rulers.

RACKS AND TRAYS. "Those of bamboo are used in almost as many ways as baskets. Bamboo racks, baskets, and trays constitute important items of equipment required for many large-scale industrial and commercial pursuits in the Orient. In the silk industry, the mulberry leaves are brought from the field in bamboo hampers, while the silkworms are hatched and spend the whole of the caterpillar stage on bamboo feeding trays. As a fitting finale, they are placed, when mature, upon racks fashioned from bamboo in a form suggesting treetops where, in the wild free state, their ancestors spun their cocoons. The shape of these spinning racks is cleverly designed, however, in deference to the requirements of space economy."[140]

RAFTS. In Colombia and Ecuador, large rafts of *Guadua angustifolia* deliver this bamboo to markets and formerly were a principal means of transport for people and mountain agricultural produce. Since bamboo grows well along rivers, the most natural and ancient means of transporting it has been to float it downstream. In the Orient, up to ten thousand culms may compose a single raft of bamboo, which is generally considered the archetype of all Chinese vessels:

> A length of bamboo cut in half longitudinally and floated on water gives a striking model of the constructional principle of all Chinese craft. It is not necessary to insist upon the sea-going bamboo sailing raft of Thaiwan as the only ancestor of all junks, for many other forms of

> bamboo raft are regularly using Chinese rivers at this day. One of the most interesting is the Ya River raft of Szechuan, which moves both up and down 100 miles of intractable waterway between Yachow and Chiating, carrying Tibetan trade. This *Chu-fa chhuan* must be one of the lightest draught general cargo-carriers in the world, for its depth below the waterline when loaded (with a cargo of 7 tons) is often as little as 3 inches and never exceeds 6, owing to the buoyancy of the bamboos. In length the rafts, which are quite unsinkable, vary between 20 and 110 feet, and are built throughout of the culms of the giant bamboo *Dendrocalamus giganteus,* which grows as high as 80 feet with a diameter of as much as a foot. The bow is narrowed, and bent upwards in a curve by heating, so that the raft can slide over rocks which may be almost level with the water surface.[141]

Worcester (1966:113–5) provides interesting details on the history and construction of the Ya River raft, with a photograph of a model (plate 12) from the London Science Museum's collection of twenty-seven models of bamboo and wood rafts, junks, sampans, houseboats, inflated skin rafts, trawlers, lifeboats, etc. (137–8).

Emperor Wu Ti (140–86 B.C.) built a floating castle 600 feet by 600 feet, garrisoned by two thousand men and some cavalry horses as well. Floating villages with room for two hundred families were reported by a seventeenth century official of the Dutch East India Company. (Cf. Rudofsky's valuable *Prodigious Builders,* 146, 373.) As the floating life becomes more common in a more crowded world, China's long history of buoyant architecture will be more consulted. More attention will be paid to bamboo's role in river transport and shelter, to its place in bankside industries of paper pulp and plybamboo, and to its help in holding banks stable against the constant current. (See *junks,* above.)

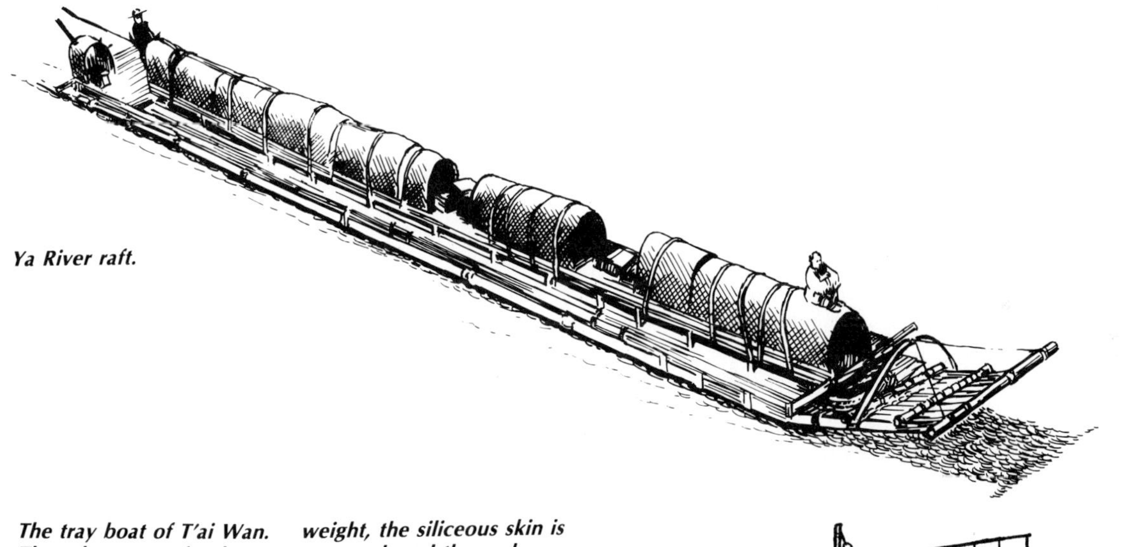

Ya River raft.

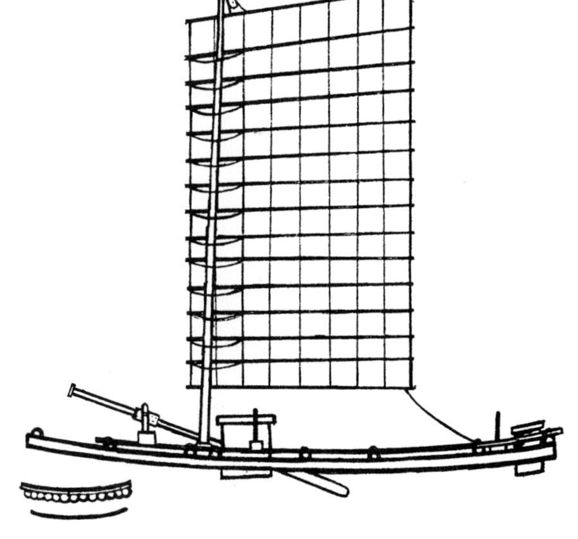

The tray boat of T'ai Wan. These boats are simply small rafts, measuring 12 to 40 feet long, with a draught of a few inches, built of about a dozen large bamboos lashed together with the tapering ends forward. The foremost end of the raft is narrower and more upturned than the after end.

The culms used grow to a height of about 60 to 80 feet with a maximum circumference of 17 inches. This species has a large core and is an extremely light wood. The bamboos are very carefully selected, for they must be of uniform size. To prevent cracking and to reduce weight, the siliceous skin is removed, and the nodes are hardened over a slow fire. The raftmen say that this method of treatment also increases their "arresting power" when afloat. A vain effort to keep out the water is made by securing a small bamboo between each of the larger ones and its neighbors. Across the fore and aft poles eight slightly smaller curved poles are lashed athwartships at intervals to suit the construction. The after ends of the main bamboos, which are usually, though not always, broader than those in the bow, are painted red, black and green.

RAYON. A bamboo rayon factory in Pakistan produces more than 5,000 tons of high quality rayon a year with technical assistance from Japan.[142]

RECORD NEEDLES. "Very satisfactory phonograph needles have been manufactured from bamboo slivers. They are heated in oil at 340°F and tumbled in barrels containing sawdust, which removes excess oil and polishes the slivers. They are then ready to be pointed and used."[143]

RITUAL OBJECTS. "As well as being a food, the shoot, like the culm, may house a god under certain conditions. In the valleys among the Chekiang hills between rice fields one often passes small shrines in which instead of the accustomed idol one sees a dried bamboo shoot. On examination, this shoot reveals very unusual characteristics. The most noticeable are the oblique joints which give a zigzag effect. It has been found that occasionally in a grove of *Phyllostachys pubescens* (*mao chu:* 'hairy bamboo'), which is very common in this region, a shoot comes up with this freak characteristic. The Chinese are quick to discover it and because of its strangeness they think it possessed of preternatural powers."[144]

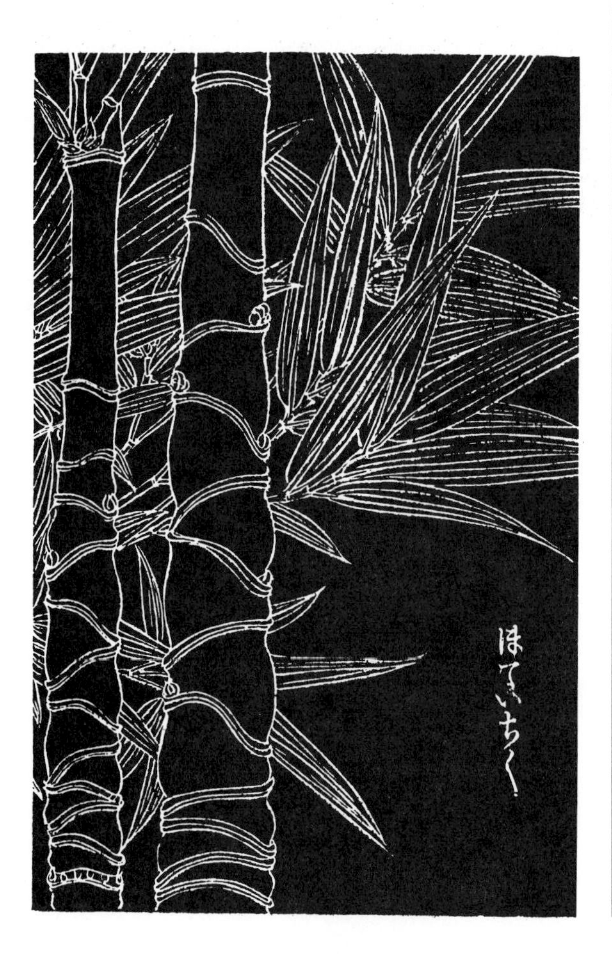

Roofing of split bamboo culms reportedly inspired the first clay tiles. Japanese woodcut.

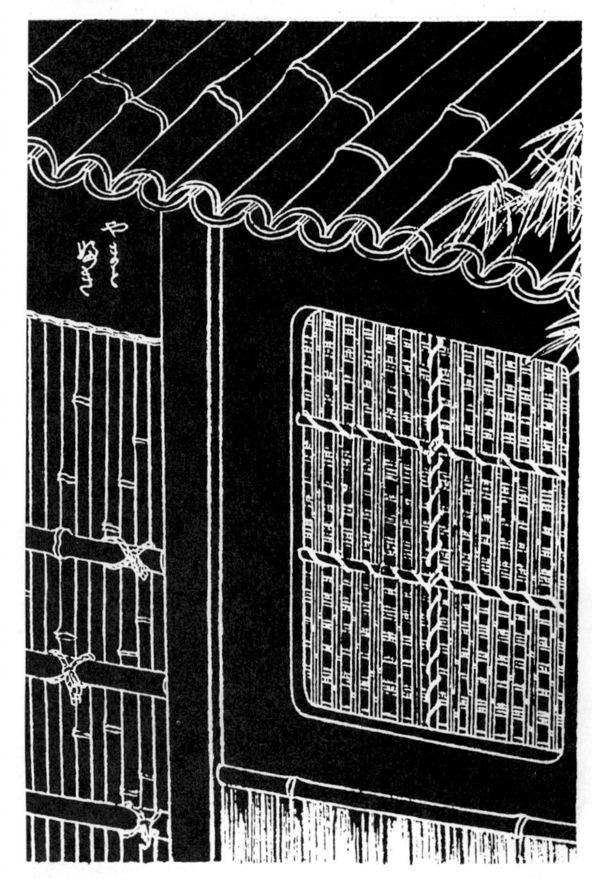

ROOFING. Shingles, thatch, and tiles. "In the dry zone around Mandalay, split bamboo may be woven into long narrow mats which are put on the roof like overlapping shingles. In some other places bamboos with diameters of 3–5 inches are split in half lengthwise, cut into two-foot sections, and laid on the roof like Chinese tiles."[145]

> The earliest type of roof material was no doubt the fully grown hollow bamboo stem split in half longitudinally, convenient lengths being then laid in rows with the concavities alternately facing outward and inward. (Bamboo and reed matting must also have been used from high antiquity, and still frequently forms the covering for sheds and boats.) This very corrugated arrangement was afterwards carried out in half-burnt grey tile[146]

Thatch of sasa bamboo leaves is said to last a long time in Japan. Tsuboi recommends late August harvest there: "if gathered while the leaves are tender and green, they will decay quickly."[147]

S Sailcovers,
sails
sailstays,
sake,
salt well drilling,
sandals,
scaffolding,
scales,
scarecrows,
scoops,
scratchers for backs,
screens (room dividers, papermaking),
scrubbers,
sedan chairs,
shades,
shakuhachis,
shavings (for pillow and mattress stuffing and for caulking for boats),
sheaths for knives and swords,
shields,
shingles,
ship design,
shoehorns
shoe soles,
shoots for food,
shovels,
shuttles for weaving,
sieves,
silk industry,
skewers for cooking,
ski poles,
slide rules,
sluices,
snow fences,
spears,
splints,
spouts for watering cans,
spray guns for citrus insect pests,
springs (for carts, toys, gates),
stakes,
staves,
sticks (incense, rhythm, roasting for shish kabob),
stilts,
stools,
string,
sugar,
sunning floors for drying coffee and other produce,
swimming pools.

SAILS. Bamboo mat-and-batten sails are unique to Chinese area ships: "The sail is made by weaving together thin and narrow strips of the outer parts of the stems of bamboo and is divided into sections grasped by bamboo battens. Thus the sail falls in tiers, ready to be hoisted. A large mainsail in a grain ship needs ten men to hoist it, but for the foresail, two suffice."[148] The widespread use of bamboo matting sails necessitated this form of frame and led naturally to the balance lug shape. The aerodynamic importance of tautness is considerable; yet such a design, which doubtless arose because of the easy availability of a material so light and at the same time so strong, never arose in any other culture. The battens have at least five other uses: they permit precise and stepwise reefing; they allow immediate furling of sail, which falls into pleats; their setting system obviates the need for cloth or canvas as strong as on other sails; and they act as ratlines, giving access for the crew to any desired part. Above all, they are a complete protection against tearing and carrying away; a Chinese sail may have half its surface full of holes and yet draw well. The sail never jams. As Audemard thought fit to emphasize, this system "avoided the sending of men aloft to take in reefs, always a dangerous operation in bad weather."[149]

Thatch boat in China. Building is a bulky business; cutting thatch on or near rivers to float the roof downstream reduces transport problems.

The sail is stiffened by battens of bamboo, each of which connects with, and indeed forms part of, the sheet, thus keeping the sail very flat: this, then, is the secret of the Chinese sail. In seagoing vessels and in craft used in very open inland waters the sail is peaked. In narrow waters it is flat-headed, narrow and very tall, so as to catch the breeze over the banks of the river, creek or canal.

The sail design of junks could be a technology appropriate in areas presently ignorant of it but blest with small navigable waters and bamboo. Building models of a junk is an interesting exercise for schools. Bamboo sails have also been used traditionally in windmills (q.v.).

SAKE. "A *sake* can be brewed from bamboo seed, which though rather sharp to the tongue does not otherwise differ from ordinary *sake.*"[150]

SALT WELL DRILLING. This is one of a number of industries that would virtually come to a halt without bamboo: "Possibly there is no more indispensible article in China than bamboo. One missionary made a list of bamboo uses, unfinished with 440 separate uses. If it were uprooted from the soil of China, it would be worse than losing a right hand. So many and varied are its uses around a salt well that one wonders what would happen if it were suddenly cut off. The geomancer uses bamboo tickets to decide the location of a new well; the mechanic, a bamboo rule to measure the land; the

Joining bamboo pipes in Szechuan, 1637.

priest, bamboo joss sticks for the opening rites. Bamboo ropes haul up the logs for the derrick frame, splice them together, and are tightened by bamboo wedges. Bamboo binds the drum over which the bamboo cable runs into the well, carried by a wheel edged with bamboo. The band first put into the well to carry the drill is of bamboo, as is the cable later used to carry the bamboo brine pipe, fastened to the bamboo cable by hemp. The brake used on the windlass that winds the cable is of split bamboo, and it runs on strips of bamboo lashed by bamboo rope to the wooden windlass. The water buffalo is harnessed with bamboo to the windlass to draw the cable out of the well, while the driver 'persuades' the awkward beast with a bamboo whip. The rope with which he is tied by the nose is of smaller bamboo, his stable divided into stalls made of old bamboo cable, and the sides of the buffalo barn made of old bamboo rope twisted about the upright posts. The well coolie wends his way home by the light of a taper made from an old bamboo cable.

finely and wrapped around the pipes to prevent their splitting in the hot sun. Bamboo hoops support the great brine vats which hold the brine before running into the boiling pans, through split bamboo supported by old bamboo cables.

"Bamboo matting separates the boiling rooms; attendants sleep on bamboo baskets which later carry the finished product, or enjoy bamboo sprouts while they watch the pans (often boiled with bamboo firewood) and scoop the refuse off the top with woven bamboo skimmers. The coolie carries his load of salt to market with a bamboo carrying pole, and his tally is recorded with a bamboo stick. His sun hat is made of bamboo, finely woven, to keep out the rain, while the mat on which the peddler spreads his wares is of bamboo. When it is impossible to use the bamboo further in the industry, it is given to the labourers who sell it, repair their homes, or boil their rice with it. The expert boiler, watching the pans, blissfully smokes a bamboo water pipe, while his wife sews shoes with soles of bamboo leaves, by light of a vegetable oil bamboo lamp. He dips the brine from pan to pan

"The brine runs to the boiling pans through bamboo pipes, supported by bamboo pieces split

Brine conduits in bamboo piping at the Tsu-liu-ching salt fields, Szechuan, 1944.

with a bamboo dipper and strains it through a bamboo sieve. Bamboo guy ropes hold the derrick secure when the great gales blow, and the wheel at the dizzy height of the mighty derrick is trussed with bamboo. . . . The subject is not exhausted, as there are many other uses to which bamboo is put daily, but the reader will readily see that, if bamboo were taken from the market here in Tzeliutsing, it would paralyze the salt industry."[151]

SANDALS. "In Southern China the sheaths of a large thorny species *(Bambusa sinospinosa)* are torn into narrow strips to serve as the weft of coarse sandals."[152] The usefulness of these sandals on slippery heights is remarked by a Chinese woman, a botanist at Harvard's Arnold Arboretum, recalling mountain treks in western China in her youth, "Climbing the Trails of the Giant Panda." Somewhat like those monsters of myth with hide so tough that it could only be cut by their own claws, the panda is sought shod in sandals made from the culm sheaths of its favorite food—bamboo. Perhaps the rarest animal in captivity, the most expensive according to the *Guinness Book of Records,* it lives in an ancient land mass that predates the Himalayan uplift in the early Tertiary period. Called the "white bear" by the Chinese who have known it some four thousand years, the panda is actually a cousin to raccoons, first heard of in the West in the nineteenth century and first seen in the Chicago zoo in 1937.

In the early 1940s, the British were offering fellowships for study in England to anyone who could catch one. "When a giant panda is ready for shipment to England, a fellow in any branch of the biological sciences may be chosen to accompany the animal for a year's research in British institutions."[153] Miss Hu, anxious to study at the Royal Botanical Garden at Kew, set out to capture a panda from the bamboo forests in northern Szechuan: "At Ya-chow we hired thirty-one porters to transport our supplies and equipment. One of the most important items among our supplies was two hundred pairs of sandals made of bamboo fibers. In a country where leather shoes are rare and climbing boots are unheard of, these bamboo sandals are indispensable for high-level climbing. They are relatively inexpensive and comparatively strong, and they do not shrink when wet. Most important, the rough ends of the tough fibres extending from the outside of the soles hold the steps in slippery places. In the mountainous region of west China, even the recognized highways between county seats are steep and narrow trails, and the byways are still narrower, often only steps cut on the hard rocks of steep mountainsides, with roaring torrents a few hundred feet below. A slip may mean death. The trails in the land of the giant panda are only occasionally climbed, by hunters or medicine diggers. The morning mist, the frequent rains, and the melting snow render these paths wet and slippery all the time. The bamboo sandals are the conventional footwear for climbing trails in this region."[154]

Pandas, whose lives are lashed more firmly to bamboo than any other animal, are presently threatened in their native mountains by the flowering and death of bamboo species in their area. A modern papercut of a panda with its totem plant, from Fushun, China.

From Miss Hu's further descriptions of the grandeur of the terrain, it sounds so much more inviting than a British zoo that I am happy to report that her two hundred sandals permitted her and companions to climb the trails of the giant panda . . . without ever so much as glimpsing the beast. "With the demand of many zoos throughout the world and the favorable foreign exchange, the capture of a specimen means a fortune to each of the cooperating hunters. Yet year after year experienced hunters have gone hunting in vain. Likewise, in my expeditions, I have not seen any giant pandas."[155]

SCAFFOLDING FOR SKYSCRAPERS. "A typhoon once struck two tall Hong Kong buildings, both protected by scaffolding: one of bamboo, the other of steel. The bamboo held firm, but the steel collapsed in a contorted heap. . . . You can sit by a twenty-story office window and watch the Bamboo Men operate. Suddenly you will catch a glimpse of a waving pole. Then a little man will appear, scampering up to the top. After tying more poles, he will disappear skywards. These are the men who have recently built a 4,000-seat theater in just over a week . . . They are the aristocrats of the local labor force and can command over $500 a month. In a typical operation—like repainting the exterior of an office building—trucks move in first to dump their loads of weather-beaten bamboo poles on the sidewalk. . . . The workers pick through the poles carefully, rejecting an occasional piece because it is cracked or weak—or simply because they don't fancy it. Next the heavy base poles are swung up and lashed to the window ledges or other projections. Soon the framework is towering over the street, and the workers are swinging along it, secured only by the tough bamboo strips and their own confidence."[156]

And for supertankers: "Sixty thousand tons of freighter, a beamy heavyweight of the high seas, takes shape inside bamboo scaffolding at Aioi. Workers put the bulk-cargo carrier together by joining prefabricated interlocking units. Devising such techniques to produce ships far more quickly and cheaply than European or American yards, Japanese builders now launch almost half the world's new vessels—and the biggest, including the largest ship afloat, the fifth-of-a-mile long, 209,000-ton supertanker *Idemitsu Maru.* Japanese designers now plan leviathans of more than twice that tonnage."[157]

SCALES. A passion for cricket fights dates in China at least from the T'ang dynasty (A.D. 618–906). Many volumes, more weighty by far than their subject, devote themselves to the best diet and training. In the Sung dynasty, the Sun of Heaven himself neglected his vast empire, lost in this pygmy passion, bent above a 4-by-8-inch Kingdom of Crickets, the minicolosseum where the mortal combats were staged. The combatants entered by sliding doors at opposite ends of the bamboo arena. A screen separating them was lifted, and the battle of the bugs began, ending only with the death of one and the triumphal return of the other to his jade residence for a victory banquet of fresh shrimp. The most interesting element—for our purposes—in this cruel absurdity is the set of scales used to weigh the crickets, like prize fighters, before the match. The accuracy required for so delicate a task can be appreciated. The material chosen was bamboo. Then in World War II, when radio parts became scarce in China, they simply made them from bamboo, drawing on a sophisticated technology sharpened for centuries on such tiny triumphs as cricket scales.[157a]

Screen, Fu-chou, Fukien Province, China, 19th century.

SCARECROWS. Those of bamboo are made in various countries in a variety of ways. "One often sees, in the more tropical parts of the Orient, scarecrows made from large stiff culm sheaths. The sheaths are either suspended by a short cord from the tip of a bamboo pole thrust into the ground at an oblique angle, or simply impaled upon a short stick set upright. As the sheath swings about in the breeze, the pale, polished, inner surface and the dull outer one reflect the light differentially, exaggerating the effect of its motion."[158]

"If we have in Europe ugly scarecrows and such like for driving away the flocks of predatory birds from the young sowings and cornfields, so has the Malay also his own invention for the same purpose: long bamboo halms, at the end of which is fixed a bamboo wind wheel, moved by the slightest breeze with an ugly rattling noise which scares away the numerous rice thieves, finches and small parrots that swarm on the ripe fields. This noise, very disagreeable to the ear, continuously interrupts the stillness of tropical nature. The Javanese often places in various parts of his field many bamboo sticks, from which are suspended pieces of cloth and other light articles, and connects all these sticks with a bamboo or rattan string. The man who keeps the ends of these strings in his hand pulls them from time to time, concealed like a spider in a little bamboo house erected for this purpose on high posts."[159]

Sedan chairs have been constructed since ancient times in China of lightweight bamboo frames and carrying poles, and used to carry persons of rank or riches. The common people used them only to carry a bride from the house of her parents to her groom. In funeral processions, an empty chair was carried ritually to even the poor man's grave.

SEDAN CHAIRS. These have been constructed since ancient times in China of lightweight bamboo frames and carrying poles. Employed to carry persons of rank or riches, the common people used them only to carry a bride, concealed from view, from the house of her parents to her groom. As if to acknowledge all are equal on the other shore, in funeral processions an empty chair was carried ritually even to the poor man's grave.

SHIP DESIGN. Oriental ship design holds in the hull a ghost of the grove: "It is clear that the ships of East Asia cannot be genetically explained on the theory of the simple floating hollow log. Bamboo is their ancestral material, not wood . . . the Chinese hull is an elongated structure as full of transverse bulkheads as the stem of the bamboo is of those partitions which botanists call septa [nodes] . . . this construction sets the Chinese ship apart from the ships of the rest of the world."[160] (See also *rafts* and *fishing.*)

SILK. "Bamboo played an important though inconspicuous part in the history of European industry. When Justinian reigned at Constantinople (A.D. 527–565), the court reserved a monopoly of the silk trade and its manufacture, the looms being worked by women in the Imperial Palace. Until then, the silkworms that feed on the leaves of the white mulberry were confined to China; those which haunt the pine, the oak, and the ash were common in the forests both of Asia and Europe, but as their education is more difficult and their produce more uncertain, they were generally neglected, except in the little island of Ceos, near the coast of Attica. The Persians had the monopoly of the trade in Chinese silk. This was a matter of deep concern to Justinian, who endeavored to procure the raw material for his looms through his adventurous Christian allies, the Abyssinians, who at that time were a naval and commercial power. His negotiations failed, the Abyssinians declining a competition with the Persians, whose proximity to India must give them an overwhelming advantage. Another expedient, however, presented itself. Two Persian monks, who had long been resident in China, traveled to Constantinople, a giant's journey, and proposed to the emperor that they should endeavor to introduce the eggs of the silkworm into Europe. The offer was accepted and liberally encouraged by Justinian. The two monks returned to China, and by smuggling the eggs in the hollow of a cane contrived to elude the vigilance of the Chinese, and made their way safely to Constantinople with their precious treasure. It is not too much to say that in that fragment of bamboo were carried the future commercial fortunes of Lyons, of Genoa, of Spitalfields, and all the other great manufactories of Europe, for from those eggs were descended all the races and varieties which stocked the Western World. But the pity of it is that we have not the record of the travels and adventures of those two Persian monks! This memorable importation is assigned to the year A.D. 552."[161]

A floating-sights water-level proto-theodolite used in 1957 for surveying irrigation projects. The bamboo with septa (nodes) bored as sights floats on the convex meniscus of water in a rice bowl. This method is probably very old as well as very practical.

SPLINTS. An Austrian doctor working several years in eastern Nicaragua noted the appropriateness of bamboo for splints to set broken bones in village settings where you often have to make do without the manufactured variety available in hospitals. Medical uses of bamboo are many, from tongue depressors to equipment for the handicapped. (See *crutches.*)

SUGAR. "For the first time in history, the *Dendrocalamus strictus* bamboo forests of Chanda, in the Central Provinces [India] began last March [1900] to exude a sweet and gummy substance in some abundance which was found very palatable to the natives in the neighborhood, who have been consuming it as food. The occurrence of the manna at this season is all the more remarkable since the greatest famine India has known is this year visiting the country, and the districts where the scarcity is most keenly felt are in the Central Provinces. . . . The manna occurs in short, stalactiform rods, about an inch long, white or light brown in color, more or less cylindrical in shape, but flattened on one side where the tear had adhered to the stem. It was pleasantly sweet, without the peculiar mawkish taste of Sicilian manna, soluble in less than its own weight of water; the solution in repose deposited white, transparent crystals of sugar. Analysis showed 2.66 percent moisture, .96 percent ash, .75 glucose, a small quantity of nitrogenous matter, and the remainder consisting of a sugar which from its solubility, melting point, and crystalline nature appeared to be related to if not identical with cane sugar. The bamboo and sugar canes belong to the same order of grasses, and perhaps it is not unnatural to expect them to yield a similar sweet substance which can be used as food; but it is a coincidence that the culms of the bamboo, hitherto regarded as dry and barren, should in a time of great scarcity afford sustenance for a famine-stricken people.

"None of the natives questioned recalled seeing the substance before. They believed it to be 'bamboo manna.' The material is soluble in water, but insoluble in alcohol, ether, and chloroform. Crystals of sugar remain on evaporation which melt at about 166° C, and a little above this temperature assume a brown color and consistency of barley sugar."[162]

SWIMMING POOLS. The owner of a construction firm living across the road from the USDA groves 9 miles south of Savannah, Georgia, told us that Thomas Edison had used bamboo to reinforce a swimming pool at his home in Florida. According to this professional, the pool was in good condition around 1980 and was then roughly eighty years old. (See *water storage.*) Edison had already demonstrated the more delicate capacities of bamboo's durability as the incandescent element in light bulbs (q.v.).

T Tabasheer,
tables,
tallies,
teahouses,
tea strainers,
tea whisks,
tents and other temporary structures,
tortures,
towers (for water, windmills or rope-making),
toys,
trailers,
transport,
traps for ambush or fish or rats or tigers,
textiles,
thatch,
theatrical uses,
theodolite,
thole pins,
threshing boards and machines,
tiles (roof and floor),
tinder,
tipis,
tokens,
tongue depressors and cotton applicators in hospitals,
toothpicks,
torches,
trays,
tree guards,
tree houses,
trellises,
trestles,
triumphal arches,
trolley poles (England),
troughs for feeding and watering chickens and other animals,
trumpets,
tubs,
tug and tracking cables,
twine.

TABASHEER. (tabaschir, tabashir): Used as variously as it is spelled, "famous for any and all ailments" (Porterfield), good for coughs, asthma, a reknowned love potion, an antidote to poisons of all descriptions, it has long been in the Orient a medicine of almost mythical powers. "It is employed in western India to cure paralytic complaints and as a stimulant and aphrodisiac. . . . It plays a great role in Chinese medicine, and large quantities are exported especially from India to that country and Arabia. Tabasheer is also used in polishing, a quality it owes to its silicious composition."[163]

Silica is apparently responsible also for its reputation as an effective antidote: "nowadays an artificial form is used internally to neutralize toxic agents by absorption."[164] *Melocanna baccifera* and *Bambusa arundinacea* are cited by Porterfield and Kurz as species especially abundant in deposits of tabasheer, but it is found in many other bamboos as well, in chunks up to an inch thick.

Jones (1966) and associates at the University of Melbourne in Victoria, Australia, studied the composition of tabasheer through an electron microscope. An abridged version of their findings: Bamboo absorbs large quantities of dissolved silica, and solid silica is deposited in its cell walls. This silica has been identified as opal by optical and x-ray techniques, and particles known as opal phytoliths of the same dimensions as cells can be isolated from the plant tissues. Large masses of silica found in the hollow stems of bamboo are called tabasheer. Known anciently but only recently examined by modern techniques, it is a porous, hydrous silica composed of roughly spherical particles about 100 angstroms in diameter, linked together in chains. As determined by the immersion method in sodium light, tabasheer has an index of refraction of 1.427

+/− 0.002; its specific gravity is 1.93, appreciably less than that of common opal, which tabasheer resembles in other ways: it is milky in appearance, translucent, and shows conchoidal fracture. Roughly parallel bands of different optical densities can be seen within the tabasheer in transmitted light. The chemical composition of tabasheer by weight is: SiO_2, 85.82 percent; H_2O (−100°C), 5.87 percent; Na_2O, 0.002 percent; K_2O, 0.039 percent; CaO, 0.007 percent; and MgO, 0.004 percent. Tabasheer contains more water than is commonly found in opals but smaller amounts of alkalis and alkaline earths. Comparison of tabasheer with opal phytoliths, silica gel, and opal of inorganic origin under the electron microscope revealed basic but subtle differences that the interested reader can pursue in "An Opal of Plant Origin," *Science* vol. 151, no. 3709:464–6.[165]

TEMPORARY SHELTERS. "If temporary shelter is required by the native or traveler in the jungles, nothing is so convenient as the bamboo; and how quick do they finish such a temporary house! A few hours' patience and the traveler is comfortably housed for the night, having not only shelter above him, but also his table, chair, and 'bali-bali'—bedstead—all made of bamboo."[166] These remarks suggest reflecting on bamboo for refugee camps in the wake of war, quake, flood, or whatever man-made or natural disaster, when instant shelter is required.

THEATRICAL USES. Bamboo's use is related to its convenience for temporary structures. Easy to carry, easy to assemble and disassemble, it is much used by traveling performers throughout the world in many ways that deserve study by schools. A report from China: "Itinerant theatrical troupes employ bamboo structures of a distinctive architecture, tall and narrow, with the walls often covered with gaudily decorated mats, and surmounted by ornamental devices of traditional rococo design. The floor, which is elevated several feet above the ground, is made of thin wooden planks laid on bamboo beams and held in place by thin strips of bamboo bound down by bamboo thongs. The top-heavy structure is held erect by means of long bamboo braces, to which is often added the security of bamboo guy ropes."[167] (*Fiesta assistant,* above, explores similar uses; see also *transport,* below.)

TRANSPORT. Rafts, ox carts, sedan chairs, carrying poles, wheelbarrows, tug cables, punting poles, sails, docks, and even bikes have been made of bamboo. Its lightweight strength has anciently associated bamboo with travel and transport, as vehicle or accessory, axle or anchor, caulking material or bilge pump . . . a complete list would begin to repeat other entries. The next time you walk hilly terrain, carry a bamboo staff. Bridges of bamboo we have seen earlier. In some mountainous areas, bamboo stairs are also constructed at critical points of the path: "When the path goes over very steep and slippery ground, the bamboo is used to form steps. Pieces are cut, about a yard long, bamboo pegs are driven through holes made at each end, and a ladder or staircase is quickly made."[168]

TRESTLE. "A trestle of bamboo, 125 feet high with a span of 400 feet, was erected near Miyaroshita, Japan, and used until a steel railroad bridge could be completed."[169]

U

UMBRELLAS. "The most common umbrella, found in all parts of China, has a bamboo handle, spring, and ribs, and is covered with oiled paper," which may also be made of bamboo.[170]

V

VALIHA. The valiha is the national instrument of Madagascar, an island with roughly one-fiftieth (2 percent) the land mass of its neighbor Africa but nearly three times as many genera of bamboo. (*Decaryochloa, Perrierbambus, Cephalostachyum, Hitchockella, Hickelia, Nastus, Pseudocoix,* and *Schizostachyum* in Madagascar; *Oreobambus, Oxytenanthera,* and *Arundinaria* in Africa.)[171] With this richness of bamboo to choose from, it is not surprising that the valiha, a bamboo

zither of sorts, should be central to the island's musical traditions. A record is available that includes photographs and detailed descriptions of the valiha and its musical history.[172]

VIOLIN. "There is also a kind of Chinese violin called the *hyi ieng*. . . . It consists of a 2-inch thick bamboo joint, 3 to 4 inches long, closed at its extremity by a tightly stretched snake's skin. To this is inserted a bamboo handle about 2 feet long, to the upper end of which are fixed the two strings resting on a bridge on the snake's skin. A piece of split bamboo is used as a bow."[173]

It is actually a mistake to try to understand or judge these instruments in relation to Western counterparts. Calling them "a kind of violin" or "zither of sorts" is a convenient shorthand, but it does more to confuse their reality than to clarify it. Best to scrub your ears of all memories of Mozart and try just to hear how they sound. (See Bamboo Discography for a brief listing of records, p. 333, and the Hornbostel-Sachs system of instrument classification, p. 74.)

Whip.

W Wagons, walking sticks, walls, war, water jugs, water pistols, water storage, waterwheels, waxes, weapons, weaving shuttles and looms, weirs, well sweeps, wheelbarrow, whetstones, whips, whistles, wicks, windbreaks, windmills, wine storage, winnowing machines, writing brushes.

WAR. The art of war has many ancient links with bamboo. "Fine spears and arrows as well as bows and shields used to be made of bamboo: so are torches, conical military hats, criminal beaters, and splints for binding up wounded limbs."[174]

Research on bamboo often increases in wartime, the forceps of history which drag new technologies into the world. Scarcities in World War I inspired the Chinese to build railroad trestles of bamboo cement. In Vietnam, U.S. military interest in bamboo rose, as it had also in World War II, when the U.S. government funded investigations into bamboo ski poles to replace the supply of *Arundinaria amabilis* cut off from China. Fighting in the Alps, you discover the best ski pole shafts are built of this species, growing only in a very narrow range of southern China. So you go investigating, quickly, the properties of bamboos of abundant known supply in territories ally to your purposes. McClure (1944) did that, from the United States to Brazil.

Bamboo cement receives a spurt of interest in modern wars because you need to build landing strips and other installations in the middle of the jungle where you find no iron and much bamboo. Soldier shelters, from semipermanent barracks to armatures for field tents, are rapidly constructed from bamboo when the material and labor skilled in its use are available. Revetments of bamboo and earth can protect from heavy bombardment.

Bamboo ambush is a genre in the art of war in Southeast Asia: "The so-called 'rangyoos' are thin bamboo pegs sharpened at both ends which are put in oil and slightly burnt in fire. Such pegs are put vertically in the ground hid in grass. They cause very dangerous wounds and, in wet weather, can penetrate also the moistened soles of shoes. In the campaign of the Dutch against the Boogginese of Boni [Celebes] in 1859, the Dutch soldiers all carried bundles of such rangyoos, but the Boogginese were not such fools as to run into them, nor had the Booginese rangyoos any effect upon the Dutch troops. All the Malayan tribes, and the hill-peoples of Assam and Burma use similar pegs, and larger ones are employed against cavalry, placed singly and obliquely in the ground in high grass, or crosswise and tied. It is a very common custom with Malays and Burmans to place strong bamboo poles across paths in long grass or dense jungle, fixing them firmly at the one end while the bamboo is tightly strained and fastened at the other end in such a way that it immediately unbuckles as one steps on it or only uncautiously touches the pole, thus striking with all force against the legs of the passersby or the passing enemy. The people of Arracan and Tenasserim have, for catching tigers, a similar method. The bamboo pole is then vertically planted in the ground and strained downwards by means of a strong rope terminating in a large noose arranged so that the tiger, which preys upon a bait laid for him, must pass and touch the noose, when, of course, he is at once launched into eternity."[175]

WATER STORAGE. Bamboo-cement tanks have been used as an alternative to dirty, distant, and insufficient water stored in small earthenware jars or costly aluminum tanks. The Adaptive Technology Group (ATG) in rural Thailand, with help from the civil engineering department of Chula University, evolved feasible designs for bamboo-cement rainwater storage tanks that proved to be four times cheaper than galvanized steel tanks of equivalent size. Easy to build using local materials and local technology, independent of outside supplies or assistance, the tanks are an exercise in self-reliance, a demonstration that the rural villagers can autoresolve their most critical problems. Five-day seminars were conducted in three villages in October 1980, and a handbook was distributed in additional areas. For drinking and cooking water, figuring daily personal consumption at roughly 5 liters, 3,000 liters or 3 cubic meters is needed for a family of five during the four-month dry season.

Construction begins by leveling and compacting the soil base. Bamboo is then woven into a square structure, with spacings of 2 square inches between strips, 20 cm wider than the diameter of tank base on both sides. Walls are next woven and tied to the base. Flint coat is painted on the bamboo before coating with a mix of cement to sand to gravel (1:2:4) 10 cm thick in the base, 3 cm in the walls, where two coaters work opposite one another, wearing thick rubber gloves. Coating hardens one day before giving a finish coat. A bamboo lid is woven, leaving a hole where rainwater can be channeled to enter and which is large enough for people to get in and clean. The tank is cured a week or two covered with straw and dampened as needed with water.

Communication of the technique: Lakdan is a village 300 km from Bangkok in northeast Thailand where 120 households suffering many years from water shortage have to fetch water from a source 2 km away since the town pond is often dry. In July the abbot of the local temple invited ATG to send a study team to investigate village conditions. Two workers visited in August and proposed an official letter from the village headman and the abbot to the Thai government requesting a large pond be dug at the low end of the village for general village use. For individual families, bamboo-cement tanks were proposed, and four workers returned two months later for five-day demonstrations at Lakdan and another village of 150 households 40 km away. Twelve young men and women were chosen by the village to participate in the construction and learn the technique to later extend to others. A four-day training was also given to thirty monks and twenty "local elites" from a large number of villages suffering from water shortage.

Training included theory, practice, and group discussions. General water problems and the technique and importance of passing on the training were stressed. The fifty participants, in groups of ten trainees per trainer, built a tank per group, 1.5 m in diameter by 1.5 m high. Five new useful water tanks for the school where the seminar was held and fifty enthusiastic and knowledgeable messengers of the bamboo-cement method came out of the seminar —which was also a demonstration to all involved of participatory solutions to community problems.

Discussions were held about applying the format to other rural difficulties. The participants were respected members of their respective communities, so their strongly positive endorsement of bamboo-cement technology would be well received. It was felt that successful training in an alternative technology solution to water storage would make the villagers more self-reliant and receptive in other areas of innovation as well. A Bangkok bank donated a sufficient quantity of money so that a bamboo water storage tank could be built in the village of each participant and help spread the message. ATG will send workers to these villages as consultants to check on progress and resolve any problems confronted in construction.[176]

Hidalgo (1978: 88–100) illustrates with thirty-eight photos the process of making bamboo-cement tanks for water storage and processing coffee and for panels that admit of various uses.

WATERWHEELS. "These and the cups with which water is raised to the rice fields are made of bamboo . . . A split bamboo pole with the partitions knocked out serves to conduct the overflow from higher rice paddies across a country path to another on the other side on the next lower level. Also, they are ingeniously fixed so as to catch water from the edges of the sluice that feeds the waterwheel of a primitive flour mill and carry it to the axle supports for the purpose of reducing the friction. As long as the water comes through the sluice, the axle of the wheel is automatically lubricated."[177]

WAXES. "While conducting experiments on *Sasa paniculata,* Tsujimoto found that the leaves of this species, upon extraction with petroleum ether, yield approximately 1 percent of crude wax. When refined with animal charcoal in ethyl acetate, the product was a hard, brittle wax melting at 79–

Oblique axis windmill fitted with mat-and-batten sails for irrigation.

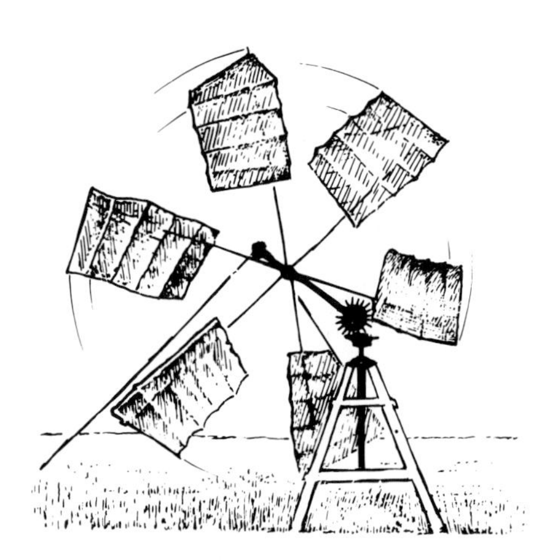

80°C and possessing, in general, the properties of carnauba wax."[178]

WHISTLE KITE. "A kind of very curious whistle is used by the Chinese for driving away evil spirits. Several holes are pierced in a piece of bamboo, two of the natural knots being left, one of which offers an opening out in a slope; to each extremity are fastened two long strips of paper from 15 to 18 feet in length and 6 to 8 inches wide. A string is attached to a groove made in the bamboo, and when there is a little wind, this wierd whistling kite is sent aloft and a monotonous whistling is produced, resembling at times the noise of a jet of steam, sometimes the sighing of the wind in trees."[179] (See, also, *flying art,* above, and pp. 83–84, 252–254.)

WINDMILLS. "Numbers of six-sail wind machines are currently in use in the salt works around the northern shore of the Gulf of Thailand. The machines are of about 6 meters diameter and use bamboo spars, rope, and wire to form a wheel which carries six triangular sails, each woven from rush or split bamboo."[180]

These wind machines drive paddles of traditional water-ladder low-lift pumps. Wind energy is receiving more attention worldwide in developing countries, where only 12 percent of the rural population has electricity, and where—according to FAO—centrally produced power from thermal or hydroelectric stations remains centrally consumed: 80 percent for urban industry, 10 percent for urban domestic consumption, and 10 percent trickling out to rural areas.[181] By contrast, decentralized energy sources such as wind generators or wind-powered irrigation pumps are directly controlled by local owners. Irrigation is the most crucial issue in increasing agricultural production, and wind machines have proven a low-cost way of lifting water with locally crafted equipment. "A small number of people are working on water-pumping windmill designs in developing countries. Most promising at present appear to be sail windmills—machines with cloth or bamboo sails rather than fixed blades."[182] Bamboo in the wind was a traditional favorite of Chinese artists for depicting the vital force of nature. Modern appropriate technologists are rediscovering its ancient Asian role in towers and sails to harness that force for serving human needs as well.[183]

WINE STORAGE. Wine kept in green bamboo a few days is said to improve in flavor. Bamboo has other associations with wine, as well. "The custom so common among the southwestern tribal peoples of drinking wine ceremonially through long bamboo

Washing fermented rice (for making wine) in bamboo baskets.

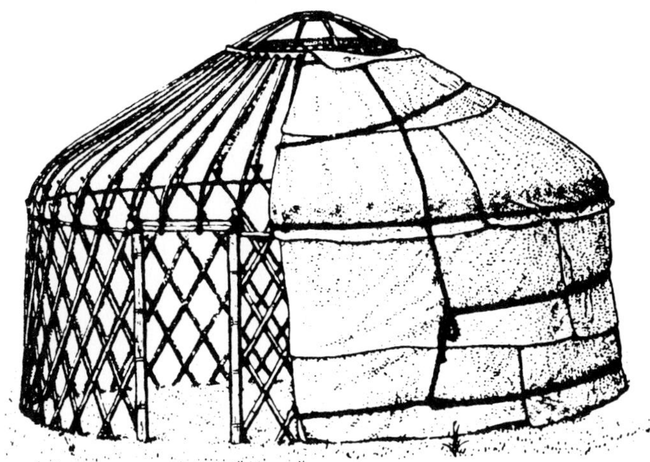

Yurt.

tubes is one which links them directly with the ancient peoples of the Fertile Crescent."[184] Satow remarks the Japanese practice of making bamboo sake (q.v.), and classical Chinese poets felt for centuries that a bamboo grove was the best of all possible places to get drunk.

X

XYLOPHONES. See page 00 and *musical instruments,* above.

Y

YURTS. Some made of bamboo are discussed in Kuehn's *Mongolian Cloud Houses* (1980:43). See also Charney's *Build a Yurt* (1974).

Z

ZITHERS. See p. 255 and *musical instruments,* above.

"This résumé of the uses of bamboo, although still meagre, may yet remove any surprise upon reading that the Radjahs of Boutan were proud in asserting that their forefathers sprung from the womb of a bamboo."[185]

CHAPTER 2.

1. Porterfield 1933:181–3.
   "During my war years in Chungking, a Chinese pointed out how history had failed to record one of man's great ages. 'Your books,' he said, 'speak of the Iron Age, Bronze Age and Steel Age. But in Asia we have the Bamboo Age.' And he demonstrated how bamboo was used in China's blockaded wartime capital: for water and sewer pipes, for tools and knives, to replace broken springs in automobiles and trucks, for building materials and even for radio parts and watch springs." From an article "The Ominous Blooms of Japan's Beloved Bamboo" by Carl Mydans, undated and without a source in our Xerox from the Smithsonian.
2. Needham 1980:iv.2:129, 142.
3. Ibid.: 145.
4. Ibid.: 153.
5. Ibid.: 188.
6. Ibid.: 272.
7. Ibid.: 313.
8. Ibid.: 333.
9. Ibid.: 355.
10. Ibid.: fig. 695.
11. VITA 1963:86.
12. *Encyclopaedia Britannica,* 3:17–8 (1961).
13. Appropriate technology is sometimes called "alternative technology," an unfortunate name in some respects because it suggests "second rate"—what you do if you can't afford the real thing. In the United States, 1 to 2 million wood space heaters are sold yearly. In 1980, they contributed more energy than nuclear power. In this case, which is the "alternative" energy source?
14. Papanek 1971.
15. Paul Ehrlich, quoted in "Ecology, Capitalism, and Communism," *Co-Evolution Quarterly,* Spring 1977:46–8. A recommended article.
16. Needham 1980:iv. 2:63.
17. Hidalgo 1974:245–51; see Leon 1956, Hidalgo's source.
18. Kurz 1876:235.
19. Varmah 1980:12.
20. Young 1961:29.
21. Laufer 1925:3–4.
22. Porterfield 1927.
23. Kurz 1876:231.
24. Munro 1868:29.

25. Darrow 1981:691. Review of *Bicycling Science* by F. Whitt and D. Wilson (Cambridge, Mass.: MIT Press).
26. Needham 1980:iv.2:63.
27. Darrow 1981:680–3.
28. Ibid.: 190–1.
29. Simmonds 1963:334.
30. Freeman-Mitford 1899:273–4.
31. Mannix 1956; Schurmacher 1949.
32. Kurz 1876:225–6.
33. Needham 1980:iv.3, sec. 28 Bridges: 191.
34. Wilson 1913,I:171.
35. Kurz 1876:228.
36. McClure 1958c:402.
37. Kurz 1876:228.
38. Porterfield 1933:179.
39. Needham iv.2, sec. 27: 64.
40. Kurz 1876:229.
41. Porterfield 1927:40–1.
42. Kurz 1876:229–30.
43. Ibid.: 230.
44. Darrow 1981:581,566; see also 100–1 for cart designs.
45. Ramaswamy 1979. (Reviewed, Darrow 1981:566.) Free to Third World groups. See bibliography.
46. McClure 1958c:405.
47. Varmah 1980:12.
48. Kurz 1876:228.
49. Hutt 1979, Introduction.
50. Kurz 1876:236.
51. Junka 1972:19–20.
52. Horn 1943.
53. Dasmann 1980: *Planet Earth 1980.*
54. Darrow 1981:569.
55. McClure 1956c:39.
56. Ueda 1960.
57. Piatti 1947b.
58. McClure 1958c:401.
59. Darrow 1981:118.
60. McClure 1958c:406.
61. Ibid.
62. Ibid.: 397.
63. Darrow 1981:485.
64. Ibid: 484. Review of *Appropriate Technology for Grain Storage*, 94 pages, by the Community Development Trust Fund of Tanzania. $2.50 from Economic Development Bureau, P.O. Box 1717, New Haven, CT 06507.
65. *Encyclopaedia Britannica* 3:17–8 (1961).
66. Freeman-Mitford 1899.
67. Aero 1980:100–1.
68. Austin 1970:130–5, 180. Seventeen photos document fan fabrication.
69. Porterfield 1927:38–9.
70. Sloane 1979:51–4.
71. Austin 1970:13.
72. Tsuboi 1913:221.
73. Ferrar-Delgado 1951.
74. McClure 1958c:401.
75. Kurz 1876:226–7.
76. Hidalgo 1974:244.
77. Dickason 1966:7.
78. Aero 1980:133.
79. Ibid.: 134.
80. Darrow 1981:494, quoting Dasmann's *Planet Earth 1980.*
81. Eckholm 1976:26.
82. Ibid.: 18–9.
83. Club du Sahel 1978:45.
84. Aprovecho 1981:6.
85. FAO 1978.
86. Darrow 1981:496.
87. Ibid.: 498. From Sanger 1977.
88. Aprovecho 1981:10.
89. See also *diesel fuel,* above.
90. McClure 1958c:403–4.
91. FAO 1976.
92. Darrow 1981:512–24.
93. Porterfield 1927:44–5.
94. Needham 1980:iv.2: 313.
95. Ibid.: 578.
96. Freeman-Mitford 1899:282.
97. Needham 1980:iv.3. sec. 28: 295,339.
98. Hunter 1977:39.
99. Porterfield 1927:8–9.
100. Hooker 1854, I:29.
101. A. H. Church, Food Grains of India, Suppl. 1901:6.See *sugar,* below, for another form of bamboo food.
102. Aprovecho [359 Polk Street, Eugene, OR 97402) (503)345-5981] offers designs for a removable "renter's greenhouse" that can be adapted to bamboo.
103. Dickason 1966:5.
104. VITA 1963:270.
105. Needham 1980:iv.3. 391, 395.
106. Worcester 1966.
107. Hidalgo 1974:195–204.
108. Newman 1974; Peterson 1969.
109. *Melocalamus compactiflorus* is also used. Dickason 1966:5.
110. U Pe Kin 1933:635–8.
111. Kurz 1876:233.
112. Ibid.: 224–5.
113. Porterfield 1927:41.
114. Junka 1972:26–7.
115. Cave 1955:A3.

116. Aero 1980:22.
117. Kiang 1973.
118. Nonesuch Steel Band record jacket.
119. Lessard 1980:201.
120. Ibid.: 201–4: "The Angklung and Other West Javanese Bamboo Musical Instruments," by Elizabeth Widjaja, Lembaga Biologi Nasional, LIPI, Bogor, Indonesia.
121. Lessard 1980:202–3.
122. May 1983 and Reck 1977.
123. Sineath 1953:55.
124. Austin 1970:184, 186, 187: photos of *netsuke* and *inro.*
125. Lutz 1975. Brush pot photos. A full page photo of an extraordinary brush pot depicting a bamboo grove complete with philosophers occurs in Austin 1970:177.
126. Kiang 1973.
127. VITA 1963:367–8.
128. Lessard 1980:53.
129. Varmah 1980:20.
130. See Narayanamurti and Bist 1948, 1963.
131. McClure 1956c:40.
132. Dickason 1966:6.
133. Kiang 1973:17.
134. Kurz 1876:240.
135. Needham 1980:iv.2.587.
136. Aero 1980:109. "The Five Punishments."
137. de Fonblanque 1863.
138. Porterfield 1927:52.
139. Ibid.: 48.
140. McClure 1958c:396.
141. Needham 1980:iv.3.394.
142. Junka 1972:27. See also Thoria 1947.
143. More in *Scientific American* (July 5, 1919), 121(6):6; Sineath 1953:55.
144. Porterfield 1927:50.
145. Dickason 1966:5.
146. Needham 1980:iv.3, sec.28 Civil Engineering: 134.
147. Tsuboi 1913:219.
148. Needham 1980:iv.3: 604.
149. Ibid.: 597.
150. Satow 1899.
151. Crawford 1926: 171–2, abridged.
152. McClure 1958c:409.
153. Hu 170.
154. Ibid.: 168.
155. Ibid.: 171.
156. Shelter 1973:75.
157. *National Geographic* (Sept. 1967), 132(3): 317.
157a. Aero 1980:60–1.
158. McClure 1958c:409.
159. Kurz 1876:230–1.
160. Needham 1980:iv.3, sec. 29 Shipping: 389.
161. Gibbon, *Decline and Fall of the Roman Empire,* Chap. X1; *Encyclopaedia Britannica,* "Silk"; Freeman-Mitford 1896:31–3.
162. Hooper 1900, abridged; Sineath 1953:17.
163. Kurz 1876:239.
164. Austin 1970:22.
165. See also "Catalyst for General Use," *Chemical Abstracts* (1947), 41:3592.
166. Kurz 1876:223.
167. McClure 1958c:401.
168. Kurz 1876:224.
169. Sineath 1953:56. See also *Scientific American* (Feb. 1917), 116(124):3, "Building a Railroad Bridge of Grass to Expedite the Construction of a Steel One."
170. Aero 1980:237.
171. Numata 1979:231.
172. OCORA, OCR 18, Valiha—Madagascar. An extensive introduction. See also the excellent *African Music, A People's Art,* Francis Bebey, 1975:48 (Lawrence Hill & Co., Westport, Connecticut).
173. Kurz 1876:234.
174. Porterfield 1927:51.
175. Kurz 1876:237.
176. Tankrush 1981. Is the author's name a pseudonym or weird coincidence?
177. Porterfield 1927:44–5.
178. Sineath 1953:17. See also *Chemical Abstracts* (1940), 34:2177.
179. Kurz 1876:234–5.
180. Heronemus 1974. (Reviewed in Darrow 1981:140.)
181. Darrow 1981:558.
182. Ibid.: 593.
183. See Chapter 3.
184. Austin 1970:22.
185. Kurz 1876:240.

3. NUDE MODEL TO THE EAST

Just here body, just now mind,
just this act will leave behind
nothing in the shoreless sea
of nothing which envelops me.

Eye just see, just taste tongue,
ear just hear, just breathe lung—
just this moment you may find
just now body, just here mind.

—Buddhist lullaby

THE ARTIST'S ALLY, THE EMPEROR'S ORNAMENT, THE PEOPLE'S FRIEND

Among painters in China, bamboo occupied as central a position as the human figure does in the West. It was the nude model for the East, providing artists with brush and paper as well as subject for their work. Manuals of countless masters contain a long tradition on the subtle nuances of depicting its moods and motions. It was a primary theme of poets as well, and centuries of moral philosophers regarded bamboo as a complete pattern for correct human behavior. It stood for constancy because it remains green throughout the year; fidelity for its patient resignation beneath the weight of winter snows; integrity because when split its parts are straight and even; purity because its heart is always empty and immaculate; rectitude since in the hurly-burly of the wildest storm, it bends without breaking, and stands up again.

Bamboo aripana. This symbolic mandala of the cosmos as a bamboo grove was painted by the village women of Mithila in NE India as image magic for a fertile, healthy family, which grows up around an ancestor like a clump of bamboo around the first shoot.

Bamboo bone.

There are bamboos 10,000 meters high if you look at their shadows by moonlight.
—Su Tung-p'o

Bamboo itself would we have writ
had we the hand to scribble it—
but only glimpses of the grove,
brief leaves of a branch we love
black the pages of this book.
And all the planet scriptures say,
"Words are wild and miss the Way."

Those who wander by and find,
wading in our shallow brook,
an inward wish to dive more deep
down the wisdom of bamboo,
pilgrims of the plant, must leap
unmeasured fathoms further to
that central stillness of the mind.

There the moon-dipped, winelit eye
of that sage crazy, Su Tung-p'o,
in China groves nine centuries ago,
drenched in imagination, saw
bamboo 10,000 meters high . . .

After unripe years of raw
and greedy study, we found how

the data addict will not know
bamboo, nor bookful botanists with loads
of learned lumber in their heads.
The busy expert will not see
the tremble of the living tree.

But a lover who believes
in hollow mind, much alone
with midnight and dawn's dew
jewels on the tips of leaves
mirroring a bamboo
morning, may hope to gnaw
the dark and empty bone
of it, and know the skin.

Bamboo was referred to as "this gentleman" in literate Chinese. With the orchid, chrysanthemum, and plum, it was one of the Four Noble Plants of Chinese garden lore. And with the plum and the pine it was among the Three Friends, a plant trinity representing Lao Tzu, born beneath a plum tree, the Buddha, who died in a grove, and Confucius. The "mother culm" in a stand of bamboo, which feeds nutrients to her surrounding offspring, was regarded as an image of altruism in Chinese folklore. In northern India, aware of the cooperative underground union of a clump in which apparent individual stems are in fact interfostering fellows of a single plant, village women painted a stylized bamboo grove on their walls as a prayer for a happy family. Mythologies of a number of early cultures in both hemispheres regarded an empty bamboo culm as the womb of the race, and in China until recently the plant was often chosen as a spiritual parent or "godfather."

> If parents are too poor to bring up their children, especially their sons, the child may be commended at the instigation of a fortune-teller to the care of a tree. The spirit of the tree henceforth becomes its patron. Because it is regarded as a prince among trees, the bamboo is preferred before all others for this kind of adoption. On this account the child, because he has become the ward of such an influential spirit, may have a better chance in life.[1]

The spirit of bamboo was called upon not only to secure the future, but also to discern it. The Kachins of the northern hills region in Burma place sections of the thin-walled *Pseudostachyum polymorphum* in a fire, and the "nat" or spirit priests divine the future by the pattern of explosions.[2]

The origin of bamboo itself has been the source of various myths. A Japanese creation myth recounts that the first grove grew where Izanagi, one of their creator gods, flung down his fine-toothed comb—and up leapt the erect Adams of bamboo.

The eminence of bamboo in Japanese culture is suggested by an honorific formerly bestowed on her emperors: "King of the Bamboo Garden" *(Take no Sonoo).* Many cultural forms are limited to an elite or loved only by the multitudes, but one characteristic of bamboo in oriental culture is its pervasive popularity among all social classes. The pampered favorite of the emperor's garden was also a standard component of popular art forms. In lantern paper cuts, for example, a folk art dating from the T'ang dynasty (A.D. 618–906), split bamboo provided the frame, pulp provided the paper, while the plant itself provided a motif for the traditional lantern, often graced by a full moon and a few culms in this "poor man's stained glass" glowing in the night.[3] As for bamboo's place in oriental music, some musicologists consider bamboo instruments and the prominence of idiophones as the two chief acoustical features of that vast culture area stretching from Korea to Indonesia.*

Wherever bamboo abounds—if traditions have not been severely ruptured by conquest—it

**Editor's note:* When Western students began examining planet music of other cultures, they found that the traditional Western division of instruments—strings, woodwinds, brass, and percussion—were often not applicable to the sound devices they encountered. Eric von Hornbostel (1877–1935) and Curt Sachs (1881–1959) suggested a more inclusive classification (the Hornbostel-Sachs system): sounds are produced in *idiophones* by the unstretched material of the instrument itself; in *aerophones* by a vibrating column of air, with a few exceptions such as a bullroarer; in *membranophones* by a skin or other membrane stretched over a resonating chamber; in *chordophones* by stretched strings; in *electrophones,* added in 1914, by electrical current. (May 1983:xii.)

Four out of eight immortals use bamboo

If you peer deep enough into the misty peaks of China's past, you'll find *Chang Kuo Lao,* a mountain hermit with a magical white mule that carried him long trips in all directions and shrank to pocket size when not needed. Chang Kuo Lao carried a bamboo drum—and drums in traditional Chinese military music meant "Forward!" Even the emperor tried to get him to stop bumming about and take a steady job in the court . . . Chang Kuo Lao pulled his mule out of his pocket and hasn't been seen since.

After he fell out of a peach tree one day and became immortal,[4] the bamboo flute of *Han Hsiang Tzu* made flowers bloom, drew beasts from caves and birds from sky as he wandered the countryside playing. Vagabond patron of musicians, itinerant intimate of wind, riding mounted on the backs of clouds, his music was not for sale—and if some villager ever offered him money by mistake, he flung it about, laughing madly, and was off on his way with the weather down the next valley.

Chung Li-ch'uan one day in the country came across a young woman in deep mourning, fanning a fresh grave. "My husband's last wish was that I wait to remarry till his grave was dry . . . I've found another man." He helped her dry the grave, and she ran off so excited she forgot the fan, which Chung Li-ch'uan kept to remind himself of human memory. Later, he burnt his own house down to set off wandering, saving from the flames only his *Tao Te Ching* ("The Book of Road Power") and the bamboo fan. According to some accounts, the fan later served to blow breath into corpses, reviving the dead.

One shoe off, one shoe on, *Lan Ts'ai-ho* in a long blue gown strolled the streets of a thousand towns, ragged, disheveled, and panhandling as she scattered blossoms from a bamboo basket of flowers, singing irresistable melodies about the crazy wisdom of letting it all go. It's ironic that this rootless gypsy became the lucky angel of gardeners and florists. She loves best a simple outdoor shrine, blessing any field where she finds her image propped up with devotion and a few flowers on a stone altar in the open wind.[5]

Chang Kuo-lao.

Chung-li Ch'uan.

Han Hsiang-tzu.

Lan Ts'ai-ho.

has stood at the upright center of the most ancient tales and cultural values, reflecting its crucial position in the local economy. With prolonged and intimate acquaintance, bamboo has everywhere been used intensely, and intensely loved—with that mellow and abiding affection which made it not the Servant of Chinese folklore, but the Friend.

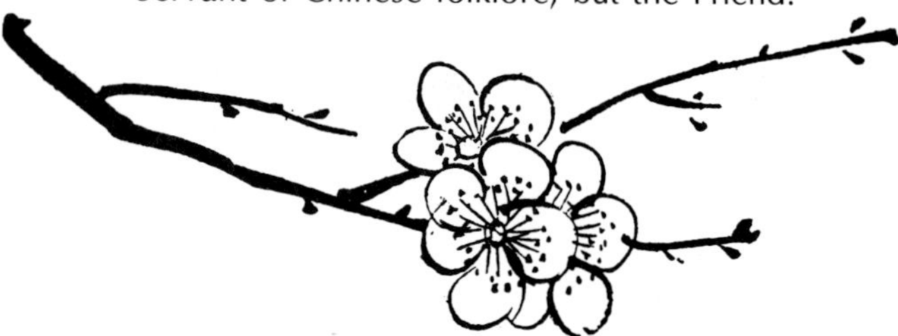

Plum blossoms, brave emblem of spring's recurrent victory over snows, provided a fresh challenge to centuries of brushes in China. "The plum trunk is a sinuous dragon with a feeling of iron power, curiously twisted at cliff's edge; combed by wind, washed by rain; blossoms drawn tenderly in mists as though holding smiles and gentle laughter on the branch. The wonder of it can't be caught in words."

Flowers.

> Flowers preach to us if we will hear.
> —C. G. Rossetti

In her format of flowers, Lan Ts'ai-ho recalls many famous saints who selected the same message. Jesus suggested lilies of the field as the ultimate tailors, more radiant than Solomon in all his glory; and the Buddha, in his "flower sermon," held one mute blossom up in reply to acres of tedious questioning disciples. Flowers are a delicate element of the tea ceremony, where their ritual use formed the root of Japanese flower arrangement. The Japanese schools themselves are blooms on one branch among many world traditions. In the San Francisco Arboretum library, the art of displaying flowers is the subject of four yards of shelved volumes, which gives some notion of the degree of human fascination with a blooming plant. We give flowers to mark arrivals and departures, births and deaths, and all small versions of them in between. The sudden, theatrical opening of petals, which then soon strew the garden path, makes them a natural metaphor for motion. In the symbol systems of many traditions, flowers stand up on their stems for that fleet and frail/durable beauty of life in which only a wandering spirit who kisses the joy as it flies will live in eternity's sunrise, and be sure to die at home.

The zero zone.

Wandering, the world's longest art form, is also perhaps its most ancient, a subtle science old as shoes. The nomadic nature of the immortals in Chinese folklore, their feet firmly planted on a cloud as you can see in their pictures, reflects their respect for the homeless. As though to acknowledge that wanderers and householders compose the world, the Wanderer and the Marrying Maiden are the only social roles dignified with a chapter in the *I Ching.* The wanderer is an archtypal nonrole in many literatures, enjoying a unique position by holding none. The Fool or Zero of the tarot pack is one famous example. "Full of intelligence and expectant dream, flute in one hand, in the other, a bamboo pole. Morning quiet . . . keen light . . . high mountains, sudden cliffs, vast masses of air . . . the spirit in search of experience! Only the sun behind you knows where you come from, where you are going, how you'll return by another road after many days."

Travel, like meditation, is a zero zone. It is institutionalized in the practice of pilgrimage, which forms a rite of all religions, like fasting, silence, solitude, or song, that handful of constants belonging to all spiritual seekers throughout the world.

Weightlessness—and a consequent ability to leap or fly long distances—is a mark of the Holy that hovers above many cultures as well. Up plus light equals Heaven. Down plus heavy equals Hell. Bamboo became the symbol of many immortals because its weightlessness was crucial for wanderers—which gods and heroes usually are. The bamboo staff and backpack-basket became the equipment of many a pilgrim or apprentice saint in the ancient East, where actual physical homelessness was regarded as the most effective metaphor to help us inhabit at last the true dwelling of our inward village, that abiding flux in which existence swims. "Leaving home" in Buddhism came to mean to adopt the Buddhist teaching, and in the Zen tradition, "travel for study" was the complementary pole to "sitting quietly doing nothing." The

Bamboo fool.

monks were called *unsui* or "cloud-water" to remind them of the easy motions of enlightened mind through which the nonopposing body also flows happily as water and floats effortless as clouds.

Image magic.

> *Ponder 10,000 volumes,*
> *wander 10,000 miles.*
> *Stick like glue to the four treasures.*
> *Drown the brush daily,*
> *grind the inkstone to dust.*
> *Take a week to paint a creek.*
> *Spend a month on a rock.*
> *Go through complexity*
> *to simplicity.*
> *Learn method*
> *to get rid of it.*
>
> —Mai-mai Sze

The journey played an integral part in the training of oriental artists as well as holy men. In the earliest days, there was less of a distinction between the path of the spirit and the path of art. For the shaman-artists of the Orient, as for their cousins in the West, the journey was a central part of their learning and power—sometimes chosen, sometimes an exile in which they found themselves, by which they broke through to a realm of power denied to people safe in their village and home.

The magical power of painting has been felt from the earliest times in a number of cultures. The cave paintings of Europe were used to assemble power for the hunt, and in the East also the primitive shaman-artist regarded his brush not as an instrument to "hold a mirror up to nature" but as a tool to control it. Witness the farmer's son in the Japanese tale who constantly drew cats: he was fit for nothing but his obsession of perpetually sketching this one subject, so his father gave him away to the local temple. "You're useless for farming. Maybe you should be a priest." But the priests also became exasperated in time with his addiction to drawing and sent the boy away. Arriving one night at a haunted monastery where the monks had fled, routed by a local demon, he sketched an enormous cat in the crusted dirt of the abandoned temple wall before retiring for the night to a distant corner of the grounds. In the morning, he found in the temple an immense rat whose carcass stretched from the altar to the main door . . . and from the huge dusty mouth above him on the wall, the fresh blood dripped.

Modern artists can readily relate to the notion of art as therapy or exorcism to drive our demons

Pilgrim. Daitokuji Temple founder Daito Kokushi, by Hakuin, early 18th century Japanese Zen master. With rain-battered bamboo hat and overcoat, staff nibbled by a thousand mountains, begging bowl for the day's refreshment, and straw mat for the night's repose, Kokushi taught Nothing to the Emperor Godaigo.

out. They can also easily share the attitude if not the brush of that court artist in China who, powerless to resist the demands of the emperor, painted the ordered landscape on the palace walls. When the royal patron arrived to inspect the finished work, the painter leapt up into it, ascending a path that entered the wilderness in the left foreground, and disappeared into the dark mouth of a cave tucked in the mountains high above the astonished monarch, beyond the reach of applause or punishment. The work was then sucked, with increasing speed, off the wall and into the cave after him.

Somewhere between this earliest age of art, in which the artist magically transformed reality itself, and our late days when we have come to believe that the business of artists is the production of objects, lies that era when bamboo leaves rustled in the foreground of oriental art attitudes and when the primary function of art was not to provide artifacts for market but to effect a subtle alchemy in the artist's ultimate artifact, his own soul.

Doors were made to open
roads were made to wind
ten thousand miles before you,
ten thousand more behind . . .

Rocks and bamboo. Chao Meng-fu (1254–1322).

GALLOPING THROUGH RULES

gem of the river
and evening—
windswept noble one
with small leaves—

in storm or calm
sunlight and rain
bending or upright
feel it from the heart,

galloping through rules
and beyond method
to the living quality
of bamboo itself[6]

For a culture in which the transformation of the artist is the real issue, the work of art is merely the by-product and proof of that growth—the tracks, not the actual animal. "Build living Buddhas, not pagodas," advises a Zen proverb reflecting the same attitude. From years of careful mastery of a rich tradition, a spontaneous rightness at last burst forth, vigorous, unhesitant, and ripe. "Wen T'ung, a sage born with knowledge, his bosom rich in hills and valleys, worked in harmony with Nature, kept within the rules—and yet roamed beyond the dusty world." An artist attempting to reproduce nature without having first entered her was monstrous and ridiculous to the oriental. "To learn to paint, still your heart, clarify your understanding, let wisdom open in you like a flower."

Ink's excellent absence.
This wise passiveness in the artist is reflected in the oriental method of composition, in which a certain emptiness invites the viewer to join as a participant in the creation. Confucious remarks that a work is finished not when the last thing has been added, but when the last thing has been taken away. "Painters nowadays," complained Yun Shou-p'ing, "consider only brush and ink. The ancients considered its absence. If you can understand that *excellent absence of ink,* your brush is close to the tao of painting."

Sweeping brushstrokes in which a fairly dry brush left the paper or silk still visible beneath were one expression of this appreciation of the excellence of emptiness, called "flying white."[7] A chance to explore absence on a larger scale was afforded by landscapes. "Mountain-water" paintings *(shan shui),* as the genre was early known in China, gave fullest scope to manifest the accord of heart and hand—soulwork and brushwork—whose harmony marked the fully mature artist. Often they were skyscapes, cloudscapes. "Clouds are the ornaments of sky, the embroidery of mountains and streams. They move as swift as horses, striking a mountain with such force you can hear the sound: such is the *ch'i* of clouds."[8] Snow, mists, or clouds often united or dominated the landscape, whose true subject was space: "Space, as it was rendered in the best of Chinese painting, might be described as a spiritual solid."[9]

The roots of clouds.
It could be said as well that the true, invisible subject of Chinese painting was *ch'i*—the breath, vitality, or life force at the sightless center of the ten thousand things. The only way to paint it was to have it, and the ultimate proof of a painter's work was the presence of *ch'i,* the lack of it the only failure. The Western genre of still life—"dead nature" in French *(nature morte)*—is the antithesis of the Chinese ideal. For the Chinese, even rocks, "the bones of mountains and the roots of clouds," were bursting with vitality. "The Book of Rocks" in the seventeenth century *Mustard Seed Garden Manual of Painting* sums up instruction in rock painting with a single phrase: "Rocks must be alive. How could a cultivated person paint a lifeless rock?"

Though rocks were pulsing with the breath of Tao, other pieces of nature expressed *ch'i* more completely. The wind especially was "a tangible and direct manifestation of *ch'i,* described in the *I Ching* as a force of Heaven visibly stirring life."[10] And of all the subjects in the long history of Chinese painting, a traditional favorite for rendering the movement of wind and making visible its felt but unseen power was—bamboo.

Bamboo in the wind.
Stems bending in a heavy storm or with a light

Bamboo in the wind.

breeze running its hundred thin fingers through the leaning leaves, bamboo served as the ultimate test of an artist's power to assemble his spirit and fling it down on white paper with black ink in one rush of energy: "When calm, paint the iris; when angry, bamboo." The very names of brush strokes used to paint bamboo leaves indicate their nervous vitality: goldfish tail, startled rook, wild goose landing, swallows in flight, stag's horns, bird's claw, and fishbone are among those described in "The Book of Bamboo."[11] The bamboo and the plum tree were favorites of the scholar-painters. Bamboo was a perfect choice for these masters of the brush. A work depicting bamboos is both a painting and a piece of calligraphy. To produce such a work, it is absolutely necessary to have a steady wrist and complete control of brush and ink, and to work in swift, sure brushstrokes without the least hesitation. Experience teaches the values of ink tones, the way of handling a dry or wet brush, and a variety of brushstrokes. The bamboo plant, like the orchid, is interpreted as having all the ideal qualities of a scholar and gentleman, the essence of refinement and culture: gentle and graceful in fair weather, strong and resilient under adverse conditions. Supple, adaptable, upright, firm, vigorous, fresh. Even the sweet melancholy from the rustle of its leaves has been translated into qualities of mind.

"Su Tung-p'o referred to bamboos as 'those dear princely joints.' His passionate admiration seems to have been shared by painters and connoisseurs to the point that amounted to a cult of the bamboo. Certainly, among all the subjects in its class, bamboo offers in painting the most direct and effective communication of the vitality called *sheng tung* (life-movement) and the breath *(ch'i)* of *Tao*" (Sze 1977:96–7).

The four treasures.

The Four Treasures in China—the brush, ink, inkstone, and paper—were traditionally held in great respect. The brush and paper have long associations with bamboo, and centuries before bamboo paper, slips of bamboo were used as a message base. The brush was said to have been invented by one Meng Tien, a general under a Ch'in emperor who reigned from 221–209 B.C. Decorations on pottery from Shang-Yin dynasty (1776–1122 B.C.) were probably made with some kind of brush, however, so perhaps the general only improved it or was maybe among the first to use the more modern name form: *pi,* composed of *chu* (bamboo) and *yu* (brush or pen stylus), which first came into use at that time.

> It is a direct representation of a hand wielding a brush in its bamboo holder to write or paint. Brush handles for ordinary use are made of bamboo, although in the past they were sometimes made of jade, quartz, gold, silver, or ivory, and tipped with buttons or knobs of these precious metals for special occasions. The plain bamboo handle has usually been the most satisfactory since its main advantage is lightness, an important factor in the balance of the brush.[12]

Hair of various animals—sheep, goat, deer, sable, wolf, fox, rabbit, weasel—have been used. Su Tung-p'o favored mouse whiskers. Chicken down, the hair of children, and bristles from a pig's neck are some infrequent alternatives. Goat, rabbit, and sheep are presently most commonly used. More rarely, plant fiber is used: sugar cane sucked dry was one Sung choice, and a T'ang critic spoke of a brush composed of mountain bamboo fiber that drew lines sharp as a slicing sword—an apt foreshadowing of the ideal in Japan centuries later, when samurai were judged by mastery of brush and sword, calligraphy and battle. Austin pictures a bamboo brush made by burying a piece of bamboo with the tip underground for several months.[13]

Bacteria eat the pith and leave the fiber to embody the artist's eye.

THE BRUSH DANCES, THE INK SINGS

> Full appreciation of Chinese painting depends a great deal on the spectator's sensibility to the tempo of the brush, following it with eye and imagination as it dots, flicks, or moves forward, sweeping, turning, lifting, plunging, thinning out, swelling, sometimes stopping abruptly, sometimes crouching to leap again. It has often been remarked that the brush dances and the ink sings.[14]

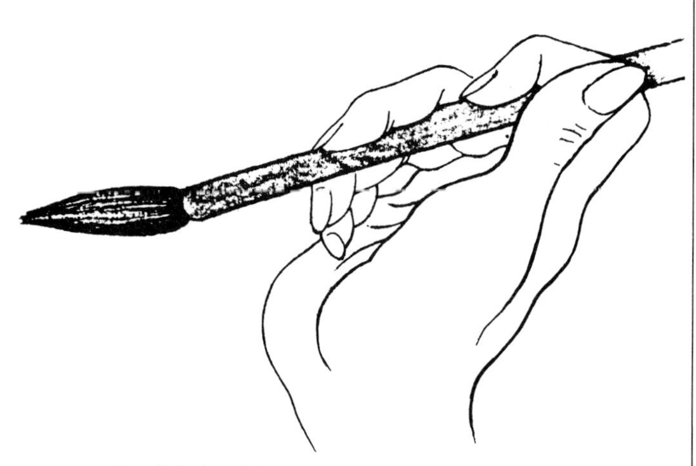

Brush in hand, hand and heart in agreement.

Sumi-e: the color of the heart.

Sumi-e is a style of painting particularly close to the nervous ch'i of bamboo leaves, trembling in the stillest air. The painter fenced rapidly with his brush on the fragile paper which tore if he paused. Black was usually the only color—called "the color of the heart" because it could convey the feeling of all the rest. "The aim of the sumi-e artist is not the reproduction of the subject matter but the elimination of the inessential."[15] Not to record every rock on the mountain or each leaf in the grove, but to capture—with a spare economy of strokes—the moment-by-moment urgency of life itself. The paper is thin, the brush is wet—a pause will soak and rip the page—quickly! quickly!

Sumi-e is the art form that most closely approximates live music, whose just-now, once-only quality admits of no corrections: Flutes have no erasers. To achieve fluid motion, the brush is held lightly, perpendicular to the paper with the tip of the thumb and the index finger, at least 5 inches above the hair of the brush, with the elbow free. The posture of the painter is important. Slumpy makes a sluggish line. He must paint with his whole body, hold himself as he holds his brush, erect, with the spine straight but not stiff.

"Ultimately, the brush becomes like a needle of a highly sensitive graph that records immediately the minutest impulse of the thought process."[16] You can't cheat. You can't lie. The simplest line measures the alertness of your mind and sincerity of your spirit. Look your heart in the eye, love your brush as you love your wife and happy family—and it becomes magic alive in your hand.

Su Tung-p'o.

Chinese painters were noted for their extravagant behavior. One, reproached by a friend for painting naked, replied, "The world is my home, my house, my trousers. What are you doing in my pants?" Among them all, none was more dazzling-various than Su Tung-p'o (A.D. 1036–1101), a many-skilled marvel of Sung dynasty China, embodiment of the oriental art ideal of the complete person who could shine in as many directions as the sun. He was hired by a handful of governments, was friend to a dozen remembered geniuses and shelter to unnumbered and forgotten poor. A city planner who inspired the largest bamboo water systems ever built (in Hangchow, A.D. 1089, and Canton, A.D. 1096), he also excelled as poet, painter, and creative drunk—a classical role in Chinese cultural history.

When he laughed, his head fell off his shoulders. When he sketched, heaven and earth fell off his arm. "Su Tung-p'o painted old trees contorted like dragons, while his wrinkled and sharp rocks were queerly tangled, like sorrows coiled up in his

Su Tung-p'o, "The Gay Genius."

Bamboo by my gate to doctor my mood . . .

breast. 'When my dry bowels are refreshed with wine, the rapid strokes begin to flow, and from the flushed liver and lymph, bamboos and stones are born.' " Su Tung-p'o fathered children as well as art. He was a family man stitched more firmly to the world than the old hermit or young monk without wife and child to send him into the fray of the human marketplace. His "Prayer for My Son" suggests the complex tissue of the heart that so doted on bamboo's "dear princely joints."

> At a child's birth, fathers generally hope for a
> smart baby.
> Wits wrecked my life, so let's hope this fellow
> turns out dull enough to get on in the world
> and enjoy his senility in high government office.

The prayer also provides an excellent portrait of office mind from an art mind, wind's-eye view. Wise counselors have declared it "off the bamboo bull's eye," but against their better judgment we include it as a sample of the toughness of mind of China's most famous fanatic or beau of bamboo.

Su Tung-p'o held office himself with his art mind firmly intact. Once while he was doing duty as judge in a small claims court, a poor defendant excited his sympathy. Su Tung-p'o called for brush and paper, quickly sketched a branch of bamboo and cluster of leaves, and gave it to the man to sell and pay his debt.

By the nature of Chinese calligraphy, poetry and painting there were much more closely allied than in the West. Most paintings included a brief poem in the margin, and, just as bamboo was a favorite subject for a sketch, poets vied with one another in praise of the plant. Su Tung-p'o, with a smile up his sleeve, once wrote a boast for bamboo departing from a story about Confucius, who reputedly became so absorbed in the melodious soughing of bamboos that he forgot to taste meat for three months and remarked to a friend: "People get thin without meat, but without bamboo they get vulgar."

Plain but good
for my food,
for my bed,
a straw mat.

Would I whine at my fate
or pine to get fat
with bamboo at my gate
to doctor my mood?

No cash? You grow thin
in the paunch and the purse.
No bamboo? You get coarse
and your thinking runs thick.

The skinny or sick
can get plump again
and fate sometimes switches
from squalor to riches—

but the dummies you meet
in palace or street
in general wax worse
from cradle to hearse.

—Su Tung-p'o

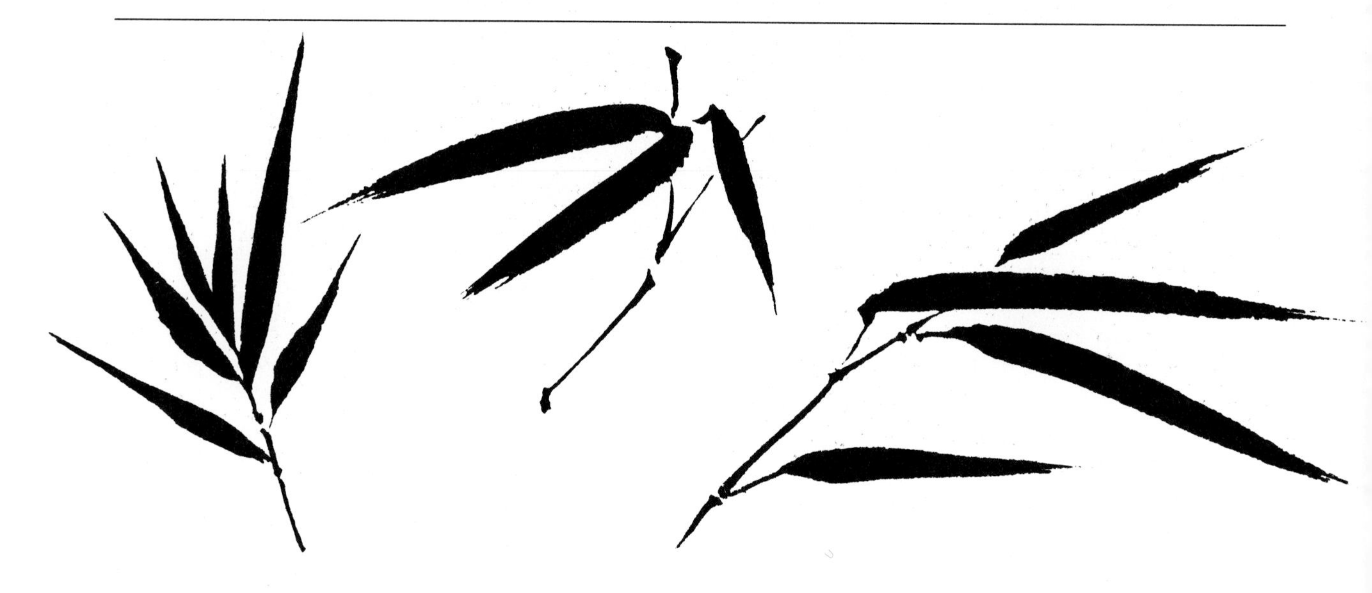

Blackening the silk.
Rich in cultural reference, placing the ultimate challenge on wrist and fluid brushwork of the ripest talent, ubiquitous in nature, coddled beauty of the rich man's garden, tough friend in a dozen daily uses to the poor, bamboo became as tenaciously rooted in the artistic traditions of China as the plant itself was anchored in her mountains. More common longer as a subject than the naked human figure in the West, manuals multiplied on painting techniques. Li K'an (A.D. 1245–1320) in his *Bamboo Treatise* compiled a classic statement of the art: "To paint bamboo, carry the whole matter first in the mind. Then seize the brush, concentrate your attention, go straight ahead and follow what you see as quickly as the hawk swoops when the hare jumps out. If you hesitate for a single moment, it is gone! It isn't making joint after joint and piling up leaves. . . . Every leaf, curved or straight, slanting down or pointing up, beaten by rain or fluttering in the wind, has its own appearance. If they're all made the same, you're just blackening the silk." (Sze.)

Some say that a woman by moonlight painted the first bamboo:

> *Empty and receptive,*
> *Lady Li traced shadows of bamboo*
> *the moonlight cast on*
> *paper of a windowscreen:*
> *the lean of each leaf,*
> *the bend of each branch,*
> *complete in the mind before beginning.*
> *Then, from the heart in order,*
> *power flows which shoves the brush to action*
> *with confidence and ease.*

This is an interesting method for the amateur artist, by moonlight, noonlight, or lamp. Arrange a paper on a wall or ground beneath a plant, and let the brush dance in the shadows. This is one way, at least, to begin the apprenticeship. There is no end.[17]

THE YELLOW BELL

> A single, eternal note sounds at the center of Chinese musical theory and forms the cornerstone of good government. This fundamental tone is called *huang chung*—"the yellow bell." Emperor Huang Ti (3000 B.C.) assigned Ling Lung, the "music ruler," the task of determining correct pitch. Ling went to the westernmost edge of the kingdom, where he cut a piece of bamboo and blew into it. The note was the pitch of a man's voice when he spoke without passion.
> —Rita Aero[18]

The voice without passion: *wu wei,* the art of not doing.
Ling then cut the rest of the scale in accord with this primal note, returned to the court, and tuned the emperor. The emperor tuned the ministers, they in turn tuned the functionaries of the state, and these then governed the people with minimal intrusion—in a voice without passion.

The mist-hid mountain, the word unspoken, the note not blown all dominate the painting, poetry, and music of Eastern art, and many Eastern theories of government reflected the same principle: *wu wei,* not doing, was an art every bit as crucial as doing. "Govern people as you cook small fish," says Lao Tzu—poking as little as possible.

Music was the nondoing of correct government: Better to move men from inside with pitch and rhythm than with external compulsion and laws.[19]

"Music is a higher revelation than philosophy or science," Beethoven contended. Ling and many other early Chinese masters of music would agree heartily. Most anciently, music was a tool, designed to achieve specific effects, in politics, medicine, and agriculture. The shaman's drum helped him travel to other dimensions from which he returned to his local physical neighborhood with healing power and news of the invisible. People of knowledge or healing used song and rhythm as a raft to other shores.

In India, the science was sufficiently perfected to create ragas that influenced not only plant growth but also even weather. Since all life is pulsing with vibration, the trick was to find sounds that resonated with nature and so massaged the natural harmonies of matter with sound. The Chinese prescribed different types of instruments for different seasons:

> In summer, silkworms' work is encouraged with the zither's silk strings. In autumn, bells were the appropriate instrument, for they were used in retreat in battles. In winter, drums—used to advance in wartime—encouraged the sun to return. In spring, when men desire trees to bud and crops to grow, the most potent instrument would naturally be one of bamboo, a plant of such vitality that it remains green even in the winter snow. The various pipes of bamboo, then, through which men's *ch'i* causes a similar *ch'i* in Nature to respond, were the instruments of spring.[20]

Modern experiments confirm the "soundness" of early oriental practice. The fundamental metabolic processes of plants, such as transpiration and carbon assimilation, were quite accelerated, and they increased over 200 percent in comparison with controls, when excited by music or a rhythmic

beat. A chapter in *The Secret Life of Plants* is devoted to the influence sound has on plant growth, a reality well documented by a number of investigators throughout the world.[21] Many an ancient commonplace is weird news to the nowaday ear. We usually think of music, for example, as of human origin—but earlier cultures believed in "the music of the spheres." And earth has its music, too.

> *The music of people is made on flutes and drums.*
> *The music of earth sings through a thousand holes—*
> *mooing and roaring, whistling and grumbling,*
> *high pitched and screaming—one call awakens another . . .*
> *Have you listened then how everything trembles and dies down?*
>
> *Such is the music of people and earth.*
> *And the music of heaven?*
> *Something is blowing on a thousand holes.*
> *Some power stands behind all this,*
> *beyond sound and silence.*
> *What is this power?*
>
> —Chuang Tzu

FLUTES CAME BEFORE FINGERS

Flutes came before people. The wind made them and played them. One culm leaning on another rubbed holes the weather whispered through; birds pursuing bugs or building nests drilled other notes in other internodes. Long afterwards, the human race came wandering down the path of spontaneous mutation and built their villages beside the groves. Bamboo moaning in the windy midnight made melodies that seemed played by spirit fingers, strange lips, a ghost orchestra of ancestors come to mumble down the long tube of dreams . . .

Weeping bamboos.
As with many inventions, the first flutes were probably the biggest ever built. Someone shaped the haphazard processes of nature into deliberate art, climbed up in the groves and experimented opening holes of different length and volume in internodes. Munro (1866) notes the persistence of this practice among the natives of Malacca, where slits in every internode produced up to twenty notes in these *bulu perindu* or "weeping bamboos," as they were called.[22] (See also *flying art, kites,* and *whistle kite,* Chap.2; see also pre-Colombian *Guadua,* Chap. 5, for Japanese and Colombian military use of this effect.)

Accordian ancestor: the *sheng*.
Birds were also early music masters to our race, and some of the earliest instruments obliquely acknowledge this ancestry:

> Sweet and delicate, the sound of the mouth organ or *sheng* is said to resemble the cry of the phoenix. The instrument itself is crafted to imitate a phoenix with folded wings: Seventeen bamboo pipes are arranged in a circle and held in a gourdlike wind chest. This type of mouth organ is very ancient—characters on bone from the Shang dynasty (c. 1550–1030 B.C.) refer to it. The pipes contain reeds cut to the diameter of the pipe so that they are free, unattached . . . Each pipe also contains a finger hole. When the hole is closed, air is forced over the reed, causing it to vibrate and produce a note.[23]

The sheng is related to the *sho* of Japan and to the most important folk instrument of Thailand, the *khāen,* generally made of about fourteen thin bamboo tubes, up to 4 inches long, each with a small free-beating metal reed. The khāen is usually tuned to seven pitches per octave, thus producing a two-octave range. "The bamboo pipes are put in two rows through a wooden mouthpiece, above which in each pipe a small round hole is cut. The instrument is held with the two hands cupped around the mouthpiece so that the fingers fit naturally over the holes. The hole of one pipe may be closed with a piece of wax, pitch, or other material to produce a drone. The sound of the khāen is often described as mournful and plaintive."[24] A number of similar instruments appear in Southeast Asia; in fact, the

Khaen.

sheng's notes drifted even farther from its original home.

> The sheng works on the same principle as the reed organ and is the earliest instrument known to use that principle. In the eighteenth century, a Chinese sheng found its way to Saint Petersburg in Russia. A German organ builder studied it there and then introduced the free reed to Europe. By the start of the next century, Western instrument makers had applied the principle to develop the harmonica and the accordion.[25]

At the same time that this Eastern bamboo mouth organ was bearing strange new Western offspring—the cultural kin to "hybrid vigor" in biology—a European missionary in the East was evolving the world's first bamboo organ, a musical mestizo of Spanish-Philippine descent.

Appropriate music: Philippine bamboo organ.
Just 11 miles south of Manila stands the world's

oldest and perhaps only Western-style bamboo organ, completed in 1821 and still playable in spite of all that a century and a half of unnumbered earthquakes, typhoons, termites, and a generous abundance of wars have cast against it. According to tradition in the village of Las Pinas, Father Cera, the local pastor, requested an organ from Europe in a letter he addressed to the queen of Spain. Neither aid nor apology arrived in reply for so long that he at last took the matter into his own capable hands. Helping himself to native material, local musical wind-instrument design, and Philippine bamboo building traditions, in three years he constructed an organ 14 feet high, with 950 pieces of bamboo, which 163 years later remains incredibly intact, in tune, and in use. As a quite effective preservative measure, the friar buried his bamboo six months in sand before starting to build. It would be interesting to know the bamboo species used.[26]

BAMBOO ORCHESTRA

In the 3½ millenia between the earliest sheng and the Philippine organ, there have been so many wind, string, and percussion instruments made of the plant wherever it flourishes that any student of bamboo music would agree that "in every bamboo bush are hidden the instruments for a whole orchestra."[27] Borrowing a line from Shelley, Freeman-Mitford (1899) called bamboo "the slave of music."

Trumpets, didjeridu, bocina, quenas, rondadores.

Some version of a bamboo trumpet, for example, is found throughout Melanesia. A large bamboo tube, closed at one end by a node in which a hole has been made, is played by buzzing the lips against the hole. Similar instruments exist throughout the world. The *didjeridu* of Australian aborigines is an impressive bamboo wind instrument made from a hollowed tube 5 to 6 feet long, 2 to 4 inches in diameter, with a mouthpiece of beeswax.[28] In Ecuador, a trumpet called a *bocina,* made from a single 5-foot internode of a bamboo locally called *tunda (Aulonemia queko),* is used for a message system in the mountains around Otavalo. Better known in the Andes are *quenas,* bamboo vertical flutes, and *rondadores,* bamboo panpipes of various sizes, both increasingly familiar with the recent adoption of Andean music by young Latinos as a musical form of anticolonialism, a regional counteroffensive to multinational electronic rock.

Valiha.

In New Guinea, the jew's harp, rattles, simple xylophones, and even a string instrument are made from bamboo. Their bamboo zither resembles the *valiha* from Madagascar, which is similarly constructed. Strips cut in the silica-rich skin of the culm, with both ends left carefully attached, function as strings. Small wooden bridges wedged under them lift the strips above the culm surface for strumming.[29]

"The Jakoons in Malacca make also a sort of guitarre consisting of a bamboo tube about a foot long, on which are lengthwise strained three or four strings which rest on small pieces of wax instead of a bridge."[30] As a flying footnote, it's worth noting that calling these instruments trumpets, zithers, or guitars roughly approximates their reality, but implies a Eurocentric—and myopic—vision of *Music of the Whole Earth.*[31]

Raft pipes and bundle pipes.

Some of these instruments, such as the bamboo "zithers" made in Nigeria and other countries, seem made in spite of limitations of bamboo. Others, especially flutes, are designed in complete agreement with the natural properties of the plant.

A scan of a handful from a single country will suggest the impossible amplitude of our theme. In China, bamboo is used in a number of flutes and

Trumpet. "La Bocina" summons workers to eat near Otavalo, Ecuador.

wind instruments in addition to the sheng discussed above. A set of bamboo pipes stopped at the lower end, the *p'ai hsiao,* is among the most ancient. Twelve, later sixteen, pipes were bound together in a circle ("bundle pipes") or in a row ("raft pipes") and blown from above across the opening. The *ch'ih,* perhaps the world's first cross flute, was a six-hole bamboo flute mentioned in a ninth century B.C. ode, played with the *hsuan,* a bone or earthenware wind instrument dating from the Shang dynasty (1550–1030 B.C.), shaped like a small barrel about 2½ inches high with a hole in the top. "Heaven enlightens the people when the bamboo flute responds to the earthenware whistle."

The *ti,* another bamboo cross flute said to have come from Central Asia around the first century B.C., is in its present form 2 feet long with six finger holes. Just below the mouth hole, a seventh hole is covered with rice paper to make a buzzing tone that accompanies the notes. Two more holes are used to tie an ornamental silk tassel to the end of the instrument.

The *hsiao* and the *yueh* are vertical flutes, the first with the mouthpiece cut through a node that closes the hsiao at the top. The yueh (meaning "foot" or "measure" or "stalk") was tuned to the *huang chung,* the "yellow bell," or fundamental note. The instrument measured an old Chinese foot in length—22.99 cm or 9.06 inches. "The distance between holes was determined purely by measurement: The center of the lowest finger hole was 3 Chinese inches from the bottom, the other holes all 2 inches between their centers. A scale requires holes spaced farther apart towards the lower end of the instrument. For proper pitches to be played on a Chinese flute, the size of the holes would have to be varied or the player would have to make adjustments in fingering and breathing."[32]

Voice of the dead.

From Egypt to India, thence to China came an instrument—the *sebi*—which around the Tang dynasty (A.D. 900) evolved closer to the shape of its present-day descendant, the *shakuhachi,* a Japanese vertical flute some 1.8 feet (54.5 cm) long with four holes on top and one below.

A Zen Buddhist sect of mendicant musicians—the *komuso,* "priests of empty nothing"—wan-

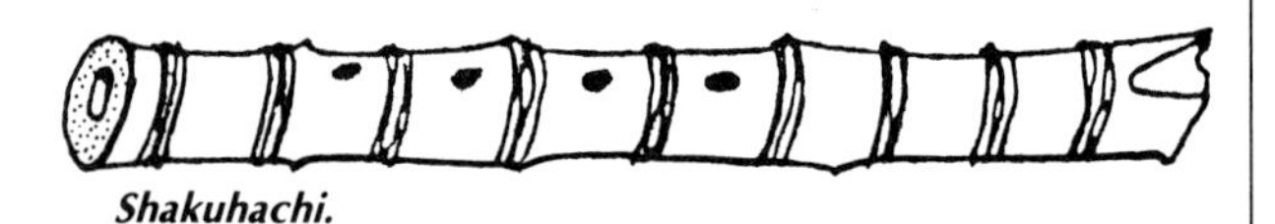

Shakuhachi.

Priests of Empty Nothing.

dered Edo playing bamboo flutes as part of their begging style, wearing bamboo baskets over their heads for anonymity. A group of them formed the Fuke sect, forged some documents making them appear more ancient and venerable than they were, and persuaded the shogun to grant them a monopoly on the flute and basket begging format. In exchange, they roamed the red-light district and other areas of old Edo as itinerant ears of the shogun, pausing at the paper windows on the dark streets playing the breathy, eerie music that, on first hearing it, Mahatma Gandhi, weeping, called "the voice of the dead." Between haunting notes, they kept their ears open inside their baskets for useful information. Buddhists are generally nonviolent, but the Fuke were often samurai stripped of their swords. Weaponless by government decree, in time they redesigned the shakuhachi to double as a club: it began to be cut from the heavy basal section of the thick-walled *madake (Phyllostachys bambusoides).*[33]

Creation by subtraction: hollow + hollow = flute.

Flute-making was a perfect embodiment of the Confucian art principle that a work was finished not by adding the last lacking, but by taking the last extra away. The simplest bamboo flute is indeed constructed by subtraction. And the new holes in the original vacancy reflect the circular emptiness of the bamboo itself. Hollow plus hollow equals flute, a trembling void.

Considering the simplicity of its structure, the

emotional complexities of its effects are all the more interesting. Characteristically described as indescribable, eerie, of a strange mournfulness, the flute echoes a basic ache of the soul stretched between the imagined and the embodied. The flute has always been the lightweight companion of wanderers, wind-whipped shepherds, homeless gypsies playing to a lonely immensity of stars. Flutes are more of the mountain and open field than of the street or living room, and when they are played in more civilized surroundings, they often seem a sound in exile, from another world or another, more inward dimension—from cosmic rather than merely social strata of our being.

The shrewd use of the bamboo flute by the "priests of empty nothing" resonated with a rich traditional alliance between the Eastern spirit and the plant. The komuso were a late, brief moment in a bamboo parable that runs for centuries through oriental spiritual imagination.

BAMBOODHISM: SOMETHING ABOUT NOTHING

> In a river mist, if another boat knocks against yours, you might yell at the other fellow to stay clear. But if you notice, then, that it's an empty boat, adrift with nobody aboard, you stop yelling. When you discover that all others are drifting boats, there's no one to yell at. And when you find out you also are an empty boat, there's no one to yell.
> —Chuang Tzu

"Yesterday upon the stair, I saw a man who wasn't there. He wasn't there again today. I wish that man would go away." Only a silly jingle—but it embodies accurately the Western mind's discomfort with nonbeing. In the skies constructed by Western theologians, there is only a Supreme Being: "I am Who am." A Supreme Nothing is almost unthinkable in the West, while our planet neighbors in the East have always been very much at home in the void. Empty among them is an honored guest. With micro- and telescopes, Western scientists have recently discovered that there is much more nothing than something. The gaps have gradually appeared so much more ample than the atoms, space so much vaster than stars, that siding so rambunctiously with being has begun to seem more fragile after all than that ease with emptiness which has, for millennia, remained a central feature and patrimony of the East.

A cup is made
of bottom and sides—
but its use
lies in emptiness.

A house is made
of roof and walls—
but its use
lies in emptiness.

The greater the road,
The greater the emptiness.
Something about nothing
makes us able to use
what's there by what isn't.

So, tell me, please,
which you like best:
being . . .
or nothingness?

—Lao Tzu

In a mind climate so amiable with absence, it was only natural that, of all the useful pieces of bamboo, the part that most firmly rooted the plant in Chinese affections was the part that wasn't there. Radiating into every corner of the culture from its hollow core, bamboo became a natural symbol for that "flexible emptiness" regarded particularly by the masters of Zen as the subtle center of spiritual development. In fact, to "trim bamboo" was a Chinese phrase meaning to become a Buddhist.[34]

Hollow design.
From a botanical viewpoint, we might remark that the abundance of bamboo derives in great measure from its capacity to leap immediately to full growth, which, in turn, depends largely on its emptiness. Instead of sanely constructing itself inch by solid inch like trees, soberly climbing into the contested forest air, bamboo sprints sunward to complete stature in about two months. We people attain our

total altitude in maybe fifteen years—say a quarter of our average lifetime. Although its wood is fully seasoned and mature in three to five years, a bamboo culm can easily stand as long as ten or more. In 1/60 its total lifetime, it is as tall as it will ever be. To pursue our human analogy, it is as if we were fully grown by our first birthday. After this initial vertical burst, bamboo unfolds branches and uncoils leaves to capture the sunlight it leapt up to get. As a survival tactic among many plant species reaching up to compete for available light, the growth pattern of bamboo is shrewdly designed. And it revolves around a basic emptiness—2/3 of its volume in thin-walled species, 1/3 in those with thick walls, never has to be born or fed. The nutrients and moisture that would have been exhausted making and maintaining this empty center can be utilized for growth of other culms.

Nine uses of void.
The utility as well as vitality of bamboo depends greatly on that part of it so fortunately absent.

> With its internal septae removed, bamboo forms a natural pipe, and this fact exerted a cardinal influence on East Asian invention. In the earliest times it offered itself as a material for flutes and pipelike instruments of music, instruments which deeply moulded the development of Chinese acoustics through the ages . . . Then it was turned to great effect from the Han onwards in the conveyance of brine from the deep springs to the places where evaporation was to take place. It also found use in piped water installations. Cut longitudinally it served for light tiles on roofs and every sort of simple channel. But bamboo tubing was used further in alchemy and the beginnings of chemical technology in the form of containers for such purposes as the descensory distillation of mercury and the solubilization of minerals. It generated the sighting-tube so characteristic of medieval Chinese astronomical instruments and fulfilled its most fateful destiny by becoming the ancestor of all barrel guns early in the +12th century.[35]

Silence has a lot to be said for it.
The useful hollows of bamboo amply demonstrated to the Orient the opulence of emptiness. The healthy respect for nothing is reflected also in the feeling that silence has a lot to be said for it. A spareness of expression in poetry and a use of silences in music, unexplored by comparison in the West, are rooted in a philosophical tradition that has always carefully distinguished explanations from reality—and preferred the latter. "Since words are given, it's good to know when to stop . . . The Path you can name isn't the Path you can walk . . . Those who know don't speak. Those who speak don't know." The *Tao Te Ching* is the briefest of world bibles and the most generous among them in its frequent warnings to beware of bibles. It was reputedly left, as the last reluctant social act of its author, at the insistence of the gatekeeper at Han Ku before Lao Tzu disappeared over the last horizon, returning to his root.

> *Hold tight to Quiet and, of ten thousand things,*
> *there's not one you won't handle.*
> *I've seen them go back. Look:*
> *whatever grows, goes back at last*
> *to its root.*
> *Return to the root means "Quiet."*
> *Quiet means accepting destiny.*
> *Accepting destiny links us with "Always."*
> *Knowing Always means "Lit Up."*
> *Not knowing it means plunging darkly*
> *from chaos to disaster.*
>
> —Lao Tzu

Quiet ears ripen. And the rustle of bamboo leaves, as in the case of Confucius, were a favorite "classical" sound, with a famous precedent in cultural history, which helped bring stillness to the heart and will.

The philosopher's womb.
The empty center of Buddhist cosmology was mirrored in the sage's empty will. Achieving this "silent will" often implied withdrawing from the human hubbub, and the example of the Buddha himself—who chose a bamboo grove as one of his most constant homes after his enlightenment—was imitated by many later would-be wise. The seven sages of the third century (A.D.) China, who retired from the hectic pleasures of court life to a bamboo grove where they could give themselves more fully to philosophy, provided a much-followed model for ages after them.

Monasteries multiplied, some even named after bamboo, providing an institutionalized format for withdrawal. The retreats became crowded with recluses, forming societies rife with the same ambitions, conflicts, and messy human motives they had been established to avoid. Frequently the more serious seekers of emptiness found themselves compelled to quit the monasteries stuffed full of "rice bags"—as the idle monks were sometimes called in disgust by the masters. These true pilgrims had left home for the temple. When the temple proved an-

Philosopher's womb. From the example of the seven sages (c. A.D. 300), the bamboo groves became the standard mind resort of would-be Buddhas, reclusive scholars bent above a prolonged task, embittered exiles in the shadow of the displeasure of the Sun of Heaven, gentlemen painters, drunk poets, and exhausted officials.

other home, they left it also to travel deeper into the mountains and the void.

The Tang dynasty (A.D. 618–906) was a golden moment to the literate Chinese for centuries afterwards. It was then that Zen Buddhism—*Ch'an* in China—attained the classic form that first embodied and later encased so much of oriental culture. Kyogen was a monk of those times who had spent twenty tense and fruitless years in quest of an elusive enlightenment. One day he decided to burn the useless sutras, settle far from the monastery in a mountain hut, and abandon his obsession as violently as he'd formerly pursued it.

More years passed . . . nobody counted them. Kyogen calmed down. Like Su Tung-p'o, he had a stand of bamboo at his gate to doctor his mood, serving as a green quotation from much of Chinese cultural history. In his threadbare poverty, their erect culms reminded him that the ultimate riches were a mind without "twisty thoughts" (as Confucius called them) and a heart without scheming. "The whole universe surrenders to the quiet mind," as some burnt scripture had once told him. Or, in the words of Chuang Tzu, "Keep your will single. Don't listen with your ears, but with your mind . . . Not with mind, but with *ch'i.* The ear is satisfied with sounds, the mind with concepts: but *ch'i* is an immense emptiness ready to receive anything."

The zero zone . . . bamboo satori.
But for Kyogen, this tangle of phrases we are now recalling would long be a small pile of cold ashes. He had been alone a long time now, good medicine to rid one of the mesh of memory that clogs the mind. Truly alone forgets *alone*—and *not alone.* It forgets even the one who forgets.

Wrapped in this emptiness of memory and desire, steeped deep in mountain silences, ears scrubbed with secret scriptures of the dew, one morning cleaning his path with his bamboo broom, Kyogen swept a piece of broken tile against a bamboo culm—*konk!* In that magical moment, the membrane popped between the total texture of reality and his imagined "me." That one hollow *klunk* echoed . . . centuries down a sudden amplitude in Kyogen's skull, hollow of its last illusion. As a Buddhist nun gratefully scribbled once of that same empty moment when the mad mind halts:

Time after time
I patched the old bucket.
Tonight, the bottom fell out.
No water. No moon.

The tale of Kyogen's lucky ear echoed also down the minds of generations of hopeful monks. Sweeping became a chief metaphor and exercise of the Zen tradition, and another legend was added to the bamboo lore in the endless attic of Eastern consciousness.

Rush is the mother of ugly.
Aware of the cumbersome credentials bamboo holds as an accelerator of inward evolution, two young meditators in China many centuries ago entered an enlightenment pact: Together they would take only a few necessary belongings and leave a farewell note to their parents, letting them know they were ascending the local mountain to an isolated grove to descend only when complete illumination was their's.

Kyogen, on the edge of enlightenment. Hsiang-yen, his Chinese name, is less familiar in the West.

The food ran out shortly before their resolve. They got bored to unknown zones of desperation contemplating the bamboo leaves. Their blankets were wet in the improvised shelter—forget it! They packed to go down, then remembered their notes in embarrassing detail and unpacked . . . This went on a few days. The next midnight, one tiptoed out —leaving another note. Late next afternoon, the second figured he could make it to the village before dark if he left right away.

In Korea, such an evaporation of aim is called, "head like dragon, tail like snake." Modern Americans aren't the only ones in a hurry, a quality of mind that rarely reaches its point of imagined rest. Rush has always been cited by spiritual masters as a particularly vicious form of unenlightenment because it negates the very process of one by one, moment by moment, patient opening of the extra eye focused on the third time zone so few inhabit: *now*.

An encompassing device of space and shelter that embodies much of oriental philosophy's attempt to relocate people in the calm and breathing now is the teahouse, a shelter genre more aped than truly entered in the West.

Tea's easy way.

Tea's way's an easy way.
Roof wide enough to keep off the rain,
bowl full enough to hush hunger—
Fetch fuel and water with your own hands,
build the fire, and put the kettle on.
Arrange fresh flowers as if still living in the field.
Suggest the cool in summer and in winter, warmth.
Burn incense. Boil water. Make tea.
Offer it at the altar, serve friends, and then yourself.
All this is done to search the subtle meaning of the Buddha's deeds.

—Rikyu (1521–1591), *TAPROOT OF JAPANESE TEA*

HAIKU HOUSING

The bloody eyelids of the Bodhidharma.
Thus we have heard: Tea's origin lies in the eyelids of Bodhidharma ("Awake to Real"), the Indian Buddhist who brought the teachings of the Enlightened One to China around A.D. 500. Dozing off in meditation one evening, he became exasperated and ripped his eyelids off, flinging them aside. Where they landed, the first tea plant grew, with eye-shaped leaves that helped wake the mind. Tea became the special favorite of meditators, scholars, artists, and government officials, and tealore became a cup 5 fathoms deep.

Bodhidharma.

The teahouse.
The teahouse is a miniature church. Brevity of statement in poetry—haiku. Economy of material in a sparse architecture—the teahouse. The hermit's hovel, the humble shack of the recluse distant from the human hubbub in the Chinese tradition became the Zen teahouse centuries later in Japan, where it was known as the empty house, "vacancy's place." The building itself is a piece of inhabitable sculpture, characteristically small and of simple construction. Natural structural elements—bamboo, rough-hewn hardwoods, and thatch—surround ritual conventions fixed by centuries of cultural experiment.

Path and garden.
The path to the teahouse is a haiku landscape, a brief quotation, in a few stones and leaves, of the long journey to the mountain hut. Step by step, the city's smell floats off the clothes. A dozen culms of black bamboo rise in bold contrast to raked white sand beside a large fieldstone. Bamboo hoops line a pebbled path, and the guest will pass three traditional fences, all using bamboo. The outside fence surrounds the property, tall enough to block vision of the house from the street. Its principal door is of wood and bamboo, made according to time-honored designs, with a narrow bamboo roof above. The "little fence"—hardly a yard wide and no more than 4 or 5 feet tall—is used to break up space or conceal objects or portions of the house from the garden and is made also of bamboo combined with straw and other thatch.

The "inner fence" serves principally to separate the teahouse from the garden. Traditionally very transparent, the bamboo is thinly spaced and simply, irregularly tied together in an apparently makeshift manner. It is a studied carelessness, however, reflecting one of the principal notions of the tea ceremony, to find beauty in the commonplace and imperfect. A story is told of one tea master whose son zealously swept the path immaculate and plucked the garden clean, removing the last offending twig. The father, inspecting these preparations for an honored guest, completed them by seizing and shaking a low branch of the tree above. Down drifted six dry leaves.

The ceremony of friendship.
The tea ceremony gives ritual expression to the central shape of friendship: on entry, all become equal . . . and tranquil. Social status, economic rank, and all worries are left at the door with your shoes.

Drinking a cup of tea,
I stopped the war.

—Paul Reps

The humble tasks of boiling water, pouring tea, in the calm hands of the master become exact and intimate gestures, steeped in quiet significance. The tea ceremony is a kitchen ballet.

Bamboo is found in a number of the unpretentious utensils used to create the event. "*Matcha* [powdered tea] tastes better when the powder is stirred into bubbles. The stirring utensil is a tea-whisk made by splitting a culm 2 centimeters in diameter into eighty fine lacelike prongs. The feat is possible with no other plant. The tea ladle used to scoop the matcha is also made of bamboo. Craftsmen in former days vied with each other to produce

Tea room. Nan-en-ji Temple, Kyoto.

Hui-neng, the Sixth Patriarch of Ch'an (Zen) Buddhism in China, cutting a culm. By Liang K'ai, early 13th century.

the most beautiful and graceful tea utensils and some of their masterworks survive today."[36]

The best tea ceremony utensils now cost from 80 to 500 dollars—so we have come full circle, from the hermit's humble hovel to the rich man's hobby.

But it is not as a superrefined pastime of an elite that the teahouse interests us. We are drawn not to the perfections of the external, for-sale form, but to the inner folds of meaning in teahouse traditions as they relate to present world building needs, especially among the poor, that is, most of us here.

The tao of housing: building builders.

The human scale, the use of locally renewable resources, a highly personal weaving of house and habitat, shelter and weather and leaf wed in one seamless garment—these design virtues of the teahouse are all the more relevant in a time when high-rise plastic pollution, exiled from vegetation and made by anonymous and ulcered others is the tao of modern housing. Another art has floundered down the sad path to industry. How can we help the art of housing come back home? The teahouse was not only self-built, but Self-built by masters who knew that the builder's main monument was the builder—"Build living Buddhas, not pagodas."

In floor and ceiling,
wall and shelf,
beneath your house
you build yourself.
Work mindfully, for
calm builds most
accurate carpenters,
who make the rest.

CHAPTER 3.

1. Porterfield 1927:50.
2. Dickason 1966:7.
3. Aero 1980:185
4. The peach is a symbol of longevity. It was said if you could taste fruit from the peach tree of the gods, you would never taste death. Han Hsiang Tzu, brought to this tree by his teacher, slipped from a branch and would have died in the fall, but he managed to bite a peach on the way down.
5. Williams 1931:151–6.
6. Assembled from phrases of Satow (1899) and Sze (1956, 1977).
7. Sze 1977: 302.
8. Ibid.: 299.
9. Ibid.: 110.
10. Ibid.: 111.
11. "The Book of Bamboo" forms a part of *The Mustard Seed Garden Manual of Painting,* published in the West as *The Tao of Painting* by Sze (1977), a paperback abridgement of a more complete text published by Sze in 1956. Citations here refer to the 1977 paperback.
12. Sze 1977:65.
13. Austin 1970:121.
14. Sze 1977:117.
15. Oi 1963.
16. Ibid.
17. A number of recent publications give details of traditional bamboo painting brushwork. See especially Sze 1956, Leong 1979, Tseng-Tseng Yu 1981, Cameron 1968. For historical background, see Cahill, van Briessen 1975, Suzuki 1971.
18. Aero 1980:168.
19. "In ancient China, music was from the beginning unmistakably linked to politics." This subject is discussed at length, May 11.
20. Needham 1980:iv.1:156.
21. Tompkins 1974:161–8.
22. See also Kurz 1876:233–4 and Lawson 1968:15 for various accounts of this practice.
23. Aero 1980:166.
24. May 1983:68.

25. Aero 1980:166.
26. *New York Times,* 5 Feb. 1969.
27. Kurz 1876:233.
28. May 1983:158–61.
29. Malm 1977:15 ff.
30. Kurz 1876:234.
31. The title of a highly recommended global survey by Reck 1977.
32. Aero 1980:112.
33. Levenson 1974.
34. Sze 1977:275.
35. Needham 1980:vol. iv.2,sec.27:64.
36. Junka 1972:26. See photos of tea ceremony utensils in Austin 1970:152–9. Sen 1979 offers over five hundred photographs on tea architecture, objects, and etiquette, noting that Japan presents an example "to be kept in mind," adapting it always to the special resources and cultural needs of new locations. Tea has no frontiers. Its path is everywhere. Its essense is "extracting essence through relationship." *Aisatsu,* the encompassing Japanese idea of "greetings," derives from *ai,* "friend," and *satsu,* "drawing out the good qualities of one another by gathering together to share time." "As iron sharpens iron, a friend sharpens the face of a friend" (*Old Testament,* Proverbs) (Sen 1979:3).

    The way of tea and its behavior is based on "the mutual contribution of the host and guest to the mood of their meeting." The tea ceremony is a "practice," a rehearsal ritually designed to ease tea's way into everyday life.

    The recent flood of tea news includes: *The Way of Tea* (Rand Castile), *Japanese Arts and the Tea Ceremony* (T. Hayashiya et al.), and *Tea Ceremony Utensils* (Ryoichi Fujioka), all from Weatherhill. See also Hammitzsch 1980 and Okakura 1906, the first book of tea published in the West. Lee (1962:395–7) provides a useful critique of the rigidities stifling the central sniff of tea: "Beginning as an informal and uncodified meeting of congenital spirits, it became a rigid cult of taste, artificially dedicated to simplicity . . . Rikyu's four requirements for the ceremony—harmony, respect, purity, and tranquillity—are understandable defenses for the conservative and cultivated few against the robust, gorgeous, and sometimes gaudy tendencies of the new dominant classes of the seventeenth century. Two all-important tea ceremony qualities, *sabi* ('reticent and lacking in the assertiveness of the new') and *wabi* ('quiet simplicity'), can also be best understood against such a historical background . . . What had begun as an informal gathering has become an exercise in studied nonchalance with overtones of repression and symbolic poverty . . . One small incense box of glazed stoneware brought almost $40,000 in the 1930s . . . That the tea ceremony represents the essence of Japanese art, or even taste, is open to question. The many other manifestations of art in the Ashikaga, Momoyama, and Tokugawa periods should place it in proper perspective as a conservative, elite, holding action." Fossilization, loss of original energy, is a constant in all cultural evolution. It provides a fallow period of neglect, prologue to appropriate resurrection. Fathy's remarks on tradition and the reactive creation of the individual that keeps tradition alive are important in this context (see pp. 116–122, "Architecture with Heart").

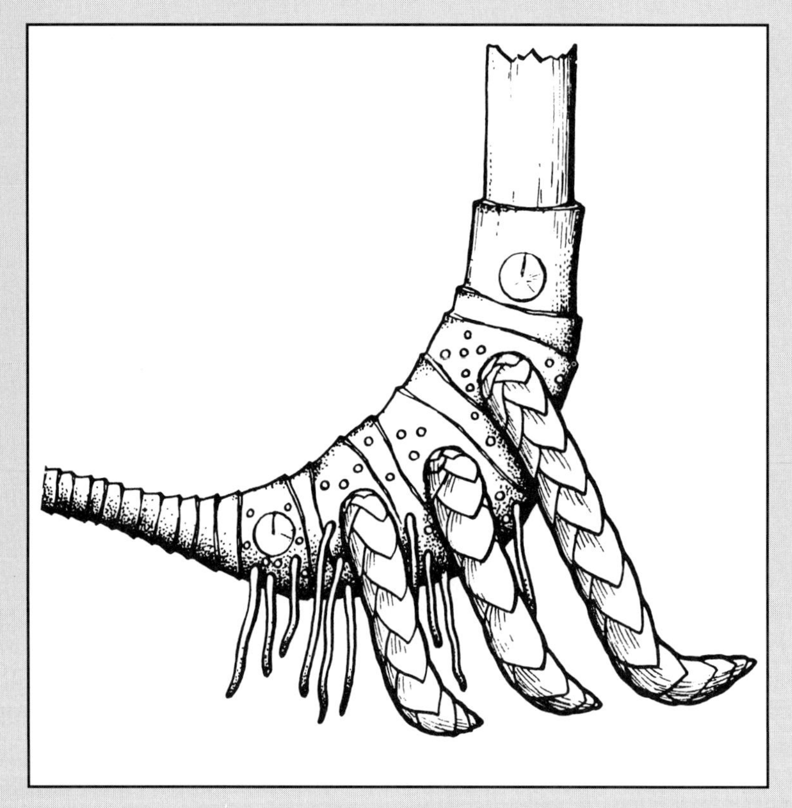

4. SHELTER: IN THE BEGINNING WAS BAMBOO

Things came before people.
People came before words.

Tall, green, and nameless,
bamboo walked down centuries
and crossed continents
in time to stand there, waiting,
naked of language,
when the first people came
to make the first village.

Lots more people means lots more houses. Bamboo outproduces any alternative source for the raw materials of construction. Growing walls and roof on the spot saves transport and materials cost, decentralizes the building business, unplugs it from banks, and humanizes shelter. The house is brought home again. You can grow your own, like tomatoes.

THIRD SKIN

Shelter is our third skin. Although everyone agrees everyone needs it, the housing shortage is ever more acute and ever more everywhere. Governments are dealing with unprecedented numbers demanding more as minimal housing, and as more line up for a roof, fewer are equipped with skills, tools, materials, land, and/or cash to weave a shelter around them of their own labor and will.

Shelter shortage.

More are moving more, which means more renters less attached to any one shelter, less experienced as owner-builder-menders permanently related to a piece of earth. Government low-cost housing projects around the world are never as abundant as needed, always beyond the means of the really poor, always a double-ugly crime zone miles wide of the cultural mark—and always erected with near zero planning and participation of the people.

Autoarchitecture forms part of the schooling of all traditional peoples. Only in fragmented modern education has building been abandoned as a central study. Are reading and writing and arithmetic more crucial than architecting our own dwellings? A world in chronic shelter shortage, an intense drama with a supporting cast of some 2½ billion, should reestablish popular architecture in its educational system from the earliest levels. And an integral architecture should plant as well as build: We should all be foresters as well as carpenters, growing the raw material as well as shaping it.

Unprecedented demand for human shelter demands in turn a deep rethinking of our cultural designs for it. Bamboo is examined first in two forms relevant to high-tech modern construction—lamination and bamboo reinforced concrete. Both are

In this sympodial rhizome of* Guadua angustifolia, *buds of the next generation form "feet" to help support the 100-foot culm growing from the mother rhizome.

appropriate in a number of ways for a decentralized village technology.

Disasters are sufficiently chronic in human experience for us to take the trouble to design their wake better. And in disaster design the need for "instant shelter" provides an intensified miniature model for the constant disaster area of present world culture, our wretched world housing, monstrously below any code of the heart. People participation has been observed as central in postdisaster designs to rebuild. That hint should be applied to the permanent emergency the homeless have become in our time.

In contrast to the sprawling, shallow, and money-dominated disorder of modern mass housing, the Japanese teahouse represents a wholistic response to the deepest psychological duties of a shelter design in touch with restrained human needs. The idea of an ultimate art, an all-inclusive art is present in many cultures. "Primitive" fiestas embody this better than any modern art form. A Hopi bean dance, for example, includes every art you could imagine—dance, song, drama, clowns, cooking, architecture, and god dolls or Kachinas: the theology and cosmology of the Hopis embodied in beautifully painted and costumed cottonwood statuettes given to children as a catechism of Hopi myth at fiestas. Elaborate jewelry of turquoise and silver, pounding on the chests of the dancers, becomes a percussion instrument.

Europe in the nineteenth century also dreamt of the complete work of art that would unite all the arts in its creation and found the opera. Architecture provided the buxom embellishments of the standard opera house, the construction of which provides few hints about mass housing: It is a glittering exception, not the possible norm.

The tea ceremony is another of these total art forms weaving many arts into a coherent unified expression. Its summation of Japanese culture, within a tiny space and intimate ritual, was effective and seminal: The simplicity of the form made it available at all social levels, and the teahouse became the dominant philosophic and design influence on what we know today as the characteristic Japanese house.

Architectural activists.

Feeding the people doesn't mean putting every spoonful in their mouths but being sure that food and fuel are available in sufficient quantity for everyone. Sheltering the people should not mean housing projects but making resources available, reawakening traditional skills, and playing midwife to new forms of old solutions so people can resume responsibility for self-shelter.

The true task of governments with respect to housing is to help create the conditions for a culture of architectural activists. This emphatically includes reforestation designs geared to the needs of a popular architecture.

LAMINATION

New forms of the ancient technology of lamination may be opening the biggest door for bamboo in contemporary industrial design. The Chinese employed plybamboo in the vanes of winnowing machines in the eleventh century, and Japanese bowmakers were laminating bamboo with wood for centuries before large-scale modern uses of the process in the West in 1907, when the first architectural members of laminated wood were made in Europe. Laminated woods were used in airplanes in World War I and later in U.S. factories, gymnasiums, hangars, and houses from the middle thirties on. Laminated bridges and marine structures followed the development of new glues and resins, with which laminated spans up to 174 feet have been made in the United States.[1]

Bats, rackets, and parquets.

The lamination process makes large pieces from small, even scrap, material without the splits and fissures common in large timbers. Within certain limits, lower grade woods can be used without affecting the strength of the final laminated product. As smaller and inferior grade trees appear on the pinched world lumber market, lamination will increase as a process with a thousand applications, and laminated bamboos will be taking over more and more of the tasks performed by laminated woods. The same rapid proliferation that has occurred in the brief history of laminated woods can

Laminated bamboo plate.

be expected in products of laminated bamboo, many of which are already on the market. Laminated bamboo plates, cups, bowls, trays, screens, bows and arrows, tennis rackets—scores of traditional objects and bizarre East-West hybrids such as laminated bamboo baseball bats have, for some time now, been exported widely from Taiwan and Japan.

Bamboo tile, parquet for floors, thin rollable bamboo veneer, as well as plybamboos of woven matting are now available, and many more objects presently made in plastic can and probably will be made of bamboo as lamination extends the natural versatility of the plant. Plastics are petroleum based and polluting, while bamboo creates lots of oxygen as well as crude cellulose fiber. Countries can't move oil fields to their back yards, but they can plant bamboo. As nonpolluting, replenishable, biodegradable substances are welcomed by ecologically conscious industries—somewhere in the uncertain and finite future of petroleum—laminated bamboo may replace plastics in many more roles.

Plybamboo.

In rural areas of the world, bamboo will continue to be used as it has been for many centuries—in unprocessed poles, cut to the desired length with a machete, flattened into boards with an ax, or woven into long mats and sold by the finished yard. But the most promising future for bamboo in urban housing lies in the direction explored by the Forest Research Institute at Dehra Dun (India). Narayanamurti (1956) summarizes—in "Building Boards from Bamboos"—seven years experimental production of woven panels of plybamboo for moulded furniture and prefabricated ceilings, walls, and floors for economical housing.

The appearance, strength, and durability of desk fronts and dressing tables, folding screens for room dividers, drawers, sewing machine covers, suitcases, wardrobes, and architectural elements manufactured were considered very encouraging. *Bambusa arundinacea, Bambusa polymorpha,* and *Dendrocalamus strictus* were the principal species used, with the first showing greatest strength.

Successful experiments were conducted producing flakeboards from shavings and sawdust of bamboo, but most work was done with bamboo mats, woven by hand or on a loom modeled after a textile-weaving machine, using strips of varied width approximately 1/16 inch thick. Mats were soaked in a resin solution and then pressed—usually three together—at 140°C. A resin solution of 15 percent and pressure of 400 pounds per square inch were found to produce satisfactory boards. "Single mat boards are flexible and very suitable for light partitions or ceilings . . . The bamboo strips can be interwoven with veneer strips or special wood shavings. Bamboo shavings sprinkled on the mats before being pressed produce a very attractive surface." (Sawdust, sand, charcoal, and cork dust were also tried for filler.)

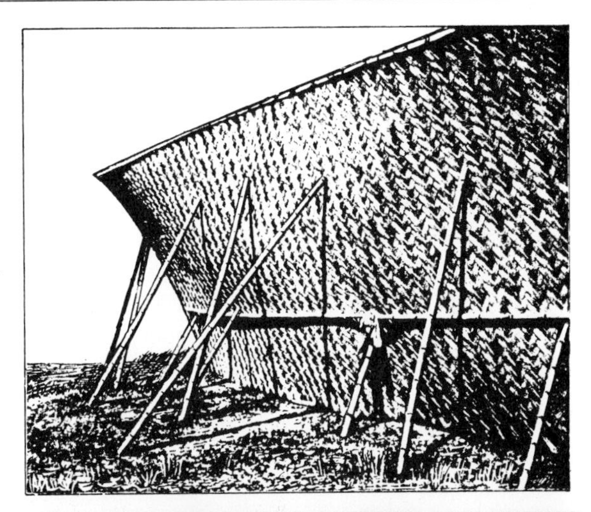

Prefab walls of woven bamboo are common in oriental countries. (A) **Muli (Melocanna baccifera),** *a major species of India. (B) Detail. Note the nodes; each strip is a whole, flattened culm.*

Boards proved quite water resistant. Specimens soaked in water were sound even after three years. Boards buried in the Institute test yards at the end of seven years were still unattacked by termites and fungi. This resistance can be further increased by adding a small percentage of pentachlorophenol to the resin solution. A building with floor, walls, and roof of bamboo boards had survived, at the time of this writing, five years with 85-inch annual rainfall and was still in good condition. The boards used had a filler of sawdust and were produced at the comparatively low pressure of 200 pounds per square inch.

Some notion of the feel of finished products in plybamboo can be had from the laminated bamboo plates produced by a similar process in Taiwan, China, and Japan, which are broadly available at low cost in the United States. They are incredibly light yet strong.[2]

Plybamboo is a relevant direction for research because its technology is simple enough to fit decentralized village production for local use. Fabrication relies on basic traditional weaving skills found in most areas where mats and baskets are made with bamboo and a wide range of other grasses, reeds, and vines. The use of plybamboo in eleventh century China suggests the low-tech, village industry feasibility of this shelter material. Plybamboo represents a revival of an ancient process wed to the common practice through many parts of Southeast Asia of weaving long prefab walls of split and flattened thin-walled bamboos. *Schizostachyum* species are used in the well-known Philippine version of this woven wall called sawale. A 1-inch-diameter bamboo, opened out flat, yields a strip as long as the culm, roughly 3 to 3½ inches wide, a large member for rapid weaving ease. In some areas of Asia, walls up to 30 feet long and 10 feet high are woven, rolled up, and delivered to the house site by ox cart. Such rural technologies provide a sound traditional base for plybamboo development, but one urgent priority in countries with abundant bamboo is the search for native resins that will free plybamboo from costly imported adhesives.

Barn swallows and other species built mud shelters reinforced with twigs and fibers. Wattle, daub, and reinforced concrete are recent evolutions of an ancient technology.

REINFORCED CONCRETE

Use of bamboo in construction has, until now, tended to diminish with increased industrialization. For example, India, with a population still largely rural, uses 150 times as much bamboo for construction as Japan—and one-quarter as much for manufacture. But with bamboo now available in a contemporary industrial format as laminated panels and structural members for the skin and frame, these figures could change greatly in industrialized countries like Japan. And the use of bamboo to replace expensive iron rods in reinforced concrete may do as much as lamination to alter the profile of world use.

Around 1918, the Chinese were the first at studying bamboo to replace iron in the reinforced concrete of railroad bridges and other constructions. It was employed by both the United States and Japan during World War II, and its usefulness to the military in Vietnam recently spurred more research in America. The industrial implications of bamboo used as reinforcing rods are enormous, and detailed research has been conducted in a number of countries including China (1920s), Germany (1935), and the United States (1943, 1968).

Until now, it has been primarily the crisis conditions of war that have intensified interest in bamboo as a possible replacement for needed metal in construction. Technologies using available on-site material are preferable in wartime to reliance on vulnerable supply lines. However, many products and processes first tried in the military have found their way into the domestic, peacetime economy. Refrigerators were first used not in family kitchens,

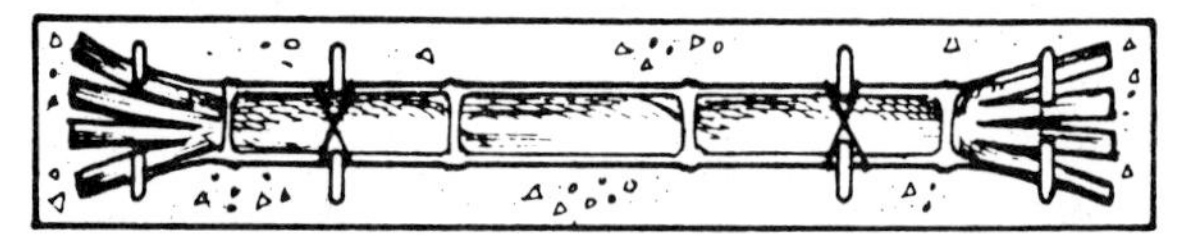

but on battleships. Bamboo reinforced concrete may follow this route from weaponry to "livingry," as Buckminster Fuller called the peaceful ending of this familiar pattern. Population growth and the rising costs of steel and transportation may soon prove sufficient to stimulate increased use of bamboo cement, especially where bamboo is locally abundant.

Fortunately for world housing needs, the areas of the most rapid population growth are also generally the areas with the largest growth of bamboo, or at least with climates friendly to its growth. Latin American governments especially, short 20 million houses now and facing the world's fastest population doubling rate, intensifying that shortage, are going to be turning to bamboo.

South Carolina experiments.
Bamboo reinforced concrete experiments have been conducted since roughly 1917 in China, Japan, and the Philippines in the East, and by Germany (1934) and Italy in Europe. Various applications have also been explored by Oscar Hidalgo (1974, 1978) in South America. The most comprehensive experiments to date in the United States were conducted by Glenn and associates (1950) at the Clemson College of Engineering in South Carolina. McClure reprints their conclusions and construction principles in full.[3] Some highlights of their summary: Bamboo reinforcement in concrete beams increases the load-bearing capacity of beams most effectively when it constitutes 3 to 4 percent of the cross-sectional area of concrete in the member. Reinforced beams bore four to five times the load of unreinforced beams of equal dimensions before failure occurred; they can be safely designed to carry two to three times the load of unreinforced beams.

In concrete slabs and secondary members, green, unseasoned, whole culms up to ¾ inch may be used but are not recommended for important concrete members. When possible, allow a drying period of three to four weeks before using. Only fully mature culms, three years or older, should be used. Alternate base and tip of whole culms because this makes for a uniform percentage of bamboo in cross section and creates a wedging effect that increases bond. "Apparently, the bamboo reinforcement of nonload-bearing members, such as wall panels and floors resting on well-compacted earth, has more to recommend it than bamboo reinforcement of load-bearing members of a structure."[3a] Before pouring, request information from Dehra Dun (next section) and consult the Glenn study in detail to avoid repeating their mistakes.

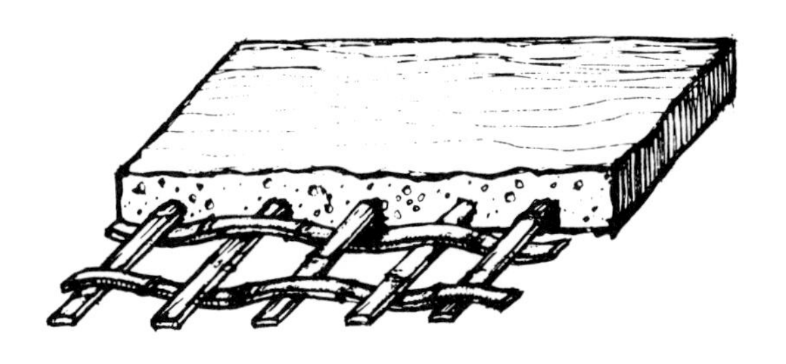

Cross section of bamboo reinforced concrete slab.

Report from Dehra Dun, India.

> To save steel in the country, a research program by the Timber Engineering Branch tested bamboo as a reinforcing material in various cement concrete precast and cast-in-situ structural components. High absorption of water from cement concrete initially resulted in dreadful cracks in the concreted components and discouraged the engineers. However, suitable, easily applied water-repellent has been found, resulting in safe structures: by dipping bamboo *(Bambusa arundinacea)* strips, 20 by 9 mm, in 80/100 grade hot bitumen and sand blasting, the bitumen acts as water-repellent, and the sand coating helps increase bond. After drying, the bamboo strings are tied into a mesh as is usually done.

These investigations showed bamboo use in place of steel reduced costs 33 percent.

A number of full-scale demonstration and use structures have been built to observe behavior under long-term loading and operation. These structures include a double-story residential block with a variety of structural components ranging from a sunshade to a grain silo. The results are encouraging and more such structures are planned.[4]

Lattice structures.
West Coast U.S. bamboo architect Jim Orjala sees the Glenn experiments as insensitive to other design possibilities with bamboo, such as lattice structures. Traditional bamboo lattices, from basketry to roofs, developed out of familiarity with the flexibility of bamboo and its strength in tension. The bamboo lattice when curved, or "prestressed," gains considerable strength in its ability to span space and resist bending due to external compressive loading.

The bamboo lattice, thinly coated with lightweight cement to aid in taking compressive

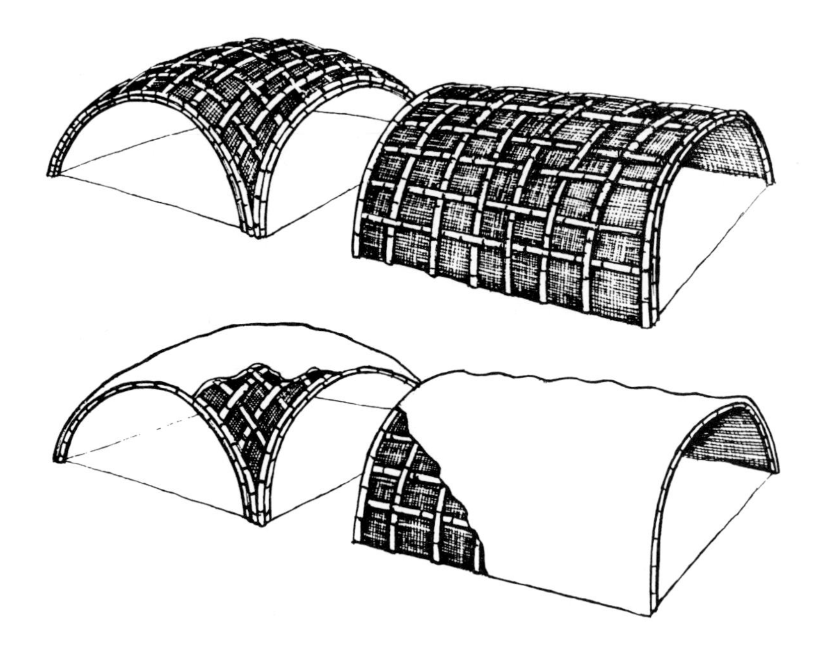

stresses, provides a structure of exceptional efficiency and economy in the use of materials. Major problems encountered in previous work with bamboo can be traced to the structural systems employed: beams, slabs, columns. Beams, the most inefficient structural members, are particularly subject to bending due to concentrated perpendicular loading. The beams tested failed ultimately due to failure of the bond between the bamboo and concrete. Physical characteristics of bamboo, in comparison to steel, make bonding with concrete difficult.

Techniques developed for thin-shell ferrocement vaults and domes offer significant potentials to alleviate these problems. The curved bamboo lattice is similar to steel mesh used in ferrocement. Vaults and domes are particularly efficient structural systems. The structural advantages of these forms lie in dispersal of stresses along the curved surface, which effectively minimizes bending. A higher percentage of reinforcement provides more surface for bonding. Distribution of reinforcement significantly reduces concentration of stresses on the bonding and allows greater resiliency of the composite material.

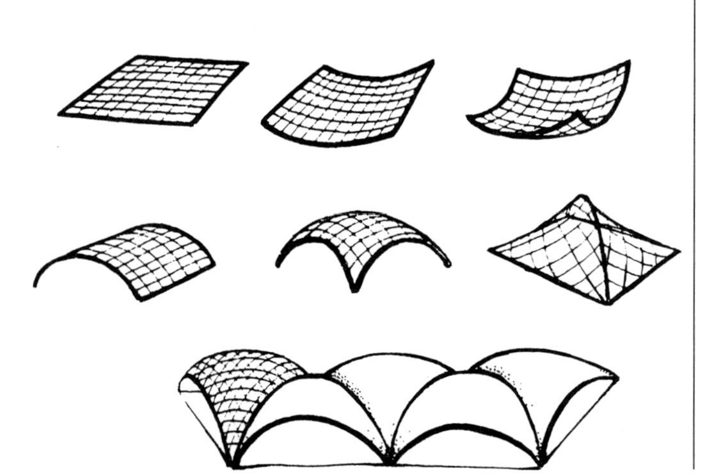

There are several other advantages to the lattice framework as well: it requires the simplest of skills and techniques for plastering. Labor and cost of formwork are eliminated. The use of bamboo in place of steel further reduces the weight and cost of reinforcement. The thin coating of cement significantly reduces the weight, quantity, and cost in comparison to conventional concrete construction.[5] See also "Water storage" (pp. 66–67) on Thailand's recent work with bamboo-reinforced storage tanks for rural areas. Hidalgo (1978: 82–127) is highly recommended for extensive how-to photos, exploring low-technology production of bamboo-cement roofing, tanks, panels, fence posts, and beams. Cables of twisted strips of bamboo, designed to increase bonding, were an innovation of these Colombian experiments.

QUAKING WITHOUT BREAKING: BAMBOO CITY PLANNING

Of the half million earthquakes rippling the earth's surface annually, some one hundred thousand can be felt. Of these, one thousand cause some dam-

Bamboo design models. Evolution of the original round bamboo hut in India. Bamboo was a structural element in the earliest shelters of Asia not only because of its abundance, but also because crude tools could work it easily. In some areas of the world where native peoples enjoy the options of large hardwood forests and groves of bamboo, the latter is still chosen for houses for the same reasons that first determined its use: ease of harvest, transport, and construction.

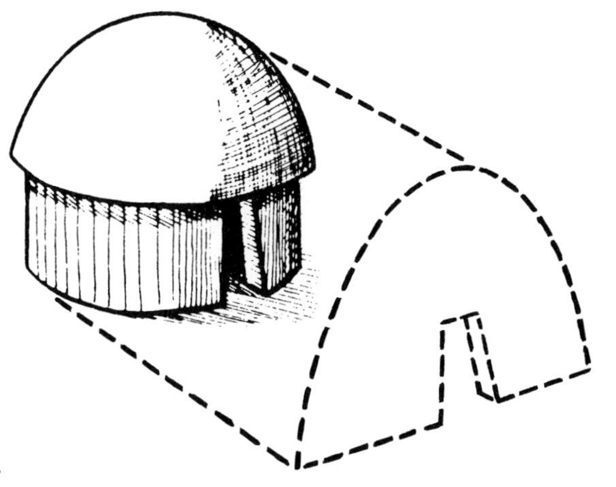

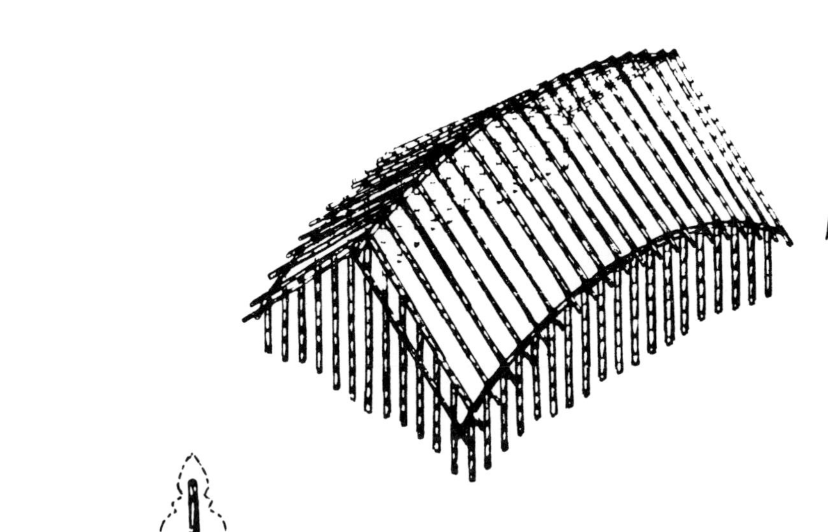

Chauchala *(India). Wood and stone became important building materials as primitive tools grew more sophisticated. This trinity of bamboo, wood, and stone reflects not only the architectural evolution of the Orient, but its social classes as well. The poor used bamboo, and the rich, wood; stone was preferred for the palaces of princes and the temples of priests. Bamboo continues to be the wood of the poor, particularly in India, where some 500,000 tons are used annually for housing.*

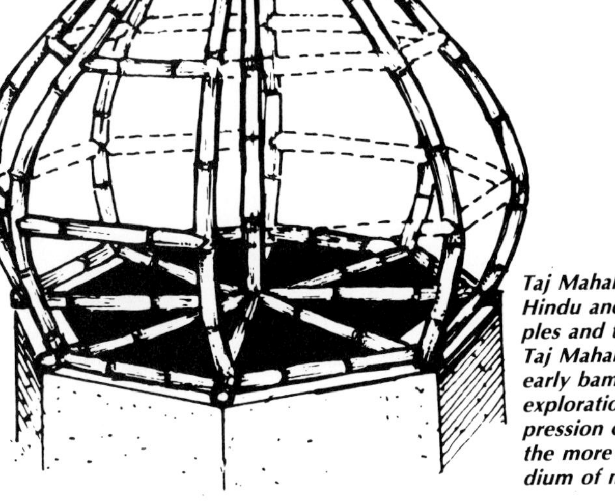

Taj Mahal. The cupola of Hindu and Buddhist temples and the dome of the Taj Mahal are examples of early bamboo architectural explorations finding expression centuries later in the more permanent medium of masonry.

age. Three times a day, a quake somewhere is tumbling things back to earth. Quake-zone city planners can ponder the Japanese tradition that a bamboo grove is the safest place to be in an earthquake. Dense growth of bamboo around a building functions in the soil as windbreaks act in air: rhizomes buffer the blow and diminish the intensity of motion. Widely planted in any quake zone, bamboos absorb much of the earth's ripple, moreover provide immediate construction material for temporary disaster shelters. "Scenario" has become a cliche of war games. "What if bomb A fell on city B?" In order to hint how bamboo development could occur, let us imagine for a moment a true lie, a tale botanically and socially possible, but which never occurred. It sketches an imagined collaboration among government agencies, mass media, the schools, and the people that is probably a precondition for any durable and deep cultural innovations with bamboo in our times.

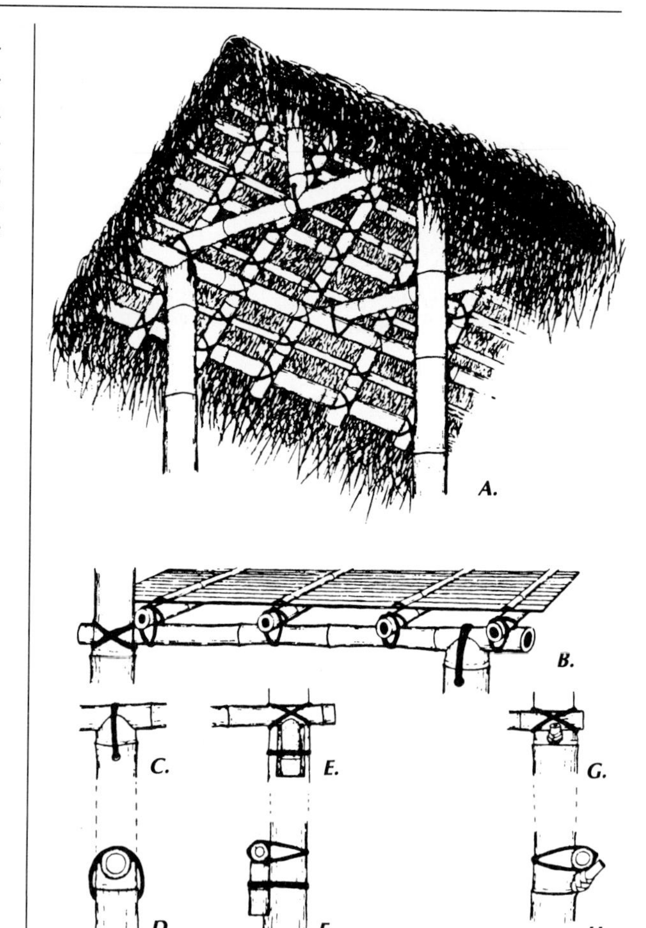

Basic bamboo construction: (A) fitting and binding culms at joints in roof and frame; (B) fitting and securing bamboo boards in floor; (C and D) saddle joint; (E and F) inset block to strengthen support of horizontal above; (G and H) use of branch stump for support.

Managua 1972: rebuilding with bamboo.
Managua provides an interesting example of how bamboo can relieve disaster in a tropical climate. In December 1972, the city "trembled like a wounded animal," as one survivor said, and the Nicaraguan people were left with a demolished capital to repair. Central Managua was in ruins, with thirty thousand dead and many thousands homeless.

Bamboo houses are common in Nicaragua, especially in the rainy northeastern part where some 50,000 acres of bamboo grow mainly along a number of rivers emptying into the Atlantic. Many Managua residents—often the oldest members of the household—had come from rural areas where rustic bamboo constructions were common. Two species of bamboo useful for building grew some 40 km south of Managua. For a number of reasons, it was natural that disaster housing was partly erected with bamboo. It was light to transport; some 700 poles fit into a truck with a 4 by 9 by 15 foot bed. It was rapid to work with, requiring only a machete to cut, split, and shape. Basket makers from the Masaya area, the craft center of Nicaragua, were brought in to show people how to weave bamboo panels. From this beginning, sprang up small workshops producing prefab panels woven to specific dimensions.

The basket makers also taught some Managua residents how to plant. The freshly cut culms of *Bambusa vulgaris* they were building with could be propagated easily from basal nodes and wasted branches. One barrio organized a larger planting in a park and playground area. It gave quick shade, extended in a few years by using bamboo arbors for chayotes, calalas, and other climbing plants, which provided food in the inner city without transport as well as large leafy roofs to gentle the oven of city air, torrid enough to inspire a popular joke: "When Managuans die and go to hell, they send home for their blankets."

Integrated development.
The municipal reconstruction offices in Managua responded to the people's spontaneous plantings of bamboo by beginning groves in completely empty areas scattered throughout the quake zone where no building was contemplated. They planted two drought-resistant species, *Bambusa ventricosa* and *Dendrocalamus strictus,* and other bamboos from the government collection of oriental tropical spe-

cies in El Recreo, Zelaya. More delicate species like *Bambusa textilis* were mulched heavily and planted under vine-bearing bamboo arbors until they were well enough established to begin shading themselves.

Urban umbrella: bamboo microclimates.
The ministries of agriculture, education, and housing were involved in the Managua bamboo project design, as was the architecture school, *Universidad Nacional Autonoma de Nicaragua* (UNAN). Articles appeared in newspapers, supporting and explaining the why and how of bamboo and urban planning. "One rhizome of *Bambusa vulgaris,* in a Colombian experiment to measure bamboo's capacity to increase and multiply, produced 200 culms in five years. Yearly growth is 2–5 percent of existing size in most tree species; in bamboo the annual increment is 10–30 percent. Managua can grow much of its needed construction material on location, saving enormously in transportation costs.

"A great deal of this new growth in bamboo is in its leaves, which do not increase the size of the culm they grow on, but nourish the underground system of rhizomes uniting the grove. These rhizomes, bamboo's energy bank, in turn produce more culms. Massively planted by students in the city's abundant empty lots, bamboo will change many microclimates in Managua, as the experience of Don Bosco and other barrios has proven, in fairly large plantings already almost two years old. Bamboo shadows the architectural madness of zinc roofs—low in altitude but not in cost or temperature. It muffles city noise, increases privacy, and screens buildings often less attractive than the plant. It can air-condition the torrid center of Managua, still virtually treeless since the quake."

City farming.
"The foliage of bamboo is evergreen and superdense. Reaching 4 inches per year in the leaf fall of some species, in others it equals the weight of the grove's new culms. Its green abundance accounts for bamboo's wide use in many rural areas of the world for animal fodder. It has been used as a supplement rich in vitamin A for chicken food,[6] and in the Andes, in both Ecuador and Peru, bamboo leaves are fed to guinea pigs, an important part of the local diet. Such uses in small animal husbandry have suggested the use of bamboo leaves for rabbit food and culms for cages in urban farming designs to increase meat production in the crowded city. Japanese yield studies found that two species measured produced 5–15 tons of leaves per hectare.

"This abundant foliage means that bamboo shade is quite tangibly cooler than the surrounding air and makes it perhaps the most intense injection of oxygen into urban exhaust that coughing moderns could plan."

Green air.
"A neglected aspect of modern urban architecture is a serious leaf/people disequilibrium. Never have so many people lived with so few leaves, nor dirtied their air more thoroughly in traffic jams trying to escape it—to the countryside, where we instinctively seek out a greater abundance of vegetation. The search is not surprising. More leaves mean more oxygen. Contemporary psychologists note that oxygen reduction in blood creates in brain cells a proliferation of negative emotions, fears, aggressions, worries—the whole configuration of antiother feelings which characterize the psychology of modern urban dwellers, sick of too much cement, too many faces, and too few leaves. Anyone who has experienced the refreshing atmosphere of dense stands of the plant will appreciate why, after enlightenment, the Buddha chose to live in a forevergreen grove of bamboo." (*La Prensa,* 14 October 1974.)

These figures are botanically accurate, but—except for the bit about the Buddha—none of the above actually happened. Nicaragua's dictator at

the time, Anastasio Somoza, pocketed most of the international aid sent for disaster relief, but even if he hadn't, it's unlikely anyone would have thought of bamboo. The reasons in favor of bamboo are multiple and complex; they cross the frontiers of too many ministries. Modern governments, fragmented into immense bureaucracies of specialists, by their very nature are ill-equipped to enact solutions that touch too subtly too many keys on the grand piano of cultural design. Ten years after the quake, Managua is still barren of trees, filled with empty lots, and experiencing a housing shortage that present building rates will require a century to meet. (The Sandinistas have, however, made tree planting in Managua a part of their platform, which aspires nationwide to an ecological sanity as bandage to a century or so of ruthless resource plunder.)[7]

Disaster designs.

Massive, merciless bombing, and resulting masses of roofless refugees, must be added to earthquakes and other natural disasters as sufficiently recurrent to deserve advance planning. Someone has statistics somewhere on how many people, worldwide, are presently living in disaster architecture. Cost and quickness are two critical factors for disaster relief; another, often overlooked, is victim participation. Flying sophisticated equipment and crews in to spray a thousand polyurethane foam domes denies the people a chance to rely on their own creative energy: first they are knocked over by the weather or war, then high-tech generosity floods them with a solution alien to the context . . . and drives home a dismal sense of helplessness.

Bamboo is an alternative. Providing victims of natural or human disasters the wherewithal to create their own solution is more creative than a finished package usually designed far from the local reality. Green belts of bamboo in and around large population centers would be one intelligent and anticipatory practice in the tropical regions of its greatest growth. In *Shelter After Disaster,* Ian Davis (1978) examines three hundred years of human response to disaster—an increasing global experience as more people, especially in poorer countries, jam dangerous sites on precarious hillsides or hurricane-prone waterfronts. The victims' ingenuity does 80 percent of the cleanup—actively recycling rubble into new homes—if foreign aid doesn't steamroll in to frustrate their efforts. Experience indicates that international agencies and relief design should foster what the people are already doing instead of pushing an alien scheme and technology on them. Low technology, by contrast, is affordable by the people, affirms a reassuring continuity with traditions, and generates local employment. (See Darrow 1981: 661–2.)

Quake-zone construction.

The queen in *Alice in Wonderland* explains to Alice the advantage of advance weeping: it leaves you free to deal with the accident calmly when it occurs so effective response to tragedy of whatever dimension is unmuddied by emotional confusion. Routine plantings of bamboo long before disaster strikes could function as anticipatory tears, permitting efficient relief when suddenly needed. Similar in intent is a current housing project in Guatemala that features a retraining program promoting quake-conscious construction using traditional materials and known skills; it results in structures familiar in appearance but safer than traditional housing in the area.[8] Along the same lines, Narayanamurti, in the U.N. sponsored *Bamboo and Reeds in Building Construction* (1972), notes a variety of ways in which conventional bamboo and other low-tech structures can be made more quake proof: "Earthquake forces that a building has to withstand are proportional to its weight and are predominantly horizontal. The heavier a building, the more likely it is to get damaged. Lightweight material such as bamboo, with a high strength/weight ratio, is therefore preferred in regions where earthquakes occur.

"Experience in different seismic regions of the world has shown that a house built of bamboo, properly lashed together, is earthquake resistant. In this respect, bamboo is somewhat superior to timber. It has the capacity to absorb more energy, and shows large deflections before failure occurs. A bamboo frame structure yields readily to vibrations during a quake and does not collapse. If such a mishap occurs, loss of life and property are less in lightweight structures.

"Construction details should be adopted at the joints of the framing members and wall panels so that the bamboo frame structure as a whole behaves as one unit against earthquake forces. Experience in India suggests that a closed-frame construction should be adopted with horizontal connecting members for the columns at foundation level. Walls and partitions should be provided with diagonal braces and anchored properly to the vertical and horizontal struts. Observations on behavior of framed structures during earthquakes in Assam have led to the conclusion that the superstructure

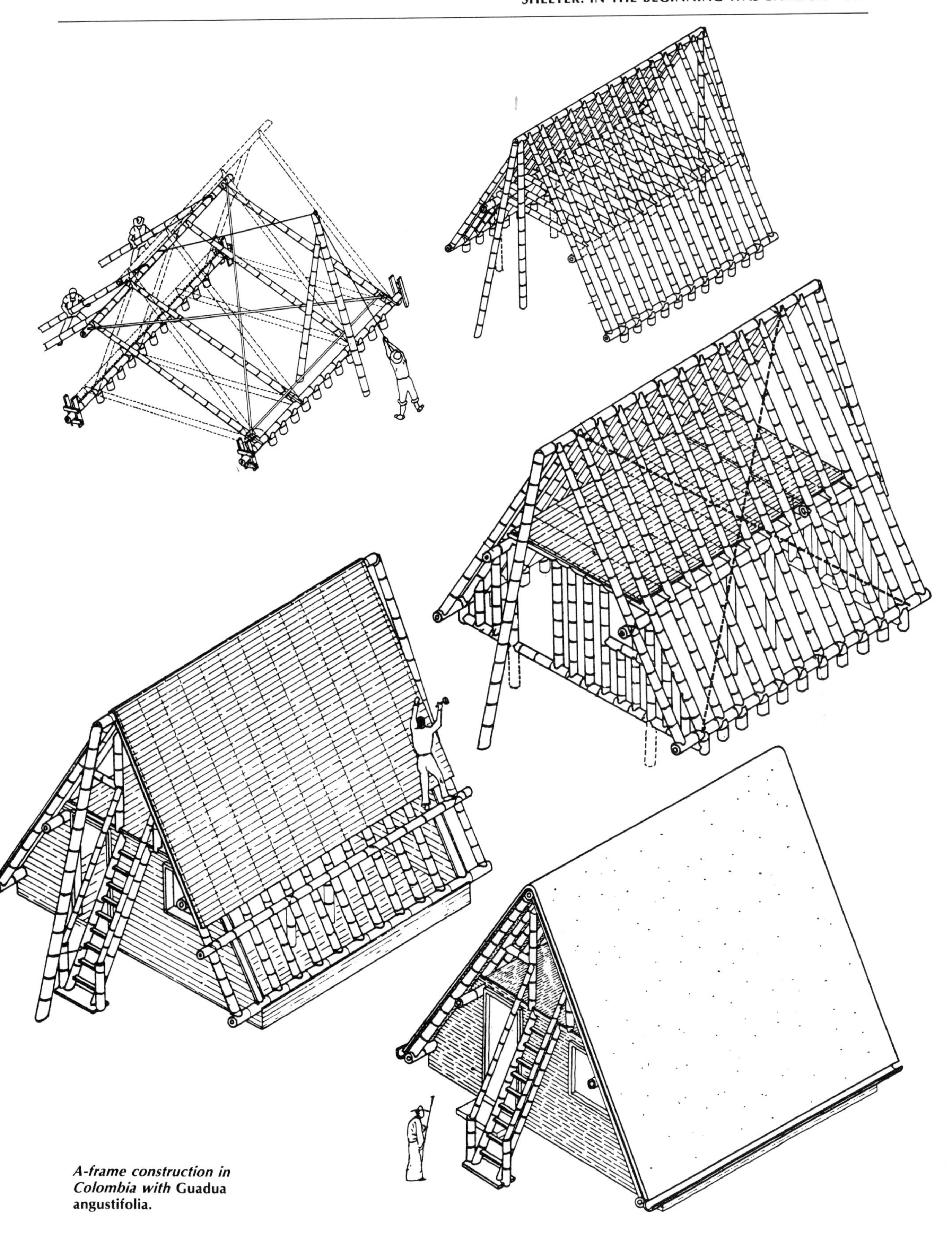

A-frame construction in Colombia with Guadua angustifolia.

should rest on a foundation of masonry. Small, one-story buildings up to 50 square meters may rest on firm ground, but in larger buildings the posts should be attached to foundations by means of pins or straps, bolts, and nuts.

"Bamboo board, matting, and plastered matting walls, being light and flexible, are suitable for seismic areas. Experience in Colombia has shown that the *bajareque* wall construction, more massive than wattle and daub but less massive than rammed earth or adobe, is earthquake resistant. Load-bearing adobe and heavy mud walls, which fail under relatively slight tensile or bending force, are the first to fall during a seismic vibration. It is recommended that a bamboo lattice should be used in mud walls to strengthen them.

"Brick masonry walls also have poor resistance to earthquake shocks, especially when a weak mortar such as mud is used. In ancient Babylonia and Ur, reeds embedded in asphalt appear to have been used as horizontal reinforcement in brick walls. Vertical steel reinforcement is now specified for brick masonry walls at corners and junctions. Vertical steel at doorjambs and a lintel band are also recommended. The use of bamboo splints or reeds in place of steel in these places should prove beneficial, especially in single-story houses.

"A false ceiling should be tied rigidly to the roof. Plaster on the ceiling should be avoided or kept to minimum thickness. Light roofing materials are advantageous in reducing the inertia force at the top of a building. Bamboo tile, bamboo shingle, and thatch are satisfactory roof coverings in this respect."[9]

How many pounds make a house?

Flexibility is one bamboo feature that helps it survive quakes; another is *weightlessness.* "Gentlemen, how much do your buildings weigh?" asked Bucky Fuller, whose geodesic domes sit more lightly on the earth and so withstand quakes and winds better than conventional urban architecture.

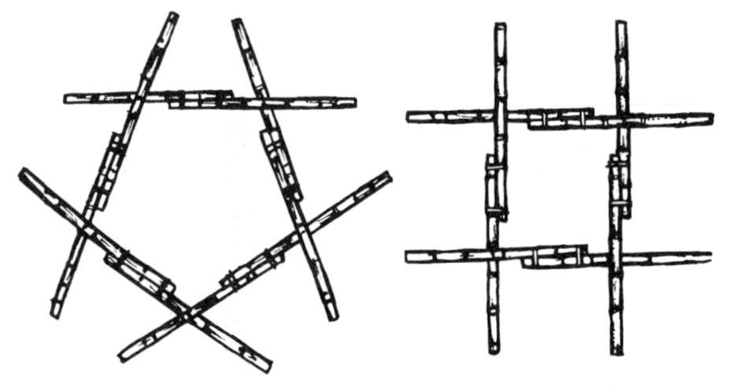

Bamboo dome members.

Bamboo polyhedra: a synergetic mylar Walden in Louisiana, c. 1942.

Fuller built some early bamboo domes with students at the school of engineering in Calcutta, India.[10]

Bamboo polyhedra, Louisiana 1940s.

An architect from Lafayette, Louisiana, Neil Nehrbass, followed Fuller's footprints in the grove, reporting on building and living in bamboo shelters inspired by the crystal shapes in nature.[11] Of the five regular polyhedra that matter finds comfortable, two are collapsible: the cube or "hexahedron," with square faces, and the dodecahedron, with twelve pentagonal faces. Three are rigid, with triangular faces: the tetrahedron, the octahedron, and the icosahedron.

These shapes—known to mathematicians since the time of Plato in 400 B.C.—show that triangulation creates greatest strength with least material; an experiment with straw—or bamboo branches—and string models can prove this fact to anyone's satisfaction.

Nehrbass designed his bamboo polyhedral shelters for low-cost mass shelter in areas of bamboo abundance. He notes that a hacksaw and drill are the only required tools; that structures using *all members* of the *same length* can be constructed without the skills of reading and measuring; that lashing is a skill available in all cultures, with a variety of local materials such as vine, leather, rope, and so on; that the shelters could be mass-produced at grove site, with a line cutting, curing, trimming, sunning, cutting to lengths, drilling, and packaging.

Total weight of a 175-square-foot shelter (roughly 13 feet × 13 feet in a square) including all framing, skin, ventilation, and door is only 60 pounds, 10 cubic feet in volume. Packaged, this complete shelter can be carried by one person and assembled in four hours without tools on site requiring no foundation but anchored to stakes in winds.

From a harvest of four hundred poles with 1½–1¾ inch basal diameter, Nehrbass cut thirty 10-foot bamboos and built one complete icosahedron, a twenty-sided figure of equilateral triangles. If you cut off the bottom, you're left with ⅚ of an icosahedron, composed of twenty-five equal members. Nehrbass built nine of these, four with 10-foot members, three with 7-foot members, and two with 5-foot members, using the different structures to compare different methods of flooring, skinning, ventilation, and doors.

At the hub a rubber ball rode the ends of five poles drilled an inch from the end and lashed with leather to form a firm but flexible joint.

Although Davis in *Shelter After Disaster* (1978) warns against confronting disaster victims with outlandish shapes that might shock them more, the ease and rapidity of construction might warrant using these polyhedras experimentally to test their acceptance and practicality. Demonstration of the form in several Nicaraguan villages inspired interest. Construction as part of a school project or as playground structures might be one way of introducing them to a community. Villagers responded to the idea of their use as kitchens since in many areas a small cookhouse or *cocina* apart from the house is characteristic of rural Latin architecture. Boy Scouts, summer camps, and tropical guerrilla fighters are others who might find the form useful.

Those interested in bamboo polyhedras should be familiar with *Polyhedra: A Visual Approach* and *Introduction to Tensegrity* (1975 and 1976), both by Anthony Pugh (University of California Press, Berkeley, CA 94720).

THE PSYCHOLOGY OF SHELTER

After nine months in the womb, a child—the new package of perfect energy—enters the home, a woven texture of human emotion that swirls between the walls of the house. This swirl is the nest that shapes a consciousness. This emotional microclimate, the family, takes the raw organism of a human being and puts a whole language on two square inches of tongue; dresses the mind, still swaddled in eternity, in local time and custom and routine. The house is the living womb of consciousness, the mold of spirit. Space is the incubator of our patterns of learning and knowing, so architecture is much more than merely the study of physical structure.

Old house and garden, Kyoto, 1880.

Bamboo Walden.

Of all houses in the United States, the White House and Walden are among the most internationally famous. The White House has become the black symbol in the world for the United States as *Macho Negro,* the dark bully of brute force swaggering down history's main street dressed in metal and bombs as if human destiny were a Hollywood western. The president has become an anti-Santa to the whole wide world. Out of his chimney swarm sooty helicopters, dangerous dragons.

Power and force: the longing to build.

Walden is the counterimage of this violent physical force, comparable to the constant juxtaposition in the Old Testament of the power of prophets to the force of kings. Walden represents the quiet rippling, through continents and centuries, of mind power, soul power, naked of excessive ownership. The tiny hut by the small Massachusetts pond, reflected in the careful mind and sometimes sassy Yankee spirit of Thoreau, moved Tolstoy, Gandhi, and other cultural movers far removed from Walden Pond. Thoreau's book reflects that instinctive, precultural longing in us all to shape our own space and shelter, to weave outside our skin some reflection of our mind. Henry's hut is perhaps the most mindfully constructed and recorded event of American architecture. The humility of its scale makes it a feasible norm for everyone and reminds us that the greatness of a person or country is not measured by meters but by motive, not by capacity to dominate but by capacity to meet.

Yankee ashram.

The foundations of Thoreau's cabin were sunk some four thousand years deep in Hindu classics,

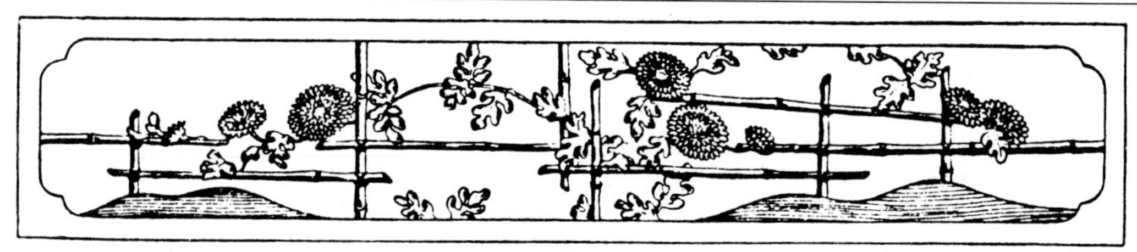

The "rama" in the space between door and ceiling was a favorite place for Japanese cabinetmakers and carpenters to display their mastery, often in bamboo only, or in wood and bamboo combined.

which he knew through Emerson. So it isn't surprising that today, a brief century and some later, his construction still feels fairly upright and sound, a house at home in the world. He was building a Yankee version of the Indian ashram or hermit's retreat, pitched just close enough to the careening track of American destiny to shout insights. Looking far back over Thoreau's shoulder, you can make out the mountain sage, familiar from both Chinese philosophy and dinner plates, living close to nature in a place remote enough from humans for a true encounter to occur on the rare occasions when someone actually showed up.

It is not fanciful to choose Thoreau's hut as cousin from American architectural traditions to the teahouse of the East. Both share the same ancestry, both are a small altar to friendship, both offer a model for a tiny, sane, handmade gesture on the edge of a ever more immense and mechanical madness . . . and both are equally dangerous if taken as a complete model to be swallowed in an uncritical gulp. The hazards of excellence were noted by Thoreau himself. "Our manners have become corrupted by communion with the saints," he warns, perhaps aware that he would in time become one of that dangerous—and endangered—species.

Perfection smothers invention.
A "climax" in biology describes a stable community of plants and animals that establishes a balance in time in a given climate and terrain after a long succession of species. In cultures, similar equilibriums occur, perfect embodiments of mind that sum up a prolonged evolution so well they stifle further development. "Man's earliest and most important victories are over his gods," André Gide declares in a long story, "Theseus" devoted in part to this theme.

Escaping the teahouse.
It's possible to approach the East on your knees, but reaching it is more likely with full leg and lung, your whole erect, creative equipment, because the East, most deeply considered, is you. As with all serious quests, the sought turns out to be the seeker. So creative spirits digest traditions to devise and invent, while dummies duplicate. A heavy-handed and exact external imitation of the Japanese teahouse and the ceremony there enacted will be light-years wide of the inner significance and possible present use in Western cultures.

Tradition only comes truly alive when we feel it from inside, sense the bones beneath the historical expression and discover relevant contemporary forms for ourselves. Blind imitation always cheats us of finding our own way. German castles reconstructed on the Hudson in New York, Zen temples carried stone by stone to northern California from Japan—these are whims that only waste precious fossil fuels in senseless transport. And "thou shalt not waste" is a prime commandment in the Zen Eclogue. Just as a scientific experiment must be re-enactable by others to be valid, part of the validity of an art form resides in the degree of its availability as a form for others.

Tight urban dwelling for our times.
The economy and simplicity of the teahouse has proved a possible model for centuries of Japanese. But what would be a true inhabitable teahouse for our time in the West? How can its miniature magnificence serve as a tight model in the space crunch of modern cities, which imposes the good necessity to design our dwelling area as neatly and efficiently as ships? We need a minimal architecture, dweller-made or adapted, with zero waste space. How can the teahouse serve as a suggestive, flexible inspiration in our efforts to roof everyone in a manner that shelters spirit as well as flesh?

Metaphysics of architecture: the Möbius strip.
Contemporary physics in the West has discovered that there are no isolated facts or objects, only "fields" and relationships. Buddhist psychology has always declared there is no "I." Only the crudeness of our observation can persist in maintaining the distinction between this swarm of sensation "inside" our skins that we call "I" and the encompass-

ing universe "outside." The Möbius strip is a Western form that reeks of the East and serves as a useful model to annihilate dualistic categories. If you do not already know the properties of the Möbius strip, cut a piece of paper an inch or so wide from a sheet of paper the length of this page. Twist it once and glue the ends together. Run your finger along one side and then one edge. Although clearly a three-dimensional figure, it has only one side and one edge. The Möbius strip is a physical model of the psychological fact that the other is "maybe us."

Minimal membrane: leaving the outside in. In Japanese architecture there is an analogous attempt to reduce the distinction between inside and outside. The house is wed to weather and garden, providing a minimal membrane that protects our frail flesh from elements without completely exiling us from leaves and air. Just as we put on shoes rather than try to cover the earth with leather, we can change our attitude rather than the event. In terms of shelter, the Japanese impulse is to put on another sweater rather than turn up the thermostat or build a bigger wall that can't be removed as easily as clothing when fine weather comes. The *shoji,* a sliding screen of paper in a light wooden frame that gives the Japanese home such flexibility in interior partitions, also helps blur the boundary between home and garden. The shoji leaves the roof without fixed wall, opening often onto a platform where the floor continues the room, but without roof, with potted plants further merging homescape and landscape. Many such directions for design can be found tucked into the teahouse. Its diminutive dimensions, natural materials, and attempt at harmony with the surrounding garden—which opens up an otherwise confining scale—are all qualities we can adopt in our shelter designs.

Appropriate nostalgia: Uncle Sam's cabin. It could be argued that nostalgia is one of the great glues of human culture, providing a reassuring continuity in change. In Yucatán, for example, you can admire the Mayan temples built in stone carved to imitate the woven pattern of the bamboo mat walls used to construct temple buildings before Chichén-Itzá and Uxmal lifted their glory above the jungles. Americans like to face cement block walls with knotty pine for the same reasons: The veneer of the old technology is used to cover and lend moral authority to the new. Marshall McLuhan, the Canadian who studied the cultural impact of our communications technologies, remarked that as a technology becomes obsolete it becomes an art form. Bamboo plays that role in the Japanese teahouse for the Japanese. Perhaps a log cabin would be a crude

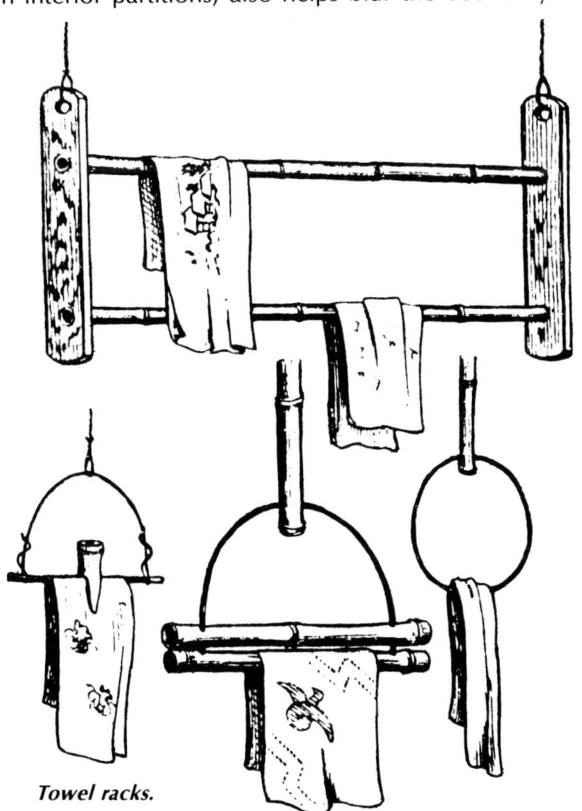

Towel racks.

Kitchen racks can be easily made with a few cuts in larger bamboos—probably moso (Phyllostachys pubescens) *in this sketch by Morse.*

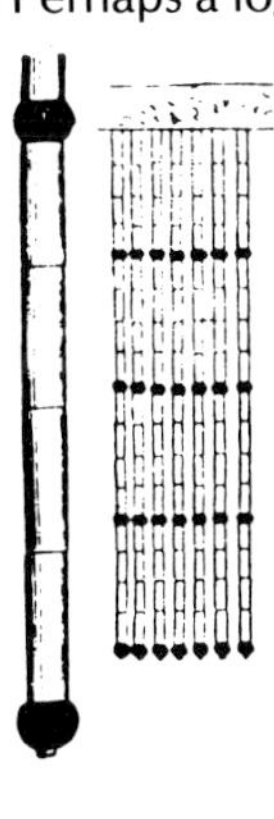

Door curtains are strung from 1-inch culm sections of very thin bamboo, or from waste branch sections of larger species. They make a cheerful clatter when passed through.

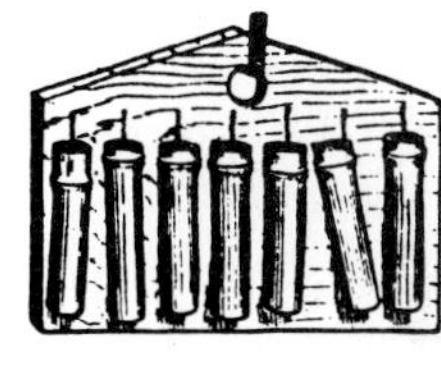

Gate rattle. Alarms of bamboo are of many varieties. In fact, the word bamboo, ***of Malaysian origin, probably derives from the early custom of throwing green bamboo sections into campfires to explode and frighten off wild animals.***

and partial equivalent for Americans: an architecture reaching deep into our psychic past as a nation, our robust infancy before we put so many layers of history on. As Japan becomes more industrialized and drifts ever further from all the things bamboo stood for, the teahouse becomes more attractive: a modest, rustic, uncluttered, nonmetallic, unmanufactured manageable human scale—everything, in short, the skyscraper in downtown Tokyo isn't.

Leisure shelter: bar bamboo, thatch therapy. It's interesting that the places built to pass a happy leisure are often so different from the architecture where we live and work. Taverns with bamboo and thatch in the decor hopefully suggest relaxed South Seas isles as far as possible from the busy sidewalk where we are. But even in the tropics, urban cantinas often choose bamboo and indigenous architectural forms—a circular shape and high-pitched palm roof—although expensive zinc, an oven in sunlight and a racket in rain, is chosen as the more stylishly "modern" roofing for homes.

Fifty billion fingers can roof the race. The universality of nostalgia suggests that instead of trying to stomp it out as a conservative idiocy stuck on the past, fixating on the rearview mirror in place of the road, we should perhaps try to find an appropriate format to replace the fake and cosmetic form it often takes—glue-on plastic veneers of knotty pine. A proper nostalgia could lend emotional energy to our path as we wend our way through inevitable changes, carrying the essence of our relevant past—like Theseus in the labyrinth, leashed gently to both his past—the entrance—and his future—the exit—through Ariadne's weightless thread.

Popular architecture, self-shelter distributing the heavy load of housing on all shoulders, could be the most sturdy and valid nostalgia, a constant in the midst of changes to fasten our future securely to our past. Participation is the key most likely to unlock the door to appropriate world housing, that enormous room for everyone—built on a small, human scale with affection, local renewable resources, and two trained hands.

The moral geometry of tea's easy way. "A social sacrament, a worship of the Imperfect, a tender attempt to accomplish something possible in this impossible thing we know as life." Okakura in his classic *Book of Tea* calls its cult a "moral geometry which defines our sense of proportion to the universe." The island isolation of Japan begat an introspective spirit, a national trait classically embodied in the Zen traditions, which trained one to see the smallness of the great—holding galaxies in the palm of the mind—and the immensity of the diminutive. The whole universe bends attentively above a blooming flower.

> *In one hair pore—ten thousand Buddhalands.*
> *Unnumbered beings there, whom Buddhas teach.*
> *How many hair pores on their face and hands?*
> *Ten thousand Buddhalands in each.*

This attitude implies that our behavior is the most intimate and true altar. The most complete art form is complete people related to others in complete friendship.

Rikyu's art of being in the world. Rikyu (1521–1591), the great grandfather of Japanese tea, was befriended and later condemned to death by his emperor-patron, the peasant dictator Hideyoshi (1536–1598). Through Rikyu's creative use of the form, multiplied by an early official encouragement, the ideals of the "cult of tea" have influenced Japanese domestic architecture immensely to the present day. The chaste simplicity of Japanese homes that at first strikes foreigners as barren is the stylistic heir of Rikyu's sixteenth century vision of "the art of being in the world."

The tearoom, built to accommodate five people, roughly 10 by 10 feet, was the main structure of a minicomplex that included, attached to it or separate: a side room *(mizuya)* where tea things were washed and arranged and a portico *(machiai)* where guests waited until invited to come along the tea garden path (the *roji*) to the house. The air of refined poverty surrounding the teahouse was partly owing to the fact that everything was scrupu-

lously clean—a virtue also inherited from Zen. "One of the first requisites of a tea master is the knowledge of how to sweep, clean, and wash, for there is an art in cleaning and dusting. A piece of antique metalwork must not be attacked with the unscrupulous zeal of the Dutch housewife."[12]

Zendo to teahouse.
The smallness of the teahouse, 4½ tatami mats or 100 square feet; the unrefined, impermanent materials; the absence of ornament or color, all these qualities contrast strongly not only with Western ideas of architecture but also with the classical Japanese temples and palaces that preceded the immense impact of this tiny shack. The high-class church-and-state official style of fifteenth century Japan was neither small, simple, nor austere. Pillars of wood 2 to 3 feet in diameter and 30 to 40 feet high supported enormous timbers framed for heavy tile. Elaborately decorated canopies, gilded baldachinos, walls full of frescoes and mirrors, statues in residence—the traditional "noble" architecture—was the opposite of everything the teahouse stood for.

The teahouse prototype was not temple or palace, but the Japanese Zen monastery meditation hall. This was a bare dormitory-laboratory for students of mind science, uncluttered and unlike the highly decorated temples of other Buddhist devotional sects, which were crowded with banners, prayer flags, images, children underfoot, and grandmothers in the kitchen cooking the banquet. In the *zendo,* or meditation place, a single flower and Buddha statue lent stark adornment to the altar, which later became the *tokonoma* of Japanese homes, the alcove where the single scroll or artifact is out on temporary display to grace the evening.

Cult of life.
The roji, or path from gatehouse to teahouse, was compared to the first stage of meditation. The keenest patrons of the early teahouses were the harried samurai and government officials who most needed to leave their hurried lives behind; the function of the roji was to sweep the town away. The tea master cleaned the house, the path cleaned the visitors. Twilight in pine needles, rough steppingstones, moss-covered granite lanterns . . . five bamboo culms complete the tranquillity, and the guest bends low to enter the door, no more than 3 feet high, where only a child could pass without bowing. The tea ceremony was the national balance, the cult of life as counterpoint to the cult of war and

martial arts in the cultural pattern of seventeenth century Japan; so samurai left their swords at the door . . . and bowed again to the picture or flower arrangement in the tokonoma before quietly taking their place. Only the sound of water boiling in the kettle breaks the silence. "The kettle sings well, for pieces of iron are so arranged in the bottom to produce a peculiar melody—the echoes of cataracts muffled by clouds, a distant sea breaking among rocks, rainstorms sweeping through a bamboo forest, or wind in pines on some faraway hill."[13]

CHAPTER 4.

1. Hidalgo 1974: 238–42.
2. The technical data of Narayanamurti's study is extensive, filling ten crammed tables and most of the thirteen pages of his valuable article. Request a copy from Dehra Dun (address, p. 305). M. Singh (1969) also contains relevant but highly technical data. See also "plybamboo" (p. 55) for more how-to details.
3. McClure 1953:7–11.

3a. Ibid.: 11; see also Limaye 1952.

4. Varmah 1980:20.
5. Orjala 1980.
6. Squibb 1953, 1957.
7. *New York Times*, May–June 1983. Ernesto Cardenal, the Sandinista minister of culture, claimed in the San Francisco Bay Area in December 1983 that the Nicaraguan armadillos like the new regime.
8. Darrow 1981:662.
9. Narayanamurti 1972:76–8.
10. Hidalgo 1974:234–7.
11. Nehrbass 1942.
12. Okakura 1906:36.
13. Ibid.: 35.

5. POPULAR ARCHITECTURE: A BAMBOO BACKBONE

The philosophy and know-how of the anonymous builders presents the largest untapped source of architectural inspiration for industrial peoples.

—Rudofsky

Of all the poverties of rich countries, one of the saddest is that the expense of housing makes it ever more difficult for young people to experience building. If we don't grow up building, we will never be fully at home in the act—or in our houses. We will set out too late to ever find our native hand. We will always build "with an accent" if we build at all.

For one countertactic to the architectural illiteracy of "advanced" countries, for rapid diffusion of the experience of self-shelter, we suggest dovetailing people's perennial need for a leakless roof with their chronic longing for a learning road. A College of Wandering and Popular Architecture, with its campus scattered in small holdings among the poor of the world, could earn its crooked Way by its declared homework: the creation of a lightly linked chain of simple shelters by the road where travelers can learn the local methods of building and gardening by direct participation with the neighboring people in these tasks. In return, the travelers can pass on to their hosts appropriate seeds from their own native lands and hints from the growing global inventory of appropriate tricks of human shelter and trades, salvaged from the technological folklore of the whole wide world. The most serious architectural question of our time, virtually unasked in the standard curriculums of schools, is: How can we roof the race in a lean and noble manner? Lean, so there be enough resource for everyone, none standing, shivering, beyond the embrace of the eaves. Noble, so that the dwellings of the peoples of the world cease being a cage and become a class and womb of spirit. How can we design a global building process in which we are all at home again in the intimate art of self-shelter? How can we make villages and towns reflect the ongoing sturdy learning and joyful creation of the housed?

A Pattern Language (Alexander et al.) is one of the few architectural manuals designed both to come to philosophic grips with these questions and to provide people with the means of answering them. It needs to be rewritten and mainly *drawn* for the villagers of the earth, but it is an excellent point of departure for rethinking our conventional notions about world and personal shelter.

To awaken respect for the immense *fact* of popular architecture—still almost invisible in our

"Beehive" house, Kenya.

"Cathedrals" of bamboo, ceremonial structures providing space for sacred or secular business of the tribe, have until now been mainly male clubhouses. They make use of bamboo's monumental dimensions for shelters that could be models for housing in a crowded world. In the framework of a Dawi, *a men's ceremonial house from the Purari delta region of New Guinea, the gable ends may be 80 feet high.*

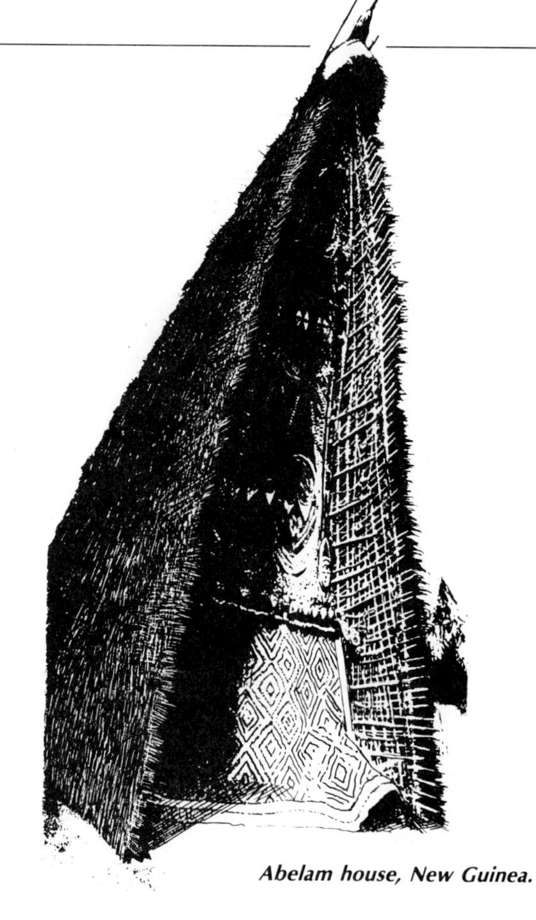

Abelam house, New Guinea.

Toradja, the Celebes.

schools—we present first of all two classic statements in defense of heartchitecture, architecture with heart, blueprints for sane planet shelter by Rudofsky (1964) and Fathy (1973) "condensed for modern readers." Their reflections serve to frame a sketch of *guadua,* the bamboo backbone of popular architecture in Colombia, now threatened with economic extinction. The precarious present position of this resource points out a primary need for any durable popular shelter designs in our crowded time: "Permashelter" for people implies revegetation. Planting is the rational prelude to building. Without replacement of resource, a design for popular architecture is a bird with one wing. Happy rafters aren't possible for long without a pact with the planet to put back what we take.

Nonpedigree architecture.
"Conventional architectural history skips most of the globe, most of the centuries, and all of the common people's nonpedigreed architecture—so unknown we have no term for it—to focus on a who's who of architects who commemorated power and wealth, a brief anthology of buildings of, by, and for the privileged. The building blight of industrial societies also emphasizes the talent of the individual architect, while the serenity of the best architecture in so-called underdeveloped countries is anonymous, a result of communal enterprise, 'not produced by a few intellectuals or specialists, but by the spontaneous and continuing activity of a whole people with a common heritage, acting under a community of experience.' "[1] [Pietro Belluschi]

Town framing: art implies finitude.
"We choose flat, featureless country, erasing flaws in the terrain with bulldozers. Communal architecture is attracted to rugged country and the most complicated configurations of landscape—Machu Pichu, Monte Alban, and the craggy republic of monks on Mount Athos are a few familiar examples of this practice. Security was doubtless one motive for this tendency, but perhaps even more important was the need to define borders. Many old-world towns are still solidly enclosed by moats, lagoons, or walls whose defensive value has long been lost, but which serve to thwart undesirable expansion. Urban derives from *urbs,* the Latin for 'walled town' and suggests that a town which aspires to being a work of art must be as finite as a painting, a book, or a piece of music. Our towns, in contrast, grow unchecked, an architectural eczema which defies all treatment. Ignorant of the duties and privileges of older civilizations, we accept chaos and ugliness as fate. We credit architects and other experts or specialists with exceptional insight into problems of living when most of them are preoccupied, in fact, with problems of business and prestige. We regard the 'art of living,' neither taught nor encouraged among us, as a form of debauch—unaware that its tenets are frugality, cleanliness, and a general respect for creation, not to mention Creation."

Vernacular anticipations of "modern" design.
"The diligence of historians, invariably emphasizing architects and their patrons, has obscured the achievements of anonymous architects, whose good sense in handling practical problems, sometimes transmitted through a hundred generations, left shapes of houses which seem eternally valid, like the shapes of their tools. Many audacious 'primitive' solutions anticipate our cumbersome technology. Many a feature invented in recent years is old hat in vernacular architecture—prefabrication, standardization of building components, flexible and movable structures, floor-heating, air-conditioning, light control, even elevators. Long before modern architects envisioned subterranean towns under the optimistic assumption of protection from future warfare, such towns existed—and still exist, on more than one continent."

The comforts of inconvenience.
"Ironically, the urban dweller periodically escapes his splendidly appointed lair to seek physical or mental renewal from 'primitive' surroundings: a cabin, a tent, a fishing village or hill town abroad. Despite our mania for mechanical comfort, our relaxation hinges on its very absence, a confession that life in old-world communities is singularly privileged. Instead of hours daily spent commuting, only a flight of stairs may separate one's workshop or study from living quarters. Environments seem to remain most alive for those who shape and preserve them themselves, largely indifferent to 'improvements.'

"Just as toys are no substitute for affection, no technical contrivance can replace 'livability' in homes. So also, the general welfare is not to be subordinated to the pursuit of profit and progress. 'Expecting every new discovery or refinement of existing means to contain the promise of higher value or greater happiness is extremely naive. It is not at all paradoxical to say that a culture may founder on real and tangible progress.' [Huizinga]

"The philosophy and know-how of the anonymous builders presents the largest untapped source of architectural inspiration for industrial peoples. The wisdom of communal architecture goes beyond economic and esthetic considerations, touching the far tougher and increasingly troublesome problem of how to live and let live, how to keep peace with one's neighbor, both in the local and universal sense."[2]

ARCHITECTURE WITH HEART

In *Architecture for the Poor,* Hassan Fathy finds our time a century without signature, eroding traditional architectures in the provinces and replacing them with International Ugly, the global nonstyle, nonsense of world building, nineteen eighty-now.

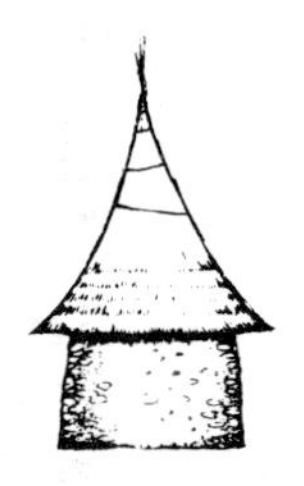

1.

Taxonomy of African popular architecture. For the thousand or so tribes of Africa, the style of individual houses (and the format of their customary arrangement in complexes) can be conveniently presented in 32 categories. The drawings by Denyer, below, deserve selective experimental embodiment on one location. Detailed descriptions will be found, by number, at the end of this chapter.

His anatomy of cultural rot led him to a renewed respect for traditions, especially in architecture. There is world relevance in his plea to Egyptian authorities to prop the main beam of traditional family architecture, which he found to be an intimate individuality in responsible dialogue with tradition.

Through a domestic architecture suffused with intense owner participation, deliberately nurturing

2.

local craft traditions of carpenter, mason, furniture builder, stained-glass maker, or whatever local excellence, the whole village becomes a homogeneous work of art, a source of both communal pride and communal livelihood. A scattered constellation of such villages, interrelated but subtly distinct through their faithful reflection of the local genius in woodcarving or other shelter-related art, creates a region with its own unique and authentic cultural character in delicate balance with local materials, climate, and land.*

"Who knows only his Bible, knows not even his Bible," Matthew Arnold warned Victorian bigots. He who knows only bamboo will never notice all the corners where it fits in twentieth century Western cultures. So occasionally, to suggest the cultural soil and site, we are driven to the art of relevant digressions. Fathy's tender analysis of the menaced village, his methods for getting out of the way of the people's own creative energy, his thoughtful defense of both tradition and the individual against the tasteless global soup of mass production and mass consciousness—these qualities make *Architecture for the Poor* an important document in designs for bamboo shelter development everywhere—even though Fathy never mentions bamboo.

Beneath its apparent subject, this book in your hands at this moment is more about those hands, which are always with you, than about bamboo, which is probably less present to your life. Oriental

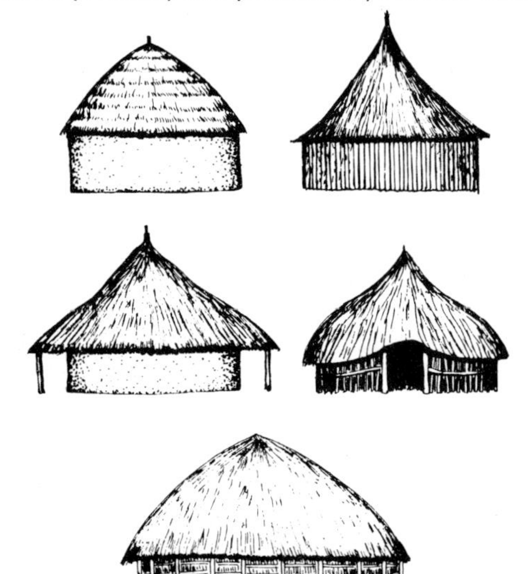

3.

**Editor's note:* Rich inspiration, especially for any Third World housing ministry or architectural activists anywhere, can be found in Denyer (1978), *African Traditional Architecture,* which extends European and Mediterranean studies of vernacular architecture to the thousand or so tribes of tropical Africa. "More architects are turning to vernacular architecture because it obviously satisfied psychological needs far better than most modern suburban settlements."[3] We have drawn on Denyer's "Taxonomy of House Forms," seventy tiny drawings of the wonder of African natural architecture, to punctuate the Egyptian experience of Fathy below. Descriptions of the thirty-two styles drawn are found at the end of this chapter.

traditions of bamboo use happen to be a very instructive example of a durable and creative relation between people and resources. Through centuries of cultural evolution, bamboo remained relevant. Change did not make it obsolete. New conditions revealed new facets of its precious friendliness to human need. It has remained a flexible mate to the quick cunning of human fingers for so long that a deeper and deeper feeling, a more mossy and ma-

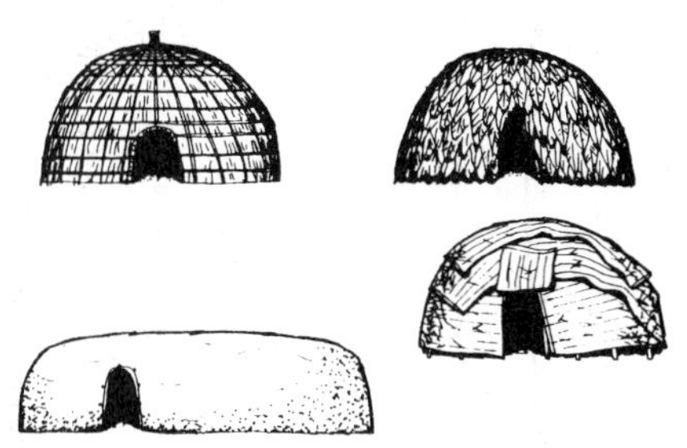

4.

ture mellowness, has had centuries to ripen for the plant in oriental affection. In our time, when three out of five marriages end in divorce, when friendships share roughly the same fatality, when dizzy obsolescence is the design norm, and the same nervous, searching, unbalanced condition prevails everywhere in art, commerce, faith, or fad; in these times, we have much to learn from the steadiness with which the oriental heart woke up beating in unbroken fondness for bamboo for 1,500,000

5. *6.*

mornings—to cast our mind's eye back only some 4 millennia to, say, 2000 B.C. And those are only the mornings that fall within the small glance of history. We must always remember that, like the 10 percent of the iceberg that's above sea level, history is only the small tip of human reality that thrusts its visible fraction above oblivion into our range of perception. This book, like most, encompasses mere history; but bamboo has been standing beside people far longer than that.

And yet, 25 percent of all people ever alive are alive in our crowded now, so history is more densely present, in a sense, than it has ever been in all history or all prehistory. Human action, love, friction, creation, death—the whole constant upheaval of our natural condition is more everywhere than any humans before us have ever had to encounter.

All cultures, like their comprising members, are mortal. But cultural designs that do not acknowledge this new density of history are doomed to a greater brevity than necessary. We are threatened with shipwreck not by the turbulence of the surrounding ocean, but by conflicts of the crew aboard who are so rocking the boat that even those quietly trying to withdraw from the fray will drown with the belligerents. We are in need indeed of cultural forms less jerky than our hyperlively human consciousness, to steady and ballast the voyage.

The shelter equation: hand + land = house.
Bamboo is one resource that fits this need. Its history contains hints for use of other available materials, sometimes with bamboo, sometimes without. Bamboo's own future history will receive similar

7.

hints, in turn, from many directions, the nearest of which lies, mostly unnoticed, beneath all our feet. Broad distribution is one result of bamboo's vitality, but bamboo isn't as omnipresent, globally, as are people—or mud. But not to get lost in mud either and forget the spectrum of earth's resources, we should keep in mind that the basic equation contains only two constants, human work and human shelter: hand + land = house. The variable, the land and its local building resources, presents a perpetual feast of options where no crude sameness reigns.

Mud.
In the 1940s, Hassan Fathy discovered that the rural housing problems of Egypt were not without exit: They had already been solved by her past. Mud brick vaults, built without scaffolding, were an ancient technique still alive in one region. Fathy set about training masons and securing commissions to explore this viable native reply to the costly and fashionable reinforced concrete, roughly seven times more expensive, in his time and place, and therefore light-years beyond the economy of the poor villager. He proposed owner participation and mud as a truly appropriate architectural reply to Egypt's housing shortage, and after a number of scattered demonstration buildings, he was given the job of building a village for seven thousand people being relocated by the government from the site where, for fifty years, they'd been settled on top of

8.

an important archaeological dig, plundering antiquities, and melting down priceless gold artifacts thousands of years old. Their path from Old Gourna to New Gourna is a walk rich with instruction.

The government bureaucracy that funded, then frustrated, Fathy's attempts to collaborate with the people in creating sane and beautiful low-cost housing is a familiar dragon that has devoured large pieces of many lives in many countries other than Egypt. His tale is tenderly told and resonates on numerous levels as the best introduction to architecture with heart we could recommend. It has nothing to do with bamboo, but everything to do with the proper spirit, directions, and problems of bamboo development in the poorer countries of the tropics. How to be a Human Being, and remain one while wandering corridors of power trying to pry loose some *piasters* for the poor. How to maintain calm in face of, in spite of, the mindscapes of government officials. These are among the topics that thread Fathy's magic carpet of architectural concern for the world's homeless, which will fly your heart to the mud villages. Be warned that it may be hard to get home again to mesh easily once more with your old assumptions.

The missing signature.

"Every people that has produced an architecture has evolved its own favorite forms, as peculiar to that people as its language, dress, or folklore, beautiful children of a happy marriage between their imagination and the demands of their countryside. No one could mistake the curve of a Persian dome

9.

10.

11.

and arch for the curve of a Syrian, Moorish, or Egyptian one. No one can fail to recognize the same curve, the same signature, in dome, and jar and turban from the same district. It follows, too, that no one can look with complacency upon buildings transplanted to an alien environment.

"Yet in modern Egypt there is no indigenous style. The signature is missing: the houses of rich and poor alike are without character, without an Egyptian accent. Style is looked upon as some sort of surface finish that can be applied to any building and even scraped off and changed if necessary. The graduate architect imagines a building can change its style as a man changes clothes. It is not yet understood that real architecture cannot exist except in a living tradition, and that architectural tradition is all but dead in Egypt today. As a direct result of this lack of tradition, our cities and villages are becoming more and more ugly. Every single new building manages to increase the ugliness, every attempted remedy only underlines it more heavily."

How international ugly invades villages.

"On the outskirts of provincial towns, the ugly design is emphasized by shoddy execution. Cramped boxes of assorted sizes in a style copied from poorer quarters of the metropolis, half unfinished yet already falling apart and set at all angles to one another, are stuck up over a shabby wilderness of unmade roads, wire and washing lines hanging over

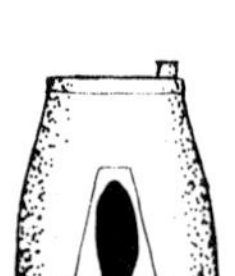

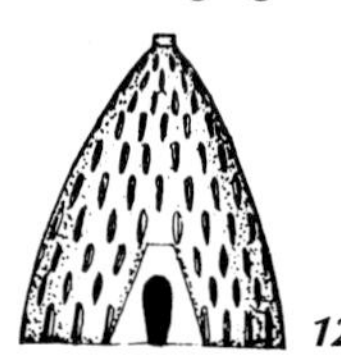

12.

dusty chicken runs. In these nightmare neighborhoods, a craving for show and modernity causes the owner to lavish money on the tawdry decorations of urban houses, while being stingy with living space and denying himself the benefits of real craftsmanship.

"To flatter his clients and persuade them that they are sophisticated and urban, the village mason starts to experiment with styles that he has seen only at second or third hand and with materials that he cannot really handle with understanding. He abandons the sure guide of tradition and, without the science and experience of an architect, tries to produce 'architect's architecture.' What he achieves is buildings with all the defects and none

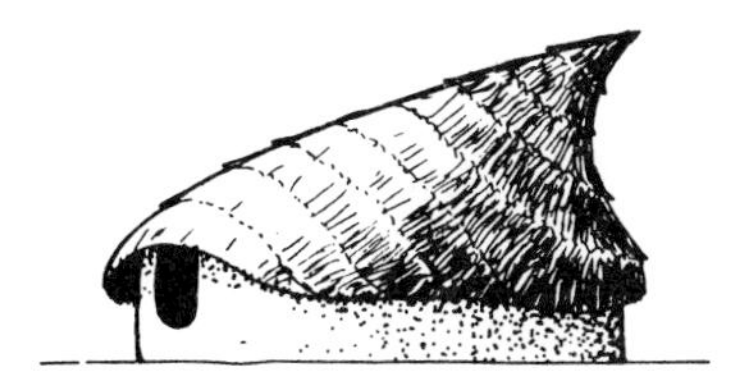

13.

of the advantages of the architect's work.

"Thus the work of an architect who designs, say, an apartment house in the poor quarters of Cairo for some stingy speculator, incorporating features of modern design copied from fashionable European work, will filter down over a period of years through the cheap suburbs into the villages, where it will slowly poison the genuine tradition. A scientific examination of this is required to reverse the trend toward bad, ugly, vulgar, and inefficient housing in our villages."[4]

The function of tradition.
"To be alive is to make decisions, and the most subtle decisions are called for when a person makes

14.

something. Habit may release a man from the need to make many less interesting decisions so that he can concentrate on the important decisions of his art, like a musician scarcely following each finger as it produces a note. Tradition is a social habit releasing the artist from the distracting and inessential so that he can give his whole attention to the vital."[5]

Indulgent innovation = cultural murder.
"Some problems are easy to solve in a few minutes, others need a lifetime; in each case the solution may be the work of one person. Other solutions may require generations, and this is where tradition has a creative role to play. Some traditions—in bread making or brick making, for example—go back to the beginnings of human society and will perhaps exist till its end. Others, though they appeared only recently, were in fact stillborn. Innovation must be a completely thought-out response to change in circumstances and not indulged in for its

15.

own sake. The individual can be happily relieved of many irrelevant decisions by tradition but will be obliged to make others equally demanding to keep tradition from dying on his hands. In fact, the further a tradition has developed, the more effort each step forward in it costs."[6]

Reactive creation.
"To willfully break a tradition in a basically traditional society like a peasant one is a kind of cultural

16.

murder. The architect must not suppose tradition will hamper. When the full power of a human imagination is backed by the weight of a living tradition, the resulting work of art is far greater than any that an artist can achieve alone with no tradition to work in or through willfully abandoning tradition. One person's effort can bring about an altogether disproportionate advance, if the person is building on an established tradition. It is rather like adding a single microscopic crystal to a solution that is already supersaturated so that the whole suddenly crystallizes in a spectacular fashion. Yet this artistic crys-

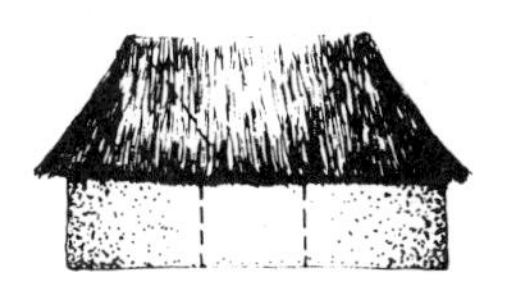

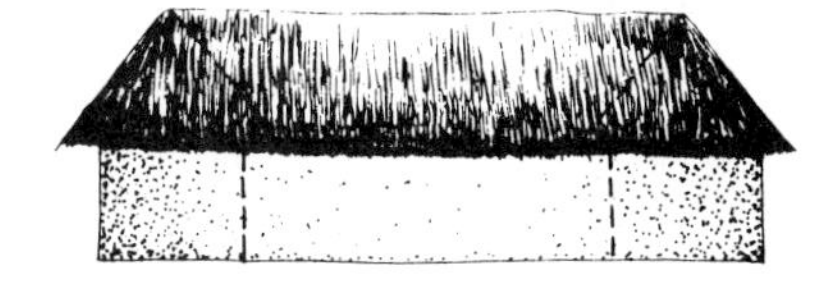

17.

tallization is an act of creative reaction and reactive creation that has to be perpetually renewed.

False synthesis in quest of applause.
"Architecture is still one of the most traditional arts. However hard the architect strains after originality, by far the larger part of his work will be done in some tradition or other. Why then should he despise the tradition of his own country or district; why should he drag alien traditions into an artificial and uncomfortable synthesis; why should he be so rude to earlier architects as to distort and misapply

18.

their ideas simply to gratify a selfish appetite for fame?"[7]

Mass solutions, mass culture death.
"If in your dealings with others, you consider them as a mass, abstracting and exploiting the features they have in common, you destroy the unique features of each. The advertiser who plays upon the common weaknesses of humankind, the manufacturer who satisfies the common appetites, the schoolmaster who drills the common reflexes, each in some way kills the soul. That is, each, by overvaluing the common features, crowds out the individual ones. Largely unchallenged, the promoters of sameness have eliminated the tradition of individuality from modern life."[8]

"In medicine no one expects the doctor when dealing with the poor to try to mass-produce operations. Why, then, when a passing infirmity like a sore appendix is honored by careful personal treatment, should a permanent necessity like a family house be accorded any less? If you chop off appendixes by thousands with a machine, your patients will die, and if you push families into rows of identical houses, then something in those families will

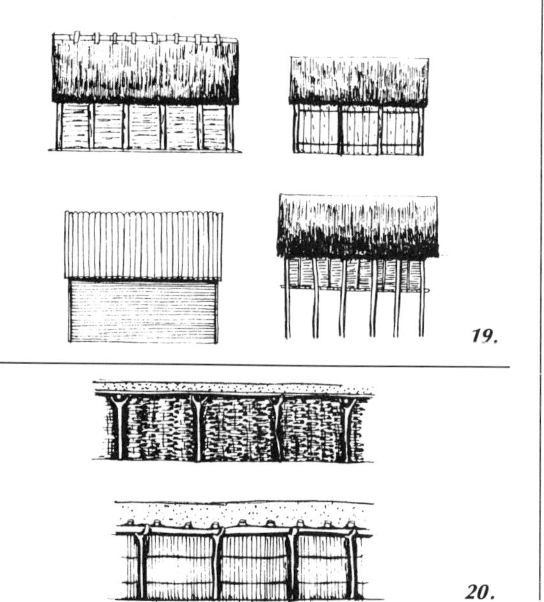

19.

20.

die, especially if they are poor. The people will grow dull and dispirited like their houses, and their imaginations will shrivel. If governments regard people as millions to be shoveled into boxes like loads of gravel—inanimate, passive, always needing things done to them—they will miss the biggest opportunity to save money ever presented.

A creative economy of housing: to create conditions for creating.
"People have minds and hands of their own. A human is an active creature, a source of initiative, and you no more have to build a person a house than you have to build nests for the birds of the air. But the government would still have a very big part to play in a building revival stemming from the individual family. It would have to create the conditions for such a revival to flourish; clearly the conditions do not exist now, or there would be no problem."[9]

Government attitudes and family housing.
"If only governments will change their attitudes towards housing, will remember that a house is the

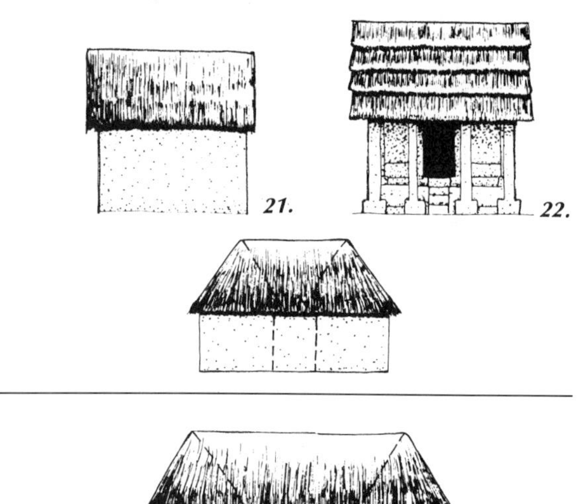

21. 22.

23.

visible symbol of a family's identity, the most important material possession people can ever have, the enduring witness to their existence, its lack one of the most potent causes of civil discontent and conversely its possession one of the most effective guarantees of social stability, then governments will recognize that nothing less will do than the utmost that can be given in thought, care, time, and labor to the making of the family house. They will recognize that one of the greatest services they can render their people is to give each family the chance to build its own individual house, to decide at every

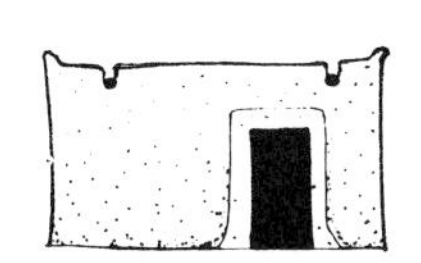
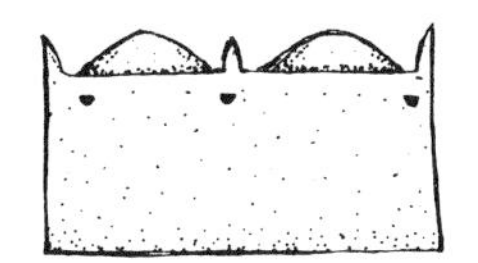
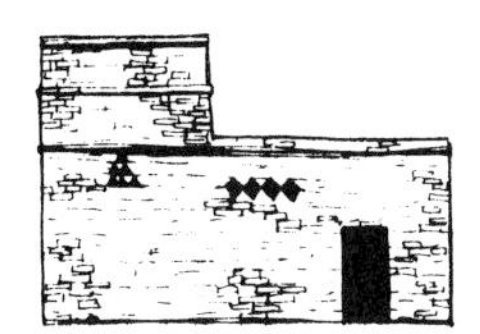
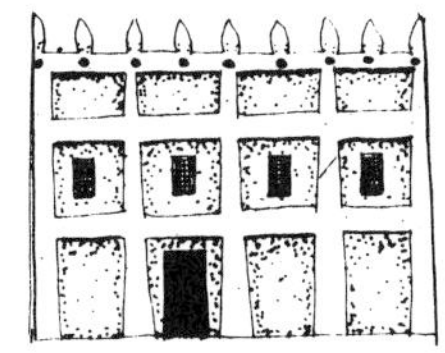

23.

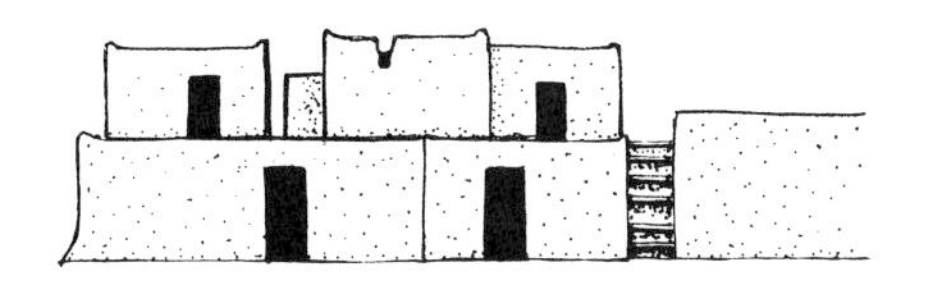

24.

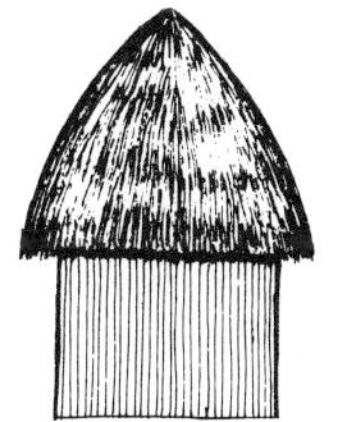

25.

26.

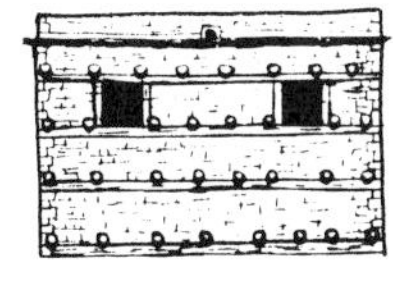

27.

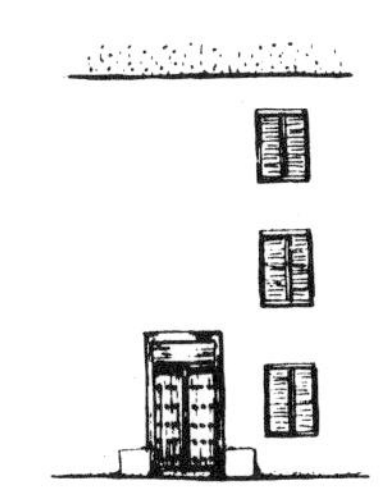

28.

29.

30.

31.

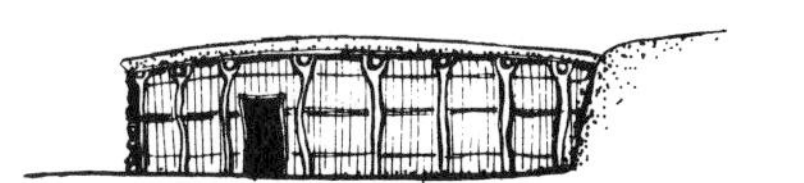

32.

stage how the house is to be so they feel the finished building is a full expression of their family's personality.

Peru 1959: Christmas birth of a bastard village. "Something happened in Peru that is a lesson to all planners everywhere. In 1959, a hundred thousand people living in the slums of Lima decided to build a complete new suburb for themselves on empty land some way outside the city. Knowing that the authorities would be unsympathetic, they planned the whole operation in secret, like a military maneuver. They divided into four groups, each to have a district in the new suburb, and each with its own leader. They drew up plans, laying out the suburb with roads, squares, schools, and churches. Then on the night of 25 December they marched, carrying materials with them. They reached their target and between 10 P.M. and midnight put up a thousand temporary houses, sited according to their plan, and each quarter had a church. By midnight the authorities had noticed what was going on, and police were rushed out to stop the squatting. Despite this, five thousand people (out of a planned hundred thousand) stayed and still live there at Ciudad de Dios, 10 miles from Lima. If five thousand people can house themselves in one night in a well-laid-out suburb planned by themselves in the teeth of official opposition, what could they do with official encouragement?"[10]

GUADUA ANGUSTIFOLIA

> Guadua is more conspicuous in the local economy of the areas of Colombia where it abounds than is bamboo in any area of the Far East with the possible exception of Java.
> —F. A. McClure[11]

Authentic architecture and misapplied modernity.

> We've always limited the history of architecture to the analysis of a few cultures, represented by the most refined expressions of genius in great religious monuments, palaces, or mansions—while ignoring the roots found everywhere, in the most elemental and humble popular architecture, born spontaneously, profoundly felt, always a clear expression of the spirit of the people.
>
> Although this has been the dominant architecture in our surroundings, we fool ourselves by ignoring it, creating increasingly narrow concepts of beauty and utility. In the name of a misunderstood modernity, we've insisted on applying new techniques, often superficially known, strange to the surroundings, and quite incompatible with the training of the local laborers, who know neither the materials nor the most elementary methods of working them. Workmen adjust slowly to complicated systems and strange ways, comprehensible only after knowing local materials which lend themselves to empirical exploration and appropriate adaptations of new techniques.
>
> Contrary to the neglect lavished on popular bamboo architecture in Colombia, we believe that only a depth study of our own traditions and local materials will make possible the erection of an authentic modern architecture in our midst.
> —Londoño: *La Guadua*

World's most durable, West's largest bamboo. Of seven superbamboos from around the globe selected for special scrutiny by McClure, only *Guadua angustifolia* is native to the Americas.[12] The world's most durable bamboo, and the New World's largest species, it is also blessed with excellent fiber for paper pulp. As skyscraper assistant in scaffolds and poured concrete forms in Colombia and Ecuador, it costs a third as much as wood for these purposes, according to the dean of an architecture school in Guayaquil. It has long been the most common material for low-cost, instant architecture at the city's edge, wherever it grew close or could be transported by raft. Overuse without replanting threatens this multihandy species with virtual economic extinction by the year 2000. With proper care on the part of the people in its regions of growth, guadua could be strategically grown as a hillside crop that also helps hold soil against erosion. It could provide cheap local building material and raw matter for small crafts as well as abundant educational material for schools.

The extended treatment of the species here provides a sample study of bamboo's cultural past, present problems, and possible futures in one region that could be applied to other bamboos in other areas of the West. This sketch of guadua is also intended as half of a diptych—those hinged, twin paintings on ancient altars—which includes the temperate oriental bamboo, moso *(Phyllostachys pubescens),* whose care we examine with care as a similar sample in Chapter 8, pp. 226–231.

From Noah till now.
"The Pantagoras tribe which populated the northern part of Caldas, in western Colombia, had a legend that only one Indian survived the Deluge and for many years led a solitary and sad life, until one day God, taking pity on him, turned a bamboo sprout into a woman and gave her to him as his mate. Since then, it is said, the bamboo has given man shelter."

Guadua constructions perch on steep hillsides, and are up to seven stories high. They are usually occupied initially on the street level, with the rest of the structure completed as needed.

Spanish invasion: forts and giant flutes.
"That is how the guadua became a necessary element of Indian survival, especially during the era of Spanish conquest, when they were forced to build guadua forts to protect their families and homes from the greed of the invaders. These forts, built around their villages, were made with a wide rampart using a double wall of guadua stuffed with dirt and rocks, leaving small openings in the lower wall for shooting their arrows and darts. On the upper wall, they would build small temporary bridges covered with guadua mats for throwing rocks or red-hot coals and shooting arrows at the enemy. To scare off the Spanish conquerors, they displayed the heads of their enemies on the upper ends of the guadua canes and drilled holes in some internodes for giant flutes. When the wind blew, the noise would be absolutely terrifying, according to some chroniclers of that era."

Armor and portable stockades.
"The Spanish soldiers not only learned from the Indians how to protect their arms and legs from the arrows by covering them with sections of guadua, but also to use portable stockades of bamboo to protect advances on the forts to set them on fire . . . It has been 400 years since the Spanish conquest. During these four centuries, the guadua has remained an essential element in the socioeconomic development of many parts of western Colombia where this plant grows wild, such as Antioquia, Caldas, Risaralda, and Valle."[13]

Vast virgin groves.
According to Oviedo, the distinguished Spanish chronicler of colonial times *(Historia Natural de Las Indias),* the fertile and well-watered region known as Quindiu, now the Department of Caldas, was once a great forest of *Guadua angustifolia* with scarcely another kind of plant to be seen. The imagination pictures, as a view from the mountain tops in those days, an immense, undulating sea of deli-

cate green, made up of millions upon millions of giant, fernlike plumes of guaduas gently swaying in the breeze; seen from within—a great cathedral with slender rays of sunlight filtering through green windows above.

Colonial uses.
The stems of guadua served a multitude of uses for the early settlers. With such versatile material it was possible, using only a machete, to fashion practically everything necessary for the home from floor to roof, complete with furniture—tables, chairs, beds—even the cradle and utensils for the table. Water was brought to the kitchen in *tarros* of guadua or in aquaducts formed from its hollow stems. Nodes with the branch bases still attached made hooks for hanging the garments in order. Outside, sturdy fences of guadua were constructed as needed to guard crops from wild animals and keep the livestock from straying afield.

Guardian of watersheds.
The land was cleared at great labor, for the guadua, a sturdy and well-anchored plant, resists destruction and decay. Repeated burnings were required to kill the stumps, and even dead they persisted eight to ten years. Corn and other crops were planted in the fertile soil among the stumps, and grasses gradually established themselves, making cultivation more difficult. The grasses in turn supplied pasture for the livestock. More land was cleared, and little by little the *guaduales* (guadua groves) were pushed back from the habitations, the fields became broader, the houses more numerous. It became evident that land the quadua inhabited by preference was the most productive, and so destroying the guaduales became the major concern of the more industrious and ambitious settlers.

As the guaduales were cleared from larger and larger areas, it was discovered that the flow of water in the streams became intermittent: flooding more violently after rains, then subsiding quickly to ever lower levels. And so the idea developed that the guadua is a natural guardian of the water supply. But instead of encouraging its growth on the hillsides where, as we know today, it would do most good, it was kept in the valleys, along the watercourses, while the rolling hills were devoted more and more to pasture, the foundation of the lucrative livestock industry.

Today, the famous Quindio region is traversed daily by many airplanes, from which it is possible to get a view denied the hardy settlers who traveled laboriously on horseback, in the shadow of the guaduas. Today only the watercourses are still shaded and fringed by the giant green plumes, while the hills are smooth pastures where tens of thousands of cattle roam and graze.

Water carriers with guadua canteens.

Weed or resource?
The guadua is still the main source of building material for farm structures and fences of the land, but it remains a problem as well—a "weed" that invades the pastures from every margin. In the overall economy, it is considered a problem more than a resource, for constant labor must be expended to keep it from invading and occupying the pastures completely. The market value is little more than the cost of felling and transporting it to the centers of trade. In most areas, guadua for fuel may be had for the labor of cutting it. And in many guaduales the rate of harvesting is so low that great culms die of old age, clog the forest, and decay without being used.

The natural habitat of the guadua appears to be confined chiefly to the area where coffee flourishes. And although some efforts have been made to establish the plant outside its natural area of distribution, the dominant preoccupation has been to limit it as an agricultural measure, so the cultivation and management of guadua plantations is an art as yet undeveloped. One landowner, when encouraged to grow guadua, replied, "I'm not going to cultivate a plant I spent twenty years of my life trying to exterminate."

Where river transport, by rafts, is lacking, costs limit the use of the guadua rather narrowly to the areas where it is produced. Guaduas do reach distant places for construction use, but the price is increased several hundred percent by the outlay for transportation and handling commissions.

Distribution.
Although its abundance has been severely reduced

Guadua rafts in Colombia are used to transport people and goods. In the Orient, bamboo rafts of up to 10,000 culms are built in a series of sections which overlap like tiles on a roof, the front of each miniraft resting on the aft of the raft section ahead.

by clearing to establish pasture lands and cultivated fields, guadua is still the most conspicuous feature of the native vegetation in much of the area. Many people say it "grows like a weed" and that no effort has been made by anyone to establish new groves. The fact that this plant grows spontaneously in all habitats in the region from riverbanks and moist ravines to steep slopes and even the tops of the hills —the most arid sites within its altitudinal range—is ecological evidence that it is "at home" here. It seems likely that this is one of the ancient centers of its distribution, if not the original one. The limits of its present distribution are not fully known. It appears to decrease in abundance as one approaches Manizales from Chinchiná. Manizales, at 2,150 meters (6,987 feet), may represent its altitudinal limit in this area. In Ecuador, the upper limit of this species is generally around 5,000 feet, roughly 1,500 meters.

Wide role in local economy.
Guadua plays a very important part in the local economy of the region. Its culms are found in almost every structure, rural or urban—residences, coffee-drying platforms, coffee and sugar cane mills *(beneficios),* corrals for farm animals, and so on—often to the complete exclusion of all other building materials. Guadua supplies the fence posts and telephone poles as well as the aqueducts that bring water from hillside springs to the rural housewife's kitchen. In fact, guadua is more conspicuous in the local economy of areas in Colombia where it abounds than is bamboo in any area of the Far East, with the possible exception of Japan. The importance of guadua in the local economy is reflected by the criteria by which a *finca* (farm) is judged. It is said that among the first questions asked by a prospective buyer are: "Has it a good water supply?" and "Does it have plenty of guadua?"

Local cultivation lore.
As with bamboo in China, the size and lasting qualities of the culms of guadua in Colombia are said to depend upon the nature of the habitat in which the plant grows. On rich, moist, lowland soils, the culms reach their greatest development in size and thickness of wood; while on less fertile, less moist soils, the culms are smaller in size with thinner wood. The latter, however, are said to give longer service under average conditions. Here, as in other Latin American countries where this and other species of Guadua grow, it is held as self-evident that a definite relation exists between the time of cutting and the durability of the culms. The belief is that culms cut in the increase of the moon *(creciente)* decay promptly and are promptly attacked by bamboo beetles, locally called *gorgoja* (*Dinoderus minutus* and other species known collectively as *"la polilla"* in much of Latin America). Those cut in the decrease of the moon *(menguante)* are considered to be endowed with the fullest degree of durability. (However, see pp. 221–222 on beetle tests in Puerto Rico and Trinidad.)

The idea of utilizing more fully the great resource that guadua offers has occurred to many progressive persons in Colombia, but the obstacles pointed out have prevented the realization of definite plans.

Guadua is a plant ideally suited to the conservation of the soil in those areas where it is native. It grows well on slopes of all gradients, is free from destructive disease and insects, and offers a product of importance in the local economy. It is important that guadua's potential be developed.[14]

FLEXIBLE SHELTER

Urban planning and the ephemeral family.
"My first interest in guadua was pictorial. These constructions were a good theme for paintings and drawings because of their geometrical structure and the way white houses with splashes of primary color leapt up from an exuberant vegetation. Then the architectural and urban solutions implied in the dwellings began to interest me. Later I was able to compare solutions seen in Caldas with the most modern ideas on the subject and they coincided in a number of ways: in the intention to have dense urban nuclei; to give the street other functions apart from mere space for traffic; to construct in the most mobile, flexible, and economic way, most in accord with our evolving world.

"The use of guadua has lasted until our own time, although it is now being replaced by more durable and expensive materials whose very durability serves to stratify existing inconvenience. They don't permit flexible adjustment to urban, social, or familial changes. Before, everyone used guadua; now its use is marginal, temporary, without proper appreciation of its possibilities. Colombia is a country in process of rapid industrialization, with all the advantages and inconveniences which this implies, particularly with respect to urban sprawl. Our cities are growing uncontrollably, through country people arriving in search of work—or in search of security lost in rural areas during bitter political struggles. Rural villagers crowd into urban slums, and Colombian government agencies in change of such problems invest in high-cost, long-lasting materials out of range of the people's resources.*

"Guadua construction, lasting roughly forty years, corresponds to the cycle of a family's evolution, beginning with a newly married couple; then a few children; then, a few years later, returning to the initial pair. Prices of guadua houses are on a level with the extremely limited means of the inhabitants. The construction is rapid and later changes can be made without great time, cost, or labor. As a rule, they are constructed by the inhabitants with help perhaps from a carpenter but never from engineer, architect, or other professional.

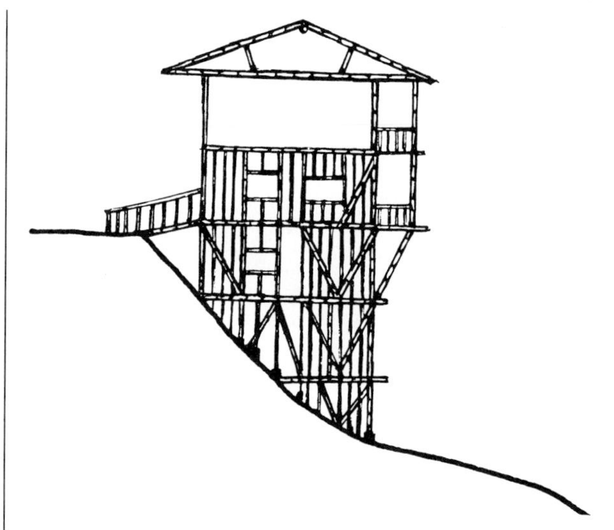

"Light and cheap, easy to change, easy to raise or remove, guadua constructions seem particularly appropriate in our changing urban environment where buildings are knocked down to widen roads, and the city is constantly rearranging itself in an effort to function more smoothly."[15]

Manizales: Bambooville.
Barrio Galan, the bamboo shantytown on the north edge of Manizales, resounds all day with the whack of hammer and hum of saw—an entire community in constant turmoil of construction, almost *breeding* rather than erecting dwarf dwellings built with all the economy of a trim ship, overgrown doll houses by U.S. standards of excess, homes that are homemade with affection and friends and jauntily concluded with a splurge of red paint on the stairway bannister or a bright blue door that swings open easy to the knock of strangers.

How many townpeople in Zonesville, industrialized cultures, have the luxury of building their own homes? Nearly a century and a half ago Henry Thoreau was asking in *Walden* how often we come across people busy with this basic task of raising their own roof. An experienced ornithologist whose word is not to be taken lightly, he maintains that if we all wove our own nests, we would also sing more, like the birds.

"The teeth fall out before the tongue," as the Chinese proverb says: an irony of architecture now in earthquake zones like Manizales is the collapse of "permanent" concrete and steel structures too

**Editor's note:* A by-product of this nonsolution is to erase the still-fresh capacity for self-shelter that new city dwellers bring with them from the countryside. Instead of encouraging and preserving and extending this capacity in the curriculum of the school system, country people are taught to regard their village skills as obsolete and definitely inferior to the International Ugly replacing them. With the eagerness of all immigrants to do as Romans when they get to Rome, they dutifully forget their traditions with haste and embarrassment instead of exploring how they could be dovetailed into urban building needs.

rigid to give, while the "temporary," seemingly insubstantial, bamboo "slums" remain, flimsy and flexible, rolling with the earth ripples.

And then bamboo, after the disaster, helps put the big city downtown back together again—becoming scaffolds, forms, and supports in the poured concrete architecture downtown, where cement floors are suppled and lightened with *esterilla* bamboo boxes: big 2 by 4 by 8 foot cartons that honeycomb the structure and give it more give.

Balloon frame bamboo.

"The topography of western Colombia is often quite steep: as a consequence, houses with one or two floors on the street might have up to seven floors in the rear, depending on the incline of the land. For quake or steep land constructions, guadua is excellent. A light material, it doesn't require heavy foundations, sometimes just rests on big rocks . . ."

Walls.

"Walls are formed of guadua poles, 30 cm between centers, with diagonals to add resistance, creating a system similar to American 'balloon frame' construction. A 'cage' quite capable of resisting seismic movements and landslides is thus created, which is rarely seen to collapse completely.

"This structure is covered with opened guadua, flattened out, creating an excellent surface to receive plaster, of horse manure, mud, sand, and cut straw, which makes an air space reducing noise and moderating temperature changes. A solid wall is produced by strips of guadua tied on horizontally, every 15 cm on both sides of frame, the space between then filled with earth and chopped dry hay. Lime is used for the final coat."

Unified streetscapes through wide use of few materials.

"The general use of a few materials in all buildings regulates dimension and appearance of urban complexes, increasing homogeneity, so absent in conglomerate modern urban architecture. Walls, covered with lime, serve as a neutral background for colors used with great freedom of combination and imagination in doors, windows, and woodwork generally, individualizing each house without de-

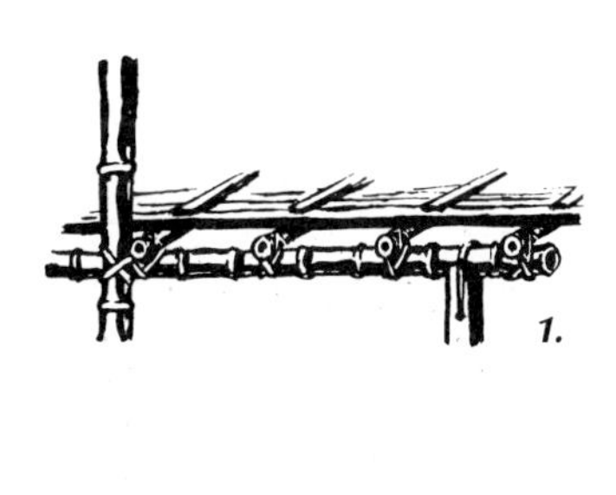

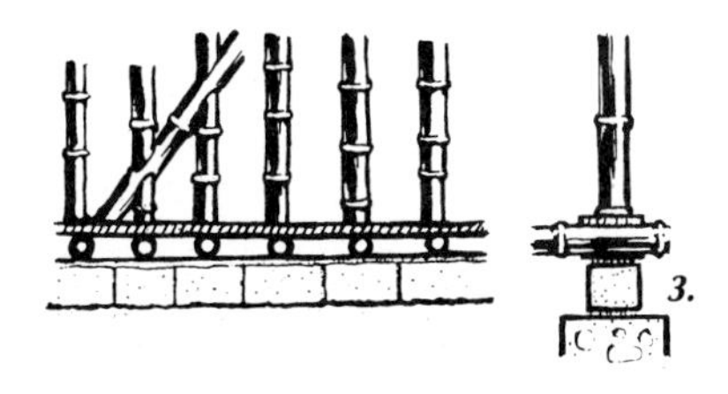

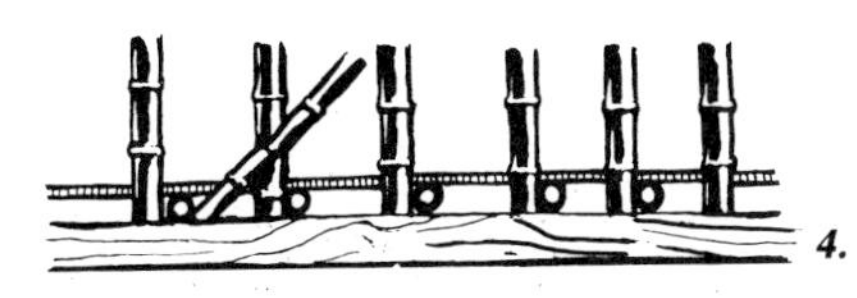

Guadua construction, Colombia. 1. Flooring. 2. Detail. 3, 4. Foundation, framing. 5. Windows greatly diminish the load-bearing capacity of the walls. Unplastered esterilla *(bamboo board) provides a thousand tiny horizontal windows, slivers of light which give a wonderful glow inside. 6. Window and door.*

1.

2.

3.

4.

Guadua houses under construction in Manizales, Colombia, by CRAMSA, a local corporation dedicated to appropriate technology.

tracting from the effect of the whole.

"The fences—made with many weaves of guadua—complete the scene, delicately indicating boundaries without breaking the flow of outside spaces as stone walls would."[16]

Modern times: guadua high rise.
"Taking advantage of guadua's qualities, the campesinos of Caldas have developed excellent construction techniques suitable not only to erect large structures like gigantic plants used by coffee growers to dry and store coffee, but more importantly, in the construction of urban dwellings. The best examples of the latter can be found in Manizales, the cradle of guadua architecture in Colombia, where magnificent homes built half a century ago have been only recently modernized by changing their facade to mortar cement, more in keeping with the new brick and concrete buildings that have been slowly replacing them.

"The arrival of the concrete era that gave such impetus to the construction industry in Colombia also brought new ways to use guadua: temporary retaining walls under protecting eaves and in the construction of secondary frames, used to support the slabs of the building while the concrete is cast and set. Large boxes or cases made of guadua boards are used to reduce the cost and weight of concrete slabs. They are placed over wooden or guadua frames between the spaces that separate the steel framework which reinforces the slabs.

"Once these boxes are in place, casting slabs can begin, by filling spaces between them. At the same time, the cases are covered with a layer of cement that will form a thin slab and become the ceiling foundation. Later, this will be covered below by a ceiling often made of braided guadua panels or *esterilla*—guadua boards—covered with mortar. After building construction, guadua scaffolds are built to work on the facade."

The death of bamboo.
"Guadua has been the cheapest and most readily available construction material, widely used by the poor in building their homes. Some 6 million Colombians in town and country—mostly in marginal areas of big cities—are living in homes partially or totally built with guadua. Many of these neighborhoods are on extremely steep terrain where, in order to build along the road, it's often necessary on the downhill side to build a 5–8 story guadua structure supported on rocks placed directly on the ground. Although in these areas some

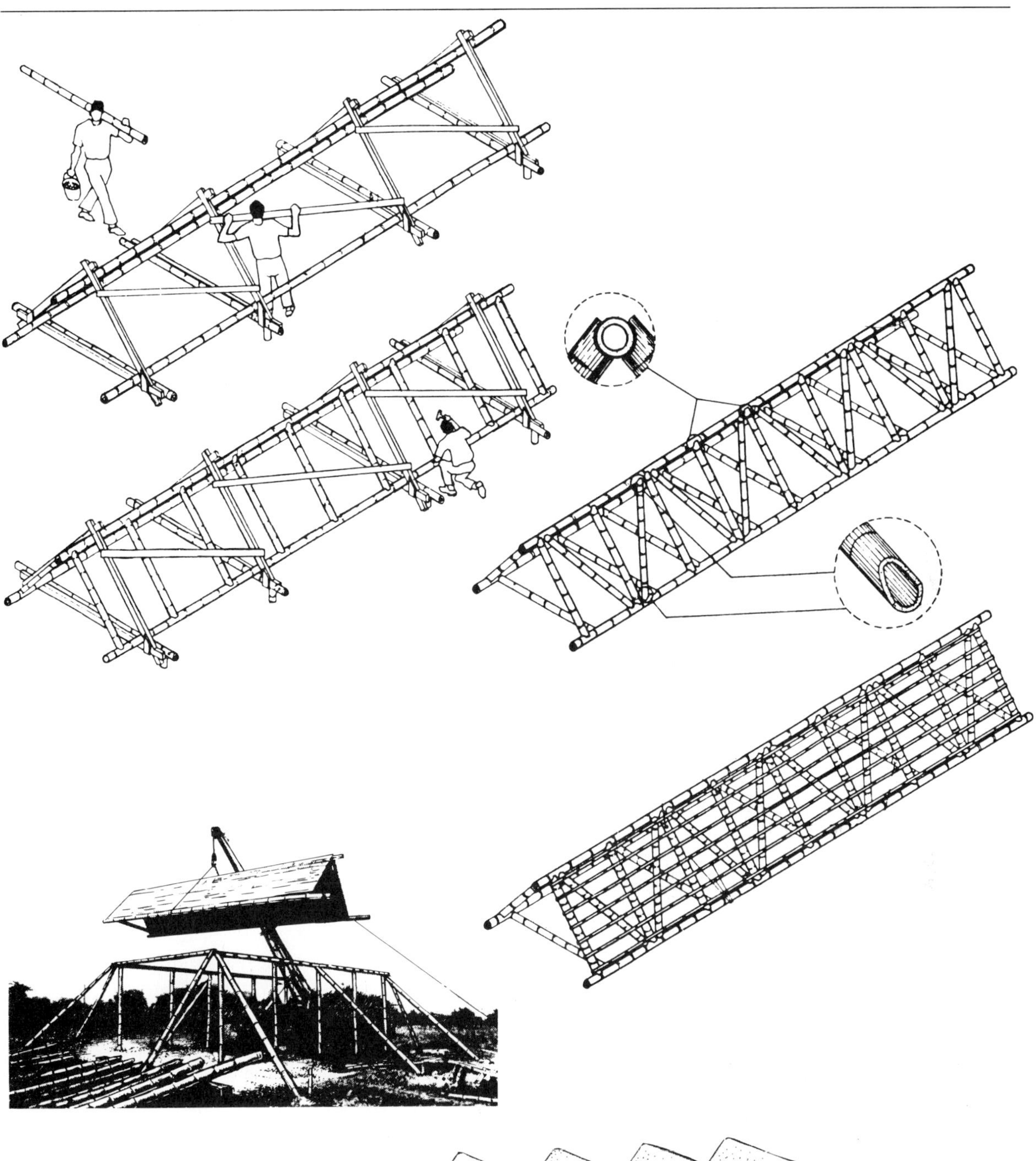

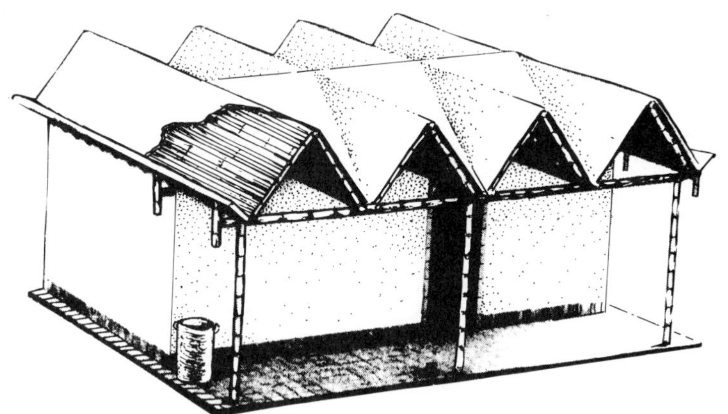

Triangular sections of a crinkle roof guadua structure are easily assembled on the ground and lifted into location by crane. A pulley on a branch overhanging the site could do as well. The finished structure can serve many community roles from school to clinic, and can easily be extended as needed.

A few thousand miles north of its natural distribution,* Guadua angustifolia *displays its characteristic squat nodes in the St. Louis, Missouri, Climatron in Tower Grove Park.

excellent guadua structural systems were developed, it was here also that the common aversion to guadua was born, stemming from misuse people made of it in huts constructed with little technical know-how or esthetic consideration—and with unappealing results that have made guadua synonymous with impoverished squalor.*

"There is a complete lack of interest in guadua cultivation, as well, and large groves are leveled off for coffee, sugar cane, and other more profitable crops—or to satisfy the growing demand of the construction industry which presently uses thousands of tons of guadua, apart from what is used for low-cost housing and other purposes. *As a result of intense exploitation, we foresee guadua's commercial extinction in Colombia before the end of this century.*

"Has anyone considered what will happen once the guadua is gone? What material will the campesinos and the poor people use to build their homes? There is no doubt that they will have to resort to agricultural and industrial refuse. The rural areas and the poor urban neighborhoods that once used guadua will then be a terrible spectacle of poverty, full of huts built with left-over scrapwood, cardboard, and metal sheets. Then, too late, through its absence, the real value of guadua will be seen."[17]

Extinction rhythms: antiextinction.

Extinction of species is a work-in-progress of humanity now.[18] Biologists used to say that extinction was happening at fifty times its preindustrial pace: now there are estimates that between 500,000 to 2 million species of plants and animals will disappear by the year 2000 with our assistance . . . roughly 70 a day at a conservative count, or one species every 20 minutes. That's the rent the rest of the planet has to pay for being neighbors to our race. What tactics can we adopt between now and the time that governments decide that this is a ruinous rate which, even from the perspective of successful "exploitation" of nature, we can't afford?

Life is a thin, breathing blanket on the earth in which the interdependent weave of species is not always obvious until we begin to unravel the fabric. "Useful to people" is a very crude criterion to apply, tainted by the same homocentricity that accelerates extinction around us to begin with. But guadua is a plant of major importance not only to us, but also to the biosphere. Defending its right to be here is an act of ecological as well as economic importance. The time may come when *Colombianos* will lament its passing as deeply as the plains Indians in the United States grieve the extinction of the buffalo.

"Regroving" the areas of guadua's distribution is high on the list of bamboo tasks in the West. International agencies such as the United Nations should assume responsibility for this and other bamboo errands of critical urgency—until that day when we have a United Creatures with ambassadors from the land of bamboo and other species to make people more conscious of their duties to the rest of creation here with us, with equal right to life, liberty, and the pursuit of sunlight.

The man who loved trees.

In 1913 a young Frenchman walking in a remote and barren area of southeastern France stumbled across a widowed shepherd who was dedicating his still-vigorous energies—he was then fifty-five—to planting one hundred acorns each day in a belief that the region, devastated by charcoal makers and others, needed trees. Elzeard Bouffier had been planting since 1910, and of his 100,000 acorns, 20,000 had sprouted, of which he expected 10,000 to survive whatever rats or unlucky weather might cast against them.

The young Frenchman, Jean Giono, was soon drafted into five years of World War I and all but forgot Bouffier, whose work he had "considered a hobby, a stamp collection." But in 1920 Jean returned and found Elzeard tending bees instead of sheep which would threaten his shoulder-high stands of young beech and birch trees, "delicate as

**Editor's note:* Harvest and curing methods, critical for longevity of building or basket, are little known. See pp. xx–xx.

Guadua flower.

young girls," that he had planted along with his acorns.

"I had seen too many men die during those five years not to imagine easily that Elzeard Bouffier was dead, especially since at twenty, one regards men of fifty as old men with nothing left to do but die. He was not dead. As a matter of fact, he was extremely spry . . . The oaks of 1910 were then ten years old and taller than either of us. It was an impressive spectacle. I was literally speechless and, as he did not talk, we spent the whole day walking in silence through his forest."

Bouffier's hobby was at that point 11 kilometers long by 3 wide. "When you remember that all this had sprung from the hands and soul of this one man, without technical resources, you understand that people can be as effectual as God in realms other than destruction." In 1913, Jean had found dry, desolate villages, "five or six houses, roofless ruins like an old wasps' nest, gnawed by rain, the tiny chapel with its crumbling steeple . . . And over the carcasses of the houses, the wind growled with unendurable ferocity, like a lion disturbed at its meal." By 1920, a grayish mist covered the mountaintops like a carpet when seen from a distance, and close up there was the sound of running water: "Creation seemed to come about in a sort of chain reaction. As we went back towards the village, I saw water flowing in brooks dry since the memory of man. Some of the dreary villages had been built on sites of Roman settlements, traces of which still remained. Archaeologists, exploring there, found fishhooks where, in the twentieth century, cisterns were needed to assure a small supply of water."

At seventy-five in 1934, Elzeard built a stone cabin about 12 kilometers from his original cottage to begin a new forest of beech. By 1935, when Jean visited again, the desert of twenty years earlier was covered with trees 20 to 25 feet tall. "Peaceful, regular toil, the vigorous mountain air, frugality and, above all, serenity in the spirit, had endowed this old man with awe-inspiring health. He was one of God's athletes. I wondered how many more acres he was going to cover with trees."

A friend of Jean's in the French forest service assigned three rangers to guard the forests, "and so terrorized them that they remained proof against all the bottles of wine the charcoal burners could offer." In World War II, cars in France were operating on *gazogenes,* wood-burning generators, and cutting was tried on the oaks of 1910 but was abandoned as uneconomical due to distance. Elzeard was unaware of the attempt and the failure, busy planting 30 kilometers away, ignoring the second world war as he had ignored the first.

Jean visited his friend for the last time in 1945, when Elzeard was eighty-seven. There was now bus service between the Durance Valley and the mountains, and only the name of a village convinced him he was traveling the same region he had first traversed on foot in 1913. "The bus put me down at Vergons. In 1913 this hamlet of ten or twelve houses had three inhabitants, savage creatures, hating one another, living by trapping game, all about them nettles feeding on the remains of abandoned houses. Their condition had been beyond hope, nothing to await but death—a situation which rarely disposes to virtue.

"Everything was changed. Even the air. Instead of the harsh dry winds that used to attack me, a gentle breeze was blowing, laden with scents. A sound like water came from the mountains; it was the wind in the forest. Most amazing of all, I heard the actual sound of water falling into a pool. I saw that a fountain had been built, that it flowed freely, and—what touched me most—that someone had planted a linden beside it that must have been four years old, already in full leaf, the incontestable symbol of resurrection.

"Besides, Vergons bore evidence of labor at the sort of undertaking for which hope is required.

Hope, then, had returned. Ruins had been cleared away, dilapidated walls torn down, and five houses restored. Now there were twenty-eight inhabitants, four of them young married couples. The new houses, freshly painted, were surrounded by gardens where vegetables and flowers grew in orderly confusion, cabbages and roses, leeks and snapdragons, celery and anemones. It was now a village where one would like to live . . . On the lower slopes of the mountain I saw little fields of barley and rye; deep in that narrow valley, the meadows were turning green."

Two years later, in 1947, Elzeard Bouffier died as peacefully as he had lived, in the hospice at Banon. In 1953, forty years after his initial visit, Jean returned again to the region, which he found glowing with health. "On the site of the ruins I had seen in 1913 now stand neat farms, cleanly plastered. Old streams, fed by rains and snows the forest conserves, are flowing again. Their waters have been channeled and on each farm, in groves of maples, fountain pools overflow on to carpets of fresh mint. Little by little the villages have been rebuilt. People from the plains, where land is costly, have settled here, bringing youth, motion, the spirit of adventure. Along the roads you meet hearty men and women, boys and girls who understand laughter and have recovered a taste for picnics. Counting the former population, unrecognizable now that they live in comfort, more than 10,000 people owe their happiness to Elzeard Bouffier."[19]

Minimal ecoethic.

There is a hidden irony in much high-tech development that is exemplified in a computer's use of paper, which is horrendous and immense. In poorer countries, waste computer paper is just replacing banana leaves as the cheapest way to wrap a roadside restaurant meal, which is a rough index of how much paper pulp our slim machines devour. The evolution from pen to typewriter to Xerox and computer printouts can seem a further and further leap from our original wilderness; but, in fact, the more sophisticated and high-tech we become, the more our communication system and general cultural style tends to go back to trees and erase them. The more artificial and humanized our environment, the more apparently distant from nature, the more it in fact depends on a natural balance already overburdened by our extravagant use.

A minimal ecoethic for our time would be for people to replace at least renewable resources they personally consume. Tree planting is the most tangible, ceremonial, and appealing activity through which to move culturally in this mending direction. Our schools and mass media should work toward creating popular awareness of the importance of revegetating to keep pace with human use and to correct former abuse. Reestablishing forest communities of all sorts, including bamboo where relevant, will cure planet and people in a single act.

Recording our thoughts about the matter knocks still more forest down. If it does not inspire planting, the gross effect of this book will be a balder globe.

Planet pact for happy rafters.

New nest beside this ancient tree,
let these woods be good to me.
In return, I vow to make
twice the lumber that I take
and plant for every daughter and each son
a tree each time our earth has run
its journey once again around its sun.

Whatever dryads dwell within these trees,
or hidden wood nymphs, braided with the breeze,
gnomes tunneling beneath the forest floor
or elves and brownies dozing near my door—
this treaty is directed, by its scribe,
to all earth spirits, of whatever tribe:

Grant my get my vegetating style,
dirt on their knees and taste for toil,
that I may leave a planet yet
more leafed and handsome than I met,
less faked, less driven, less afraid,
facades and faces looking more homemade
when what was gardener a while
joins again the garden soil
and these bones, at length, are laid
their length beneath a greener shade.

***A Taxonomy of Housing:* Notes on structures and locations.**

1. Round plan, free-standing; diameter less than height; walled with mud and/or stone; often with stone foundations; thatched roof (conical or trumpet-shaped); arranged in clusters of buildings, usually on the ring pattern, with buildings part of enclosing wall or fence. (Sudan, Tanzania, Nigeria, Cameroon, Mali, Senegal, Guinea, Togo, Benin, Central African Republic.)

2. Round plan, free-standing; diameter approximately equal to height; roof of poles leaning against central framework; poles sometimes encased in dry stone work at base; thatching of grass or turf. (Ethiopia, Tanzania.)

3. Round plan, free-standing; diameter equal to or greater than height; walls of mud and/or wattle, bamboo or palm fronds; thatched conical roof (convex or concave profile); often with verandah full or part way round; arranged in clusters of buildings within surrounding fence, hedge, or wall. (Kenya, Zaïre, Nigeria, Guinea, Cameroon, Tanzania, Liberia, South Africa, Ethiopia, Zambia, Ghana, Sudan, Mali, Senegal, Ivory Coast, Sierra Leone.)

4. Round, oval, or rectangular plan with hemispherical or lozenge-shaped profile; basic framework of hoops; covering of skins, mats and/or thatch of grass, leaves or mud over brushwood; can usually be dismantled; often found in association with cattle kraals; usually arranged symmetrically. (Tanzania, Kenya, Cameroon, Zaïre, Namibia, South Africa Swaziland, Chad, Nigeria, Somali Rep., Ethiopia, Niger, Mali, Lesotho.)

5. Rectangular plan, free-standing; framework of one to four parallel arches strengthened by horizontal cross-pieces resting at ends on poles between forked posts; covering of plaited mats; very often used as portable tent; large version sometimes immobile. (Niger.)

6. Rectangular plan tent; framework of two to four rows of parallel forked sticks surmounted by horizontal cross-pieces; occasionally arches instead of middle sets of poles; covering of skins or blankets under tension. (Niger, Sudan, Ethiopia, Somali Rep.)

7. Round plan, free-standing; conical roof and no walls; framework of straight sticks (guinea-corn stalks, bamboo); sometimes thatched. (Nigeria, Tanzania, Sudan, Ethiopia.)

8. Round plan, free-standing; framework of flexible poles embedded in ground at base and tied at top under tension; known as "beehive" type; usually slightly convex profile; thatch sometimes of banana leaves but more usually of grass or reeds, either stepped or plain; sometimes low perimeter wall inside building; sometimes central support; often divided internally by partitions; same design as house; often with porch. (Sudan, Tanzania, Ruanda, Uganda, Nigeria, Chad, Niger, Ethiopia, Kenya.)

9. Round plan, free-standing; two storeys high; walls of roughly dressed stone set in mud mortar; wooden lattice windows; drip course between each story; slightly domed mud and pole ceiling, thatched roof. (Ethiopia.)

10. Round plan, free-standing; two storeys high; walls of small round boulders set in mud mortar; second storey reached by external stone staircase; within, walled courtyard with two-storey entrance porch; thatched roof. (Ethiopia.)

11. Round plan, free-standing; flat roof; walls of mud or mud and straw; flat roof of poles and mud and straw; found in tight clusters, usually built into surrounding wall; painted and incised decoration on walls common. (Granaries often had thatched covers.) (Mali, Upper Volta, Ghana.)

12. Round plan, free-standing; "shell" mud roof and no walls; slightly convex profile; sometimes embossed patterns on exterior; arranged in clusters within surrounding wall. (Cameroon, Ghana, Chad.)

13. Oval plan, free-standing; asymmetrical peaked thatched roof supported by conical mud pillar and mud arch; walls of mud and wattle. (Nigeria.)

14. Oval plan, free-standing; mud and/or wattle walls; thatched saddleback roof with semiconical ends; sometimes on stilts. (Liberia, Guinea Bissau, Senegal, Tanzania: coastal areas and lake shores; Ivory Coast.)

15. Round or oval plan, free-standing; corbelled stone construction; untrimmed sandstone blocks, doleritic boulders or trimmed doleritic slabs. (Lesotho, Botswana, South Africa.)

16. Round plan: one, two or three storeys in height; built coalescing to form "tower" houses; walls of puddled mud; flat roof and upper floors of poles, straw and mud. (Upper Volta, Benin, Mali.)

17. Crown plan (concentric circles), free-standing; central court or impluvium; mud walls; thatched saddleback roof. (Senegal, Guinea Bissau, Ivory Coast.)

18. Square plan, free-standing; conical roof; walls of mud or mud and palm fronds; thatched roof of grass or reeds. (Cameroon, Nigeria.)

19. Rectangular plan, sometimes free-standing, thatched saddleback or lean-to roof; walls of planks, bamboo, cane, matting or cane and matting; walls sometimes plastered internally; roof thatch of palm leaf mats, reeds, bark, palm fronds, sometimes on stilts. (Tanzania, Nigeria, Cameroon.)

20. Rectangular plan; often arranged contiguously around a central square open kraal; walls of wattle or stone and mud; flat or wagon-shaped mud and wattle roof supported on forked uprights just outside walls or on walls; can be known as "tembe" style. (Tanzania, Uganda, Ethiopia.)

21. Rectangular plan, free-standing; thatched saddleback roof; buildings often arranged facing across a small court with some of the sides facing court open or pillared; walls puddled mud or wattle framework plastered over; relief murals common form of decoration. (Nigeria, Ghana, Togo, Benin, Ivory Coast.)

22. Rectangular plan; thatched saddleback roof; units built round court or impluvium having continuous roof; walls of puddled mud or mud and wattle; sides facing court or impluvium sometimes open or pillared. (Nigeria.)

23. Rectangular plan; mud brick walls; flat or vaulted mud roof reinforced with wood or palm fronds; sometimes two-storeyed; buildings arranged within walled courtyards, sometimes forming part of courtyard wall. (Nigeria, Mali, Upper Volta, Mauritania.)

24. Rectangular plan units; one storey high but built coalescing and on top of one another; mud brick or puddled mud walls; flat mud roof reinforced with wood and palm fronds; sometimes found with style 3 built on top. (Ivory Coast, Mali, Upper Volta, Ghana.)

25. Square plan; free-standing; walls of poles or palm fronds and mud; hipped roof thatched with grass or reeds. (Zambia, Zaïre, Cameroon.)

26. Square plan; free-standing; thatched hipped roof framework of flexible poles embedded in ground at base and tied at apex under tension; slightly convex profile; thatch of grass; often with elaborately carved door frames. (Zaïre, Angola.)

27. Rectangular plan, free-standing; walls of roughly dressed stone set in mud mortar; reinforced with horizontal wooden beams and short round cross-pieces; flat roof of mud and poles. (Ethiopia.)

28. Rectangular plan, free-standing, with stone rubble and cement walls; thatched roof multi-storey; often with elaborately carved wooden doors. (Tanzania, Kenya.)

29. Rectangular plan, free-standing; hipped roof; thatch of palm leaf mats sometimes with two long sides lapped over other two; walls of wattle and mud; sometimes with carved wooden door posts; sometimes on stilts. (Kenya, Tanzania, Nigeria, Benin Rep.: coastal areas; Zaïre: lake shores.)

30. Square plan; tall pyramidal thatched roof; thatch of broad leaves. (Zaïre.)

31. Cave houses; caves often artificially enlarged; sometimes small courtyard in front surrounded by mud wall or fence; wattle and daub wall sometimes built across mouth of cave. (Tanzania, Kenya, Zaïre, Cameroon, Sudan, Chad.)

32. Underground or semi-underground "dug-in" buildings; rectangular in plan; sometimes with excavated passageway in front; walls of stone in mud mortar, wattle and mud, mud bricks or turfs; flat or slightly wagon-shaped roof of earth, mud and poles supported on many rows of forked uprights. (Tanzania, Ethiopia, Upper Volta.)

CHAPTER 5.

1. Rudofsky 1964.
2. Rudofsky 1964; condensed from the preface.
3. Denyer 1978: 4.
4. Fathy 1973: 19–21.
5. Ibid.: 22–4.
6. Ibid.: 24–5.
7. Ibid.: 25–6.
8. Ibid.: 26–7.
9. Ibid.: 31–3.
10. Ibid.: 33–4.
11. McClure 1950b.
12. McClure 1966:147–201.
13. Hidalgo 1980:1.
14. McClure 1950b.
15. Castro 1966, *passim.*
16. Ibid. The Castro material consists of strings of captions and brief introductions to a book of guadua photographs from western Colombia.
17. Hidalgo 1980:3–5. The movie from which the Hidalgo material in this chapter derives is available from CIBAM in Spanish or English. Address, see p. 305.
18. For an extended treatment, see *Extinction,* by Paul and Anne Ehrlich (1981), Random House.
19. Giono, Jean: *The Man Who Planted Trees and Grew Happiness;* reprinted in Brand 1980:79–80.

6. THE PLANT — BAMBOO BEHAVIORS

*"Plants are people, just like us.
You see them, they see you.
The earth isn't blind
and the mountains aren't foolish."*

—Jose Valdez,
Mexican farmhand

TROPICAL AND TEMPERATE SPECIES

Grasses have proved the most useful plants on the earth for people, providing us with such common crops as wheat, corn, rice, barley, sorghum, oats, sugar cane—and bamboo. Bamboos are perhaps the most primitive subfamily of grasses, which includes some 76 genera embracing some 1200 to 1500 species. These are broadly divided into "sympodial" and "monopodial" species, according to their rhizomes, a rootlike system of growth underground.

Sympodial (clumping) bamboos.

Sympodial bamboos are clumping, frost-sensitive tropical species, some enduring temperatures slightly below freezing. In sympodial bamboos "shooting"—the growth of culms—occurs typically from summer to autumn or at the onset of the rainy season. It is apparently controlled by moisture levels; and in warm regions with frequent rainfall throughout the year, the growth of sympodial bamboos may be virtually continuous. The rhizome is thick and short, with asymmetrical internodes more broad than long. Side buds give rise to further rhizomes, with culms growing from the ends in tight or fairly open clumps, depending on the length of the rhizome neck. The midculm branches, swollen at the base, resemble the rhizome form; often they are bearded with root primordia (in rudimentary stage of development) on the dominant branch. Cross veining in leaves—called "tessellation"—is usually difficult to see.

Monopodial (running) bamboos.

Monopodial, free-standing bamboos are hardy to a few degrees below 0° F and thrive in climates with pronounced but not severe winters, enduring even annual seasons of snow. The onset of shooting, typically in spring, is apparently controlled by temperature. The lengthy rhizomes are segmented like the culms above them, with symmetrical internodes more long than broad. The culms, growing from side buds at alternate nodes of the rhizomes, bear branches without root primordia, unswollen at the base. Tesselation is usually clearly visible in the leaves.[1]

RHIZOMES

The rambling rhizomes of monopodial bamboos may nudge their way through up to 20 feet of soil

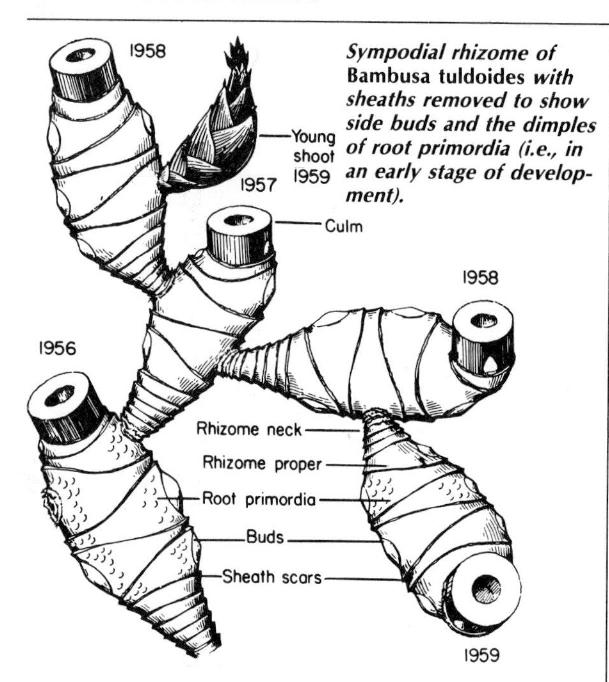

Sympodial rhizome of **Bambusa tuldoides** *with sheaths removed to show side buds and the dimples of root primordia (i.e., in an early stage of development).*

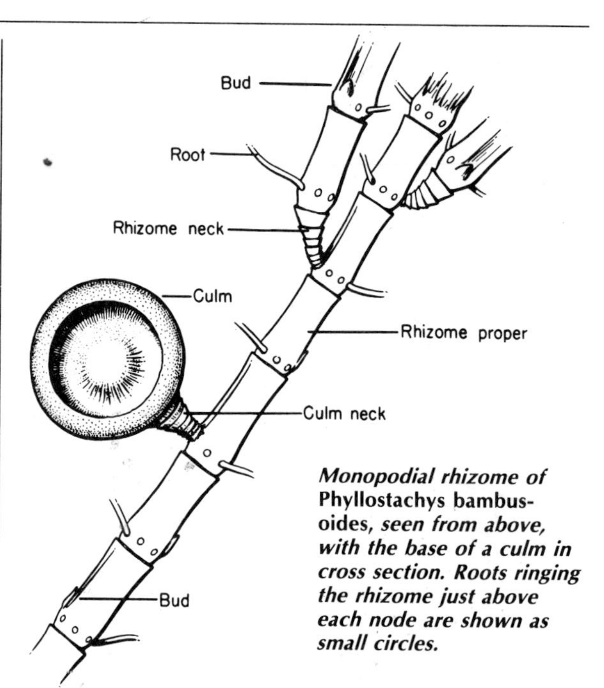

Monopodial rhizome of **Phyllostachys bambusoides**, *seen from above, with the base of a culm in cross section. Roots ringing the rhizome just above each node are shown as small circles.*

in their single spurt of growth. There is no increase in size after this, although the rhizome remains fertile for up to ten years, producing culms and other rhizomes that tangle underground in such profusion they hold firm in earthquakes and effectively diminish erosion on the steep slopes where bamboos, preferring a well-drained soil, grow well. The excavation of a quarter acre in Kyoto revealed rhizomes totaling weights up to 19,000 pounds in large species and lengths up to 53,000 yards in small bamboos.[2] That's roughly 40 tons per acre in large species, and 120 miles per acre in dwarf bamboos. Some 80 percent of this growth occurs in the top foot of soil, which accounts for bamboo's traditional reputation as guardian of the earth's fertility, stitching her hills together like an anxious grandmother.

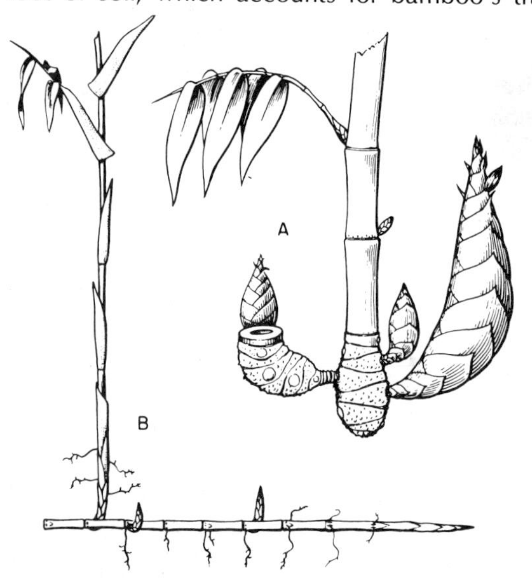

(A) Sympodial rhizome of **Bambusa beecheyana.**
(B) Monopodial rhizome of **Arundinaria amabilis.**

A healthy and vigorous young rhizome is a yellow-ivory color. Its nodes are key points of growth, giving rise to a ring of roots to gather nutrients and buds to extend the community of culms above ground and rhizomes below. The nodes store nutrients as well for distribution to the most active centers of new growth in the grove. Like the culm nodes above, each rhizome node is snugly wrapped in a protective sheath. The rhizome tip is pointed and tough. If a monopodial rhizome encounters an obstacle too unyielding to pierce or too large to pass around, it will sometimes rise briefly above ground and dive below again until it ends its season of growth—and in some species, it bends up to form a terminal culm, exuding moisture before it to soften the soil. In a shooting grove where the ground is dry enough, a small circle of moisture indicates the spot where a new culm is about to break through the soil, its tender and rapidly multiplying tissue heavily wrapped in tough overlapping sheaths that effectively protect extending stem and branch buds tucked on alternate sides of each node.

GROVE FORM

Surface growth reflects rhizome habit underground.

The form a stand of bamboo assumes above ground depends on the nature of the rhizome below. The open grove of free-standing culms is characteristic

A botany experiment

Plant a hardy bamboo rhizome in 6 to 8 inches of soil in the bottom of a deep and narrow glass tank. A pipe to water the soil directly is placed in one corner, then the tank is filled with sand. Watering through the pipe keeps the soil moist and the sand dry. As the rhizome bud eventually swells and grows, the sand before it becomes damp.[3]

of most hardy bamboos native to temperate zones. The tight-clumping habit of more southern species supposedly evolved from this earlier rhizome form as an adaptation to harsh tropical sunlight and seasons of prolonged drought followed by torrential rains. A more crowded grove provided denser shade and a thicker blanket of moisture-preserving mulch from leaf fall as well as less rhizome surface to dehydrate in those long dusty months typical of many climates at the fat middle of the globe. The tendency to spread out or stay put is particularly critical when considering bamboo plantings in limited land. It should be noted that there are hardy "running" species whose rhizome growth is so lazy that they, in fact, form fairly compact clumps extending in a leisured creep: *Pseudosasa japonica* or

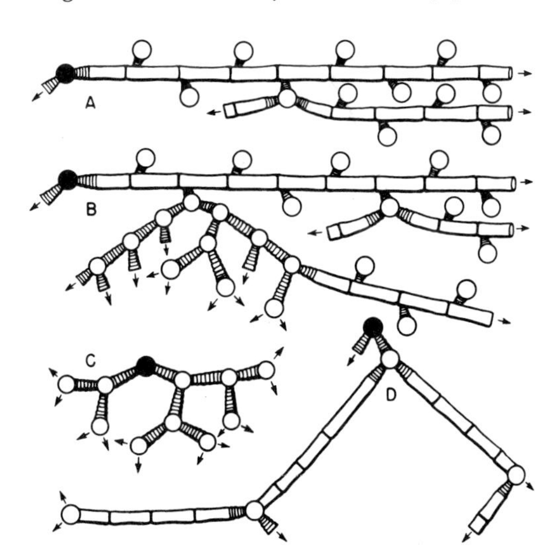

Rhizome types determine the clumping or running habit of a grove. (A) Monopodial rhizome with single culms somewhat openly spaced. (B) Monopodial rhizome with tillering culms produces scattered clumps above ground interspersed with solitary culms. (C) Sympodial rhizomes with short necks produce congested groves. (D) Sympodial rhizomes with long necks produce open groves, with culms up to six meters apart in some species. Temperate monopodials tend to scatter; tropical sympodial bamboos, to clump. The numerous exceptions to this rule are determined by the rhizome system of each species.

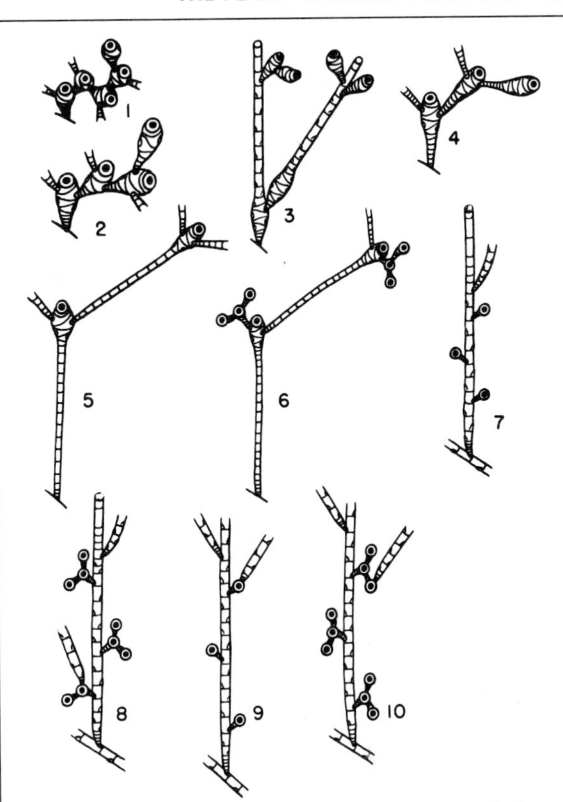

Rhizome systems of various species, seen from above. Open ends indicate further horizontal growth; a single circle (3, 7) signifies an occasional terminal culm at the end of a rhizome; a double circle, the cross section of a culm base. (1) **Bambusa pachinensis:** *a short-necked sympodial rhizome. (2)* **Bambusa tuldoides:** *a sympodial bamboo with rhizome neck shorter than the rhizome itself, which grows horizontally and then curves up into an erect culm. Produces compact tufts of culms generally characteristic of tropical species. (3)* **Arundinaria pusilla:** *distinguished by the tendency of the rhizome to form terminal culms. (4)* **Sinarundinaria nitida:** *an elongated rhizome neck that gives some sympodials an open grove form.* **Bambusa vulgaris** *is a common example. (5)* **Melocanna baccifera:** *long-necked sympodial rhizomes a meter or more long create open groves of widely spaced culms. (6)* **Yushania niitakayamensis:** *long-necked sympodial rhizomes with tillering culms produce scattered clumps united underground. (7)* **Shibataea kumasasa:** *monopodial rhizome with terminal culm and solitary culms from side buds, which tend to tiller in old age. (8)* **Arundinaria tecta:** *monopodial rhizome with terminal culms as well as culms from side buds. These culms, from their own underground basal buds, give rise to other rhizomes and tillering culms. (9)* **Phyllostachys bambusoides:** *the most broadly typical form of the monopodial rhizome. Side buds produce new rhizomes or solitary culms that do not tiller except when damaged or grown under unfavorable conditions. Terminal culms on rhizomes occur only rarely, as when rhizomes emerge on a steep hillside or encounter an obstacle such as a large rock that drives them above ground. (10)* **Chusquea fendleri:** *a South American bamboo bridging both rhizome forms. Swollen sympodial rhizomes occur in tillering clumps from side buds of a long rhizome monopodial in form.*

Phyllostachys aurea are common examples. By contrast, judging only by the grove formation above ground, one would regard some "clumping" bamboos as running, temperate species, owing to the uncommon length of their rhizome neck—up to 6 meters in the extreme case of the giraffe of bamboo tropical rhizomes, a *Guadua* species from Peru.[4]

For concerned gardeners, the style and rate of spread are considered under the descriptions of individual species below.

ROOTS

Survival value of roots in the air.

Bamboo roots are the only portion of the plant not growing in segments of nodes and internodes. They are fibrous, roughly cylindrical, thin, and do not increase in diameter with age. The largest roots of the largest species may be up to a centimeter thick and a meter long. In bamboos, as in other grasses, roots are primarily "adventitious," that is, not occurring at the usual location. Instead of growing from the primordial root, they extend from nodes of rhizomes and basal underground nodes of culms. In a number of sympodial bamboos, root primordia —in an early, dormant state—ring the swollen base of dominant branches at midculm nodes. Many species, principally sympodial bamboos, also bear root primordia at nodes as high as midculm, always diminishing in size at successive nodes in the direction of the tip. *Bambusa vulgaris, Dendrocalamus asper, Gigantochloa verticillata, Chusquea pittieri* are sample species with aerial roots. The genus *Chimonobambusa* is distinguished among monopodial bamboos with several species —for example, *Ch. quadrangularis*—bearing dry, brittle rootlettes up to midculm or higher. Aerial roots presumably anticipate the possibility of the culm being prematurely felled by storms or other falling plants. The roots are there, ready to help establish new culms and rhizomes at a number of nodes.

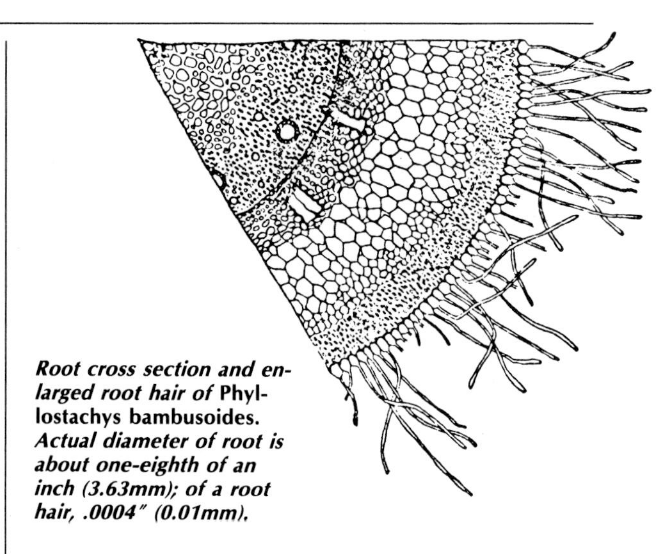

Root cross section and enlarged root hair of* Phyllostachys bambusoides. *Actual diameter of root is about one-eighth of an inch (3.63mm); of a root hair, .0004" (0.01mm).

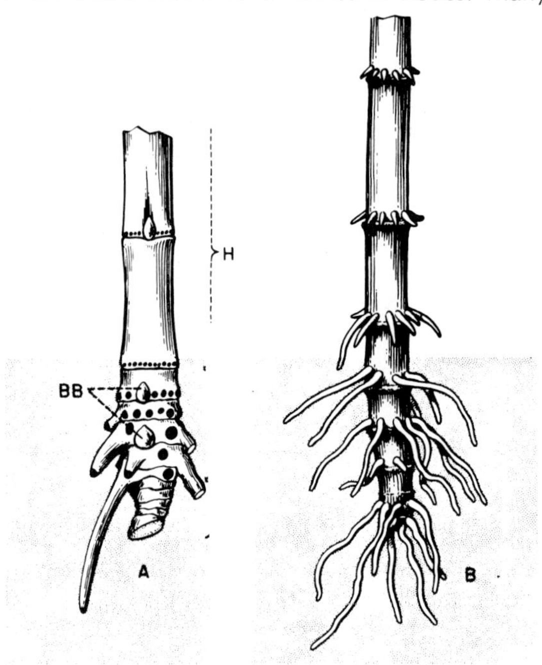

(A)* Phyllostachys nigra. *Sheaths removed from culm base to show developing young roots; BB, basal buds. (B)* Chimonobambusa quadrangularis *(square stem bamboo) is ringed above culm nodes by spiney aerial roots (*adventitious *roots) diminishing in length at each higher node and finally disappearing.

Whatever their survival value, the visual effect of culms is always enhanced by nodes or branch bases trailing roots. They provide an interesting contrast—in color, texture, and shape—to the polished and erect stance of culms. In some species, such as *Dendrocalamus asper,* they are long enough to suggest a bearded and venerable ancient. As short spikes ringing the culm in the square stemmed *Chimonobambusa quadrangularis* they provide a formal frieze bristling with vigor, an orderly crownlike adornment for a species already distinguished by its angled canes.

CULM

Visible stems: fingers of a hidden hand.

The growth of the bamboo culm, the most spectacular of the plant's visible behaviors, varies greatly depending on species, soil, age of stand, and climate. Larger species may average between 3 to 16 inches in a day, with a recorded growth for *Phyllostachys edulis* in Japan of 47.6 inches in one 24-hour period.[5] Night growth may be two to three times the growth by day in some species; in others, the reverse is true. Sympodial bamboos complete their growth in 80 to 120 days. Monopodial species reach 93 percent of their total height within one month, then taper off with a second, slower period of roughly equal duration. After this the culm hardens and matures, drops branches that may sag eventually with a bushy abundance of foliage but does not increase in height or diameter: The size reached by the eighth or ninth node to emerge from the soil in a shooting culm will be roughly the girth of the developed stem. "Like most of the plants in the great subclass of the vegetable kingdom called the monocotyledons, there is no provision for growth in thickness once maturity has been attained."[6] This abrupt growth and abrupt halt is less surprising when we remember that a culm is not a plant, but a piece of plant.

Bamboo family.
Most people, until told otherwise, unconsciously regard individual bamboo culms as trees, each a separate, living whole. In fact, of course, the stems are airy branches of a single life whose structural foundation is underground, invisible. Apparent individuals are members of a system of intersurvival, a tuft of giant grass: sharing, cooperative, striving for the common good of the grove, gathering food and drink in root and leaf, storing it in rhizomes, to be sent then to the "bamboo kids," *takenoko,* as the new shoots are affectionately known in Japanese. From each according to its vitality, to each according to its need. In many cultures, this collective

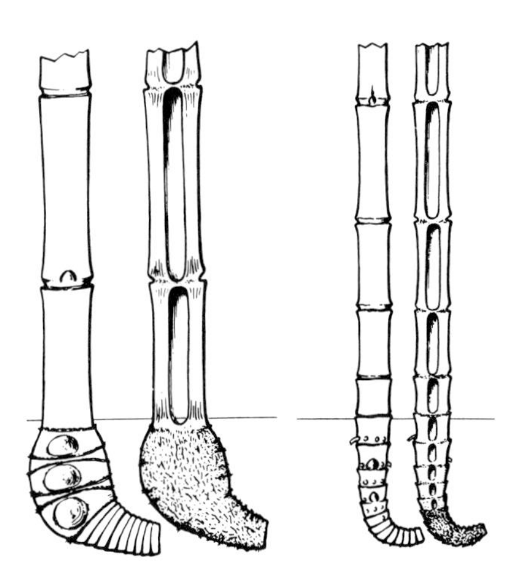

Lower culm of **Dendrocalamus latiflorus** *(left) and* **Arundinaria fastuosa** *(right).*

Japanese buyers estimate the height of culms by diameter at eye level multiplied by a number that varies according to species, but which is generally around 60. Because taper is gradual at the base in most bamboos, eye-level diameter is roughly the same for a tall adult or small child.

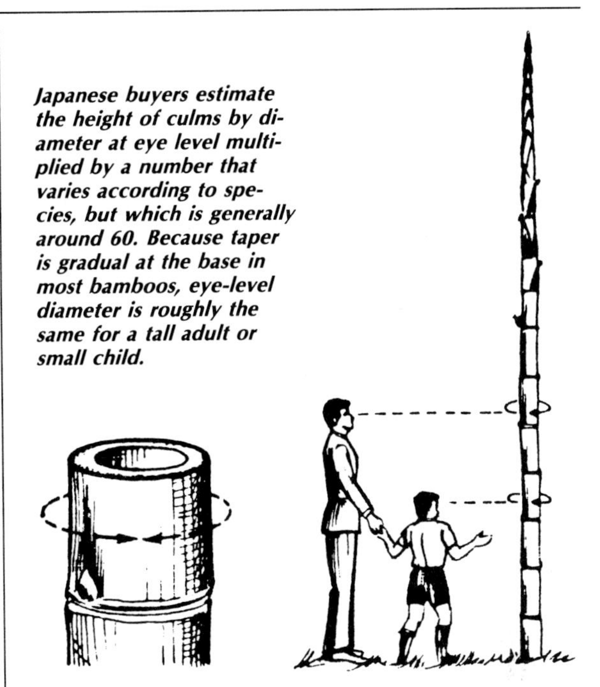

pattern of bamboo's growth was taken as a rebuke to egotism and an image of familial unity.

Grove maturity affects culm size.
Culm growth depends not only on species and climate and soil, but also on the maturity of the grove. When a new grove is established, from planting of rhizome or seed, the gradually extending system of rhizomes becomes more rich in the food necessary to produce culms. Thicker and taller stems are produced each year in greater numbers until the maximum stature and productivity for the species is reached under specific local conditions of weather and soil fertility. The grove will then continue to produce plants of maximal size until it flowers.

Longevity.
Culms may be short-lived in some species or cut back by harsh winters in the case of ornamentals introduced in climates colder than their native homes. A life of five to ten years is not uncommon, and some species—*Phyllostachys bambusoides,* the Japanese "timber bamboo," for example—produce culms living as long as twenty years. Size is as variable as duration: from pygmy *Sasas* in northern Japan 6 to 10 inches tall to *Dendrocalamus* species up to 120 feet and a reported 90 meters in a giant culm of *Dinochloa andamanica.*[7]

A highly simplified sketch of bamboo heights would reveal maximal growth in equatorial regions, diminishing towards the poles—not surprising since every 10° C fall in temperature supposedly reduces

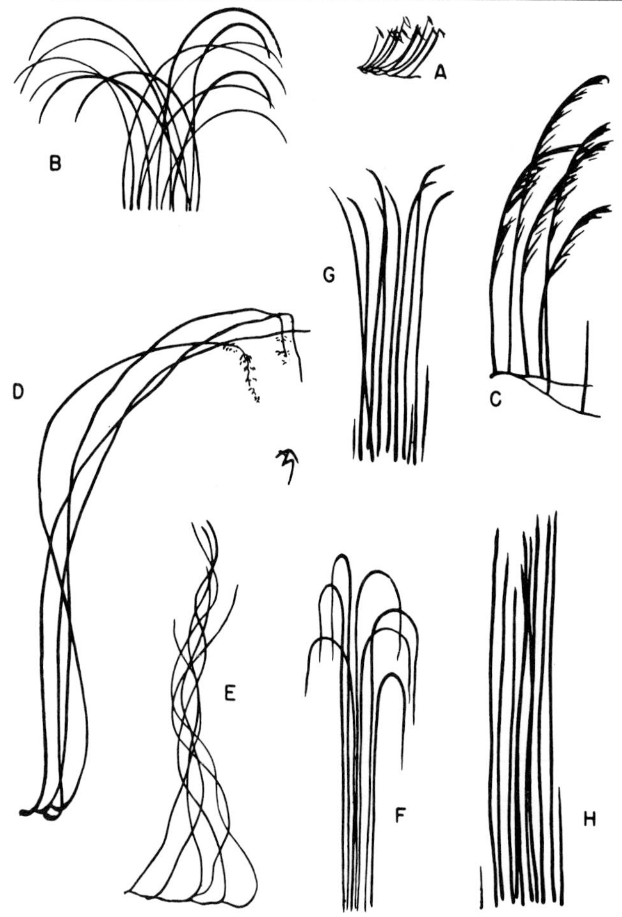

Culm stance: diagrammatic examples of some typical postures of a grove. (A) **Sasa palmata. *(B)* Sinocalamus beecheyanus. *(C)* Phyllostachys nigra. *(D)* Schizostachyum hainanense; *(E)* Dinochloa scandens. *(F)* Sinocalamus affinis. *(G)* Bambusa textilis. *(H)* Arundinaria amabilis.**

chemical processes in plants by 50 percent.[8] In form, the culms of some species are ramrod erect but more often nodding much or slightly at the tip. Other bamboos are "clambering," specially equipped to climb other plants.

Culm stance.

Culm stance also helps distinguish genera. Erect in *Pseudosasa japonica,* arching in most *Bambusa* species, plumelike in most *Phyllostachys,* in *Schizostachyum hainanense* it is clambering. In *Dinochloa* species the individual culms are more aggressive in their attitudes and gyrate like morning glories, like vines that, finding no external means of support, twine around each other. This is commonly seen in *Dinochloa,* where the twisted whirlwind of a large, unsupported clump of this bamboo suggested the genus name that means "eddy grass" or "whirling grass." Other forms include the whiplike *Dendrocalamus flagellifer,* stiff and erect for two-thirds its length, ending in a long, drooping tip. Many *Bambusa, Schizostachyum,* and *Gigantochloa* have upright culms with arched tips. Perhaps most commonly, the culm stance is strictly erect as in *Bambusa tuldoides* and all *Arundinarias.*[9]

Branching.

Culms of very young plants of most species are typically branching or bud-bearing at all nodes. Mature culms of developed groves are sometimes branched throughout, but more often they are branchless and even budless at lower or even midculm nodes. In some species, such as *Bambusa textilis,* nodes may lack branches up to three-quarters of the culm, which increases both their usefulness and the difficulties of propagation. Sometimes —as in *Chimonobambusa quadrangularis*—lower to midculm nodes may be ringed by short, thornlike roots. In some species the difficulties of harvesting bamboo are greatly increased by wicked thorns—hard and sharp dwarfed twigs—at the culm base and higher. Various species of *Guadua*—a much used Central and South American bamboo—have thorns that can shred denim to rags in a day's harvest; cut branches left on the ground can readily puncture a stout boot. *Bambusa arundinacea*—"giant thorny bamboo"—is another species as armed and dangerous as it is desirable: In spite of its bristly defenses, it supplies nearly half the raw material for India's bamboo paper pulp production, which is the largest in the world.

Internodes.

Internodes generally increase in length to midculm, then decrease to tip. Much more rarely, internodes sometimes also increase in diameter as well from base to midculm. An interesting feature of culm anatomy in bamboos is that their constituent fibers reflect the grosser form: Their length and thickness correspond to the internode length and culm diameter. They increase in length from base to midculm and then shorten, like the internodes, to the tip. Their width, like the diameter, decreases continuously with culm height. Both length and width are greatest in the inner portions of the wall.

Fiber grass. Composite materials reflect bamboo anatomy.

From a builder's viewpoint, the architecture of the bamboo culm presents an optimum configuration: fibers of greatest strength occur in increasing concentration toward the periphery of the plant. The center is hollow. These long, slender, elastic fibers are ideally suited for the skeletal framework of the plant because bamboos are shallow rooted and extremely tall. With 80 percent of the rhizome in the top foot of soil, and the rest rarely extending below a yard, the larger species are anchored in earth with foundations roughly 2 to 3 percent of their height. Only extreme resilience in the culms could keep them from being easy victims of the wind. In fact, the reverse is true: Bamboos became a proverbial symbol for resilience in adversity as oriental peoples noticed, they only bent in storms that broke the oaks, then stood up again, green and living, when the wind died.

"The essential basis for the remarkable properties of bamboo has only been revealed by recent work on two-phase materials such as fiberglass. Composite materials combine substances of high and low tensile strength and modulus of elasticity so that they can absorb loading stresses, which would rupture the weaker component, and at the same time isolate imperfections in the individual units of the stronger component. Bamboo does just this; in bamboo strong high-modulus fibers of cellulose are combined with a low-modulus plastic matrix, lignin. Hence its multifarious uses in Chinese culture, where also the two-phase structure was dissociated by retting to give the high-tensile-strength fibers alone. We shall find later on a parallel to this combination in the pattern-welding of sabers, where wrought iron and steel were forged together. Nowadays we have high-tensile-strength glass fibers embedded in a matrix of plastic resin, as described by Slayter, with a clear realization of the historical background of these triumphs of modern technology."[10]

Strength of steel.

The bamboo culm combines a certain gentle capacity to yield with a strength that rivals steel.

> The Puerto Rico Experiment Station made strength tests of *Bambusa tulda,* a good bamboo but probably not our strongest; 52,000 pounds per square inch were required to break it. Walnut, according to these tests, required 20,000 to 22,000 pounds per square inch, while the steel commonly used for reinforcing concrete required about 60,000 pounds. Certainly such an easily grown plant as bamboo with culms of this strength can have wide usage . . . In hurricane countries their great strength makes them the best possible planting material for windbreaks . . . In the First World War, when steel was difficult to obtain in China, bamboo was used in building construction to reinforce concrete.[11]

Tough skin.

The durability of bamboo culms in the grove is partly owing to their bones, their fiber; but their longevity also derives in part from the unusual toughness of their skin. If you walk in a grove of giant bamboos not cleared by a regular harvest, you find the new culms of most species a dazzling green, hard as your fingernails. Their smooth emerald surface looks made that morning by a god from Oz, but the crusty grandfathers of the grove—who have survived long enough to gradually break down the vigorous resistance of living culms to insects and fungi—are an amazing palette of grey-white silvers and russets and golds. Dozen-colored lichens and mosses embroider them like old fence posts. William Porterfield, a Western botanist resident as a teacher for many years in Shanghai in the 1920s and 1930s, describes in detail the gradual process resulting in the noble ruin of a grove, witnessed in full glory only where bamboo enjoys prolonged neglect.

> The smooth polished exterior of bamboo poles is a remarkable, almost unnatural characteristic. No finish applied by human hands is so smooth and hard. The immediate cause is the secretion by the epidermis of wax and silicon. The waxy coating is the basis of the polish, while the silicon compound is responsible for most of the hardness. Some idea of the quantity of silica contained in bamboos may be gathered from the fact recorded of one species, *Bambusa tabacaria,* that it will emit sparks when struck with an axe . . . Even parasites find it difficult to effect an entrance so long as the epidermis is left intact. In time, however, the combined action of moisture, weak mineral acids, mildews, and microorganisms succeed in forcing an opening in the protective armor of the culm . . . The scar or ring just below the raised ridge of the joint proper is a favorable place of attack from fungi and bacteria. As a tender shoot the joint is enclosed in a protective sheath-leaf. Later, as each section in order attains its full size, the sheath-leaves fall off, leaving behind a scar which, like the leaf-scars on trees, becomes thoroughly corked over as a protection against the loss of internal moisture and the invasion of pests. But this does not always take place without defect, nor can the chance always be avoided of attacks by boring insects in the wake of which spores and penetrating mycelia find their way into the tissues and eventually the joint. The toughness of the skin, however, is attested by the fact that in the bamboo forests pieces of cane are found with the wood entirely rotted out, leaving only the epidermis as a shell.[12]

BRANCH COMPLEMENT

Uses for identification.

Branches emerge from culms on alternate sides, just above nodes, from single buds just below the middle of the sheath that covers them. In all bamboo genera—with a single exception—the frequent abundance of branches ramifies from a single bud at the base of each branch complement. Only the South American genus *Chusquea* bears a principal bud flanked by two to many smaller ones that in some species (for example, *Ch. pittieri*) completely ring the culm. Some buds break in ascending order on the culm as successive internodes complete their growth. All *Phyllostachys* species and the only bamboo native to the continental United States—*Arundinaria gigantea* and *A. tecta*—follow this pattern. *Semiarundinaria viridis* breaks branch buds at midculm, and the process extends to higher and lower nodes simultaneously. Others break buds from the top down, with a period of dormancy lasting up to several weeks in the case of such species as *Bambusa textilis.*

Most young immature bamboos bear branch buds at all nodes. In more mature plants, basal buds may remain dormant indefinitely or even be lacking entirely up to ½, ⅔, or even ¾ the height of the culm. *Bambusa textilis* and *Arundinaria amabilis* are extreme examples of this condition, which enormously facilitates harvest and use—and consequently increases the value of the species. Anyone who has ever removed branches for basketry or other crafts will appreciate the advantage of a species virtually without them in that portion of the culm with thickest walls and widest diameter.

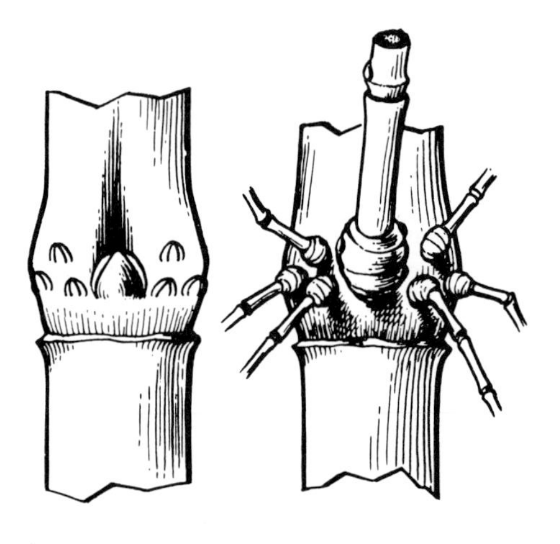

Branch buds (left) and complement (right) of the genus* Chusquea. *Multiple buds depart from the norm of all other bamboos, in which branch complements of up to 35 branches spring from the base of the dominant branch(es) emerging from a single bud.

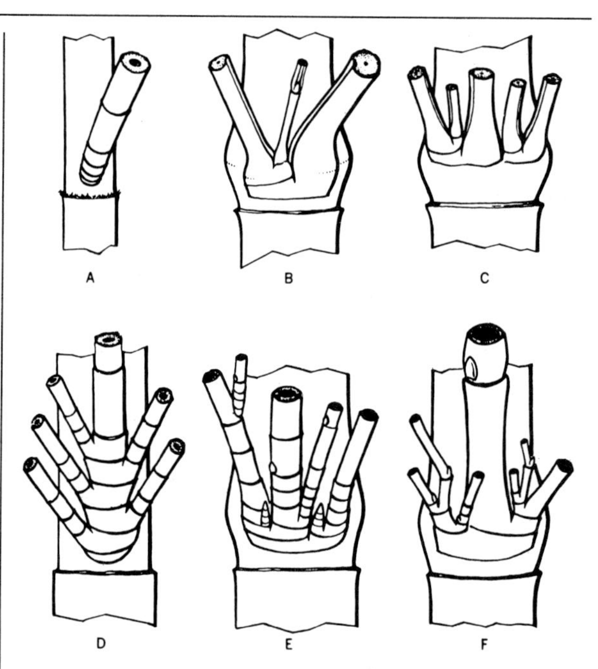

***Proliferation of branches from the base of midculm branch complements in six species. (A)* Sasa palmata. *(B)* Phyllostachys dulcis. *(C)* Shibataea kumasasa. *(D)* Arundinaria tecta *var.* decidua. *(E)* Arundinaria simonii. *(F)* Arundinaria fastuosa.**

Midculm branch complements are usually most typical of a given species. Lower branches are usually less completely developed, and branches at the upper nodes are frequently less distinctive of the genus and species.

Number per node.

The ramification in some genera is characteristically minimal. Species of the genus *Sasa* are distinguished always by a single branch at each node without basal buds on it to produce others. *Phyllostachys* species typically bear two unequal branches above a prominent "supranodal ridge" with, occasionally, a skinny and diminutive third emerging from a basal bud on the smaller of the original pair. The bud of some bamboos—such as *Melocanna baccifera* or the known species of *Schizostachyum*—may branch before even bursting through the prophyllum (the sheath at the first node of a branch) as a tuft of slender, subequal branches, with the primary branch not noticeably greater in size.

In other genera, such as *Bambusa, Dendrocalamus,* or *Gigantochloa,* the primary branch remains strongly dominant, and the branchlets emerging from its basal nodes are progressively thinner and shorter. In species of these genera, the dominant branch tends to resemble the culm itself

Identification.

Branches per node in various genera provide one gross preliminary form of identification. One branch per node tells you the bamboo is *Sasa, Sasaella,* or *Pseudosasa.* One branch at lower and upper nodes and three at middle nodes indicate *Pleioblastus, Semiarundinaria,* or *Sinobambusa.* One strong branch, surrounded by a dense cluster of short, slender, subequal ones is unique to *Dinochloa.* Two branches, unequal, with often a very small third between them are the infallible sign of *Phyllostachys.*[13]

in miniature, with the swollen base of the branch appearing like a reduced rhizome. The basal internodes of the branch are solid and the proximal nodes—those nearest the culm—are generously ringed with root primordia. *Bambusa tulda, Bambusa textilis, Bambusa vulgaris, Gigantochloa apus,* and *Dendrocalamus asper* are some common examples.

Dwarf branches—thorns.

In some genera with scandent culms—*Chusquea* and *Dinochloa*—principal branches may be nearly as thick and long as the culm itself, an obviously useful feature for vinelike climbers. In self-supporting culms, there is often a correspondence between the lengths of culm internodes and the lengths of primary branches growing from their nodes. Some genera of bamboos—*Guadua* and *Bambusa*—produce dwarfed, sharply pointed branches, or thorns, usually more densely at lower portions of the culm.[14]

SHEATHS

Protective packaging on rhizome, culm, and branch.

Every node of rhizome, culm, or branch bears a sheath that protects it during growth. The rhizome sheaths are sharply pointed shields enveloping the meristematic tissue shoving through the soil, smooth and shiny on the inner surface—like culm and branch sheaths—permitting the growing rhizome to slide forward with minimum friction, like a piston in its cylinder. Of all sheathing organs, rhizome sheaths vary least in form from node to node. The blade—at the tip of the sheath proper—is negligible in size, and there are no auricles, which in culm sheaths provide an important hint for distinguishing between species.

Culm sheath for species identification.

The culm sheath is a more complex structure and,

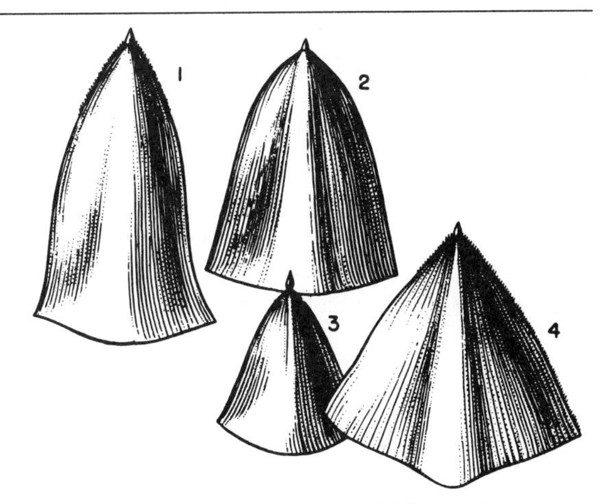

***Rhizome sheaths. The extreme reduction of the sheath blade is typical of all species. (1)* Semiarundinaria viridis. *(2)* Sasa kurilensis. *(3)* Arundinaria variegata. *(4)* Phyllostachys bambusoides.**

like the branch complement, is most characteristic of each species at midculm nodes. It is a primary means of distinguishing closely related species in a genus, depending on its size, color, shape, and the form of the blade, auricles, ligule, and oral setae at its tip. These appendages may be lacking, reduced, or much altered at base and tip of the culms. Basal sheaths approach the form of rhizome sheaths; at culm tip, they take the form of a leaf sheath with a stalked blade. Midculm sheaths are consequently those examined for species identification.

The culm sheaths serve to distinguish bamboos by their behavior as well as their form. *Sasa* sheaths

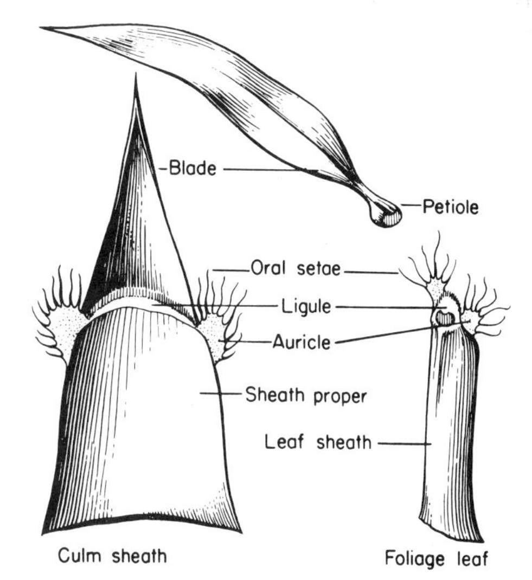

Culm sheath and foliage leaf sheath of* Bambusa *genus, not drawn to uniform scale. In species without auricles, oral setae are sometimes called shoulder bristles. Clues to species identification include the shape, size, color, and hairiness of culm sheaths; their duration on the culm; and the forms of auricles, oral setae, and blades of both clum and leaf sheaths.

Bamboo leaf.

are persistent, *Phyllostachys* deciduous, that is, tending to fall off quickly. *Arundinarias* vary from species to species. Some bamboos—for example, *Guadua angustifolia*—have sheaths that fall readily from upper nodes while remaining for years below. In some species of *Chusquea* one edge of the sheath is fused to the sheath surface some distance up the internode in such a way that the branches burst through the sheath tissue, rupturing it in a characteristic way. Some species all but lose the sheath, which hangs by a small portion at the center of the base, fluttering in a certain manner that serves as a clue to its identity (for example, *Arundinaria fastuosa*).

Branch sheaths repeat culm sheaths in a miniature form, which becomes smaller and more generalized in successive orders of branches. They may be persistent or deciduous. In clambering species whose dominant midculm branches approach the culm size in diameter and length, the culm and branch sheaths are virtually identical *(Chusquea simpliciflora, Melocalamus compactiflorus).*

Leaf sheaths cover the tips of all culms and branches. Their blades, the familiar leaves or foliage of bamboos, are "the plant's principal source of elaborated food."[15] The leaf sheath is distinguished from other sheathing organs by a blade much larger than the sheath proper, and by the *petiole* or stalk that connects the leaf base to the sheath tip. Leaf sheaths are much longer than their respective internodes and consequently overlap, each nearly covering completely the sheath above.

LEAVES

Abundance, tessellation.

Bamboo's vitality is in part owing to its leaves, which are usually plentiful and range from tiny to immense. Hair fine, scarcely 3 mm wide on *Arthrostylidium capillifolium,* leaves are 5 meters long by ½ meter wide in *Neurolepis elata* from the Andes.[16] The annual leaf fall of bamboos is commonly 4 inches deep, and in many species roughly equals in weight the year's growth of new culms. Each leaf is tightly rolled up during development, with one edge, not the tip, at the center of the roll. The parallel secondary and tertiary veins of all leaf blades are tessellated—connected by transverse veinlets, which are weakly to strongly visible on one or both surfaces in hardy, monopodial species and obscured to superficial view in sympodial bamboos by overlying tissue. Foliage is commonly

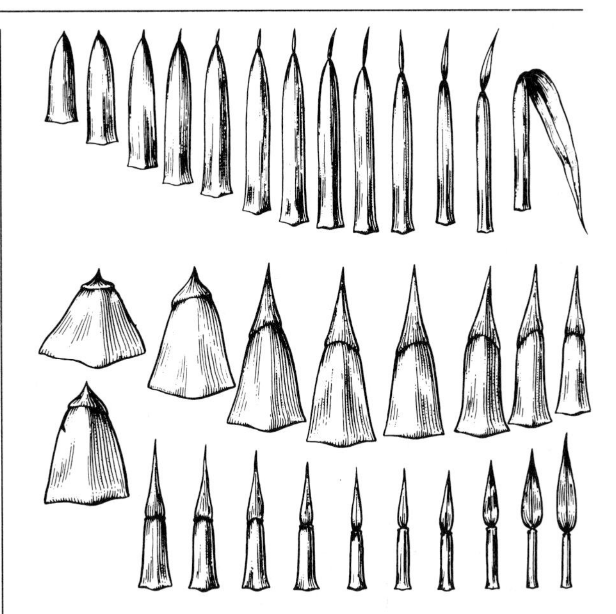

Culm sheaths of successive nodes from a single culm of* Pseudosasa japonica *(above) and* Bambusa pachinensis *(below). Midculm sheaths are used for purposes of identification. Basal sheaths resemble the rhizome sheaths, while at the uppermost nodes the culm sheath, with much more prominent blade, resembles the leaf sheath and blade.

renewed annually from leafy twigs but is diminished or stopped when flowering starts. Extreme drought causes some species such as *Bambusa vulgaris* or *Dendrocalamus strictus* to shed leaves completely. Bamboo's autumn is generally in spring. The leaves from the last season fall gradually and are inconspicuously replaced by fresh fabricators of nutrients for the grove at the time when the plant needs all the energy it can get to send shoots speeding up.

Urban oxygen injection.

When in a grove, look carefully and you will find

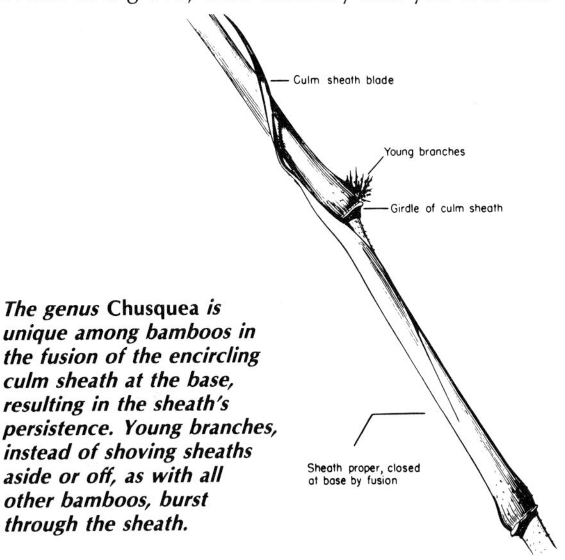

The genus* Chusquea *is unique among bamboos in the fusion of the encircling culm sheath at the base, resulting in the sheath's persistence. Young branches, instead of shoving sheaths aside or off, as with all other bamboos, burst through the sheath.

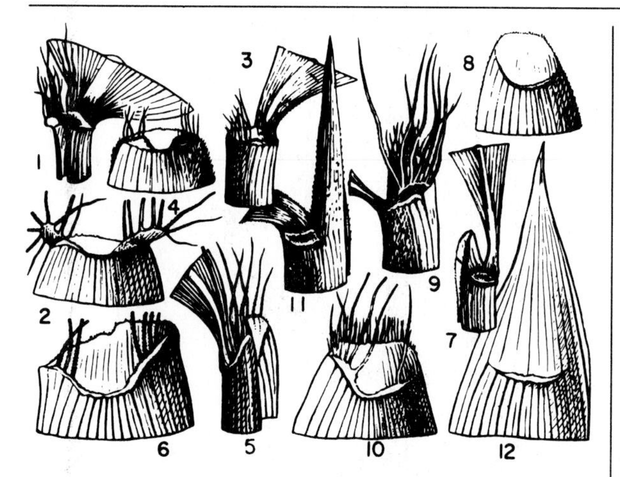

Leaf sheaths of various bamboos, with inner and outer views of the tip showing various forms of auricles and ligule. (1, 2) **Bambusa shimadai.** *(3, 4)* **Sasa tsuboiana.** *(5, 6)* **Arundinaria variegata.** *(7, 8)* **Phyllostachys pubescens.** *(9, 10)* **Phyllostachys formosana.** *(11, 12)* **Shibataea kumasasa.** *All are drawn at the same scale.*

that, in a cluster of leaves there will often be one which, even without the least breeze stirring, quivers with an astonishing rapidity, much faster than your eye can register. The effect of these scattered, extra-eager leaves is to give the whole culm a constant tremble of vitality, even in the stillest air. The evergreen foliage of bamboo is, of course, the chief reason for its popularity as an ornamental. Dense and rapid shade, a visual buffer that increases privacy and diminishes the effect of architectural errors, a muffler to mute the roar of traffic, an injection of oxygen into urban smog—the leaves of bamboo are one of its best pieces . . . and its biggest: One culm of Japanese moso—*Phyllostachys edulis*—had a total of 63.8 square meters of shimmery surface shared by 88,762 leaves.[17]

Appearance: A simple experiment.

We return often to the leaves of bamboo in the course of this book; we can leave them for the moment with some words of Alexander Lawson, whose experience in cultivating hardy ornamental

Leaf blade of **Phyllostachys pubescens,** *in cross section, before opening. Developing leaves of all known bamboos are rolled up tightly, with one edge at the center of the roll.*

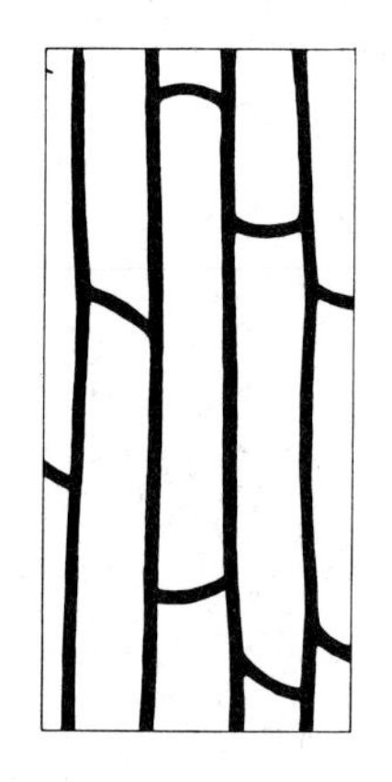

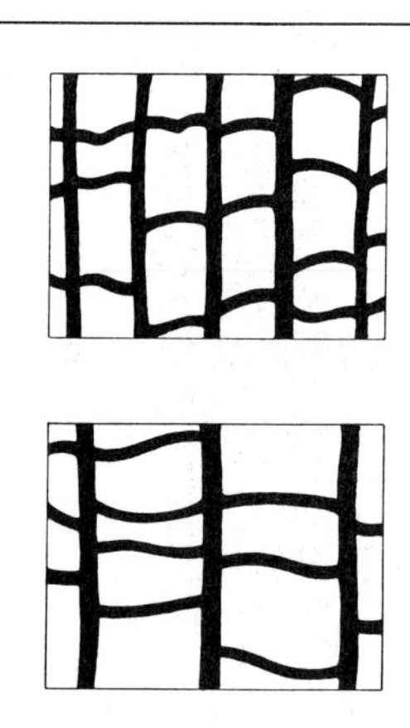

Tesselation, the mosaic pattern of longitudinal and transverse veinlets in a leaf, occurs in all bamboos but is a visible characteristic of hardy, monopodial species. It is hidden from the unaided eye by tissue in sympodial bamboos, whose leaves are often more tough and leathery than those of northern species.

bamboos derives from many years as head gardener at one of the largest European collections in Uplyme, East Devon, where over fifty species are established:

> The upper surfaces of the leaves are invariably smooth and glistening, and after a shower of rain they flash and flicker like mirrors when the sun catches them. The undersurfaces of the leaves are always somber in color and often coated with minute hairs. The dull matt effect is produced partly by innumerable minute peglike projections protruding from the lips of the many *stomata* and are designed to prevent globules of moisture from adhering and choking the entrances of the tiny breathing passages. Immerse a leaf in water and it will be seen that whilst the upper surface retains its smooth and glossy appearance, the underside acquires a glistening silvery coating, formed by the numerous bubbles of air trapped and held by the pegs. The leaf can be left immersed in water for a considerable period, and even agitated from time to time, but the pegs will continue to hold onto the air bubbles and will do so until the cell structure of the leaf begins to decompose. This simple experiment could provide an interesting discussion point for school plant biology classes.[18]

FLOWERING

Cycles: Flocks of flowers.

The flowering cycle of bamboos is one of their most unusual, disputed, and botanically mysterious characteristics. It varies from 1 to 120 years "according to numerous but weakly documented records" on the widely varying habits of different species.[19] Some flower annually; others sporadically at frequent but irregular intervals; but the pattern most peculiar to bamboos is "gregarious": Crowds of plants within groves, in crowds of groves stretched across continents, if connected genetically—alerted by some unriddled mechanism in the cells, some clock or calendar that dates the age of the genetic stock—hundreds then thousands, then hundreds of thousands of plants suddenly stop making culms and rhizomes and start making tiny flowers, as inconspicuous, individually, as they are rare. Gregarious flowering may occur over small areas or cover hundreds of square miles. It has been seen beginning in one area and spreading in a specific direction, requiring a few years to cover the entire flowering zone, continuing sometimes as long as from five to fifteen years, with fruit production ranging from abundant to zero. Many species are partly or completely sterile. Common among bamboos, gregarious flowering is characteristic also of a palm, the Talipot *(Corypha)* of south India, but the occurrence is virtually unknown among other plants.[20] Many variables, considered as possible causes, have been dismissed:

> Studies in Japan have failed to show any relation between flowering time and age, size of clump, thickness of stem, soil fertility or moisture, exposure to sun, climatic factors or locality differences . . . In India . . . physiological disturbances caused by injury, cutting, or prolonged hot, dry weather have been mentioned as possible stimulants to flowering, but there seem to be numerous exceptions. Obviously more study is needed on this question.[21]

A blooming mystery.
Researching a phenomenon whose ample rhythm spans several lifetimes requires the patient—almost fanatically patient—collaboration of generations of students of bamboo behavior. "In a test of their blooming cycle, *moso-chiku (Phyllostachys mitis)* seeds were sown in 1912 by Tokyo University in its Chiba experimental grove and by Kyoto University in its Kamigamo experimental grounds. Neither have bloomed yet."[22]

Death of groves.
The flowering of many species is followed by the death or severe setback of groves. The larger species are generally the most important economically, and these take the longest to reestablish themselves from seedlings. *Dendrocalamus strictus,* for example, may require twelve to thirteen years to reach

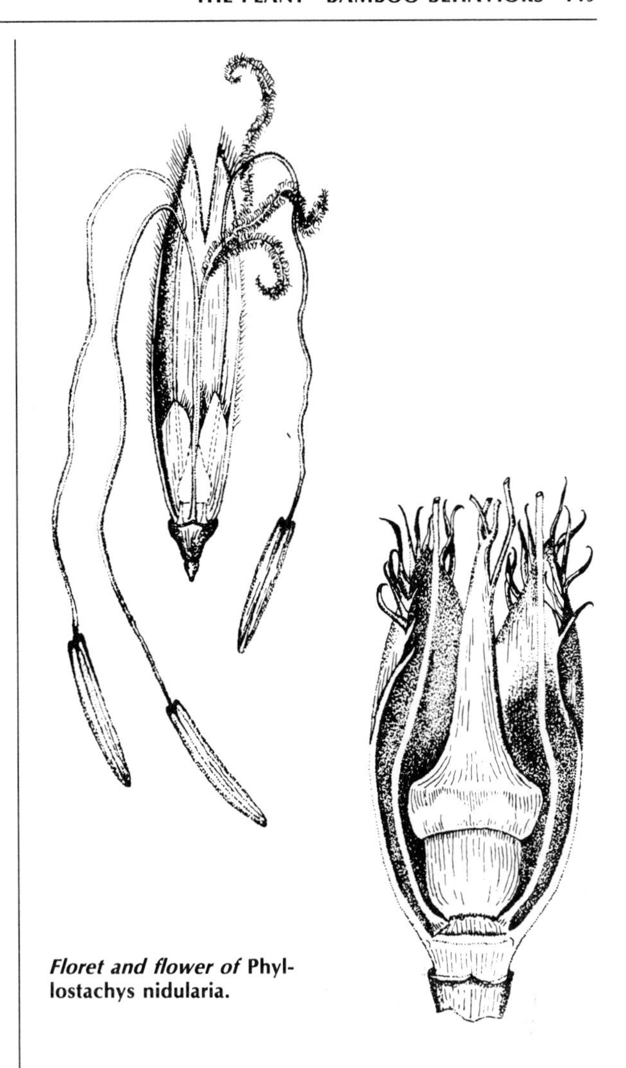

***Floret and flower of* Phyllostachys nidularia.**

mature statures again in a grove after flowering. Aided by conscious human care, this can be reduced to six years. If a given species is important in the local economy, the cycle becomes crucial, and in parts of India people's lives are even counted in *kutung* units of bamboo flowerings. In Japan, when the principal commercial species for baskets and construction, *madake (Phyllostachys bambusoides)* flowered in 1844–46, its excellent wood became so scarce that one culm was worth three bags of rice. In the case of species producing an abundance of seeds, flowering in India has sometimes helped relieve famines occasioned by prolonged drought, which some observers felt in fact triggered the flowering itself. (*Dendrocalamus strictus* and *Bambusa arundinacea* are the species involved.)

Famine relief and rodent explosion.
Although bamboo seed has often relieved human hunger, it has often been bad luck for farmers be-

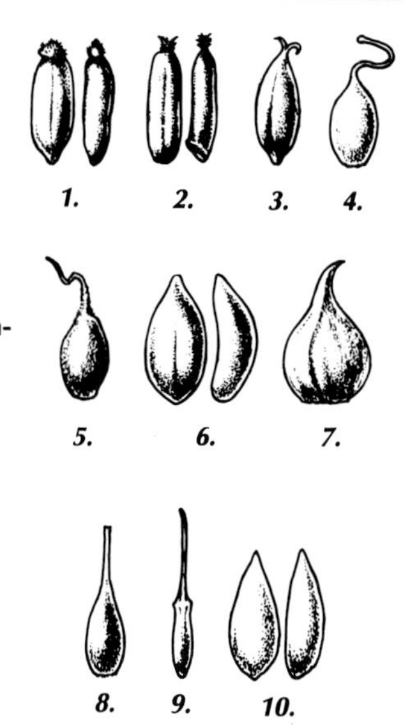

Seeds of ten species. (1) **Bambusa glaucescens,** *dorsal (back) and lateral (side) views, times 1½. (2)* **Bambusa longispiculata,** *dorsal and lateral views, times 1¼. (3)* **Chimonobambusa marmorea,** *dorsal view, times 2½. (4)* **Dendrocalamus asper,** *times 4. (5)* **Dendrocalamus strictus,** *times 2¼. (6)* **Guadua aculeata,** *dorsal and lateral views, times 3. (7)* **Melocanna baccifera,** *times ⅓. (8)* **Ochlandra travancorica,** *times 1. (9)* **Phyllostachys pubescens,** *ventral (front) view, times ¾. (10)* **Pseudosasa japonica,** *ventral and lateral views, times 2¼.*

cause of the swarms of rats attracted to the groves. In June 1855, a species of *Sasa* bloomed in a mountainous area of Japan in Yamagata Prefecture, and the hordes of wild rats that arrived to eat the seeds also feasted on a number of surrounding crops. About 350,000 rats were captured. In 1966, "all the sasa in a highland area of 10,000 hectares in central Ehime Prefecture on the island of Shikoku bloomed and bore seeds. An estimated 170,000 rats gnawed at the *hinoki* (Japanese cypress) growing there, causing it to wither and die."[23] In southern Brazil, following the heavy seeding of a species of *Merostachys,* similar invasions—*"ratadas"* as they are locally called—periodically occur.

Seed scarcity the norm.

Because of the spectacular nature of such reports, a general notion that bamboos seed heavily has become widespread. McClure, on the basis of over forty years of personal observation of flowering plants of twenty recognized genera and examination in dried collections throughout the world of preserved flowering specimens representing the fifty-five remaining known genera, comes to the contrary conclusion that:

> The incidence of maturation of fruits is relatively low in the majority of known bamboos; abundant yields occur in relatively few species, principally bamboos not under cultivation. However, this impression should be disciplined by a recognition of the diverse nature of the evidence. In bamboos that have a long flowering period, the number of fruits discoverable at a given moment is usually very low. This suggests that either the set of fruit, or its maturation, is relatively infrequent, in both time and space. Mature fruits more or less promptly fall to the ground. Wild creatures, particularly certain birds and rodents, harvest them either from the plant or from the ground. In North Carolina the destruction of fruits by insects is so high in *Arundinaria tecta* that to get mature seeds Hughes had to dust the inflorescences regularly with DDT. Of *Guadua amplexifolia,* which flowered in 1954 in Puerto Rico, Kennard says, "In spite of many thousands of flowers produced and pollen apparently viable, fruit set was very low. Examination of thousands of spikelets yielded only 1,003 fruits." Only 1 percent of the seeds that were planted germinated.[24]

Taxonomy: Based on reproductive structures.

Since the time of Linnaeus, classification of plants in taxon, or genetically related groups, has been based primarily on reproductive structures. Systematic procedures traditionally grant little taxonomic value to the vegetative parts of the plant. In the case of bamboo, these parts are frequently bulky and/or require special treatment to preserve. Since many bamboos flower so rarely, the difficulty of precise differentiation of species is greatly increased. Since vegetative production generally ceases during flowering, portions of the plant useful for identification such as the fragile, rapidly deteriorating culm sheaths are usually not available for collection at the same time as reproductive structures and are therefore rarely found in herbarium specimens with flowers from the same plant.

Rare flowering creates a bamboo babel.

These are a few of the reasons why "seasoned agrostologists recognize the tentative nature of even the most conscientious and sophisticated treatment of bamboo classification."[25] Like the amorphous monsters that roam many mythologies, bamboo—as known to science—is constant only in its change of shape. *Sinocalamus,* established as a genus by McClure in 1940, was knocked down soon after by the same author, who concedes that the efforts of other botanists to play Adam—assigned in an earlier garden the task of naming plants —will prove equally ephemeral. In some 200 years since 1789, when the first bamboo genus—*Bambusa*—was described on the basis of a species now called *Bambusa arundinacea,* "seventy-five effectively published generic names and over a thousand specific names have been attached to plants generally recognized as bamboos." Many of these names, however, have proved deciduous, and

Bamboo flowers.

many more "are either 'illegitimate' or destined to fall into the limbo of synonymy."[26]

Confusion of names.

It is a crowded limbo. Lawson's *Bamboos* (1968) devotes some hundred pages to detailed descriptions of ninety-five species and cultivars, and then spends seven pages listing their synonyms. The flowering of umbrella bamboo in 1976 permitted Thomas Soderstrom, curator of botany at the Smithsonian Institution, to identify it as *Thamnocalamus spathaceus,* after it had, for "a hundred frustrating, flowerless years, twice been proclaimed a new genus, twice been classified as an existing genus, and four times received new species names."[27] Bamboos as a whole have been treated by some botanists as a separate family of plants, by others as a subfamily of grasses, and by still others as a tribe of Gramineae: The confusion of species merely reflects the uncertainty that surrounds the question of the correct classification of the group as a whole.

Flowers hold hints for genetics and geogeny.

The tiny flowers of bamboo—apart from their taxonomic value, which is worth their weight in botanists—may hold hints relevant to big questions in other fields of science that are not obviously related to their frail stamens: In the article referred to above, unmasking at last the true *who* of umbrella bamboo after a century of conflicting wrong guesses, Soderstrom concludes with the following speculations:

> These newly established taxonomic relationships are of more than just theoretical interest. For example, our new taxonomic studies suggest that the Sino-Himalayan bamboos of the genus *Thamnocalamus* may be related to a bamboo species native to South Africa. Long-distance seed dispersal is not characteristic of bamboos, so the only likely way these now-widely separated species could have originated as a single group is if the regions in which they now grow were once close together. It happens that Africa and Asia were once part of the single land mass Gondwanaland. If the now-dispersed *Thamnocalamus* group originated before Gondwanaland split apart, then *Thamnocalamus* must be a very old genus indeed.

Flowering of bamboos may offer hints on genetics as well as geogeny:

> The peculiar timing mechanism in bamboo cells is also a feature with broad implications. That even a young bamboo specimen grown from a cutting will flower and die if its parent plant is at the end of its predetermined life cycle is a fact of more than passing interest. It may well be that in their possession of an inner timing mechanism, virtually unique among plants, bamboos may offer a handy tool to study aging and genetics.[28]

SEEDLING

Seeds resemble wheat.

The seed of *Melocanna baccifera,* the important and prolific bamboo from east Bengal and Burma, is atypical both in its size—visualize an avocado—and in the habit of its fruit to germinate while still hanging on the branch. More commonly, bamboo seeds have roughly the size and appearance of a grain of wheat and germinate under the litter of the grove. If the seed is viable and abundant, seedlings will literally carpet the areas around each clump or flowering culm in a grove whose reproductive hour has come.

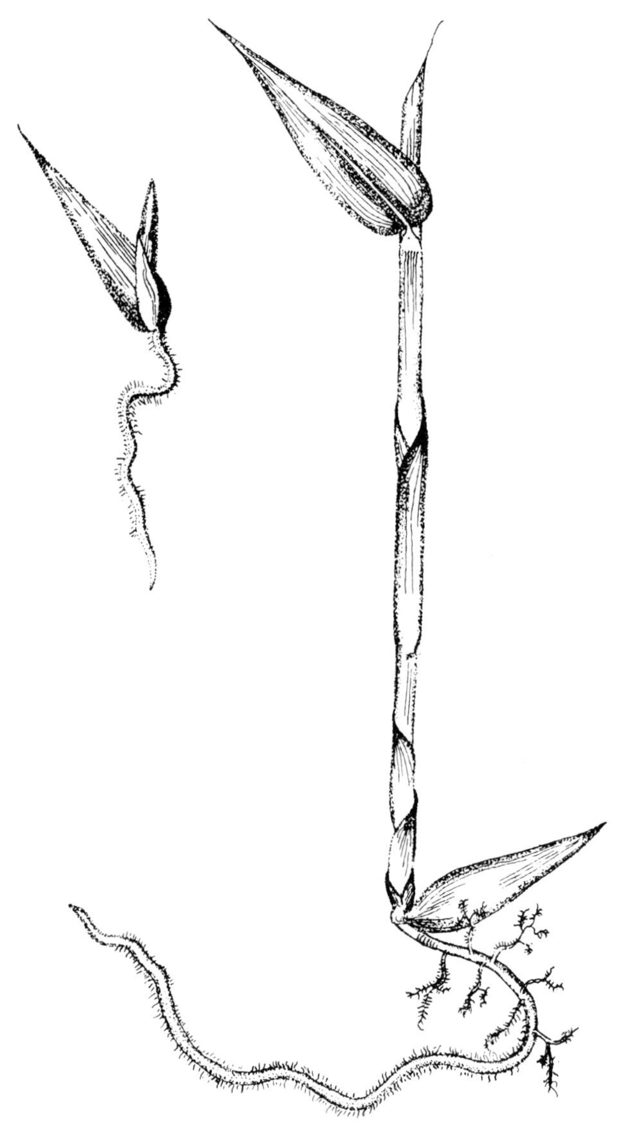

Bambusa glaucescens
seedling.

Primary root, telescopic culm.
The seedling at first consists of a primary root and primary culm. The former emerges first, producing root hairs and a few lateral branches that are soon overshadowed by adventitious roots emerging just above the lower nodes of the primary culm. The culm itself is meanwhile elongating through cell division in the meristem just above each node. (The meristem is "a body of tissue in which cell division and differentiation are active or potential."[29]) The culm "opens" like a telescope through the elongation of cells in zones of intercalary growth, that is, zones between mature tissue, which in bamboo culm, branch, or rhizome lie just above the point of insertion of a sheath. McClure notes:

> There is empirical evidence that the sheath may be the origin of substances that control, or at least influence, the process of intercalary growth and possibly also the initiation of root and branch primordia. When Chinese gardeners wish to dwarf a bamboo, they remove each culm sheath prematurely, beginning with the lowest, before the elongation taking place above its node is completed. Upon the removal of a sheath, the elongation above its node ceases. The initiation of branch buds and root primordia on any segmented axis always takes place within a zone of intercalary growth, before the tissues lose their meristematic potential, and while the subtending sheath is still living.[30]

Eighteen-foot seedling.
The primary culm of most seedlings—of *Arundinaria, Bambusa,* or *Dendrocalamus,* for example—is commonly only a few inches to around a foot high. *Melocanna baccifera* seedlings at the USDA Experiment Station in Puerto Rico, however, are said to have reached the rather astonishing height of 18 feet. In some species, the primary culm produces secondary culms from buds at the lowest nodes, usually before leafy branches emerge on the first shoot.

Proliferation through tillering from primary culm.
The rhizome generally begins to form only after the growth of several culms by tillering, that is, culms shooting from the base of the primary culm. Once a seedling has developed a rhizome, it has all the structures of a mature plant, but the rhizome's form will change with maturation, which may require three to twenty years, depending both on the species and environmental conditions. Leaves may increase in size with maturity *(Arundinaria tecta)* or, more commonly, diminish *(Melocanna baccifera).* Auricles and/or oral setae of culm sheaths may be larger in seedlings and gradually disappear in adult plants *(Phyllostachys viridis);* usually the reverse is true: Absent in seedlings, they gradually become well developed in plants of increasing maturity. Like most young creatures, bamboo seedlings can be fairly frail, easily crowded out by competing vegetation, roots drowned by too much water, or leaves burnt by too much sun.

DISTRIBUTION
The weather tolerance of bamboo is wide, result-

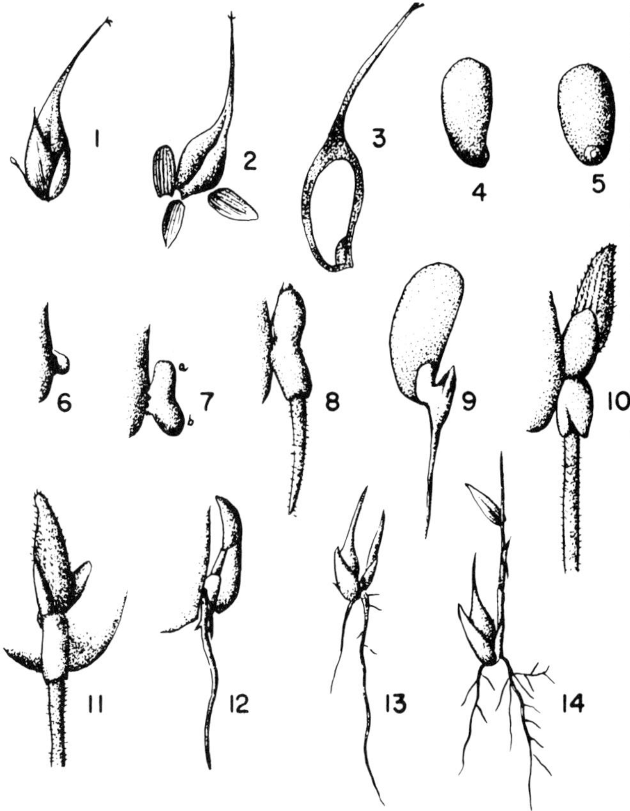

From seed to leaf: five weeks in the life of a seed of **Schizostachyum gracile.** *(1) The fruit as it falls from the plant; (2) stripped of appendages; (3) longitudinal section with the endosperm shown in white; (4) profile of germinating seed with protruding embryo; (5) front view; (6) the fourth day: part of seed, with emerging coleorhiza; (7) the fifth day: coleoptile (a) and coleorhiza (b), the primary culm and root emerging; (8) the sixth day; (9) the same on a smaller scale, with seed and the first tiny sheath of the primary culm; (10) the ninth day: side view of primary culm and root; (11) same day, front view; (12) the eleventh day: forty-eight hours later, on a smaller scale; (13) the fifteenth day: the entire plant, still nourished by the attached seed; (14) day thirty-five: leaf number one—of some 88,000 in mature culms of some bamboos, including* **moso (Phyllostachys pubescens).**

ing in a correspondingly broad distribution. Native to five continents and introduced into Europe, bamboo endures temperature ranges from −4° to 116° F and rainfall extremes from 30 to 250 inches per year. At home from sea level to 12,000 feet in altitude, it ranges between latitudes of 46° N and 47° S. "In the mountains of northern Japan, *Sasa kurilensis* is the most abundant species, growing to timberline at 1,400 m and forming pure communities in places where snow accumulates so deeply that forests are unable to establish themselves. It also occurs on the island of Sakhalin at 46° north latitude, further north than any other bamboo. The southernmost limit of bamboos has been recorded as 47° south latitude for *Chusquea culeou.*"[31]

Pygmy bamboos: fernlike understory of tropical forests.

Although bamboos are found at extremes, they occur most abundantly at low to middle elevations and, with the notable exception of Japan's 662 species, flourish most at equatorial latitudes. Australia, for example, has only 3 native bamboos, the United States, 2 and Russia, 1—while some 113 species are found in India, which lies in the umbilical of that bamboo belt stretching from Ceylon through the Malay Peninsula, southeast China, the Philippines, and New Guinea. Africa is curiously poor in bamboos, having only 3 genera, *Arundinaria, Oreobambos,* and *Oxytenanthera,* while the neighboring Madagascar is rich, enjoying some 30 species and 8 genera. Tropical America, from Mexico to Brazil, is more blessed with bamboos than any other piece of the planet with the exception of Southeast Asia and is the center of world distribution of twenty out of twenty-four known genera of the herbaceous bambusoid grasses. These understory plants of shady tropical forests, almost always smaller than bamboos—without lignified culms, with less developed rhizomes, and bearing foliage resembling ferns or palm seedlings—these pygmy

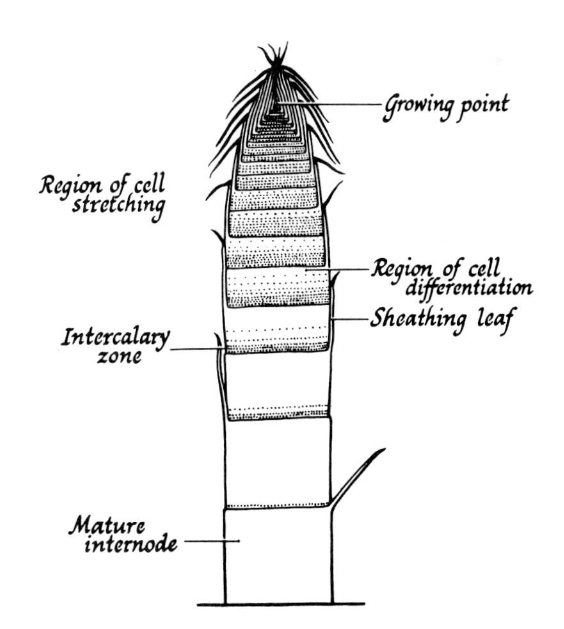

A diagrammatic section of a shooting culm, which opens like a telescope through rapid division of meristematic tissue just above the nodes. "Meristem" (fr. Gr. **meristos,** *divisible) means a body of tissue in which cell division is either occurring or potential.*

cousins of bamboos are without contemporary or past economic or cultural importance.

ENVIRONMENT

We conclude this brief sketch of the bamboos with a clipped consideration of some of the more important demands they make on their environment that determine their natural distribution, and the possibilities of their introduction where not native.

Moisture.

Bamboos are hungry plants and respond to generous treatment. Moisture is perhaps the foremost necessity of the ideal environment. A good way to observe this is to measure every day a growing bamboo shoot, keeping accurate account of the weather conditions at the same time. It will be found that the curve of growth day by day follows closely that of the percentage of moisture in the air. On rainy or cloudy days with a great deal of humidity, the growth is most rapid; on dry days with a hot sun or a coldish wind blowing, growth is very slow. Bamboos are found most abundantly along the waterways, among the rice paddies, in coastal regions where they can be bathed by sea mists, in the tropical forests and jungles, but never in dry places. All this indicates that moisture is the most decisive factor in determining the distribution of bamboo.

Shelter from winds.

An analysis of the conditions in which bamboo grows best in a natural state will show at once that shelter is also of some importance. Protection on the north and east seems to be of great assistance. The cold winds are usually from the north and northeast, and where temperature is also an object, protection from cutting winds is an advantage and in some places a necessity. The young shoots when they come up are extremely tender, and if in the spring when they first appear they are whipped about as they grow taller and battered by the older canes around them, growth will be much hampered. A southerly exposure is best; breezes from the south are warmer, less violent, and more moist.

Warmth.

The fact that most bamboos occur in the tropics shows that warmth is desirable. It is usually the case that in the presence of warmth and moisture, the work of mineralizing decaying leaves and production of soil goes forward at a rapid rate. A loamy or alluvial soil is ideal for bamboos. The hardiness of some varieties arises chiefly from the rich soil provided for their growth. Slightly sandy or clay soil is not fatal, but small canes and pale leaves indicate conditions are not right.

Surrounding flora: Companion of conifers.

Bamboo is also governed somewhat in its distribution by surrounding flora. Fairchild says it will not grow with oak or chestnut, due probably in part to root excretions. Persimmon trees, on the other hand, do not check shoot production, and some have observed the frequent association of bamboos with conifers.

Taming weather with trees.

"Tree growth helps stabilize air currents, forms windbreaks saving tips of shoots, and moderates surrounding temperature. Metabolic activities of leaves absorb a great deal of the sun's heat. At the same time, transpiration goes on. Added moisture in the air cuts down evaporation of water from the soil where tree roots also help hold moisture. During the rain, a small portion of the precipitation adheres to the leaves, about a sixth, depending on the amount of leaf surface; about a third constitutes surface run-off; while the remainder sinks into the soil. Of this the humus, like a moist blanket over the soil, holds a part; the roots absorb another part, to be made into cellulose or transpired; and the remainder drains off underground."[32]

BAMBOO ADAM

Did all the twelve to fifteen hundred species now swarming the face of the earth begin from one kind of bamboo? Botanists think sympodial bamboos were an outgrowth of monopodial species, an adaptation to climates with prolonged periods of drought. So tropical species probably came from temperate bamboos, but the details of the evolution of the species are not known. If all the bamboos presently inhabiting Asia, Africa, Australia, North and South America grew their way there from some Adam Bamboo, it is further proof of the extraordinary vigor of the plant's flexible survival design. Of course, human migrations were often the most rapid means of travel for many useful "village bamboos," as some botanists call these species symbiotically linked to the earliest human hobos. Bamboo bums were among the first wanderers because bamboo builds a mobile village.

But even without human feet beneath them, rhizomes of *P. bambusoides,* for example, travel about 12 feet a year. At that rate, they could make a mile in some 450 years. In 12.5 million years, they could circle the earth. In a world some 4,000 million years old, that's fairly fast.

Big Brother.

Apparently there are no records or samples of bamboo fossils. Nobody knows for sure how long bamboos have been striding the planet. But they are among the most primitive grasses, and many species of other genera of Gramineae have been discovered in Eocene and Oligocene beds that date back some 60 million years,[33] enough time for a leisurely monopodial rhizome traveling 2 miles per millennium to wrap five trips around the globe.

Dennis Breedlove, a botanist at the California Academy of Sciences in Golden Gate Park in San Francisco, says that bamboo has been here somewhere between one to two hundred million years (100,000,000 to 200,000,000) or some 200 to 400 times as long as the total lifetime of the human race. Known as "the brother" in Vietnam, bamboo is definitely a Big Brother far more experienced than we are at survival. And sometimes in the groves, watching and wondering how bamboo could best fold its windy green arms around our race, you realize that this big brother is also watching you. Even Walter Hawley, for twenty years director of the USDA groves near Savannah, an eminently matter-of-fact and unmystical man, a perfect paragon of prose, confessed to us with some hesitation, by the groves of *Phyllostachys arcana* in the station, an occasional uncanny sense that the groves knew and greeted him. Looking up at the sunny tips of culms whose rhizomes he'd shipped around the world from Savannah for twenty years, Walter became pensive, especially recalling the loving visits of McClure to the plants he'd nursed from China to the soil where we stood, visits which somehow intensified the feeling that the groves were not only conscious creatures but also *friendly.* "It's funny . . . you come around a corner sometimes and it feels like they're waving at you, trying to say hello . . ."

Country people are often much more in touch with the emotional life of plants and animals than the normal modern urban dweller surrounded mainly by metal and glass and cement, and for some the awareness of plants is not a question to be entertained but an indisputable fact of their experience.

As Jose Valdez explained to us while harvesting a grove of *Bambusa vulgaris* in Santa Maria del Oro (s.w. Mexico): "Plants are people, just like us. You see them, they see you. The earth isn't blind, and the mountains aren't foolish."

CHAPTER 6.

1. McClure 1966:209.
2. Ueda 1960.
3. Lawson 1968:30.
4. McClure 1966:27.
5. Ueda 1960.
6. Simmonds 1963:335.
7. McClure 1966:283. *D. andamanica* is a climber. Exceptionally large self-supporting bamboos are occasionally reported also. "Culms are known of 140–150 feet in length, and Zollinger measured a *Gigantochloa aspera* of 170 feet." Kurz 1876:251.
8. Isaachsen 1946:81.
9. McClure 1952b:18.
10. Needham 1954–1980:vol. iv.2; sec.27 : 61.
11. Lee 1944:127, 129.
12. Porterfield 1927:28–30.
13. McClure 1952b:19.
14. McClure 1966:49–61.
15. Ibid.: 69.
16. Soderstrom 1979a.
17. Ueda 1960.
18. Lawson 1968:36.
19. McClure 1966:288.
20. Simmonds 1963:335.
21. Huberman 1959:40.
22. Junka 1972:22.
23. Ibid.: 22.
24. McClure 1966:86, abridged.
25. Ibid.: 281.
26. Ibid.: 280.
27. Soderstrom 1979b:22.
28. Ibid.: 27.
29. McClure 1966:307.
30. Ibid.: 123.
31. Numata 1979:232.
32. Porterfield 1927:24–7.
33. Lawson 1968:37.

7. SPECIES: TEMPERATE AND TROPICAL

A new enthusiasm has sprung up, and there is a perfect craze for hardy bamboos. The infection, moreover, is spreading in the New World as well as the Old. There is many a Sleeping Beauty only waiting till some lover shall carry her off from her mountain fastness to awaken under the faint but kindly rays of an English sun.[1]

CRACKJAW NAMES

In *The Bamboo Garden,* Freeman-Mitford suggests that some plants may die of despair, done in by the "crackjaw names" foisted upon them by dusty botanists. Certainly students of bamboo must skirt a similar desperation, trying to remember not only the lists of names but also the shifts in the lists. Bamboo description is an art still in its infancy. We have often included some of the more relevant synonyms, the Latin names useful for reading about bamboo, and the vernacular names useful for talking about it with local people who often know more about species native to their area than can be found in books. In compiling these notes we have relied, in addition to personal experience, primarily on the following authors, whose titles may be found in the bibliography: Freeman-Mitford 1896, Lawson 1968, McClure 1944, 1957, 1966; Young 1945a–b, 1946, 1961. Sources are generally included with each species.

Our objective has not been a dry uniformity of treatment. Descriptions of most garden species of primarily ornamental interest have focused on physical features of the plant. In commercially valuable species, we have often considered use more than appearance. Dimensions given are usually the maximal heights and diameters recorded for the species and obviously will not always be reached under less than optimal conditions. Recommended *Growth Zones,* in the U.S., when recorded in USDA or other publications, have been included. We should stress that these notes are not primarily intended for field identification or for distinguishing between closely related species. Such descriptions, highly technical, are fairly boring to browse. Our aim has been more to give a rough notion of appearance, use, and lore of a fairly broad spectrum of the most common and important bamboos,* focused on, but not limited to, those species available in the United States. It is useful to remember that bamboo's form in all species reflects its environment: "It frequently happens that the same species is found in widely differing conditions of

**Starred species are samples of prominent ones, useful to get to know for a start.*

climate and soil, so changed in character and appearance to puzzle even the elect and completely bewilder the profane."[2]

The *Journal of the American Bamboo Society* (Vol. 1, No. 1, Feb. 1980, 2–11) provides a useful survey of USDA bamboo introductions from 1898–1975. Some 189 identified species and variants are included, together with a large number of introductions in which only the genus is named.

Names in English for common species.
The best way to make at least a small path of meaning through the following forbidding list of Latin names is to become familiar first with English names for the most common species on the market and in the literature. Some oriental names have passed into English usage: madake, moso, muli are a few.

Arrow Bamboo	Pseudosasa japonica
Black B.	P. nigra
Buddha's Belly B.	B. ventricosa
Calcutta B.	Dendrocalamus strictus
Castillon B.	P. bambusoides, var. castillonis
Chinese Goddess B.	B. glaucescens, var. riviereorum
Fish Pole B.	P. aurea
Giant Timber B.	P. bambusoides
Golden B.	P. aurea
Guadua	Guadua angustifolia
Hairy B.	P. pubescens
Hedge B.	B. glaucescens
Henon B.	P. nigra, var. henonis
Japanese Timber B.	P. bambusoides
Kumazasa	S. veitchii
Madake	P. bambusoides
Makino B.	P. makinoi
Medake	A. simonii
Metake	Pseudosasa japonica
Moso	P. pubescens
Muli	M. baccifera
Narihira B.	A. fastuosa
Punting Pole B.	B. tuldoides
Simon B.	A. simonii
Sweetshoot B.	P. dulcis
Thorny B.	B. arundinaceae
Tonkin Cane	A. amabilis
Yellowgroove	P. aureosulcata

TEMPERATE BAMBOOS

For the sake of a preliminary clarity, let's begin with an overall description of the three main genera of hardy bamboos normally encountered in the West, *Arundinaria, Phyllostachys,* and *Sasa.*

Arundinaria.
Arundinaria derives from Latin *arundo,* meaning reed. The genus is distinguished by the erect, smooth culms, round in cross section, without the groove in internodes that characterizes *Phyllostachys.* The branches open in new culms from tip to base. The culm sheaths are persistent, clinging sometimes for many seasons. The rhizome is generally more actively vagabond, and much more prolific in culms than *Phyllostachys.*

Phyllostachys.
Phyllostachys comes from *phyllon,* meaning leaf, and *stachys,* meaning spike. Culms are usually larger than *Arundinaria,* with a tendency to zigzag from node to node; the internodes are grooved on alternate sides. The branches open from top to bottom, generally in subequal pairs at each node, with a smaller branch between. The culm sheaths fall off quickly, as a rule, though some species such as *P. aurea* may keep their basal sheaths. Culm production is generally much less than *Arundinaria* species; basal diameter, generally greater; the rhizome, usually less vagrant, resulting in tighter groves.

Sasa.
Sa means "thin" in Japanese; *Sasa,* "superslender," refers to the culm, skinny as a pencil. Another version: "The word is believed to be a Japanese corruption of two Chinese words, *Hsai-chu,* or small bamboo."[3] *Sasas* are the most hardy and northern bamboos, covering many mountains in Japan, their country of origin. The smallest of the bamboos, they are distinguished by their leaf size, which is usually larger than any other hardy bamboos and is emphasized by their dwarf habit of growth. The branches are usually solitary, rarely in pairs. The culms are generally curved in the lower portions, and the culm sheaths are usually persistent. The leaves tend to wither at the edges in the species with large leaves.

Arundinaria.
Arundinaria alpina: 45–60 feet by 2½–4 inches.

Native to Ethiopia, Sudan, Congo, Cameroon, Uganda, Tanzania, Kenya, Abyssinia, Tanganyika. Occurring across equatorial Africa from 8,000–10,000 feet altitude in abundant and large stands, its culms are rather thin walled. Used for basketry, house construction, furniture, and crafts.[4]

**Arundinaria amabilis:* "Tonkin cane," 40 feet by 2½ inches, 13°F.

From southern China. Culms are cylindrical, straight, with nonprominent nodes, thin walled but tough, resilient and strong. Small canes serve as plant stakes; large ones used for vaulting and ski poles and especially for split-bamboo fishing

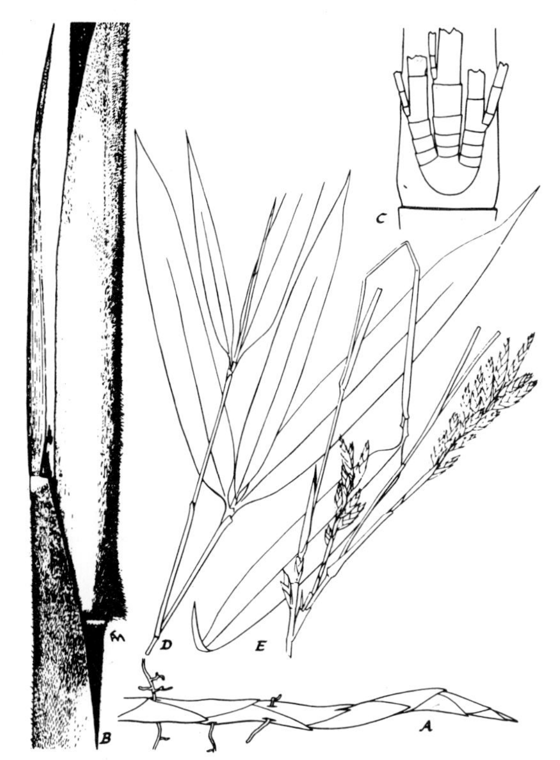

Arundinaria amabilis: ***(A) Rhizome tip. (B) Midculm node of fresh shoot, showing tip and base of hairy culm sheath. (C) Midculm branch complement. (D) Leafy twig. (E) Flowering twig.***

rods, for which this species is preferred above all other bamboos. The chief bamboo on the world market from the 1880s to the 1930s, when the Chinese revolution and World War II halted supply. Its natural advantages were complemented by disciplined grove maintenance and harvesting methods typical of the careful craftsmanship of traditional China. Selection, processing, storage, and packaging for shipping were rigorously controlled. An extensive account of this species is broadly available in a *National Geographic* bamboo cover story, October 1980.[5]

Growth zones in the United States: Gulf region, East Coast probably to Wilmington, North Carolina. West Coast probably to San Francisco area and farther north in sheltered microclimates.

Arundinaria anceps: 15 feet by ¾ inch.
A hyperhardy ornamental from India, native to the northwest Himalayas at altitudes of 10,000–11,000 feet in Sikkim and Bhutan. First introduced into England in 1865. The erect, deep green culms form an attractive hedge and, properly matured and cured, provide excellent garden stakes as well. Called *anceps* (which means two headed, that is, facing two ways or "doubtful") by Freeman-Mitford because its native land was at that time in doubt. The culms, generally straight, sometimes nodding, become heavily burdened with foliage as they mature. Initially and briefly powdered a blue white around the uninflated nodes (as much as 14 inches apart), the glossy green internodes age to a dull green brown. The culm sheaths are deciduous, although they may persist on late shooting plants until the following spring. Glossy on the inner surface, 4–8 inches long, the fresh beige green culm sheaths age to a flat straw. The thin arching branches, three to four per node in the first year beginning some yard or so above the ground, increase as the culm ages, bearing numerous twigs with leaves roughly 4 inches by ½ inch, up to 6 inches long on younger plants. Clearly tessellated, finely bristled on a single margin, midgreen above, a dull grey green below. The culms stand in small tight clumps that are separated a short distance from one another owing to the peculiar nature of the underground rhizomes. These are short, connected by long budless rhizome necks ending in a rhizome proper, from the tip of which a new culm shoots, with additional plants growing from the subterranean basal buds of this "mother" culm by tillering ("throwing out stems from the base of a stem"[6]). This species, *A. fastuosa, A. niitakayamensis, A. simonii, A. tessellata* and *Pseudosasa japonica,* are particularly recommended by Lawson[7] for canes in garden use.[8]

Arundinaria angustifolia: 8 feet by ½ inch.
Unique among bamboos for leaves with the same color on both sides; native to Japan, introduced about 1895 to England from France; the specific meaning, "narrow leaved," was given by Freeman-Mitford. "Often known to sport leaves with white variegations."[9] Rhizomes are known for their vigor and should be confined if you want limited growth.

Arundinaria auricoma: 6 feet by ¾ inch
(Arundinaria viridistriata, Bambusa fortunei, var. *aurea, Pleioblastus viridistriatus).*
Japan is the native land of this "golden haired" *(auricoma)* ornamental with quite distinctive foliage—pea green striped with golden yellow or pure yellow leaves. Named by Freeman-Mitford, introduced into Europe in the 1870s, "by far the best of the variegated species" (Lawson), *A. auricoma* is a favorite potted plant as well as an excellent garden specimen, quite hardy and, though running, not invasive. If cut back in autumn, the leaves return "fresh and sparkling" on the new canes in spring according to Lawson, who is quite vigorous in his admiration of this species. The culms are thick walled, a dark purple green, with uninflated nodes some 3–6 inches apart bearing persistent sheaths hairy at the base and tip. Branches are single, sometimes in pairs, with leaves as large as 8 inches by 1¾ inches, rough to the touch above and smooth below. Their tessellation is fairly noticeable. The species has flowered often in France, Belgium,

and England since the early twentieth century, with usually only a few canes in a stand bearing flowers. Because it is not invasive, it is ". . . a good plant for any position in the garden where it can be allowed to display its beautiful foliage to its full advantage . . . ideal for growing in the ornamental border, or in a sunny position in a rockery. The golden-yellow variegated foliage is outstanding, making this species instantly recognizable from the other yellow variegated species by its sheer brilliance . . . The golden striping varies in both width and number from leaf to leaf."[10]

Arundinaria callosa: 12–20 feet by ½–1 inch. Native to India, in the Himalayas and Khasia Hills, and Assam to 6,500 feet. Used for construction, tying thatch, crafts. Local names include *uskong, uspar, spa* (Khasia).[11]

Arundinaria elegans: 12–20 feet by 1 inch; *jilli* (Naga). From the Naga Hills in India, it grows at 5,000–7,000 feet. Used for interior walls and room dividers, walls of native huts.[12]

Arundinaria falcata: 15–20 feet by ½–¾ inch; Himalayan bamboo, *ringal, nirgal, nagre, narri, garri, gorwa, spikso, ningalo, kewi, tham, utham, kutino.* Native to India (Ravi), Nepal, Vietnam; used for construction and as a lining for roofs.[13]

**Arundinaria fastuosa:* 25 feet by 1¾ inches, −4°F
(Arundinaria narihira, Bambusa fastuosa, Semiarundinaria fastuosa).
Native to the Honshu, Shikaku, and Kyushu districts of Japan, where this elegant species is called *narihiradake* for an eleventh century mythical hero of romance, it was introduced into France in 1892. *Fastuosa*—stately—the name then given the species in Europe, suits it well: The culms are noble and erect, straight as honor, thin walled, and a deep green with a hint of purple maturing to a yellow brown. The culm sheaths are not persistent: purple outside, a quite distinctive claret on the inner surface, soon weathering to a nondescript dull light brown. "The smooth straw-colored culm sheaths often hang on in a semidetached state for a number of weeks after the new culms have completed their growth and are quite characteristic during that period."[14]

Nodes, uninflated, as much as 1 inch apart, bear two to three short stiff branches on the lower cane, more above. Leaves, four to six to a branch, are up to 10 inches by 1 inch on young canes, smaller as the culms mature. They are clearly tessellated and edged with bristles, a bright green above, dull grey green below. *A. fastuosa* flowered in England in 1935–36, in Japan in 1951, and Lawson (1968) reports several clumps sporadically in flower for many years at the bamboo gardens under his care at Pitt White in East Devon (England). The flowering canes of the species are bizarre in showing above the perfectly cylindrical lower culm typical of *Arundinarias* the peculiar features of *Phyllostachys* species. Beginning above the lowest branches, the internodes are grooved and zigzag from node to node. The species has a very slow moving rhizome and "makes a magnificent solo specimen plant."[15] A beautiful sample of *A. fastuosa* grows next to the giant Buddha in the Japanese Tea Garden in San Francisco's Golden Gate Park. "This splendid bamboo . . . is the stateliest, if not the handsomest, of the hardy bamboos."[16]

Arundinaria fortunei: 4½ feet by ½ inch
(Bambusa fortunei variegata, Pleioblastus variegatus).
A hardy ornamental from Japan, "the best of the white variegated bamboos,"[17] introduced into Belgium in 1863; it is an excellent specimen for a rock garden and an attractive potted plant. New shoots are white with green tips, culms a pale green with uninflated nodes, 6 inches apart, bearing persistent sheaths, thick and straw colored. Branches are thin and long, borne singly, sometimes paired. The clearly tessellated leaves, 8 inches by 1 inch, are covered with fine white hairs, more densely below than above, with prominent midribs. Leaves are dark green striped with white above, fading to a paler green; below, a duller green with less white striation. *Fortunei* is fairly invasive in warm climates; quite hardy but may lose leaves in a harsh winter.[18]

**Arundinaria gigantea:* 30 feet by 1¼ inch, −10°F
(Arundinaria macrosperma).
One of two bamboos native to the continental United States, its "canebrakes" once covered large areas from Virginia to Texas and provided

Arundinaria tecta ***distribution, by county, in the United States.***

an effective exit from the South for runaway slaves headed north for freedom before the Civil War. It also provided valued forage for early settlers, who found the cane an excellent indicator of fertile soils. Tests in North Carolina show that cane may be the highest yielding native range in the United States.[19] In the North Carolina coastal plain, it provided 46 percent of forest grazing in the forage types among farms surveyed in the early 1940s. Canebrakes are being drastically reduced in area by fire, uncontrolled grazing, and clearing for cultivated crops. Once killed out, cane comes back quite slowly, requiring several generations for it to naturally reestablish itself.

As with other bamboos, flowering of *A. gigantea* can mean a marked decrease in vitality or the death of a grove. "Fox Island was full of game from my earliest recollections up to 1850, when the cane which covered it went to seed and then all died. The seed grew in clusters and resembled oats, and all the animals and fowl got rolling fat from eating it. The squirrels were so fat that their kidneys were covered. This food imparted a delicious flavor to the flesh and we feasted that summer and fall. But this was the end of the cane on Fox Island, as it all died the following winter and was either carried off by high water or rotted on the ground"[20]; from a nineteenth century autobiography written in Indiana. The seed is not only eaten by squirrels; an unidentified larva bores into the seed near the base and devours the starchy endosperm, retarding prompt recovery of flowering groves.

Canebrakes were a favorite hunting ground of the Indians, as they abounded in bears, deer, panthers, wildcats, turkeys, and much small game. Canebrakes were also a good place to hide from whites, who often arrived in new territory along the rivers where canebrakes grew most abundantly. The new settlers not only pastured their livestock on cane, but also drove them into the canebreaks during winter for protection from the cold.

Canes owe their wide distribution to broad tolerance in weather and soil. They grow from sea level to 2,000 feet in the Appalachians, on all types of soil from sandy, rock cliffs and mountain slopes to muck lands *(pocosins)* and rich alluvial areas of the coastal plain—withstanding extremes of temperature from −10° to 105°F.

A smaller cane, *A. tecta,* reaches a height of about 8 feet. Some authors regard it as a separate species; others, as a form of *A. gigantea.*[21] Neither are highly regarded as ornamental plants. "There may well be situations in which a small patch would be of interest and not seriously disfiguring," is the best that Young can say for it,[22] while Lawson regretfully finds it "rather second rate . . . always sad and tattered during winter months."[23]

McClure has this to say about the confusion of species: "Small plants of *A. gigantea* are often confused with switch cane, *A. tecta,* but the two

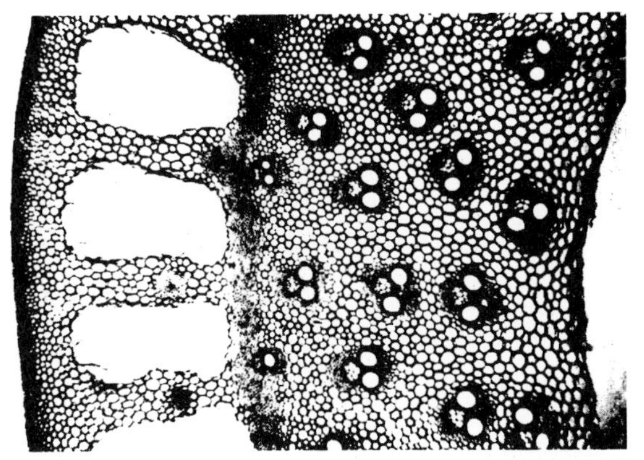

Cross section of young rhizome of* Arundinaria tecta, *showing air canals that distinguish it from* A. gigantea *(times 25).

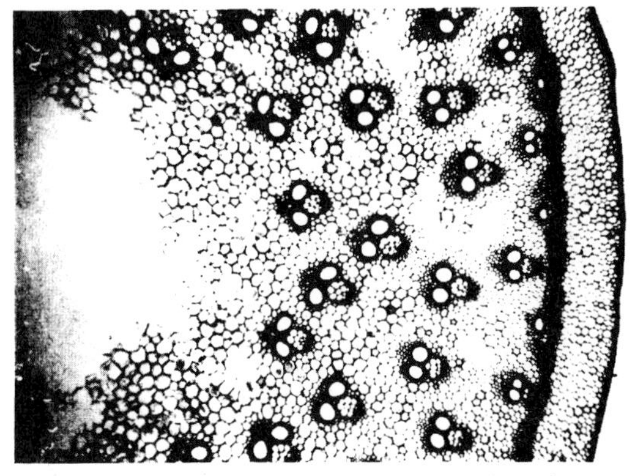

***Cross section of young rhizome of* A. gigantea *(times 25). The absence of air canals distinguishes it from* A. tecta.**

can be distinguished by the branching habit alone: culms 25–30 feet by 2–3 inches, branches short (usually much less than 1 foot) and stiffish, initiated *before* culms reach full height: *A. gigantea.* Culms 12–15 feet by ¼–½ inch, branches long and withy, initiated in the spring, several months *after* culms reach full height: *A. tecta.*"[24] McClure also reports discovering, with a 9x hand lens, air canals in the rhizomes of *A. tecta* that are absent in *A. gigantea,* further indicating they are indeed distinct species.[25]

Arundinaria graminea: 10–16 feet by ¾ inch

(Arundinaria hindsii, var. *graminea, Bambusa graminea, Pleioblastus gramineus).*

From the Ryuku Islands of Japan, *A. graminea* was introduced into England in 1877 and into the United States "from a European source, as have been most of our other oriental bamboos of small and medium size."[26] Distinguished by its narrow leaves, "grasslike," as the specific name signifies, it is one of the few hardy bamboos that love the shade. The culms, though thin walled, make very useful garden stakes. Pale green when new, they

The patron saint of **Arundinaria hindsii** *is cloud-sweeper Han Shan, the mad poet of Cold Mountain, with his companion broom.*

ripen to a dull yellow green with prominent nodes, 6 inches apart, bearing persistent, straw-colored sheaths. The branch complement, initially two or three, becomes in time numerous at each node, particularly in the upper culm. Four to six leaves per branch, up to 10 inches by ½ inch, they are sharply tapered, bristle edged, clearly tessellated, bright green above and dull green below. Flowers frequently. A fairly rapid rambler, not recommended for smaller gardens without barriers to contain its vitality.[27]

Arundinaria griffithiana: 12–20 feet by 1–1½ inches;

khnap (Khasia), *u-spar.* Native to India and Assam up to 4,500 feet. Also grows in Bhutan. Used in house construction and to tie thatch on native houses.[28]

Arundinaria hindsii: 12 feet by 1 inch

(syn. *Bambusa erecta, Pleioblastus hindsii, Thamnocalamus hindsii. Kanzan-chiku*).

Native to China, this bamboo has been grown for centuries in Japan and was introduced into England about 1875. Very hardy, grows easily in all soil types, even in deeply shaded sites. Thin-walled, olive-green culms make a good hedge—producing useful garden canes. The culms are distinguished by clusters of branches at the top giving them a top-heavy appearance. Rather rambunctious rhizome should be confined if you want to curtail travel. Often confused with *A. graminea,* but this species has coarser and broader leaves (Lawson 88–9). "As the shoots have good flavor and sprout all year round, it is cultivated as a good vegetable in southern Kyushu, where there is often a shortage of summer vegetables. Legend says it was transplanted from the Han Shan or Cold Mountain Temple in central China" (*Fuji Guidebook* 1935:35).

In Japan, this species is named for Kanzan or Han Shan, the Cold Mountain poet whose work has recently become more familiar in the West. He was often pictured with a broom and a fellow loonie-illuminate, Shih-te (Jittoku in Japanese). This species was used to make brooms, and was called "cloud sweeper" by the Japanese.[29]

Taimin-chiku, another bamboo that Satow treats under the name *A. hindsii* (var. *graminea*) has internodal lengths up to 2 feet and was early a favorite of flute makers. After a marvelously sweet flute was made from it at a certain monastery, the emperor required the area to furnish the court with culms of this "flute bamboo" as it was then called. "It is said that flutes made from stems of this bamboo grown on rocks and crags can be heard to a great distance."[30]

Arundinaria hookeriana: 15–20 feet by ½–1½ inches.

Native to India, Sikkim, and west Bhutan, an excellent potted plant, this semihardy ornamental is easily identified by its bright yellow canes with pink and green stripes. The culms are thin walled, blue green when new, ripening to their more distinctive golden yellow with green striations that show a deep pink in more sunny locations. The papery, straw color culm sheaths are moderately persistent, particularly at their base, which may remain clinging to the plant long after the rest of the sheath has weathered away. The uninflated nodes, some 8 inches apart, often show a dark blue ring, sometimes striped light green or yellow. Branches, many at each node, bear leaves quite variable in size (from 3–12 inches by ½–1½ inches) smooth blue green above, sea green below with fine white hairs at the midrib base. Not invasive. In its native lands, it grows at 4,000 to 6,500 feet but is not terribly hardy. "Usually requires some greenhouse shelter. It can be grown outdoors only in very favored areas. As a tub plant in a conservatory, it is highly valued for its wealth of delicate foliage and the uniqueness of its coloring."[31]

Arundinaria humilis: 2–6 feet by ¼ inch

(Arundinaria fortunei, var. *viridis, Bambusa nagashima, Nipponocalamus nagashima, Pleioblastus nagashima, Pleioblastus humilis).* From Japan's Honshu and Kyushu districts, this good ground cover for erosion control is called *humilis* ("low") by Freeman-Mitford for the modest stature of its dark green culms. It has dull purple new shoots with deciduous purplish sheaths drying to pale yellow. Uninflated nodes are up to 5 inches apart, bearing two to three quite long branches (3 feet). Pale green leaves (8 inches by ¾ inch) are somewhat hairy on the

dull matt green lower side. *A. humilis* is distinguished from the similar *A. vagans* by its thinner culms, longer branches, and smooth upper surface of its paler leaves. Its rhizomes are also less vagabond but are fairly invasive and can present a problem in weather friendly to its growth. "Owing to the thinness of its canes, it is often cut to the ground during very cold weather. It invariably recovers to form a green carpet before the following summer is very old."[32]

Arundinaria intermedia: 8–12 feet by ½ inch; *nigala* (Nepal), *parmick* (Lepcha), *titi nagala, prong nok.* Native to the east Himalayas of India and Nepal, Bhutan, and China, *intermedia* grows to 7,000 feet. Internodes are 15–30 centimeters. (6–12 inches) Thin-walled culms are used to make mats to cover walls and partitions.[33]

Arundinaria khasiana: 10–12 feet by ½ inch; *namlong, u-kadac namlong.* Native to the Khasia Hills of India, where it grows from 5,000 to 6,000 feet, this species is often cultivated and is widely used in wattle and daub walls of houses.[33]

Arundinaria mannii: 30 feet by ½ inch; *beneng* (Khasia). Native to India, the Jaintia Hills, and Assam, it grows to 3,000 feet. Used as withes for binding frames of houses and for firewood.[34]

Arundinaria prainii: 28 feet by ¼–½ inch; *kevva, sampit* (Naga). Native to India, the Khasia, the Jaintia and Naga hills up to 9,500 feet. Slender culms are used as lath for walls of houses.[34]

Arundinaria racemosa: 5–15 feet by 1 inch; *maling* (Nepal), *phyeum miknu, mheem, pheong, pithiu.* Native to India, Sikkim, Nepal, and China, this bamboo grows from 6,000–12,000 feet. Used for roof construction and as matting for houses.[35]

Arundinaria simonii.

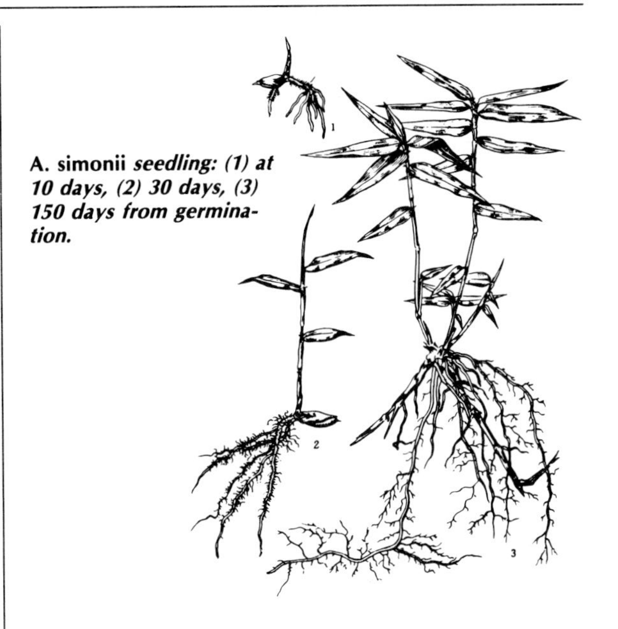

A. simonii ***seedling: (1) at 10 days, (2) 30 days, (3) 150 days from germination.***

**Arundinaria simonii:* 18–24 feet by 1½ inches, 0°F

(Bambusa simonii, Bambusa viridistriata, Nipponocalamus simonii, Pleioblastus simonii). Native to China and Japan, introduced into France in 1862 by M. Simon, then French Consul in China, the culms are thin walled but extremely strong, ideal for garden use. The species was energetically recommended as worthy of more commercial notice by 1957 Clemson College (North Carolina) tests; fishing poles, plant stakes, and shuffleboard cues were among mentioned uses. The erect, olive green culms with physical qualities resembling the excellent *A. amabilis* have uninflated nodes up to 14 inches apart. The culm sheaths are quite persistent, dull green with a hint of purple at the edges, glossy on the inner surface. They weather rapidly to a dull brown.

Branches are solitary in the first year on larger culms, then in pairs, finally a dozen or more on mature culms, growing in a distinctive steeple formation. Smaller culms may have numerous branches even the first year but always with one dominant. Three to six leaves grow on a branch, up to 12 inches by 1¼ inches on older culms, smaller on younger ones. Leaves taper to a fine tongue and are clearly tessellated. A strong green above, the underleaf is bicolored, especially towards the tip: a grey green on one side of the prominent midrib, on the other, a simple green. The species is noninvasive and makes a fine hedge. Strongly recommended for garden planting owing to its attractiveness and the useful properties of its canes for fencing, trellises, stakes, and gates. Quite hardy and vigorously abundant. "An occasional leaf may have one or more slender white stripes, and in the cultivar 'silverstripe' the striping becomes extensive."[36] Hardy to Teaneck, New Jersey, on the Atlantic coast.[37] Pacific coast, to regions where temperatures as low as 0°F do not occur more often than every three to four years.[38] In Japan, this species, known as *metake* or *Pleioblastus simonii,* covers 14,374 hectares, representing 8.5 percent of the groves and providing 9.6 percent of their harvest. After *Phyllostachys bambusoides* (61 percent) and *P. edulis* (22.4 percent) metake is Japan's main commercial bamboo species.[39]

Arundinaria tecta: 12–15 feet by ¼–½ inch.
"Switch cane" is one of two native species of bamboo in the continental United States, along with *A. gigantea.* See discussion under that species.

Arundinaria tessellata: 5–20 feet by ½ inch.
A South African bamboo used by Zulu warriors in constructing their hide-covered shields and by other forest tribes for arrows and spear handles, this species grows in such vast stands in its native lands that Bamboesberg—"bamboo mountain" in Afrikaans—is called after it. Named by Munro (1868) for the tessellation of its leaves, but this feature is much less prominent than other more distinctive characteristics of the species; the quite persistent white culm and branch sheaths are its most unique identification, not to be confused with any other hardy bamboo. The crowded lower nodes distinguish it as well, resembling somewhat those of *Chusquea culeou,* whose bottle-brush form in the branch complements *A. tessellata* also shares. New shoots are "densely covered with fine white hairs at first, smooth later, pale creamy green, often flushed with pale pink streaks."[40] The thin-walled, erect canes, a pale green maturing to deep green, age to deep purple, more intense in a sunny site. Uninflated nodes, 1½ inches apart at base, up to 8 inches at midculm, bear the persistent though paper-thin sheaths, a clear white the first year, ripening later to a dull creamy color. Downed initially with hair that weathers off, the 3–5 inch sheaths are glossy on the inner surface. Branches usually appear in the second year of growth. In time, numerous short twigs and branches grow at each node, bearing three to four leaves 2½–5 inches long by ½ inch wide. These are mid-green above, matt green below and slightly hairy at the base; tessellation is fine but clear and bristled more heavily on one edge than the other. "Bergbamboes" or "mountain bamboo" as it is known in South Africa, is most at home in moist ravines 5,000–8,000 feet in elevation. "Under cool conditions it cannot be called invasive."[41]

Arundinaria vagans: 3½ feet by ½ inch
(Pleiblastus viridistriatus vagans).
Introduced in 1892 from its native Japan into England, it is extremely hardy and vagabond (*vagans* means wandering). Excellent for erosion control, this species was used on large Victorian estates for game cover. The culms, bright green ripening to a deep olive, are faintly flushed with purple when shooting and bear persistent sheaths at uninflated nodes (some 6 inches apart), which are "bloomed conspicuously on the lower portion with a loose white powder."[42] Branches are usually solitary, sometimes paired, moderately long, with 6 inch by ¾–1½ inch leaves hairy on both surfaces (especially near the base below), bristle edged, with a prominent yellow midrib and clear tessellation. Mid-green above, matt grey green below, in winter they are characteristically withered at tips and margins. Distinguished from the similar *S. pumila* by absence of hairy ring at the base of culm sheaths and from *A. humilis* by thicker culms, shorter branches, and hairy leaves that are a darker green. *A. vagans* is "the most invasive bamboo of them all, and if allowed to get out of control can become quite a serious pest . . . will smother all but the toughest opponents."[43] This vitality, a problem in limited space, recommends the species as a stabilizer on steep banks or wherever erosion robs the soil.

Arundinaria viridi-striata: 1½–2½ feet by ¼–½ inch.
A dwarf ornamental Japanese bamboo with yellowish striped leaves that tend to curl in full summer sun, so semishady locations are preferred. Leaves, 2–5 inches long, smooth above, velvety below, in late summer lose the stripe that earned its species name. Grows better in more northern climates as far north as Philadelphia on the East Coast and on most of the Pacific Coast. Mulch is recommended in more northern winters.[44]

Arundinaria wightiana: 10–15 feet by 1 inch; *chevari.* Native to western and southern India and Ceylon (Sri Lanka), it is especially abundant on the Nilgiris. Used for matting and basketry.[45]

Chimonobambusa.

**Chimonobambusa marmorea:* 6 feet by ½ inch
(Arundinaria marmorea, Arundinaria kokantsik, Bambusa marmorea).
Native to Japan (Honshu district), introduced into France in 1889, "marble" bamboo is named for the unusual color of its new sheaths and shoots. The shoots are "pale green mottled brown and silvery white. Tipped and striped with pink . . . maturing to a dull purple or deep purple when grown in a sunny position."[46] The culms are quite thick walled, with prominent nodes some 6 inches apart, bearing persistent sheaths that are "fringed at the base with white hairs. Purplish at first and dotted with pinkish grey mottling . . ." aging to "dull silver grey." Branch complement is typically three, one short and two long. Sometimes up to five branches grow at a node, but always with one branch dominant. Leaves, 6 inches by ½ inch, end in a distinctive tongue, bright green above, a flat grey green below with

Chimonobambusa quadrangularis: ***rhizome and root system, with a leafy branch.***

a prominent midrib and moderate tessellation. Can be fairly invasive once established, though it makes a good potted plant. Attractive hedges of this species can be seen at the Japanese Tea Garden in Golden Gate Park in San Francisco.[47]

The genus name comes from its rare habit of shooting in late fall or winter (from Greek *cheimōn,* meaning winter).

**Chimonobambusa quadrangularis:* 10–30 feet by ¾ inch, 19°F.

Native of China, introduced in Japan many centuries ago, the species is distinguished, as its name indicates, by square culms. Thick walled, with strong, close-grained wood, they serve well as garden stakes. Initially dark green, the stems mature to a brownish green with occasional dark purple blotches. The swollen nodes, much larger than the slender internodes, are 5–6 inches apart and ringed with adventitious roots. Three to five branches, which snap off easily, are the usual complement at each node. The leaves, up to 9 inches long by 1¼ inches wide, are an olive green above, matt green below. The rhizomes wander, but the species is not terribly invasive as only a few canes are produced each year. In addition to its use as a striking garden ornamental, square-stem bamboo is recommended for tub plantings. Healthy stands of it can be admired in San Francisco at the entrance to the Japanese Tea Garden in Golden Gate Park.[48]

This bamboo owes its abnormal shape to supernatural powers, according to the Chinese traditions that surround its groves, although in different areas different occult agencies are credited with its creation. Some say Ko Kung, a famous fourth century alchemist, stuck his chopsticks—shaved square—into the ground of a certain monastery. There they rooted magically and produced this species.[49]

Phyllostachys.

These bamboos include many resistant species with excellent technical properties of importance to human economy. Natural stands in China are found mixed with both deciduous and coniferous forests: after its seedling stage, the shallow rhizome does not compete with deep-rooted trees.

Of 750 bamboo plant introductions into the United States by 1957, 200 were of this genus, whose center of distribution is in China where nearly all known species are native. Some were introduced to Japan long centuries ago, where only four species of *Phyllostachys* and their variants grow in general cultivation, but these make up the biggest part of Japan's bamboo harvest. In China, this genus is the principal source of paper pulp, a major source for construction and handicrafts, and an important source for edible shoots.

Of *Phyllostachys,* twenty-four species and eleven cultivars or variants have been introduced into the United States, mainly through the efforts of USDA plant explorers, especially McClure. "The potentialities of this group of plants in the economy of the United States, especially for paper pulp and for use in watershed protection and erosion control, remains largely untouched."[50]

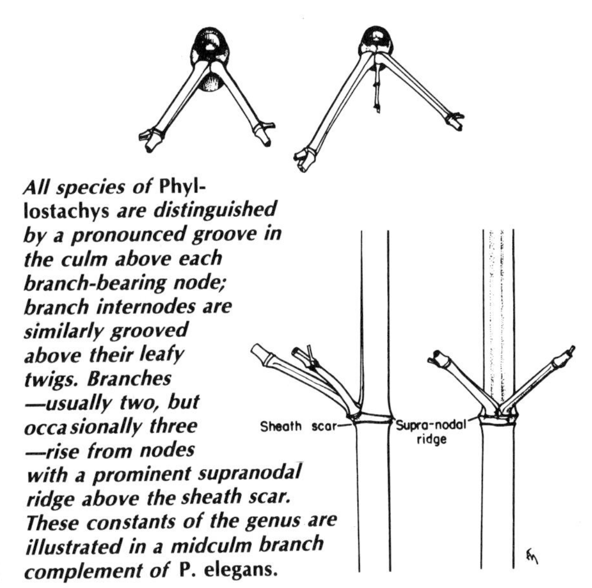

***All species of* Phyllostachys *are distinguished by a pronounced groove in the culm above each branch-bearing node; branch internodes are similarly grooved above their leafy twigs. Branches —usually two, but occasionally three —rise from nodes with a prominent supranodal ridge above the sheath scar. These constants of the genus are illustrated in a midculm branch complement of* P. elegans.**

An 1829 Japanese illustration of an unidentified variegated species whose culm resembles square stem bamboo.

Phyllostachys angusta: 21 feet 8 inches by 1 5/16 inches.

Sha Chu, "stone bamboo," in its native China, the species name refers to hard culmwood that is used in fine furniture. Nonprominent nodes are 3–10 inches apart. Midseason shoots are free of bitterness. Introduced into the United States in 1907, the species is hardy as far north as Washington, D.C., on the East Coast.[51]

Phyllostachys arcana: 27 feet 3 inches by 1 5/16 inches.

Introduced into the U.S. in 1926 by McClure from China, where its early shoots are eaten and culms are used for matting and lanterns. The specific *arcana* ("hidden") refers to "the peculiar way in which the dormant buds at lower culm nodes become partially covered by the extension of the nodal ridges." This trait and "the tall strongly arched ligule of the lower culm sheaths" serve to distinguish the species from *P. nuda,* which it resembles.[52]

**Phyllostachys aurea:* 30 feet by 1⅝ inches, −4°F.

Native to China, cultivated in Japan for many centuries where it is known as "phoenix bamboo" and "fairyland bamboo." The Latin specific "golden" can be misleading since canes are usually more green than gold, especially in shaded areas. Widely cultivated in Europe since the 1870s, popular in South America, it was the first *Phyllostachys* introduced into the United States, in Alabama, 1882. The short, swollen internodes and zigzag basal nodes of *P. aurea* are a striking characteristic of the species, and the ornamental value of this easily identified distinction is exploited in China and Japan for walking sticks and umbrella handles. Bone hard but superflexible, with the inflated basal internodes forming a natural handle to snuggle in the human hand, "fish-pole bamboo" (as the species is known in the United States) has probably landed more fish in the southern states, where it was most available, than any other single material. Excellent, too, for fan handles, clothes poles, bean poles, and plant stakes—

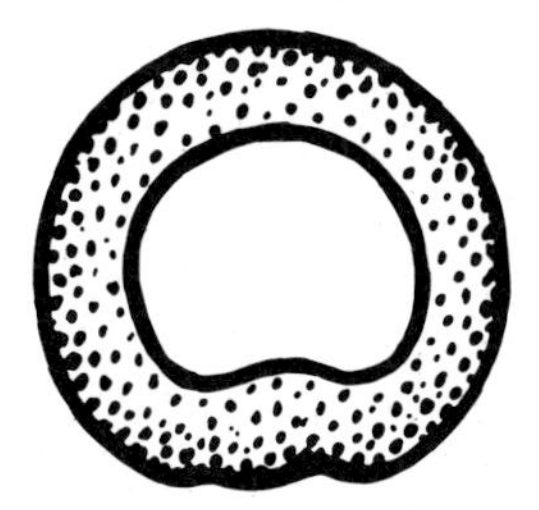

Cross section of a grooved internode of* Phyllostachys aurea *(golden bamboo).

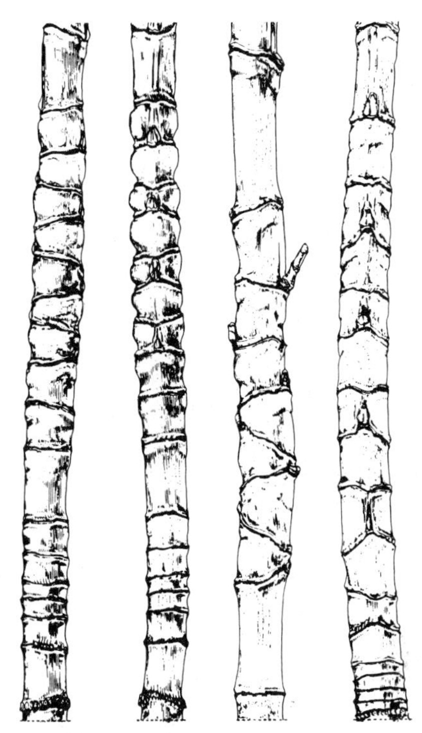

The crumpled lower culms are a distinguishing feature of* Phyllostachys aurea, *among the most common hardy bamboos in the Western garden.

branches are left on to stake tomatoes and other climbers since untrimmed canes offer more support. Pipe stems are made from branches and the tips of culms, which always stand stiffly erect in this species, even with the light source from a single side during growth.

New shoots are edible, olive green to rosy with small brown spots. The culms taper from a stout base to a slender tip, a bright green aging to a pale yellow. Only basal culm sheaths are persistent—for years—rosy, with violet streaks, maturing to a dull yellow. Nodes are crowded at the base, 1–3 inches, up to 10 inches and more in midculm. A cylindrical swelling, like an eggcup in shape, occurs below each joint in culm and branch. Stiff and upright branches are typical of the genus and occur in unequal pairs, with sometimes a third between, beginning as high as twenty nodes (13 feet) above ground in one mature 27-foot culm examined. Internodes of branches, like those on culm, are deeply grooved ("sulcate") on alternate sides above branchlettes. Leaves, two or three per twig, 4–7 inches by ¾ inch, are bristle edged more prominently on one margin, smooth on both surfaces, light pea green above, sea green below. Rhizomes are very slow moving, and the species forms fairly tight clumps. Hardy at least to Washington, D.C., on the Atlantic Coast, Pacific Coast to British Columbia.[53]

P. aurea.

Phyllostachys aureosulcata: 33 feet by 1½ inches, 0°F

(*P. nevinii*—early misidentification in United States).

Native to Chekiang Province in China, introduced into the United States in 1907, "golden-groove" bamboo, as its name indicates, is distinguished by a greenish-yellow sulcus (groove) on internodes above the point of insertion of the branch complements. New culms are rough to the touch and, in the first yard or so of some, two to three nodes are strongly "geniculate" (from Latin *geniculatus,* meaning "with bended knee"); that is, they are abruptly bent at a sharp angle in a zigzag fashion. This kink, an oddity rarely found in any other species, increases the ornamental interest of the species but decreases its commercial use. Its wood, in fact, is not of superior quality, but mature culms are recommended for fishing poles, plant stakes, and similar uses. Next to *P. aurea* and *P. bambusoides,* this species is one of the most widely cultivated bamboos among the *Phyllostachys* in the United States, owing to its ample distribution by the USDA in the 1920s as a "stake and forage bamboo." Edible midspring shoots are free of bitterness, even uncooked. When shooting, "the species is easily recognized by the pale green culm sheaths with many slender whitish stripes, a pair of prominent auricles at base of the lance-shaped sheath blade which is bristled except on sheaths near base of culm."[54] Leaves, two to three to a twig, are 6 inches by ¾ inch on new culms, smaller on older ones. Hardy to Philadelphia on the Atlantic and to the Columbia River or farther north on the Pacific Coast.

**Phyllostachys bambusoides:* 30–72 feet by 1–5¾ inches, 0°F

(*Phyllostachys quilioi, Phyllostachys reticulata,* among Japanese botanists).

This is the "Japanese timber bamboo" in the U.S. trade. In Japan, of 662 species, 61 percent of the bamboo harvest comes from *madake* (mah-dáh-kay), as this species is known there. The most versatile and preferred among all *Phyllostachys* for construction and other industrial uses, madake is excellent also for erosion control. Erect culms lean towards the light at grove margins. Mature ones are fairly thick walled; small culms of a new grove are suitable for plant stakes, rug poles, and fishing poles. Good quality split fishing rods have been made experimentally with *P. bambusoides.* Culms are durable living or dead: Stalks of madake are known to remain leafy up to twenty years in a grove.[55] Edible midspring shoots are somewhat bitter raw.

Among twelve species carefully tested during World War II for ski pole construction, this species and *A. amabilis* were found superior to

Phyllostachys bambusoides: ***The tip of a new shoot, a leafy twig, and an enlarged insert of a leaf sheath's auricles and shoulder bristles.***

Phyllostachys bambusoides, *var.* castillonis.

all others for workability.[56] Planted perhaps more widely in the United States than any other bamboo, it is most recommended by the USDA to succeed in U.S. rural economy since its use spans construction and edible sprouts.

Two other species resemble madake: *P. vivax* culms are less straight, more thin walled, earlier shooting, and faster growing. *P. viridis* has pigskin dimpled internodes that can be felt with sensitive fingers or seen with a 9x lens.

Madake is recommended in the Gulf region, the East Coast to Washington, D.C., West Coast from central California north, where winters are cool but not below 0°F.

Four distinct horticultural forms of *P. bambusoides* are found at the USDA Plant Introduction Station near Savannah, Georgia, where rhizomes of many hardy bamboos are free for the digging in February each year.

ALLGOLD. Sometimes listed as *P. sulphurea.* Culms, much smaller than madake type, are bright yellow at sheath fall, with lower internodes sometimes thinly striped with green. A white or cream stripe is found also on occasional leaves.

CASTILLON. "Undoubtedly the most impressive member of the Phyllostachys for ornamental purposes."[57] The 30-foot bright yellow culms show green slashes on grooves of culms and branches. "Golden brilliant bamboo," the Japanese call this cultivar, which they consider a spontaneous mutation of *P. bambusoides.* Highly prized as an ornamental in Japan and China, it was introduced into European cultivation in France by 1886. From there, rhizomes found their way to the collection of Henry Nehrling in Florida at some date prior to a 1916 USDA introduction directly from Japan. Culms are considerably smaller than the species type, and the edible late spring shoots are less bitter.

Castillon grows to full size in an area 25 feet across at least, after eight or more years. Initial planting should be 6–8 feet apart each way. Recommended in the United States in the northern Gulf region, East Coast probably to Norfolk, Virginia, and milder parts of the West Coast. Hardy to 5–0°F. A few degrees lower severely injures the plants.[58]

SLENDER CROOKSTEM. 48 feet by 2¼ inches, about 0°F. Many culms of this cultivar are curved near the base, and they are more slender for height than species type. It was introduced in 1925 into the United States from China by McClure. Culms are preferred to species type for harvesting nuts. Dark green above, brilliantly glaucous below, the leaves of this bamboo once seen in a slight breeze are not soon forgotten.

Zones: northern Gulf region, East Coast to Norfolk, Virginia, milder parts of the West Coast.[59]

WHITE CROOKSTEM. Introduced into the United States by McClure from Dragon Head Mountain, Kwangtung, China. Like slender crookstem, serpentine curves bend base of culms, which differ in plentiful deposits of white powder that obscure the green of old culms almost completely.[60]

Phyllostachys bissetii: 22.5 feet by ⅞ inch.

Entered the United States from Szechwan, China, in 1941. "One of the hardiest Phyllostachids under observation" at Savannah, Georgia, USDA groves, and one of the first to shoot in spring.[61]

Phyllostachys congesta: 25 feet 4 inches by 2⅛ inches.

Arrived in USDA Savannah groves in 1907 from Chekiang, China. Midseason shoots have only a slight bite. Strongly tapered, stiffly erect culms with nodes 3½–10½ inches apart are similar to *P. nigra* var. henon but are more tapered and less coated with powder.[62]

***A flowering twig of* Phyllostachys bambusoides.**

Phyllostachys decora: 23 feet 7½ inches by 1¼ inches.

Mei Chu, "beautiful bamboo" in its native China, this species entered USDA Savannah groves in 1938, sent by McClure from Hoi-wai Monastery in Kiangsu. Midseason shoots.[63]

**Phyllostachys dulcis:* 40 feet by 2¾ inches, −4°F.

Culms with prominent nodes are strongly tapered, often strongly curved. Shoots lean towards light during most active growth, then straighten up. The early spring shoots are among the most esteemed in central China where the species is native, but the culms are weaker for industrial purposes than many other species. May produce culms of maximal size in seven to eight years under favorable conditions, but drought retards this bamboo. Endures 0°F or less but won't prosper in frequent low temperatures. McIlhenny rated it the most rapid and prolific of bamboos at the Avery Island, Louisiana, groves. Recommended zones: Gulf, East Coast to Washington, D.C., West Coast except extreme north and south. *P. dulcis* grows in USDA Glen Dale, Maryland, groves, where −4°F is the minimum temperature.[64]

Phyllostachys elegans: 32 feet 5 inches by 2¼ inches.

Distinguished from others of this genus by small lance-shaped leaves especially hairy on the lower surface. Rusty brown freckles appear on culms in direct sun; internodes are 1–12 inches long. It was sent from Kwangtung (1936) and Hainan (1938), China, by McClure to USDA. *Fa chuk,* the Cantonese name for this species meaning "flowered" or "embroidered bamboo," refers to the marking on the culm sheaths. In Kwangtung, it is known as *man sun* or *man chuk,* "elegant shoot," or "elegant bamboo," a beautiful ornamental with superior flavor and quality sprouts and relatively poor culmwood.[65]

Phyllostachys flexuosa: 30 feet by 1¾ inches, −8°F.

Native of China, the culms of this bamboo are either straight or flexuous and zigzag. Midspring edible shoots are acrid raw. The medium quality wood is good for all standard bamboo uses for poles of this dimension. Hardy to −8°F with minor foliage damage.

Zones: East Coast to Philadelphia; West Coast possibly to Vancouver.[66]

Phyllostachys glauca: 33 feet 6¾ inches by 1⅞ inches.

From Kiangsi, China, to the USDA in 1926 via McClure. The early midseason shoots are relatively free from bite and bitterness, even raw.[67] (Culms of this species, misnamed *P. flexuosa,* have been tested by Glen in 1954.)

Phyllostachys makinoi: 60 feet by 2⅝ inches.

Introduced by USDA from Taiwan in 1951, this species has stiffly erect midseason shoots. During World War II, culms were imported into the United States for rug poles and as (an inadequate) substitute for *A. amabilis* in split-bamboo fishing poles, when supply of *A. amabilis* was cut off from mainland China. One of the most important commercial bamboos for scaffolds and constructions in Taiwan, curiously undistributed in the United States.[68]

Phyllostachys meyeri: 35 feet by 2 inches, −4°F (estimated).

Introduced to the United States from China in 1907 by F. Meyer. Tests show mature culms of "Meyer bamboo" are among the finest, strongest, and most versatile of the *Phyllostachys* genus. Excellent for fishing poles and nut harvesting; the longer poles are used for deep-sea fishing, and larger ones make good heavy plant stakes. Branched tops of all culms make excellent supports for vines such as lima beans and sweet peas.

Zones: Gulf region, East Coast to Norfolk, Virginia, most of West Coast.[69]

Phyllostachys nidularia: 33 feet by 1½ inches.

"Big node bamboo," *Taai Ngaan Chuk* in its native China, has culms that are strong and useful whole but do not split well: The lower internodes are frequently semisolid to solid. "Vigor and widespread occurrence in China suggest this species is worth trial for soil and water stabilization on hills or levees and for production of shoots, which are both among the best for flavor and the earliest on the spring market in Canton, where its shoots with chicken is a favorite dish."[70] *Pat Sun Chuk,* another Cantonese name for the species, meaning

Phyllostachys nigra ***(black bamboo) is a favorite ornamental, with excellent culm wood prized for cabinetwork and furniture by the Japanese.***

"writing brush bamboo," derives from the shape of these shoots, thought to resemble a brush. Another name, *So Pa Chuk,* "broom bamboo," presumably suggests use: The big nodes offer a good grip on the handle, perhaps.

**Phyllostachys nigra:* 26 feet by 1¼ inches, 0°F.

Native to southern China, anciently introduced to Japan, in 1827 black bamboo became the first hardy oriental bamboo to ramble in Western ground. Its color, striking stance, and hardiness soon made it a fairly common pleasure on a privileged lawn: "All sun-worshippers who wing their flight southward with the swallows, know the Black Bamboo as one of the chief ornamentals which grace the gardens of the Riviera."[71] The distinctive culms are first green but weather usually by the end of the first season to a fairly or very solid black. Some culms, more rarely, mature to a deep purple black. Late midspring shoots are edible. Culms are thin walled but durable. "The wood is hard enough for substantial cabinet work, and it is very cleverly used by oriental cabinet makers for decorative panels and inlays, as well as for interior finish."[72] If growing as an ornamental, remember that culms darken well in direct sunlight.[73]

Phyllostachys nigra, cultivar Bory: 0°F.

This bamboo has larger, more erect culms than the typical form, and it develops a few scattered purplish to brown spots in place of the black spots that ultimately cover lower internodes more or less completely in black bamboo. Bory has extremely limited distribution in the United States.[74]

Phyllostachys nigra, var. *henonis:* 54 feet by 3½ inches, −5°F.

Henon is among the hardiest and handsomest of Phyllostachids, native to southern China and long cultivated in Japan, from whence it was introduced into the United States in 1909. Henon bamboo is a graceful giant, thin walled, but strong. Culms subjected to comparative tests during WWII in pound resistance and fracture characteristics showed somewhat superior performance to *P. bambusoides.* Late midspring shoots are edible but remove bitterness with a few changes of water when boiling. Freeman-Mitford calls Henon "the most beautiful member of a beautiful family . . . which droughty summers do not parch and ice-bound winters do not starve."[75]

Botanists believe Henon bamboo to be the true biological species and the smaller form to be of garden origin. But the smaller was described first so botanical etiquette requires considering Henon as the cultivar. Its culm is not black, contrary to its Latin name. "Looking upward from within the clump, one sees a canopy of green foliage supported on innumerable slender pillars. Young culms are rough to the touch, until the short stiff hairs fall away from the surface; freshly divested of their protecting sheaths, they show a surface bright green, perceptibly dimmed by a thin film of white powder. In age, the powder, now off-white, all but obscures the original green, or where sun and rain beat upon the culms and swaying leafy branches sweep their surfaces, a rich lustrous golden hue may be revealed.

"As compared with mature culms, the young shoots are quite distinctive and infinitely more spectacular in their appearance, which they derive principally from the enveloping sheaths. These are an unspotted tawny hue, tinted with wine, and softened by an inconspicuous coat of spreading brown hairs. The sheath blade stands at the top, flanked at its base by a pair of well-developed dusky auricles fringed with lavender bristles. Pressed against the culm or slightly spreading, the durable and persistent sheath blade suggests a bright, krislike [wavy] dagger. As its functions are completed, each sheath, in turn, takes on a light tan color, beginning along the upper margins as the tissues dry. Presently it falls to the ground, revealing its polished inner surface and leaving behind the

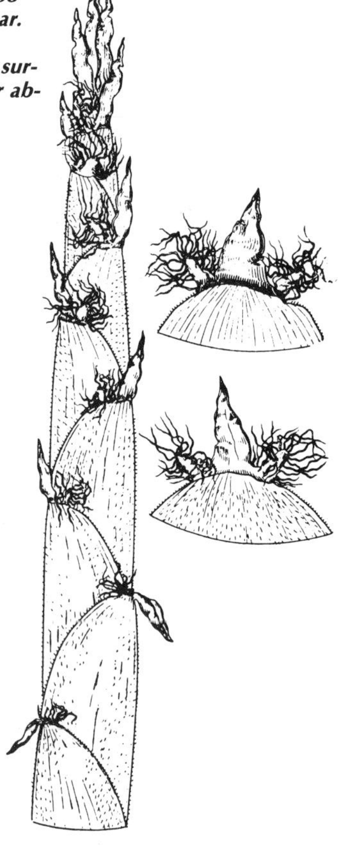

Shoot of henon bamboo* (Phyllostachys nigra, *var.* henonis), *with sheaths showing inner adaxial surface (above) and outer abaxial view (below).

fresh and spotless green internodes of the young culm."[76]

Zones: Gulf, East Coast possibly to Philadelphia; West Coast probably to the Columbia River or even farther north.[77]

Phyllostachys nuda: 34 feet by 1⅝ inches, −8°F.

From Chekiang Province, China, introduced in the United States by Frank Meyer in 1908. Its edible shoots are excellent, though a trifle bitter when raw. Mature culms are not highest industrial quality but are locally useful for fishing poles, plant stakes, and all-around farm use. This species is quite hardy, perhaps the hardiest *Phyllostachys.* It flourishes near Glen Dale, Maryland, USDA Plant Introduction Garden at the land of George Darrow.

Zones: Gulf, from southeast Texas eastward; East Coast to Maryland; West Coast probably to Vancouver. At −8°F *P. nuda* suffered minor leaf damage near Washington, D.C.[78]

Phyllostachys propinqua: 23 feet 4 inches by 1¼ inches.

Sent by McClure in 1928 from Kwangsi, China, to the USDA. It looks like *P. meyeri* but lacks the narrow line of white hairs on the base of culm sheaths of that species. Midseason shoots.[79]

**Phyllostachys pubescens:* 38–70 feet by 3–7 inches, 3°F

(*Phyllostachys edulis; Phyllostachys mitis* by earlier botanists).

Moso, as this major commercial species is known in Japan, was introduced there from China around 1738. From there it arrived in Europe about 1880 and around 1890 reached the West Coast of the United States. This species covers some 2 million hectares of China's present 2.9 million hectares of bamboo forest and represents roughly 22 percent of the bamboo harvest in Japan.[80]

The culms are larger in diameter for their height and more tapered than most bamboos. "The largest and handsomest of the giant hardy bamboos . . . with foliage more feathery and attractive than any other species of the genus."[81] Although the quality of the early to midspring shoots is a bit below many other *Phyllostachys* species, their large size makes moso the central species in the bamboo-shoot business in both China and Japan. The winter shoots of moso, on the contrary, are excellent, an esteemed delicacy.

Dig moso early as December, undeveloped, supertender.

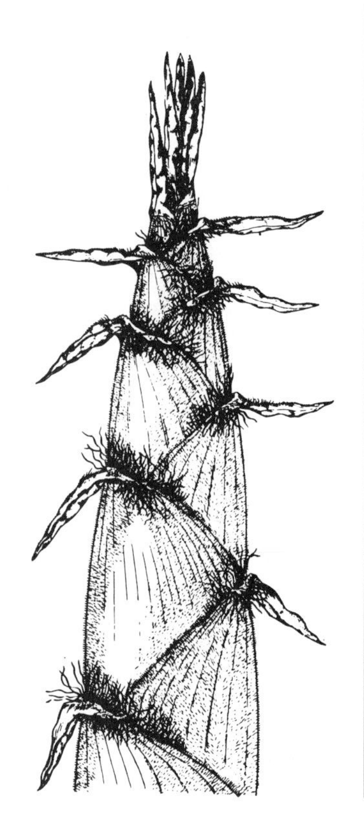

***Shoot of* moso (Phyllostachys pubescens).**

An infrequent, prized abnormality in* moso culm *development produces tortoiseshell bamboo.

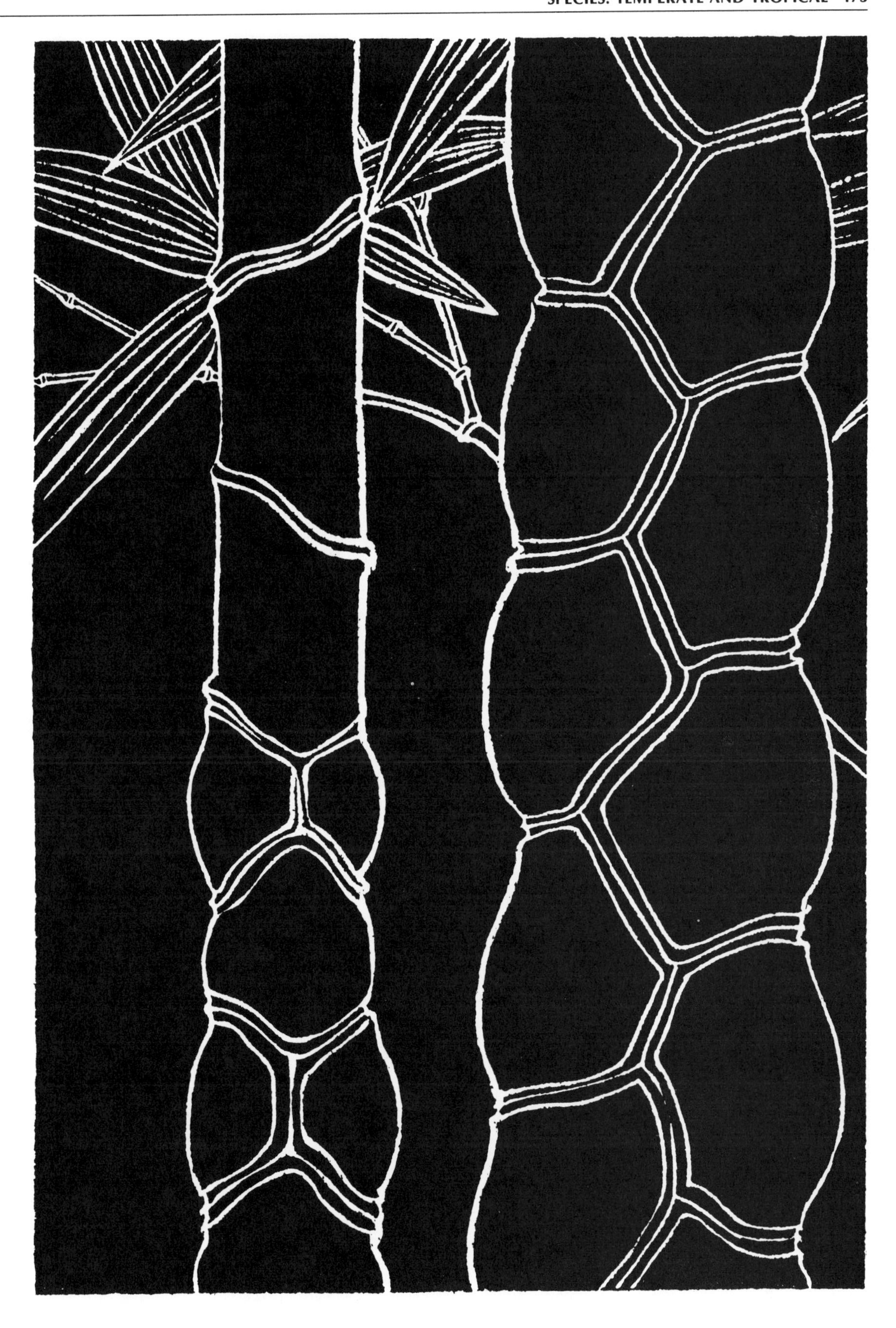

Some say the origin of the name *moso* is rooted in a tale about this esteemed midwinter dish: "This bamboo is named after Meng Tsung [Moso in Japanese], one of the twenty-four paragons of filial piety. His mother having fallen sick, craved for soup made from the young shoots of bamboo in the depth of winter, when such things are not to be had. Meng Tsung betook himself to a bamboo plantation, and wept so plentifully that the ground was softened and an abundance of young shoots sprang up."[82] Others, more prosaically, claim the name *moso* derives from *mo chu,* "hairy bamboo," the popular Chinese name given to the velvety culms. The value of moso shoots about equals the harvest of culms in a typical Japanese grove. Most canned shoots imported into the United States are of this species.

Moso is distinguished from other Phyllostachids by relatively short lower internodes in the strongly tapered, furry culms and its unusually small leaves, commonly 2–3 inches long by ⅜–½ inch wide. Perhaps moso's small leaves suffice for this giant which towers above most competing vegetation. By contrast, the diminutive *Sasas,* the smallest bamboos in the world—which are found in northern Japan—have the largest leaves among hardy bamboos to gather as much sprinkled sunlight as possible at shady ground level. (See *S. palmata* and *S. tessellata,* the latter with leaves up to 2 feet long, 4 inches wide —nearly ten times the size of moso leaves.)

Moso is the most difficult of the *Phyllostachys* to propagate. New plantings grow slowly, fail often when soil and water conditions are less than the best. Hardy on the East Coast to Wilmington, North Carolina, on the West Coast perhaps to the Columbia River.[83] For extensive information on moso use and cultivation as practiced in Japan, see Chapter 8.

"*P. pubescens* is a plant of consumate beauty. The great, broadly arched culms heavily burdened with masses of tiny leaves suggest giant pale-green ostrich plumes. Though the wood is relatively soft, the culms are much used in the Orient and are perhaps second only to *P. bambusoides* in their versatility. Used for heavy construction, cables, well-drilling equipment, tubing for piping salt brine, water, and gas, children's toys, mahjong tiles, and tobacco pipes (from dwarfed culms). The rhizomes are commonly used to make walking sticks and umbrella handles.

"Abnormal forms of this species with slanting nodes and pot-bellied internodes in the lower culm, called Tortoise-shell bamboo, are much prized and even venerated in the East. Living, they are transplanted to gardens where with proper care they can last a number of years but without reproducing their rare character. Dead, they may be enshrined behind an incense bowl or used to other decorative or magical purpose."[84]

Freeman-Mitford believes that the zigzag deformity of such culms is owing to "the repressive action of stiff soil,"[85] but he fails to explain why only this species would react to repression.

Phyllostachys purpurata: 18 feet 4 inches by ¾ inch.

Commonly arched and slender zigzag culms, with rather long internodes (up to 14¼ inches). Culm sheaths characteristically purple, for which the species is named. Sent to the USDA in 1938

***Midculm branch complement of* Phyllostachys purpurata.**

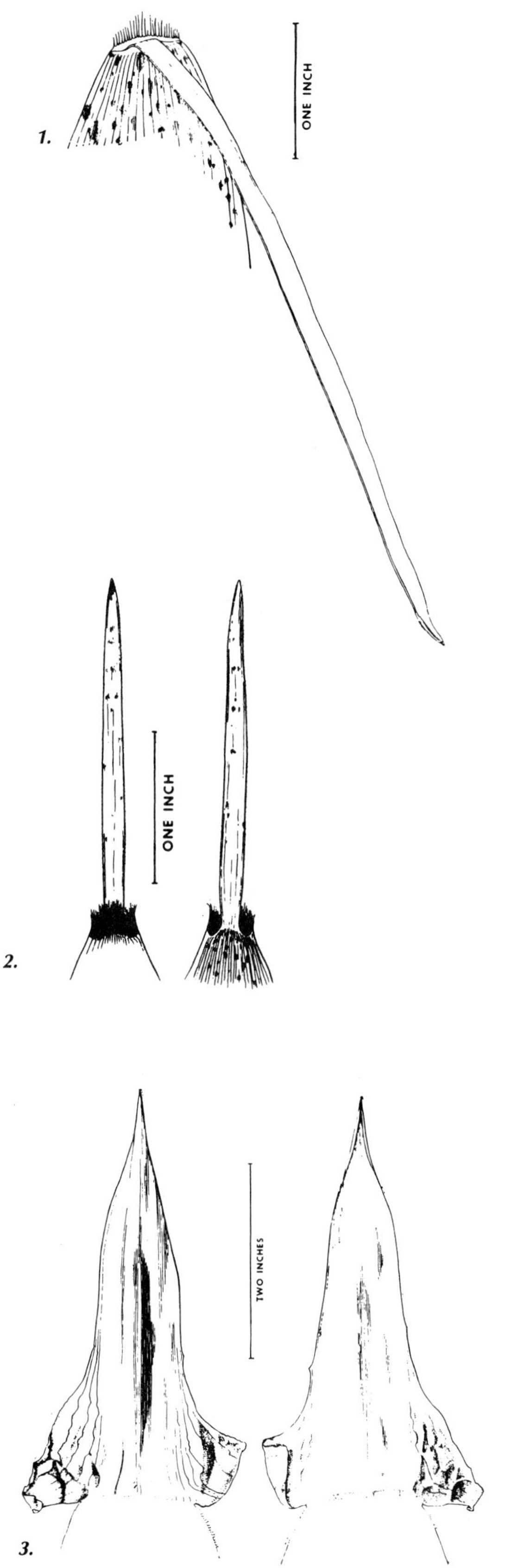

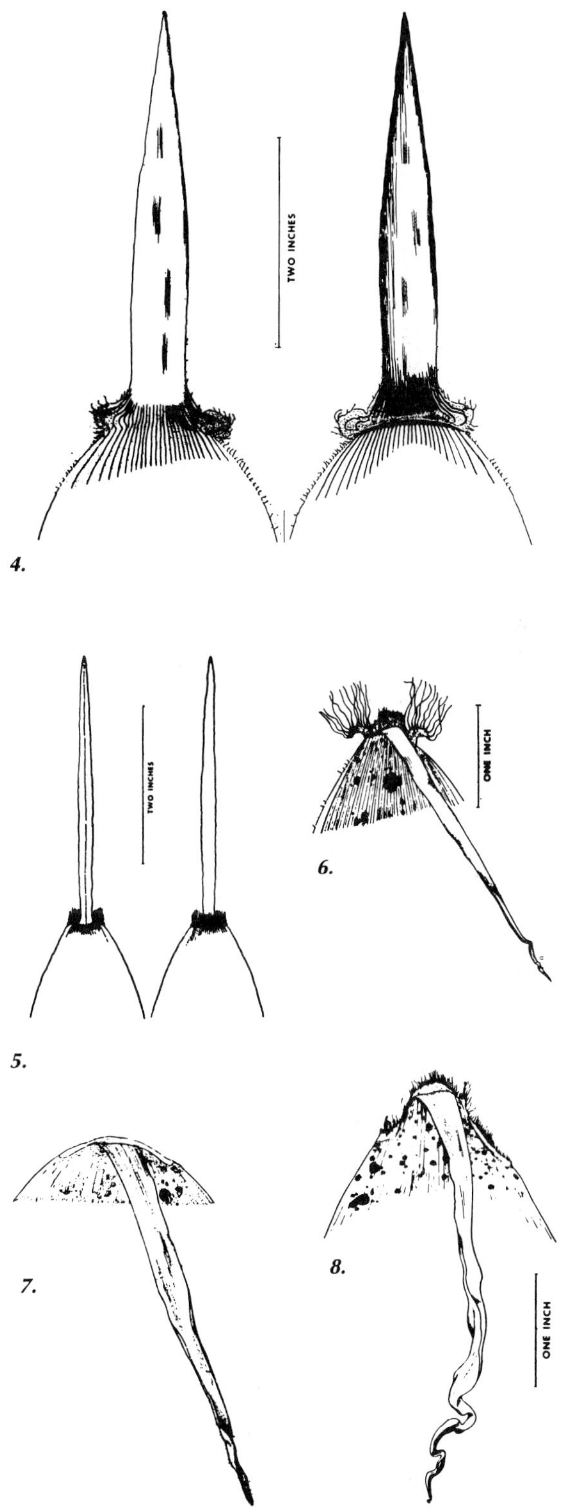

Phyllostachys *culm sheath tips of eight species. (1)* P. aurea. *(2)* P. flexuosa. *(3)* P. nidularia. *(4)* P. purpurata. *(5)* P. rubromarginata. *(6)* P. viridi-glaucescens. *(7)* P. viridis. *(8)* P. vivax.

from Kwangtung, China, by McClure. Cultivar *Solidstem:* 9–12 feet culms with the basal section solid 4–6 feet or more. Midseason shoots with slight bite.[86]
Cultivar *Straightstem:* 33 feet 4 inches by 1½ inches. Nearly double the height of typical *P. purpurata.* More stiffly erect, nodes less prominent, sheaths wine colored but suffused with green.[87]

**Phyllostachys rubromarginata:* 31½ feet by 1¼ inches.
A very hardy and plentiful super propagator from China. Among twenty-seven bamboo species grown successfully in Alabama experiments over thirty years, this bamboo produced more dry pounds of wood and more poles 15 feet or longer per acre than any other. It also proved among the three hardiest of all tested species and showed "remarkably high survival" (86 percent) of rhizomes in plantings. Its production and survival were not only greater but significantly greater than other bamboos tested.[88] In a personal communication (Savannah, November 1980) Walter Hawley, director of the USDA Savannah, Georgia, hardy bamboo groves for twenty years, stated that this species doubled the rhizome production of any other species of this genus.

Long slender culm internodes—16 inches and more—are a noticeable character trait of *P. rubromarginata,* which accounts for its use as shepherd's pipes. Culms are erect to strongly curved. In its native China, it is cultivated mainly for edible shoots, which emerge late in the season. Other culm uses include stakes and basket material. The culm sheaths make covers for the block-printmaker's barren. Unlisted by nurseries, or rare, and virtually unmentioned in the usual lists of hardy bamboos, this species is a case of mysterious neglect.

Recommended zones: southeast Texas eastward, East Coast possibly to Philadelphia, West Coast as far north as Vancouver.[89]

Phyllostachys viridi-glaucescens: 35 feet by 2 inches, −4°F (estimate).
Native of China, grown in Europe since 1846 and in the United States since 1937. The erect culms have prominent nodes. The poles have not been tested for technologically important properties but appear to be of excellent quality, well worth trial growths. Early spring shoots are edible, almost without bitterness even when raw. A vigorous and hardy species, curiously rare in cultivation, unlisted by nurseries.[90] "Recognized primarily by the brilliant blue green sheen on the underside of leaves."[91] *Zones:* Gulf region from southeast Texas eastwards; East Coast possibly to Philadelphia or farther north; West Coast at least to the Columbia River.[92]

Phyllostachys viridis: 50 feet by 3¼ inches, −8°F.
First introduced into Europe from China around 1880, the slightly curved culms of *P. viridis* are distinguished by minutely dimpled "pigskin" internodes that can be felt with sensitive fingers or seen with a 9x lens. The edible midspring shoots are quite good, almost without bitterness even raw. Mature culms are said to have excellent technological properties.

Zones: Gulf region from southeast Texas eastward; East Coast possibly north to Philadelphia. West Coast to Vancouver. Near Washington, D.C., at −8°F this species suffered leaf damage only.[93]

**Phyllostachys vivax:* 70 feet by 5 inches, 0–5°F.
Introduced into the United States by Frank Meyer in 1907 from Chekiang Province, China, this species resembles *P. bambusoides* but is a more rapid and stronger grower, sending up shoots (free of bitterness) two weeks earlier in spring. Extremely vigorous, it reaches maturity more quickly than *P. bambusoides,* whose culms are much thicker walled. *P. vivax* "completely shaded and killed the growth of *P. bambusoides*" when the two were planted side by side, according to E. A. McIlhenny's experience at his Jungle Gardens in Avery Island, Louisiana, the largest bamboo groves in the continental United States. A strikingly beautiful species with elegant drooping foliage, easily recognized at a distance once known.

Zones: southeast Texas eastwards; East Coast to Norfolk, Virginia; West Coast probably to the Columbia River.[94]

Sasas.

The *Sasa* species are distributed in several genera by different authors. The *Guidebook of the Fuji Bamboo Garden* reckons *Sasaella* species at 132, *Sasa* species at 405, *Pseudosasa* at 3.[95]

**Pseudosasa japonica:* 18 feet by ¾ inch, 0°F.
(Arundinaria japonica, Bambusa metake, Sasa japonica).
Abundant in central Korea and Japan, its origin uncertain, *P. japonica* was introduced into France in 1850 and around 1860 became the first oriental bamboo species introduced in the United States.[96] The supererect culms of *metake* or "arrow bamboo," as this species is known in Japan, are straight as its name implies and skinnier than you'd imagine possible to support the orderly mass of dark green leaves growing four to twelve in a cluster from the semierect branches at the upper nodes. The culm sheaths, often as long as the internodes or longer, are quite persistent. Branches usually single, sometimes two or three at the uninflated nodes, roughly 10 inches apart. Leaves are 5–13 inches long by ⅝–2 inches wide, glossy above, glaucous below, with a conspicuous yellow midrib. Easy to grow, slow to spread, quite hardy, with abundant and attractive foliage, *P. japonica* is among the

most common and popular ornamental bamboos in a number of Western countries. "For hedging and screening purposes, this species is without equal, and the mature canes, when ripened properly, are eminently suitable for garden purposes, even though they are fairly thin walled. It also makes a fine tub plant."[97]

Large groves can be found at least as far north as Washington, D.C., on the East Coast, where the Benedictine monks of Saint Anselm's Abbey (Brookland, D.C.) are happy to share their abundance. Hardy on the Pacific Coast to the Columbia River or more.[98]

"When planted on the west and north of a peasant's cottage, this bamboo grows thick and bushy and forms an excellent shelter against the wind. Its stems are an indispensable household article, used for raising well buckets and for fences. For catching shell fish *(Pinna japonica* and *Myaarenaria)* the tallest specimens are selected, cut in late autumn and stored during the winter in a smoky place. In early spring they are bent over a fire and an iron hook affixed to the end. With this instrument the bottom of the sea is dredged. Cut into lengths of 6 or 7 feet, it is plaited together to form a fish stew, which floating in the sea serves to keep cray fish and so forth alive.

"It bears the cold well, is very easy to cultivate, grows in soil half earth, half stone, and flourishes exposed to the violence of seashore waves. Plants growing on hillsides or river embankments expose their rootstocks, and they hang in the water without suffering any loss of strength or luxuriance. These qualities render it of great use in the construction of *kase* (groins) as a protection against floods. The flow of a side current is obstructed by planting bamboos on the banks of a large river or at the waterline of a dike where it is feared the water may break through. When they begin to grow thick and close, the inner face is stopped up with straw, earth and stones, vegetation, or the bark of trees. Such *kase* are absolutely necessary as a protection against floods.

"Smaller canes are used by the common people plaited together as ceilings, for the framework of mud walls of houses, for the frames of round fans, for all sorts of baskets, the ribs of umbrellas, and many other purposes."[99]

Sasa disticha: 2–5 feet by ¾ inch, 10°F.
(Arundinaria disticha, Bambusa disticha, Bambusa nana).
Native to Japan, from where it was introduced into Europe in the late 1860s, "dwarf fernleaf bamboo" is named for its distinctive foliage that is fernlike in sunlight but more open in shade. *Disticha,* "arranged in two rows," is the name given by Freeman-Mitford for the paired leaves that distinguish the species in a sunny site. Culms are bright green, often shaded purple. The sheaths, downy at first, are not persistent.

Pseudosasa japonica.

Uninflated nodes are 4–6 inches apart; branches generally one per node but sometimes in pairs. Leaves are somewhat downy, 2½ inches by ¾ inch, with fine bristles on both edges, bright green above, a dull flat green below. *S. disticha* has active rhizomes that can be fairly invasive.

Zones: hardy as far north as Philadelphia on the Atlantic Coast and most of the Pacific Coast.[100]

**Sasa kurilensis:* 10 feet by ¾ inch.
The most important *Sasa* economically is *S. kurilensis,* which abounds even in deep snow farther north than any other bamboo. Culms are 3+ meters high, 1–2 centimeters (⅜–¾ inches) in diameter, yield 45 tons per hectare (about 18 tons per acre) for particle boards, according to the Experiment Forest at Hokkaido University. The species "prevents erosion, but disturbs regeneration of trees and hinders foresters in summer work. . . . In 1978, about 600 tons of *Sasa* shoots were used for food in Japan, and 87,000 bundles of culms were produced for agricultural, horticultural uses, and bamboo works in Hokkaido Prefecture."[101]

In Hokkaido, *Sasa* species represent 100 million tons of vegetation, one sixth of the total wood growing, so industrial use is being

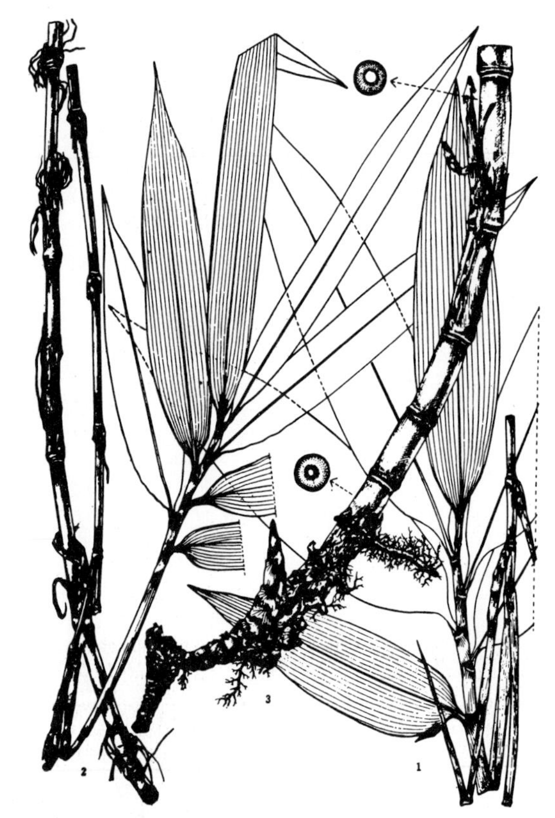

Sasa kurilensis, *"Nemagari-dake" in Japan, braves more northern latitudes than any other bamboo in the world.*

researched. Fiber dimension makes *Sasa* pulp better for thick paper and fiberboard than for thin papers. Cost of gathering and bundling *Sasa* for pulp products is an economic block. Research on pulp production dominates the intense *Sasa* studies of the last thirty years, but a possible cancer cure from *Sasa* leaf extract has recently stirred keen interest. "This species often covers whole mountains. On January 23, 1902, 211 soldiers of the Aomori Regiment started for Mt. Hakkoda, but they lost their way, and owing to the violent snow storm strayed by accident into a vast growth of this species and almost all of them were frozen to death on January 25." (Fuji 1963:64.)

**Sasa palmata*: 7½ feet by ½ inch, 0°F.
(Arundinaria palmata, Bambusa palmata).
Native to Japan, *S. palmata* covers huge areas in the mountains there, and its thick-walled culms are used as a pulp source in manufacturing hardboard. Introduced into Europe in the 1880s and into the United States in 1925. New culms are a brilliant green, often curved at the base, a powdery white below nonprominent nodes roughly 6 inches apart; internodes later covered with brown splotches. Culm sheaths, a greenish white turning straw colored, hang on until ragged. Branches are single at some mid and upper culm nodes, with one or two secondary branches often growing later from the primary branch. Leaves 4–15 inches long by 1½–3½ inches wide, bright pea green above, silvery below, thick and leathery to the touch, clustered "palmately"—like a hand and hence the name—on the tips of culm and branches. Only *Sasa tessellata* among the *Sasas* has larger leaves (up to 24 by 4 inches). According to Young, *palmata* is a "strikingly handsome . . . outstanding ornamental."[102] Leaves wither at the tips and margins in very cold weather, replaced by fresh bright greenery in the spring. Aggressively invasive. Confine it unless you want it everywhere and your neighbors agree. Hardy as far north as Boston on the Atlantic Coast, to the Columbia River or farther on Pacific; does not thrive as well in long southern summers.[103]

Sasa pumila: 2½ feet by ¼ inch.
(Arundinaria pumila, Bambusa pumila, Nipponocalamus pumilis, Pleioblastus pumilis).
Native to Japan (Honshu, Shikaku, and Kyoshu districts), an extremely hardy and hyperinvasive dwarf species, useful in erosion control. The only smaller bamboo is *Sasa pygmaea* (10 inches by 1/16 inch). The slender, dull purple culms—dull green when new—are distinguished by a thick white bloom beneath the uninflated nodes, 6 inches apart, where single, sometimes paired, branches bear five to seven leaves up to 7 inches long by ¼–¾ inch wide. The dark green leaves are sharply tapered, hairy on both surfaces, bristled at the edges, clearly tessellated, a dull matt green on their lower surface. The moderately persistent culm sheaths, purple fresh, straw colored dry, are covered by a ring of hairs at the base in those at upper culm nodes. This last feature serves to distinguish this species from *A. vagans* with which it is often confused. *Pumila* means "little." The species was named by Freeman-Mitford when introduced into England in the late nineteenth century.[104]

Sasa pygmaea: 6–18 inches by 1/16 inch.
(Arundinaria pygmaea, Bambusa pygmaea, Pleioblastus pubescens, Pleioblastus pygmaeus Sasa variegata, var. *pygmaea).*
The world's tiniest bamboo, native to Japan where it carpets the forest floors of Honshu, Kyushu, and Shikaku provinces, this speedy midget among dwarves moves horizontally with a rapidity that equals the vertical vitality of other bamboo species. The solid bright green diminutive canes are flattened and purply towards their tips. Inflated, bristled nodes bear persistent sheaths that are dull green when fresh and weather to a "dull muddy straw."[105] Branches are solitary, sometimes paired, bear three to seven leaves 5 inches long by ¾ inch wide, bristly at edges and covered with hairs on both sides that are finer below than above.

Leaves are a bright green on the upper surface, dull silver green beneath, and tend to wither at the tip, edge, and midrib. As aggressive as it is attractive, *S. pygmaea* needs to be confined. Two subvarieties exist with smooth ("glabrous") rather than hairy leaves.[106]

Sasa tessellata: 6 feet by ½ inch.
Native to China, this extremely hardy semidwarf species was in 1845 the first *Sasa* to set root in English soil. It is distinguished both by its leaf size —up to 24 by 4 inches—and the tessellation, or veins crossing the leaf which form small irregular squares, clearly visible to the lensless eye. The small, bright green culms, top-heavy beneath the generous foliage, bend gently, "giving the clump a somewhat rounded appearance, reminiscent of a small leafy haystack."[107]

The culm sheaths, pale green when fresh, straw colored dry, are quite persistent. The uninflated nodes, spaced roughly every 6 inches, typically give rise to a single branch, sometimes two. The enormous leaves—one third the length of the culm—are a bright and shiny green above, dull grey green below. In warm climates the rhizome is vigorously rampant, slower to move in cold.[108]

"*Sasa tessellata* . . . has the distinction of bearing probably the largest leaves of all known bamboos."[109] Soderstrom and Calderon, in their excellent "Commentary on the Bamboos," mention a species with leaves eight times longer: "The leaf blades of *Arthrostylidium capillifolium* are hairlike and no wider than 3 millimeters, while those of *Neurolepis elata* from the Andes may reach a length of 5 meters and width of ½ meter."[110]

Sasa variegata: 2–3 feet by ¼ inch.
(Arundinaria variegata).
Native to Japan, "dwarf whitestripe bamboo" is popularly named for the leaves clustered at tips of culms and branches, which die back in colder winters but return in midspring with the conspicuous stripe that gives this species its most distinctive feature. Leaves are 2–6 inches long, 1 inch or so wide, with fine hairs on the lower side. An attractive ground cover, this bamboo is "especially effective in artificial clump form when the rhizomes are confined by a metal or concrete curb sunk into the ground."[111] A 55-gallon oil drum, cut in half with ends removed, makes a pair of beds.

Zones: grows well on the East Coast as far as New York and on most of the Pacific Coast. Should have winter mulch in more northern areas. In more southern locations, its growth is less abundant.

**Sasa veitchii:* 3–4 feet by ¼ inch.
(Arundinaria veitchii, Bambusa alba marginata, Bambusa veitchii, Phyllostachys bambusoides, var. *alba marginata, Sasa albo-marginata).*

Sasa veitchii: ***rhizome and flowering culm, with leaves showing the white wither at the edge that distinguishes this popular ornamental.***

Kumazasa is the common native name of this hardy dwarf from Japan (Shikaku, Honshu, and Kyushu provinces). The thin-walled culms are pale green and covered with a white bloom when shooting; they mature to a purplish green and finally to a dull purple. Inconspicuous nodes 3–4 inches apart bear persistent sheaths with a tuft of coarse hairs at the tip when new. Branches are usually solitary, sometimes paired, often as long as their parent culm, and bear glossy leaves that are bristle edged, thick and smooth, up to 10 inches long by 2¼ inches wide, a deep green above and a dull grey green below. The midrib, a yellow green, is prominent; the tessellation fine but clear, with eight to nine pairs of secondary veins. The most conspicuous feature of the species is the brownish white winter withering of the margins and tips of leaves, which "far from detracting from the beauty of this plant, in fact enhance it, giving it an appearance of variegation."[112] This characteristic withering accounts for three of the five synonyms listed.

Zones: hardy on the Atlantic Coast to Norfolk, Virginia, or farther and most of Pacific Coast. Does better in the north than in the longer southern summers.[113]

Sasa veitchii, var. *nana.*

Similar to the species but with smaller and more sharply pointed leaves on which the withering is even more conspicuous. It is less invasive, and Lawson recommends it as preferable to *S. veitchii.* Common in France and England.[114]

Shibataea.

Shibataea kumasaca: 2–6 feet by ¼ inch, 10°F.

(Bambusa kumasaca, Bambusa viminalis, Phyllostachys kumasasa, Phyllostachys ruscifolia, Sasa ruscifolia).

A widely cultivated Japanese ornamental with distinctively broad leaves, the genus is named for K. Shibata, a well-known botanist who named many species of bamboo. The culms, pale green maturing to dull brown, are almost solid, half round or triangular, with short zigzag, grooved internodes and prominent nodes some 1½–3 inches apart. The culm sheaths are fairly persistent, purple fading to straw color, as long as the internodes or longer and fringed with fine hair. Typically two but occasionally up to five short, skinny branches, ¼–½ inch, bear a single terminal leaf, sometimes adding a lateral leaf also in the second season. Leaves, 2–4 inches by ½–1 inch, smooth above, with tiny hairs below, bristled at both edges, show quite prominent tessellation. Dark green fading to dull yellow green above, grey green below, "the leaves are damaged at 10°F, but as new leaves appear in late spring the plant can be grown over a considerable extent of latitude. . . . This unique plant, with its broad leaves resembling those of

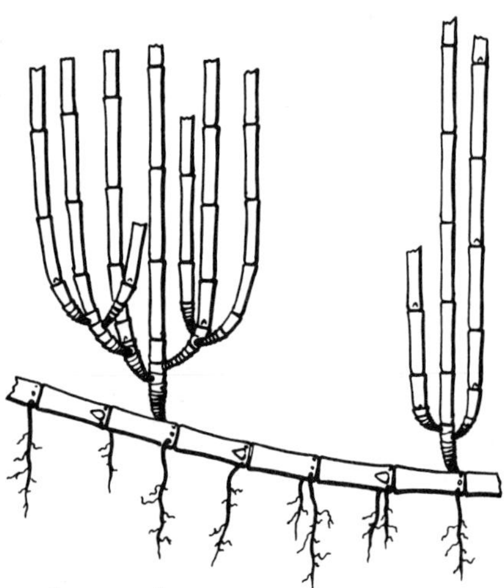

***Tillering culms are characteristic of* Shibataea kumasasa.**

Flowering branches of* Shibataea kumasasa, *the world's smallest bamboo.

the genus *Ruscus* of the lily family, its odd branching habit, and extremely slender culms looks little like any ordinary bamboo. The untrained observer would seldom recognize it as a bamboo or even as a member of the grass family."[115]

Seldom invasive; requires much moisture. Can resist severe frosts according to Lawson's English experience. Hardy at least to Norfolk, Virginia, on the Atlantic and most of Pacific Coast.[116]

TROPICAL BAMBOOS

Bambusa.

**Bambusa arundinacea:* 95 feet by 5½ inches (near Miami, Florida) 27°.

(Bambusa spinosa).

In its native India, this species is second only to *D. strictus* in paper pulp production from bamboos. It is more valued for quantity than the quality of pulp. Yields vary from 1 to 14.5 tons of dry internodes per acre (2.5 to 36 metric tons per hectare). This range indicates the importance of site and soil for culm abundance. The average yield of a large area supposedly approaches the higher figure.[117]

Under its favorite skies in the Asian tropics, it grows even larger than the figures reported for Florida above. Rather thick walled, with soft wood, often somewhat crooked, the culms shoot in fall in south Florida, then commonly drop branches in the spring. Of all bamboos, this "giant thorny bamboo" is most bristling with defense. At lower nodes a primary, almost vinelike, thorny branch is surrounded by short branches modified into thorns. These eventually form a very effective barrier against wild animals, who might otherwise devour the young shoots. The shoots are edible though somewhat bitter before boiling in at least two changes of water.

Groves die after flowering, producing abundant seed said to have relieved famine in India.[118] Seeds of this species were studied in Puerto Rico (Mayaguez USDA Experiment Station) to discover best ways of preserving viability. (See Chapter 8, pp. 218–21.) In 1789 this species was the first bamboo described in conformity with the conventions of Western botanical description fixed by Linnaeus (1707–1778). The abundance of vernacular names for it in Asia indicate its importance in many local economies: thorny bamboo, *berua, kata, koto* (Assam), *ily, mulu* (Malay), *bans, behor bans* (Bengali), *mulkas, vedru* (Telugu), *mundgay* (Bombay).[119] Young offers a description with warning that this species can tolerate only several degrees of frost.[120]

Bambusa balcooa: 50–70 feet by 3–6 inches; *balku bans* (Bengali), *baluka* (Assam), *boro-bans, sil barua, teli barua, wamnah, beru, betwa.*
Native to India, Assam, lower Bengal, and Bihar. "Best and strongest for building purposes" in this region (Gamble). In Vietnam and Indonesia, used for basketry, crafts, house construction.[121]

Bambusa beecheyana: 40 feet by 4 inches, 20°F.
Native of southeast China, beechey bamboo is a semihardy species forming a fairly open clump with quite tapered, bright green culms, often elliptical in cross section. A rapid grower under favorable conditions, it is one of the chief sources for edible shoots in southern China, where earth mounds are made each year around the base of the clumps to shield the new culms from sunlight, which quickly turns them hard and bitter. Beechey prospers in southern portions of California and Florida where temperatures rarely drop below 20°F. Used also for house construction, farm equipment, basketry, and paper pulp.[122]

Bambusa blumeana: 30–60 feet by 3–4 inches; *buloh duri* (Malay), *kida* (Semang), *bambu duri, bambu gesing, pring ori, pring gesing* (Java), *haur chuchuk* (Sudan).
Native to Java and Malaya, cultivated in Sumatra, Borneo, India, Indochina, and the Philippines. This thick-walled species is used in construction, farm equipment, crafts, and paper pulp.[123]

Bambusa burmanica: 55 feet by 10 inches.
Native to Burma, Thailand, India, and Malaysia, where it is used in house construction, rafts, and ladders.[124]

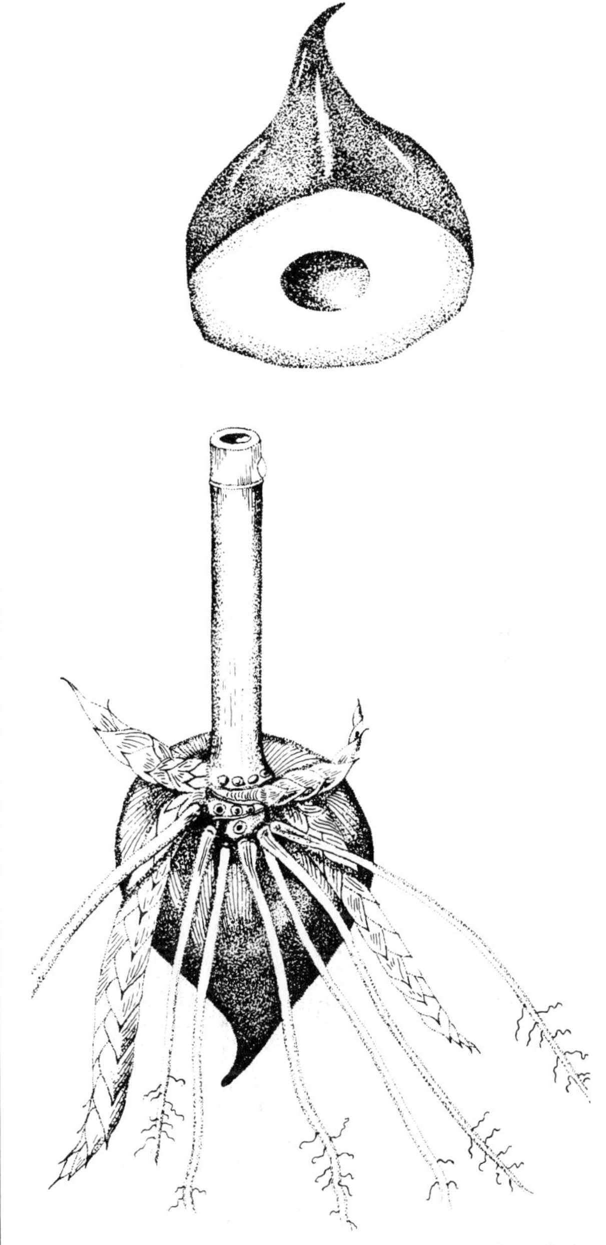

Melocanna baccifera: ***fruit in cross section and a shooting primary culm.***

Bambusa dissimulator: 50 feet by 2½ inches.
The whole culms of "durable thorny bamboo" are used in heavy construction, fences, and so on in southern China where it is native. Grows in fairly compact clumps of fairly straight culms whose internodes are fairly long (10–12 inches), and sometimes fairly hairy. The nodes are only fairly prominent, branches at all (or almost all) nodes; sometimes, at basal nodes, a few weak thorns, which never pose a problem in harvest.

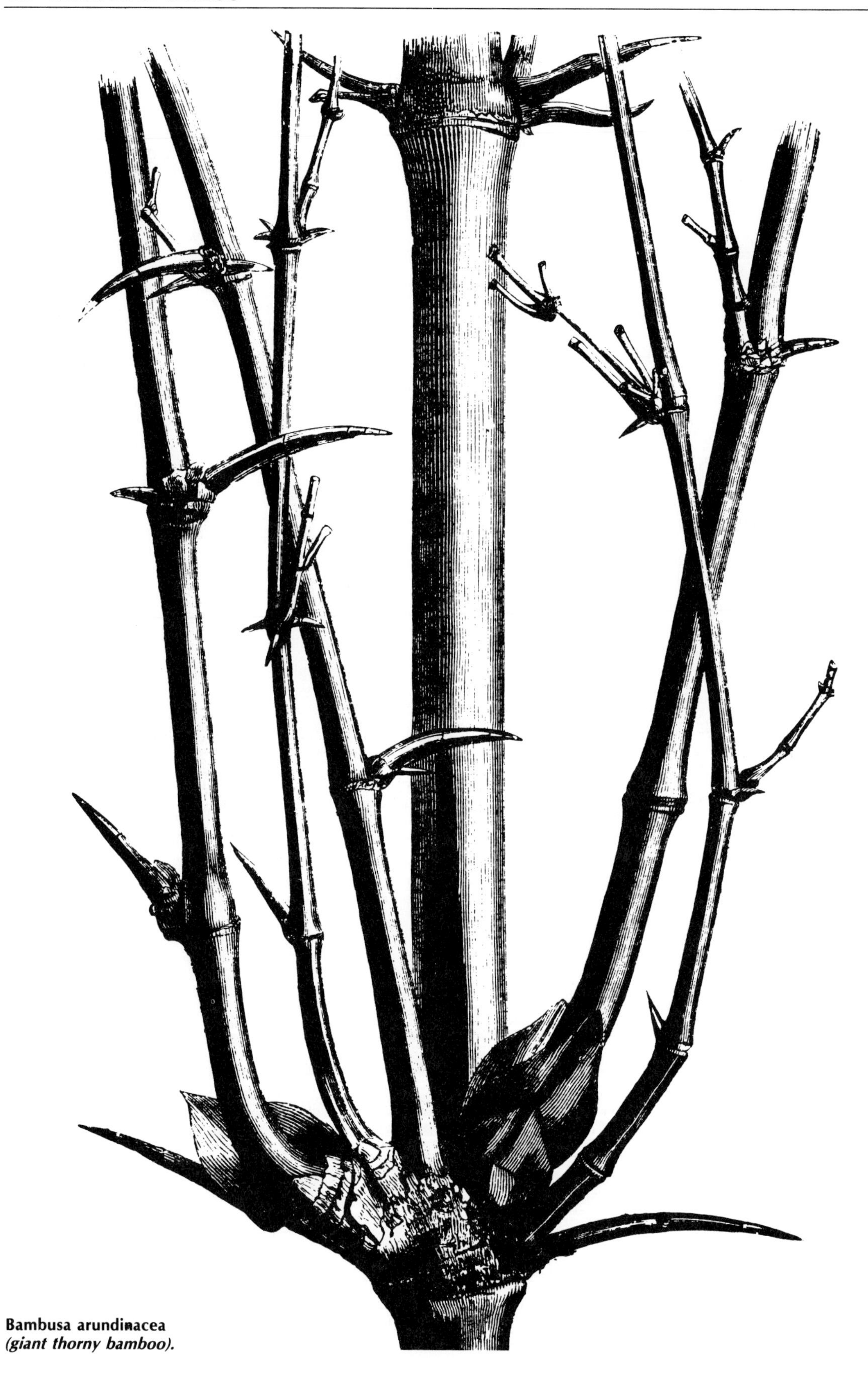

Bambusa arundinacea
(giant thorny bamboo).

An abundant bamboo, with hard wood 1 cm. thick. Sometimes striped white on lower culm.[125]

Bambusa dolichomerithalla: 65 feet by 2–4 inches.
Native to China. Internodes nearly 2 feet long, used in crafts.[126]

Bambusa edulis: 65 feet by 5 inches.
Native to China, used for edible shoots, construction, and paper pulp. Internodes 6–20 inches, walls, ½–¾ inch thick.[127]

**Bambusa glaucescens:* 10–50 feet by 1½ inches, 17°F
(Bambusa multiplex, Bambusa nana, Bambusa argentea).
There are eight forms of "hedge bamboo," probably the commonest species found in the southeast, in Gulf states, and in southern California. Among introduced species, *B. glaucescens* is second only to *B. vulgaris* in distribution throughout the tropics. Culms are rather slender for their height, thin walled and usually arching. New shoots appear around midsummer but usually wait until the following spring to extend branches and shed culm sheaths. Branches at a node vary from a few to as many as thirty-five. Shoots are bitter and rarely eaten. Often too arched for poles, too small for significant volume as paper pulp—in spite of good fiber dimensions—the main role of *B. glaucescens* is as an ornamental windbreak or hedge. Hardy to 17°F or less, this species is among the most frost resistant of clumping bamboos. Culms are pale green and thin walled in the typical form, long known in the U.S. nursery trade as *B. argentea,* in reference to the silver surface of leaves underneath, a trait common to all forms of *B. glaucescens.*

Cultivar *Alphonse Karr Young:* 25–40 feet by 1½ inches 16°F. Differs from the type form in its bright yellow, green-striped culms, with sheaths showing white to yellow stripes remaining visible when the brown green fresh sheath fades to a pale brown. Named in honor of a nineteenth century French botanist-novelist, its culms shoot in early midsummer.

Fernleaf Young: 10–20 feet. Named for the ten to twenty tiny leaves per twig that distinguish this cultivar from the species type. This characteristic is sometimes lost with favorable conditions of moisture and soil.

Cultivar *Silverstripe Fernleaf Young:* resembles fernleaf but has white-striped leaves. A rare variety.

Cultivar *Stripestem Fernleaf:* Differs from fernleaf only in its fresh culms, which at first are light reddish to yellow, striped with green.

Var. *rivierorum Maire:* 13 feet by ⅜ inch, 15°F.

"Chinese goddess bamboo," native to southeast China, resembles fernleaf in foliage but is distinguished by solid internodes. Culms more or less sinuous, internodes somewhat long, the surface strewn with stiff hairs that leave their mark after they fall off. Nodes are hairless and slightly inflated. Culm sheaths are persistent but tattered after dry, with fragile sheath blade and auricles usually absent. Tiny leaves (1–1¾ inches) on delicate branches and gracefully curved twigs. A favorite garden bamboo that also does well as a potted plant, and is more shade tolerant than most bamboos. Much used in bamboo bonsai by Chinese gardeners where, with skillful treatment, it becomes a perfect miniature a few inches tall, all parts in flawless proportion.[128]

Bambusa glaucescens, ***var.*** **rivierorum:** ***(A) Rhizome and culm base. (B) Leafy twig about to flower.***

Cultivar *Silverstripe Young:* 35–45 feet, 15°F. The largest cultivar of *B. glaucescens,* with few to many white-striped leaves, one or two thin white stripes on culm internodes, and persistent broad brown stripes on the sheaths. This is the commonest form of the species in south and southwest United States.

Cultivar *Willowy Young:* 20 feet by ¾ inch, 15°F. Thin culms, drooping markedly at the top of plants which reach full height for this cultivar. Internodes are solid at the base and thick walled above so the willowy tendency doesn't occur in shorter culms. Thin branches and twigs bear narrow leaves that are uniformly green, like the culm. The culm sheaths dry to a dull straw color.[129]

Bambusa khasiana: 30–40 feet by 1½ inches; *serim, tyrah* (Khasia). Native to India, the Khasia and Jaintia Hills, Assam, and Manipur to 4,000 feet. General uses in construction.[130]

**Bambusa longispiculata:* 59 feet by 3 inches. Native to India, introduced into the United States in 1931 by seeds. Puerto Rico USDA Experiment Station in Mayaguez reported 6 acres (McClure 1944) with culms totaling some 72,000 lineal feet per year. No known U.S. commercial exploitation of this species, which occurs mainly as an ornamental in south Florida. The arboretum in Lancetilla, north-central Honduras, the Experiment Gardens in Summit, Panama (Canal Zone, Atlantic Coast), have been reported as growing this species, which reaches recorded heights of 59 feet by 3-inch diameter in the El Recreo, Zelaya, Bamboo Garden in eastern Nicaragua.

Culms are fairly straight, cylindrical, with long internodes 10–12 inches at base and 2 feet or more at midculm. Quite similar in appearance to *B. tulda* but usually with fewer lateral branches and often with white stripes on the green culms. The many tiny hairs below each node are shed when cured, leaving scarcely visible pits. Branches are absent or slightly developed in the middle third to half of the culm. The culms are rather thin walled, roughly ½ inch at the base of the larger plants.[131]

Bambusa malingensis: 25 feet by 1 inch. Native to southern China, growing in tight clumps of stiffly erect culms with fine-grained and durable wood; formerly popular for stems of opium pipes, also used for light construction and basketry. The internodes, perfectly cylindrical, are up to 1½ feet long; the nodes, uninflated. Branches, present at almost all nodes, are rather slender.[132]

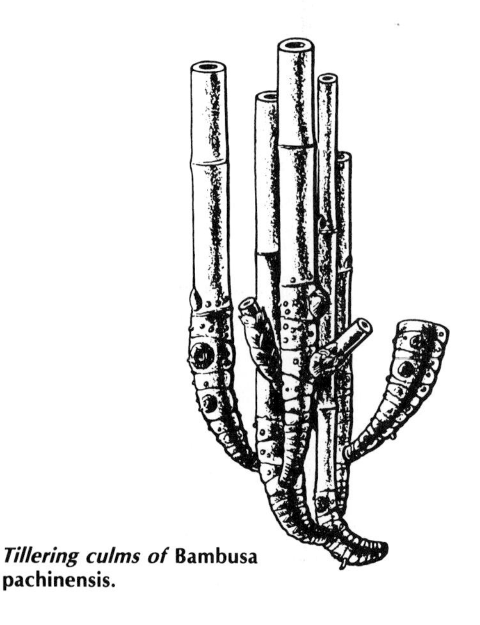

***Tillering culms of* Bambusa pachinensis.**

Bambusa nutans: 20–40 feet by 1½–3 inches; *pichle, bidhuli, nal bans, mukial, makal, mahlu, mahl, paoshi-ding-ying, jotia, deo-bans, wa-malang, sering-jai.* Native to India, lower Himalayas from the Jumna to Assam, east Bengal, and Sikkim to heights of 5,000 feet. Culm wood is strong and straight, hard and much esteemed. General construction uses and paper pulp.[133]

**Bambusa oldhami:* 55 feet by 3¼ inches, 20°F.
Native to south China and Formosa, Oldham bamboo is a large and popular ornamental common in southern California and Florida that grows in fairly open clumps. Culm walls are rather thick but soft. Used for pulp and edible shoots. Long misidentified in U.S. nursery trade as *Dendrocalamus latiflorus.*[134]

Bambusa pachinensis: 6–35 feet by 2 inches. Native to China, an ornamental garden plant also used for crafts, nodes up to 80 cm apart, walls up to 3.5 cm thick.[135]

Bambusa pallia: 65 feet by 4 inches. Native to India, northeast Thailand, north Burma, Vietnam, and Laos. Used for basketry, paper pulp, farm equipment, crafts. Nodes up to 3 feet apart.[136]

Bambusa pervariabilis: 50 feet by 2 inches. This species is closely related to *B. tuldoides* (q.v.), but is reputed less vigorous and prolific. It has more erect culms with thicker walls and more durable wood than *B. tuldoides* and more prominent nodes with more branches. The internodes are often quite hairy, as are the culm and branch sheaths. The variability of shape and degree of hairiness of the sheaths is what earns this species its name. In China, *B. pervariabilis* is much used in heavy construction and for punting poles.[137]

Bambusa polymorpha: 75–90 feet by 6 inches, 28°F.
Native to Bengal and Burma, introduced into the United States from India in 1924, *B. polymorpha* forms dense clumps of bluish waxy culms free of branches in their lower portions in mature stands. Branches, twigs, and leaves are notably slender, the latter smaller and thinner than most tropical bamboos except the fern-leafed forms of *B. multiplex.* Leaves measure 1½–7½ inches long by 5⁄16–7⁄16 wide. The culm sheaths have very prominent auricles, are extremely stiff and densely covered by fine stiff hairs that rub off easily, leaving the sheath a silver grey. Nodes are 4–6 inches apart at base, 30 inches or so at midculm.

"Because it is so highly valued for its timber, the outstanding decorative horticultural value of *B. polymorpha* has been overlooked. Growing in tight clumps, its white culms unbranched for 30 feet or more, with its abundant small evergreen leaves, it is a bamboo of outstanding beauty for planting at moderate altitudes."[138] "The best kind of bamboo for walls, floors and roofs of houses," according to Gamble.[139] McClure mentions several locations where the species can be found: McKee Jungle Gardens, Vero Beach, Florida; near the United Fruit golf club near Tela, Honduras, "a fine row of some thirty clumps"; and an even larger planting at the Experiment Gardens, Summit, Canal Zone, Panama.[140]

Bambusa stenostachya: 30–75 feet by 2–6 inches.
Native to China; used for house construction, paper pulp, crafts, and farm equipment.[141]

**Bambusa textilis:* 50 feet by 2 inches, 13°F; *wong chuk, mit chuk* (China).
An outstanding bamboo from subtropical southern China, introduced by McClure in 1936 to the United States. From Savannah USDA groves in Georgia plantings were established in Puerto Rico, Guatemala, Nicaragua, Panama, Ecuador, and Peru. Culms erect and nodding slightly at the tip, thin walled but quite strong, conveniently free of branches up to three quarters of culm length, an uncommon height among bamboos, with almost no swelling at the nodes.

In China, the culms are the best known material used for baskets, mats, rope, hats, towing cables, garden poles, and fences. When promptly and properly cured, *B. textilis* is rarely attacked by the powder post beetle. Among eleven species tested in Puerto Rico (Plank 1950), the relative susceptibility to attack by beetles in one-year-old culms varied from 44.2 percent in *B. vulgaris vittata* to 0.3 percent in *B. textilis.* This resistance to the dreaded *polilla* is what principally recommends this species for widespread distribution in Latin America, along with an exceptionally strong fiber permitting superfine splitting and weaving of durable craftworks.

This species is also a favorite ornamental, a happy union of beauty and utility: "exceptionally handsome . . . foliage distinctive and attractive."[142] Also recommended for erosion control.[143] *B. textilis* grows in many experimental stations in Latin America. Plantations should be extended in preparation for wider rural distribution. Propagation is, unfortunately, sometimes more difficult than in many other bamboos, but established groves are vigorously productive. One of the hardiest of clumping bamboos, *B. textilis* suffered only minor leaf injury at 13°F in central Florida.

**Bambusa tulda:* 70 feet by 3 inches, 27°F; *tulda, jowa, dyowa bans, mak, makor, kiranti, matela, pake mirtanga, wati, wamuna, wagi, nalbans, deobans, bijuli, jati, jao, ghora, theiwa, thaikwa.*
Native to India (central and eastern Bengal), first introduced into the United States in 1907 and since distributed to major agricultural stations in a number of countries in the Caribbean and Central America, *B. tulda* is a thick-walled species, growing in fairly compact clumps, with straight, often sinuous culms and fairly dense, straight-grained wood. Culm walls, sometimes solid at basal portion, are usually ½–¾ inch thick. Three large branches are the common complement at upper nodes, bearing large leaves 3–10 inches long by ½–1⅛ inches wide. Lower branches are usually without leaves, short and thornlike. In Puerto Rico, split fishing rods have been made of the culms, which White (1948) reports as more in use than any other bamboo at the Mayaguez station. In McClure's 1944 study of numerous species as substitutes for Tonkin cane *(Arundinaria amabilis)* in the manufacture of ski pole shafts, *B. tulda* was rated superior by one manufacturer to all other tested species.

"The Puerto Rico Experiment Station made strength tests of *Bambusa tulda,* a good bamboo but probably not our strongest; 52,000 pounds per square inch were required to break it. Walnut, according to these tests, required 20,000 to 22,000 pounds per square inch, while the steel commonly used for reinforcing concrete required about 60,000 pounds."[144]

**Bambusa tuldoides:* 55 feet by 2¼ inches, 19°F; *chaan-ko chuk, yao-chuk* (China).
Widely distributed both in cultivation and the wild in its native subtropical southeast China, "punting pole bamboo" was introduced into the West in Brazil around the 1840s by the Portuguese, probably from Macao, their colony in southern China. By 1870 it was widespread in southern Brazil and later crossed into Argentina. Introduced elsewhere in Latin America, it is known as *bara Brasileña* in El Salvador. Plants from China via the United States were later established in Puerto Rico, Guatemala, Honduras, Nicaragua, Panama, and Peru.

The moderately thick-walled culms, erect and nodding slightly at the tip, have a hard wood, nodes slightly inflated with branches at all of them, and promptly deciduous sheaths. Prolific in culm production, *B. tuldoides* is one of the main bamboos of economic importance around Canton, China, although its wood is inferior to the related *B. pervariabilis.* Used in China for punting small craft, weaving heavy crates, mats, and baskets, frames for temporary structures, scaffolding, carrying poles, and fences. In Puerto Rico, this species was found adequate (but inferior to *B. tulda* and *B. longispiculata*) for furniture and split-bamboo fishing rods.
"Peripheral wood of internodes of mature culms of *B. tuldoides* showed a tensile strength exceeding 40,000 pounds per square inch."[145]

Bambusa ventricosa McClure: 55 feet by 2¼ inches, 17°F.

Native to southern China where it is commonly cultivated as a potted plant owing to a distinctive feature that earns this species its common name: "Buddha's belly bamboo." When dwarfed, the internodes become short and swollen, in a manner suggesting the prominent paunch of popular paintings and statues of the Buddha in Chinese iconography. Newly propagated plants or plants in poor or dry soil also display this characteristic, which disappears in groves flourishing under favorable conditions. New shoots thrust up in late summer or early autumn; three larger branches often with two or more smaller ones are the usual complement at each node, which may be reduced to a single branch per node in dwarfed culms. "For indoor culture, direct sunlight is necessary for a good part of the day and good light for the remainder. Buddha bamboo in Florida has withstood a temperature of 17°F without damage, but at 13° it lost all leaves, and the culms were injured severely or killed." (Young)

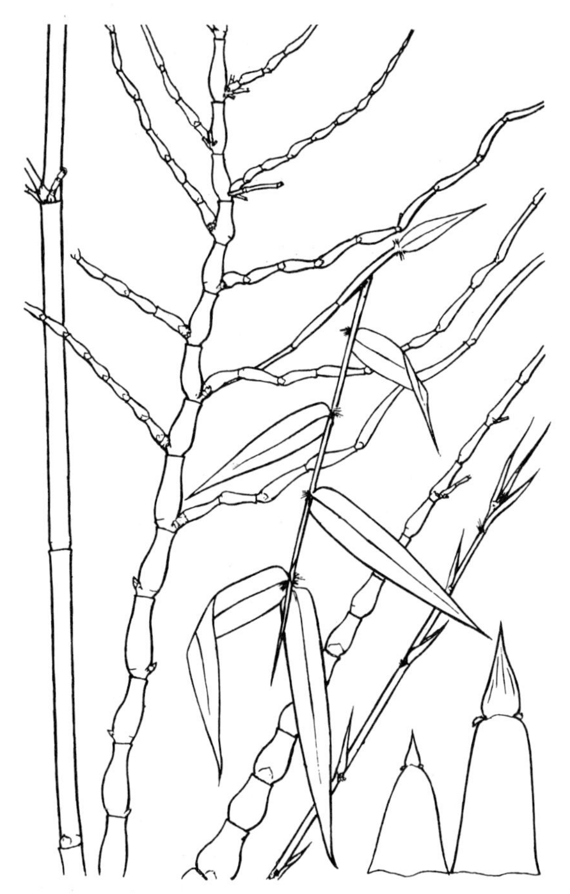

Bambusa ventricosa ***(Buddha's belly bamboo) tolerates arid climates. Abundant foliage is recommended for cattle browse.***

Sometimes internodes are solid or semisolid, but larger culms tend to follow the hollow norm of most bamboos. The glabrous (hairless) sheaths may persist until decay, but when drying they roll in along the edges in a characteristic way. The leaves remain plentiful and green even under very dry skies so McClure recommends this species for livestock forage in areas with little rain, cautioning not to overgraze until the plants are well established. In addition to extensive use as a potted ornamental, the culms of small dimensions serve for walking canes, and trimmed rows of the living plant are suggested for hedges. McClure (1938) devoted a detailed article to this species.[146]

Bambusa vulgaris: 20–60 feet by 4 inches, 32°F.

The origin of the most common bamboo in the world is unknown. Madagascar, Java, and Sri Lanka are some conjectures for the point of departure of this globetrotter that now covers more scattered acres than any other bamboo, principally owing to its easy propagation and broad adaptability to a range of weather and soil from sea level to 4,000 feet. Either the first or among the earliest bamboos introduced into Europe from the East, *B. vulgaris* was widely known to European botanists by the late 1700s as a hothouse plant.

The clump habit is quite open since rhizomes may travel as much as 2½ feet before turning vertical to form a culm. The erect or suberect culms have rather prominent nodes from which the sheaths fall quickly. Branches grow at all but a few basal nodes, which are often ringed with roots—a characteristic of the species that may be found as high as midculm nodes in some plants. Each branch complement has one heavy central branch flanked by smaller and smaller pairs, all holding abundant leaves that fall off partly or completely depending on the severity of the local dry season. Shoots appear in late summer and early fall in Florida but wait till spring to shed sheaths, drop branches, and uncurl their leaves. In warm climates with sufficient moisture, shooting may be more or less continuous throughout the year.

Among eleven tested species in Puerto Rico experiments, *B. vulgaris* was the most starchy and therefore least resistant to beetle attack, which severely limits its use in construction or fine handicrafts—but proper harvest and clump curing of culms three years or older can reduce beetle infestation 90 percent.[147] Industrial use is promising in spite of this drawback since *B. vulgaris* rates very high among almost one hundred species of bamboo available in the West for paper pulp. Feasibility for pulp is enhanced by the rarity of flowering, which is never extensive in groves, by its quick comeback even after a clear-cut harvest, and by its abundance. "At the *Centro Nacional de Investigaciones de Cafe* [Chinchina, Colombia], a single propagule of *B.*

Bambusa vulgaris, ***the world's most widely distributed bamboo.***

vulgaris produced, in five years, over 200 stems, the largest of which are 4 inches in diameter and over 40 feet tall."[148] In a short-lived Scottish paper company that operated in Trinidad during the 1930s, 1,000 acres of *B. vulgaris* on a three-year cutting cycle had a 4-ton-per-acre-per-year yield of pure, dry cellulose pulp, which a longer cutting cycle could have increased significantly.

The type form of the species has green culms. Clumps give rise, through mutation, to yellow culms with or without green stripes. This form seems more common in Central America on the mainland, while green culms predominate in frost-free parts of the United States and in the Caribbean—at least on the islands of Cuba, Jamaica, Trinidad, and Puerto Rico, where its introduction dates from the early nineteenth century. The preferred banana prop of United Fruit for many years in Central America, *B. vulgaris* owes its distribution in Jamaica to spontaneous rooting of culm segments used as yam props there.

In World War II, the corps of army engineers recommended *B. vulgaris* for tropical revetments (see Chapter 2, pp. 27–28). Their choice was determined by the same natural properties that make *B. vulgaris* a spectacular agent for erosion control.

"Puerto Rico has many winding roads through the mountainous interior. Landslides are a problem both above and below the hillside roads. The Department of the Interior has used plantings of bamboo with considerable success to maintain fills and steep road embankments. The bamboo generally used is the common species, *Bambusa vulgaris.* It develops large thick clumps, makes a rapid dense growth, and planting material has been readily available. However, many of the species of bamboos introduced by the Federal Experiment Station at Mayaguez in recent years will serve equally well, and, in addition, will furnish culms valuable for industrial purposes."[149]

(The species referred to are *B. textilis, B. tulda, B. tuldoides,* and *B. longispiculata,* commercial oriental species that require 50–70 inches of rainfall annually.)

In spite of vulnerability to polilla—the bamboo powder post beetle—"the people's bamboo" *(vulgaris)* is much used in Latin America for many ends: houses, fences, pipes for small irrigation, and containers for water or other liquids, parts of ox carts, and basketry. Proper harvest and clump curing (q.v., Chapter 8, pp. 219–223) of mature culms three years or older should be popularly published by governments in schools, newspapers, agricultural extension pamphlets, or whatever means are available.

B. vulgaris culm segments were used commercially for pots to start plants in Honduras at an estimated cost of a penny a piece in the early 1940s. "Shoots are eaten in Java; in

A shooting clump of **Bambusa vulgaris.**

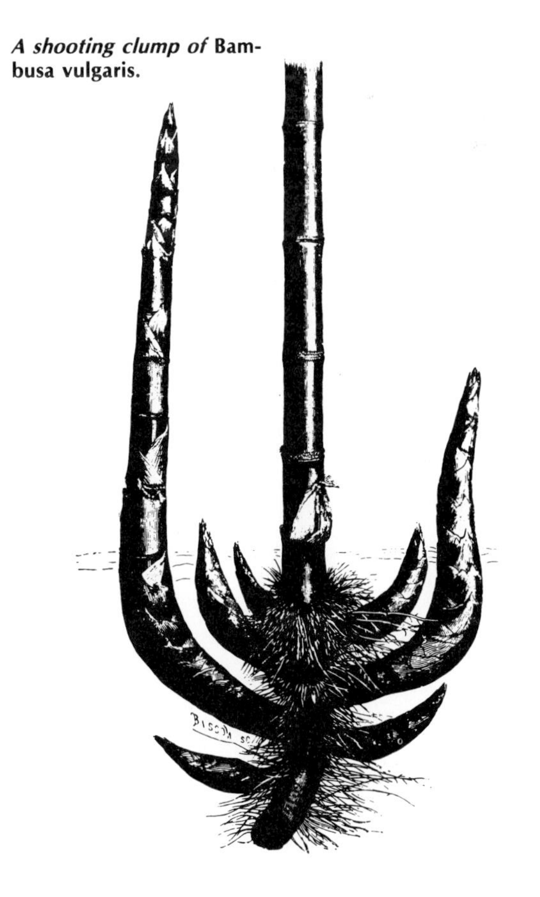

Nicaragua they are canned for the market by Chinese residents."[150]

B. vulgaris, cultivar vittata, as the variant with green striped yellow culms is called, is usually somewhat smaller than the type and supposedly less cold resistant. In some areas—El Salvador is one—it grows larger than the type. The boldly beautiful yellow culms with green stripes have made the *vittata* variant a garden favorite in many locations. It is a treasured ornamental in southern China, where it is known as "painted bamboo" and "jade-striped golden bamboo."

The flock of names surrounding the plant around the world is one indication of its popularity. It is known variously as *bambu* or *bambu amarillo*—yellow bamboo—in Latin America; *buloh minyak haur, buloh tutul, buloh gading, aur gading, pau, po-o, pook* (Malay), *jajang ampel, jajang gading, pring ampel, pring legi, pring tutul* (Java), *awi ampel, awi gading, awi haur, awi koneng, awi tatul* (Sudan), *auwe gadieng, auwe kunieng, bambu kunieng, bambu kuring-kuring* (Sumatra), *pai mai* (Siam).[151]

Bambusa wrayi: 45–65 feet by 1 inch.

Native to Malaysia, used for basketry, farming equipment, umbrella handles, drainage, fences, pipe stems, canes, and blowguns. Internodes 6 feet or more. Simmonds mentions an internodal length of 7 feet 1½ inches in this species—the longest on record among bamboos.[152]

Cephalostachyum.

Cephalostachyum capitatum: 29 feet by 1½–2 inches.

Native to Sikkim and Bhutan, used for paper pulp and basketry.[153]

**Cephalostachyum pergracile:* 30–40 feet by 2–3 inches;

tinwa (Burma), *latang* (Naga), *madang* (Sinpho). From India and Burma, introduced into the United States in 1925, established at Mayaguez, Puerto Rico, and Experiment Gardens, Canal Zone, Panama. Solid at base of culm, to ¼ inch walls in a culm 1¾ inches in diameter measured 6 feet above ground. Used as building material and in basketry. Nodes not prominent except below branches. Internodes up to 18 inches. "After *Dendrocalamus strictus* perhaps the most abundant of all species within its native range."[154]

Cephalostachyum virgatum: 40 feet by 2¼ inches.

Native to Burma, Thailand, India, Vietnam. Used for paper pulp and basketry.[155]

Chusquea.

**Chusquea culeou:* 18 feet by 1½ inches.

The most southern bamboo on the planet, 47° S, from Chile, *C. culeou* is reported as the most hardy species in the collection of the Royal Botanic Gardens in Edinburgh, Scotland. The creamy white sheaths stand out sharply on the bright olive-green new culms, without branches the first season, which soften to a yellow brown. The culms depart from bamboo's hollow norm, being solid from rhizome to tip, a characteristic of the entire *Chusquea* genus, one of the largest genera of bamboos in the world with over eighty species distributed from central Mexico to this species at Lago Buenos Aires between Chile and Argentina. Numerous short branches, 18–30 inches long, circle two thirds of every node, giving the culms the appearance of a bottle brush. Nodes are fairly prominent, 4–6 inches apart, with a grey-white bloom beneath them the first season, which disappears the second year. A rapid grower—Lawson reports 6 inches overnight in England as common—the species is also noted as being fairly drought resistant. The fact that C.

Fairyland Bamboo

This brush pot on the desk, 3 inches in diameter, holds scissors, pencils, rubber cement, and a hypodermic for injecting ink into the plastic capsules—presently irreplaceable in Nicaragua—from which this book is leaking, page by page. Cut from perhaps 50 feet up a culm of *Dendrocalamus asper,* the brush pot doubles as a paperweight to keep preceding pages of these bamboo musings from blowing away in the mischief wind at this moment rippling leaves of a potted *Bambusa vulgaris* by the door. We planted branches the basket makers routinely discard, and within three months we had saleable plants.

A burgundy hue stains the lower internodes of a freshly shooting miniature culm, thinner than the pen in hand describing it. The stem of this, the commonest of all world species, is characteristically wine colored beneath the sheaths of basal internodes of shooting culms. Otherwise, culms are lemon to honey yellow, randomly broken by streaks of green that are an inch or more wide in large culms but much thinner on this potted miniature. The green stripes are typically grouped in two to four of varying widths. The smallest on this tiny culm are thread-thin, hair-fine lines, all but absent, frail as a cobweb in dreams. If it were up to a human hand to draw them, they could only be left by the sure wrist and hyperdelicate brush of mouse whiskers, reputedly favored by Su Tung-p'o, the most renowned bamboophiliac in China's long history of eminent artists stricken with fanatic affection for the plant. Culms yellow as the road to Oz, streaked vibrant as the Emerald City—is it any wonder that, among its hundred names, the Chinese called this species "fairyland bamboo"?

Chusquea culeou, ***the world's most southern bamboo, with conspicuous "bottle brush" branches that nearly circle culm nodes.***

culeou is a remarkably hardy sympodial species makes it an ideal ornamental bamboo for limited areas in colder climates where a clumping bamboo is desired.[156]

Chusquea longipendula: 30–40 feet.
Native to Bolivia; used for house construction and paper pulp.[157]

Chusquea pilgeri: 26 feet by 2 inches.
Native to Colombia; used for houses and farm equipment.[158]

Chusquea ramosissima: 18 feet by ½ inch.
Native to Argentina, Paraguay, Uruguay; used for crafts.[159]

Chusquea simpliciflora: 30–40 feet.
Native to Panama, Peru; used in houses and paper pulp.[160]

Dendrocalamus.

**Dendrocalamus asper:* 100 feet by 8 inches.
An East Indian bamboo that grows in a fairly open clump, *D. asper* is cultivated in Java for edible shoots and construction. The walls are thin —¾ inch at most—but the culms are strong, durable, and generally straight, though leaning somewhat in the periphery of a clump. An extremely impressive species visually, all nodes in the lower culm to a considerable height are heavily bearded with adventitious roots, which occur also at the swollen base and first few nodes of branches. Leaves—characteristically larger on younger plants in all bamboo species—vary from 5 inches by ¾ inch on mature culms to 18 inches by 3 inches on culms of a younger grove. The immense shooting culms of an established grove are the soft color of a newborn fawn, a furry silver brown, and velvet to the touch. Minimum temperature endurance is not reported, but *D. asper* is fairly cold resistant and is planted in Java up to altitudes of 6,000 feet. "In Algiers this species is reported to have withstood many degrees of frost."[161]

Dendrocalamus brandisii: 60–120 feet by 5–8 inches;
kyellowa, waya, wapyu (Burma), *wakay, waklu.* Native to India (eastern slopes of Pegu Yoma and

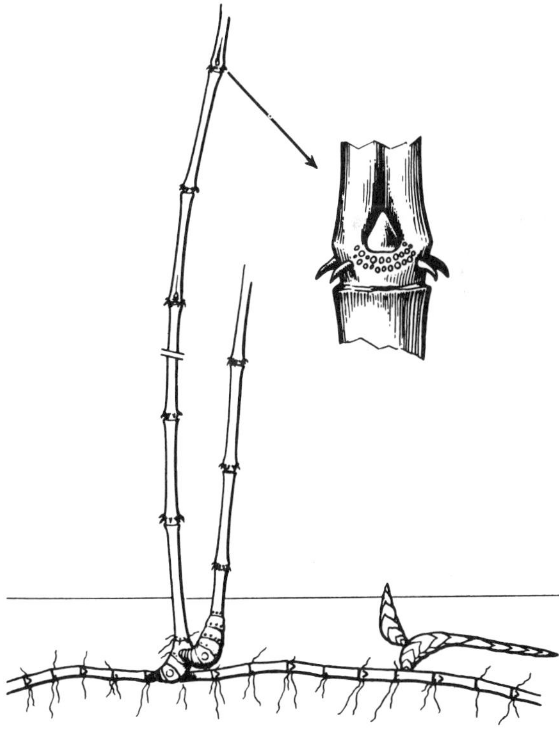

Chusquea fendleri ***is distinguished by rhizomes combining the long runners of monopodial bamboos with the swollen, clumping form of sympodial species. The enlarged node shows another feature unique to the genus*** **Chusquea:** ***a cluster of smaller buds around the principal bud of a potential branch complement. Note also the spinelike aerial roots, a survival tactic preparing fallen culms to dig in for nourishment at every node and multiply new culms and rhizomes.***

Seed of* Dendrocalamus asper, *the bamboo on the cover of this book. Seeds of different species vary greatly in size and form, often resembling wheat. Bamboo homebrew is fermented from them.

Martaban Hills to 4,000 feet), Burma, Vietnam, Laos, Thailand. Its thick walls are used in house construction, farm equipment, paper pulp, furniture, and basketry; used also for edible shoots.[162]

Dendrocalamus giganteus: 80–100 feet by 8–10 inches;
wabo (Burma), *worra* (Assam). Native to India, Calcutta northwards to Tenasserim, Burma, Sri Lanka, Thailand, Madagascar. Used in construction, furniture, crafts, farm equipment, for paper pulp and edible shoots.[163]

Dendrocalamus hamiltonii: 40–60 feet by 4–7 inches;
wabo-myetsangye (Burma), *chye* (Dehra Dun), *tama* (Nepal), *pao* (Lepcha), *kokwa* (Assam), *pecha* (Bengal), *tonay* (Mikis), *wanoke* (Garo). Native to India and Burma, northeast Himalayas, Assam Valley, Khasia Hills, Sylhet east to upper Burma and west to Sutlej, Thailand, Laos, Vietnam. This species, the most common bamboo of the Darjeeling Hills, Terai, and surrounding areas, is also much cultivated. Culm wood is somewhat soft and thin walled. Used for houses, farm equipment, paper pulp, and basketry.[164]

Dendrocalamus hookerii: 50–60 feet by 4–6 inches;
seiat, ussey, sejasai, sijong, denga, ukotang, patu, tili, kawa ule. Native to India, upper Burma, and Sikkim to 5,000 feet. House construction and pulp for paper are the principal uses of this species.[165]

Dendrocalamus latiflorus: 80 feet by 8 inches.
Native to China, with nodes up to 2 feet apart and culm walls over an inch thick, used in house construction, furniture, basketry, crafts, paper pulp, and as edible shoots.[166]

Dendrocalamus longispathus: 60 feet by 3–4 inches;
khang, ora, wa-ya, talagu. Native to India (east Bengal), Burma, Thailand. Internodes up to 2 feet; culm walls, ½ inch thick. "Not highly esteemed as building material, but used when better kinds are not available."[167] Used in house construction, furniture, and paper pulp.[168]

Dendrocalamus membranaceus: 70 feet by 4 inches;
wa-ya, wa-yai, wa-mu, wapyu (Malay). Native to India, Burma, and Thailand, in moist forests at low elevations. Uses include construction, basketry, crafts, farm equipment, and paper pulp. Internodes 9–15 inches, wood ¼–⅜ inch thick.[169]

Dendrocalamus merrillianus: 70 feet by 3–4 inches;
bayog (Ilocus), *kawayan-bayog* (Pangasinan). Culms tall and thin, thick walled and strong. Used for paper pulp, furniture, and construction.[170]

Dendrocalamus sikkimensis: 50–60 feet by 5–7 inches;
pugriang (Lepcha), *wadah* (Garo Hills), *tiria, vola* (Nepal). Native to India, Sikkim, and Bhutan to 4,000–6,000 feet; general construction uses.[171]

**Dendrocalamus strictus:* 100 feet by 5 inches, 22–116°F;
male bamboo, *bans, bans kaban, bans khurd, karail, mathan, mat, buru mat, salis bans, halpa, vadur, bhiru, kark, kal mungil, kibi bidaru, radhanapavedru, kauka, myinwa.* The world's primary volume use of bamboo is for paper. India is the main country making it, mainly from this species. "Calcutta bamboo" yields hard dense wood in slender, thick-walled to solid culms in close stands of poles ranging from quite curved to fairly erect. Of all bamboos, this is the most drought resistant—an important quality that germ plasm could, through hybridization, pass on to synthetic bamboos.

Because of the immense volume of its use, *D. strictus* is the most carefully studied bamboo in the world. Since the 1870s, when the Indian Forest Service at Dehra Dun began their long efforts to describe its most effective cultivation and use, it has been under constant scrutiny by the most active bamboo research center in the world. These studies serve as a lengthy model for the design of research with other bamboos elsewhere.

"This is the most widely distributed of all Indian bamboos, occurring in deciduous forests throughout the greater part of India, common in drier types of mixed forest throughout Burma and found typically in hilly country to 3,500 feet, sometimes almost to the exclusion of tree growth but usually forming an understory to or mixture with deciduous trees.

"Within its habitat, it is frost hardy. In the abnormal frost of 1905, it withstood the effects better than almost any tree species. In the abnormal drought of 1899–1900, it escaped damage although other species of bamboo and many tree species suffered severely."[172]

A minimum of 30 inches, maximum of 200 inches average annual rainfall; maximum shade temperature 116°F down to 22°F, atmospheric humidity is said to be one of the determining factors in the distribution of the species. It flourishes where the relative humidity is low, in interior tracts beyond the influence of sea breezes. It can stand drought better than any other bamboo though frost and drought may be very harmful to young plants.

D. strictus grows on practically all types of soils with good drainage, not on water-logged or heavy soils such as pure clay or clay mixed with lime. Localities with sandy loam overlying boulders are the best, on cooler aspects of hilly ground.[173]

"*Dendrocalamus* species, after germination, produce a grasslike seedling the first year. The plumule, which appears as a conical bud covered by sheathing scaly leaves, develops rapidly into a thin wiry stem bearing single leaves, alternate at the nodes, the leaf bases covering the stem. Fibrous roots form at the base of the young shoot, and successive pointed buds appear on the rhizome. These buds form short rhizomes which curve upward to produce an aerial shoot. This process of rhizome and shoot production continues for several years, and as the previous year's shoots die down, new rhizomes grow deeper and new shoots appear each succeeding year. The clump is formed by these short rhizomes, sometimes as congested, sometimes as open clumps, depending on the species. *D. strictus* requires twelve to thirteen years under natural forest conditions, or six years in artificially established stands, to form a mature clump."[174]

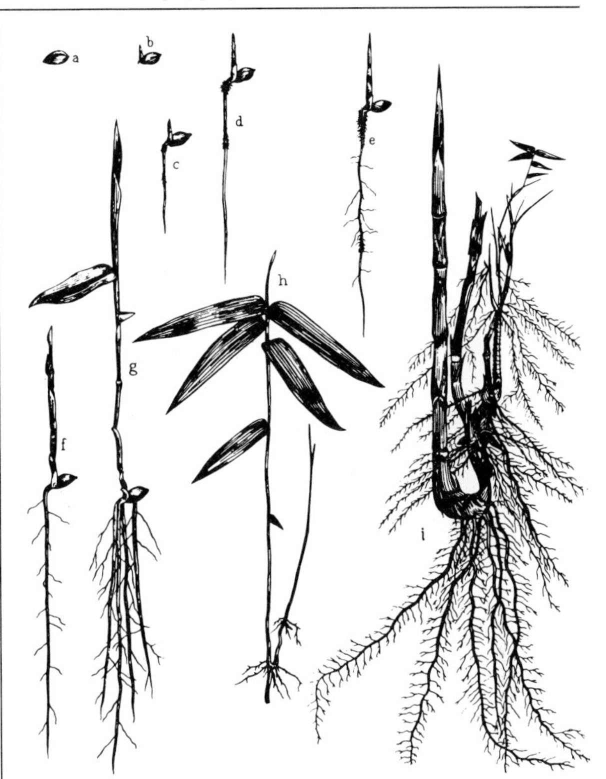

Dendrocalamus strictus: ***The seed (a) pokes up a primary culm (b) and primary root (c). Initially, the root dives further beneath ground level than the culm rises above it (d, e, f). With the first leaf (g) the visible plant begins to surpass in bulk the hidden infrastructure. In a one-year seedling (h), the primary culm is already dead, but a more vigorous and leafy heir rises from a basal bud. In a fourteen-month old nursery seedling (i) each successive rhizome burrows deeper in the soil to yield sturdier culms.***

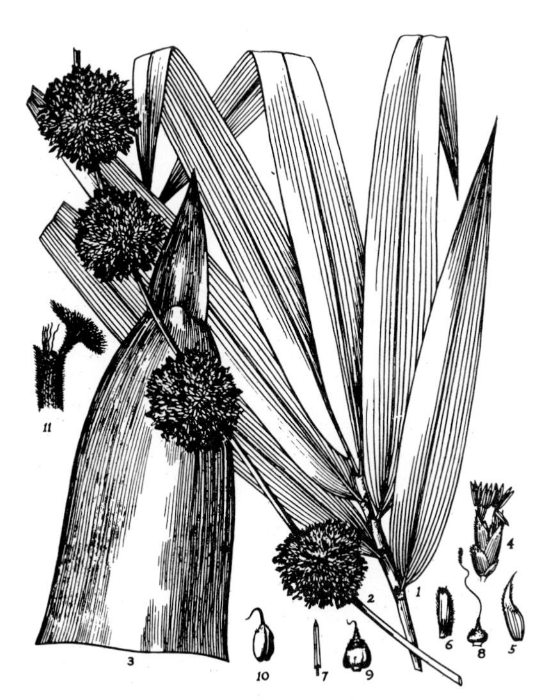

Dendrocalamus strictus, ***the world's most studied bamboo. (1) Leafy twig. (2) Flowering branch with heads of pseudospikelets. (3) Culm sheath. (4) Spikelet. (5) Lemma. (6) Palea. (7) Stamen. (8) Gynoecium. (9, 10) Fruits in different stages of development. (11) Top of leaf sheath and base of leaf blade.***

"A certain amount of overhead cover is necessary during the earliest stages of development of bamboo seedlings before they actually form clumps. Natural regeneration of bamboo is conspicuously absent from areas entirely exposed to the sun . . . A judicious amount of shade minimizes frost effect . . . The largest number of bamboos is obtained when the light falling on bamboo crops is at the maximum. But . . . a certain amount of overhead cover improves the quality of the bamboo at the expense of its quantity.

"Wind is unfavorable in direct proportion to intensity. In windy areas, more side branches develop and congest the grove."[175]

Introduced a number of times from India into the United States by seeds, *D. strictus* is fairly common in southern California, Florida, and Puerto Rico. A search for elite strains is suggested by McClure who includes this species among seven "elite bamboos," selected species from around the world.[176]

Dinochloa maclellandii: 100 feet by 2 inches. Native to Thailand, Burma, Vietnam, Laos, Cambodia; used for crafts and production of paper pulp.[177] Internodes to over 3 feet.

Gigantochloa.

**Gigantochloa apus:* 65 feet by 4 inches; *bambu apus, bambu tali* (Malaya), *delingi apoos, delingi tangsool, delingi pring, pring apes, pring apoos, pring tali* (Java), *awi tali* (Sunda), *pereng tali* (Madura). Of Malayan origin, "tabashir bamboo" is the most useful and widely planted bamboo in Java, growing up to altitudes of 5,000 feet. "Because of their strength and durability, the culms are best liked for building houses and bridges, and for all kinds of woven work. In each roof, item of basketry, and each tie where strength is required, this species is used. It is the only bamboo available in Java from which strong lashings can be made, and the only species used for the framework of rattan furniture" (Heyne 1950). The young culms are called *bamboo talie* (string bamboo) because their durability and excellent splitting qualities make them valued for string, rope, withes, and cordage. The young shoots are buried three to four days in mud to remove bitterness and a narcotic quality they reportedly possess, and then they are cooked in the usual way to be eaten as a vegetable or pickled. Full grown, the species is known as *bamboo apoos* in reference to the tabashir found characteristically in lower internodes. Tabashir is a siliceous deposit traditionally ascribed a variety of marvelous properties, which in more recent times has been patented as a catalytic agent.[178]

The clump growth is fairly open, the culms erect or arching above, with internodes up to 65 cm, cylindrical or slightly grooved above branch complements, walls up to 1½ inches thick. New culms have a conspicuous coating of white powder. The culm sheaths are without auricles and are covered in roughly oval form by small brown hairs. The nodes are slightly prominent, without branches in the lower half of the culm. The leaves, a darker green above than below, are quite large, 4–18 inches by ½–3 inches wide, with unequal sides at the base. The species is easy to propagate, even from basal cuttings of large branches, especially at lower nodes, where root primordia form on the bulbous basal portion of the branch. Introduced into the United States from Surinam around 1932, it has since been established in Puerto Rico, Guatemala, Nicaragua, and other tropical American research stations. A highly recommended species.[179]

Gigantochloa levis: 65 feet by 6–8 inches; *kawayan*-bo-o, *kawayan sina, kawayan puti, boho* (Tagalog), *boko, bolo, botong* (Bisaya), *butong.* Native to the Philippines, culms straight and easily worked, used in general construction, basketry, furniture, and for edible shoots.[180]

Gigantochloa ligulata: 30 feet by 1½ inches. Native to Malaysia, Thailand, Indonesia; used for house construction and farm equipment.[181]

Gigantochloa macrostachya: 30–50 feet by 2½–4 inches; *tekserah, madi, madaywa, wanet, wabray.* Native to India (Assam, Chittagong) and Burma; used in general construction, furniture, and agricultural equipment.[182]

Gigantochloa ridleyi: 50 feet by 4 inches. Native to Malaysia; used for house construction and farm equipment.[183]

Gigantochloa scortechinii: 65 feet by 4 inches. Native to Malaysia; used for house construction, paper pulp and farm equipment.[184]

**Gigantochloa verticillata:* 80 feet by 4–6 inches.
One of the largest bamboos of Java and the most useful after *Dendrocalamus asper* and its near relative, the more durable *Gigantochloa apus.* This species is much valued there for edible shoots and house construction, particularly for posts, rafters, and interior walls. Exposed to weather, it is reputedly not very durable. Opened into bamboo boards up to 1½ feet wide, the culms are used for floors and beds, care being taken to remove the heart wood, which is rapidly devoured by bugs. The same precaution is followed with culms split to make laths, and the poles are sometimes soaked in lye to increase durability. As with most species, clump cured mature culms of *G. verticillata* three years old or more are much less attractive to the insects that attack harvested bamboo.

The culms of the somewhat open clump are quite erect, though nodding at the tip. Easy to work because of their straight grain and unswollen nodes, they are usually free of branches for twenty or more joints—½ to ⅔ of the culm—after a few small branches at the basal nodes. The culm walls are roughly ¾ inch thick at the base, the lower nodes characteristically ringed by root primordia. Internodes are relatively long—up to about 36 inches in midculm—which partly accounts for the frequent use of this bamboo in Java for carrying water. Supposedly its use to contain liquids in no way affects the taste of the liquid contained.

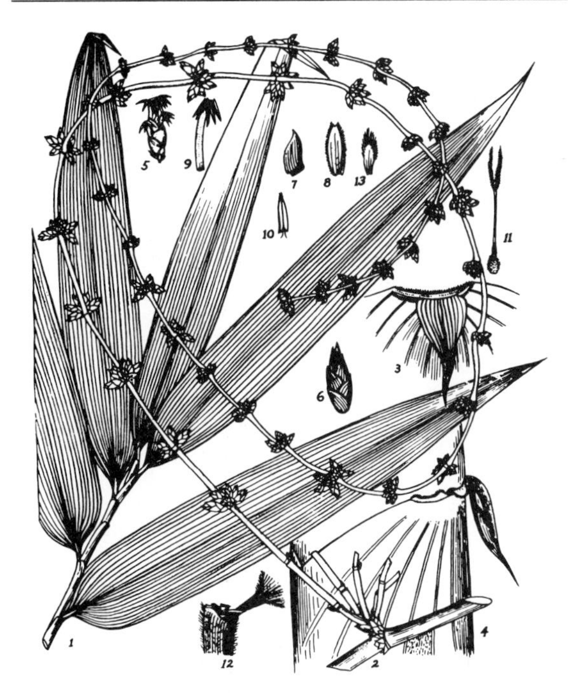

Gigantochloa verticillata: ***(1) Leafy twig. (2) Flowering branch, with tufts of pseudospikelets. (3) Culm sheath tip, dorsal view. (4) Same, lateral view. (5) Pseudospikelet. (6) Spikelet. (7) Lemma. (8) Palea. (9) Androecium. (10) Stamen. (11) Gynoecium. (12) Top of leaf sheath and base of leaf blade.***

This species is one of seven economically elite bamboos given extended attention by McClure,[185] principally for its ranking among the top six of some one hundred species tested in a survey of pulping properties for paper production of bamboos available in the West. According to its "spectacular showing" in field tests in Guatemala (at Rosario in the Polochic Valley), *G. verticillata* could yield "more than 4 tons of oven-dry cellulose per acre per year from plants spaced at 7 × 7 m."[186] Groves have been established at Coconut Grove, Florida, Lancetilla, Honduras, and at Chocolá and Rosario, Guatemala.[187]

Guadua.

Guadua aculeata: 75–90 feet by 6 inches; *tarro* (Central America). Native from Mexico to Panama, the erect culms nodding somewhat at the tip, culm walls nearly 1 inch thick at the base of clumps that are fairly open, like all *Guadua* species, owing to the long-necked rhizomes. The uninflated nodes are conspicuously ringed beneath by an area of dense feltlike hairs that wear away in time. These stand out sharply on the bright green new culms, providing—along with typically short internodes—a characteristic identification for members of this genus. The culm sheaths are triangular, with tiny or absent auricles and a sheath blade that is small and persistent. The sheaths are quite persistent at basal nodes, deciduous above, thickly covered with hairs that are rich in itches to the touch. Single branches well armed with wicked thorns are usual at basal nodes; branches are typically two at midculm and several above. Heavy construction, fences, water pipes, and water vessels are among its commoner uses. Once abundant locally in a number of Central American countries, this species has been almost completely eliminated in many locations by excessive harvesting for building purposes without replanting.[188]

Guadua amplexifolia: 60 feet by 4 inches; *cauro* (Central America). Native Venezuela to Mexico. Short internodes, semisolid in lower culm. General uses in construction. "The least desirable of the listed species for the purpose, but much used in Nicaragua."[189]

*Guadua *angustifolia:* 90 feet by 6 inches; *guadua.* "Best known and most versatile species of the genus. . . . This bamboo apparently has a relatively high resistance to both rot fungi and wood-eating insects. It has been observed repeatedly that ordinary hardwoods used in conjunction with this bamboo have had to be replaced because of insect damage while the

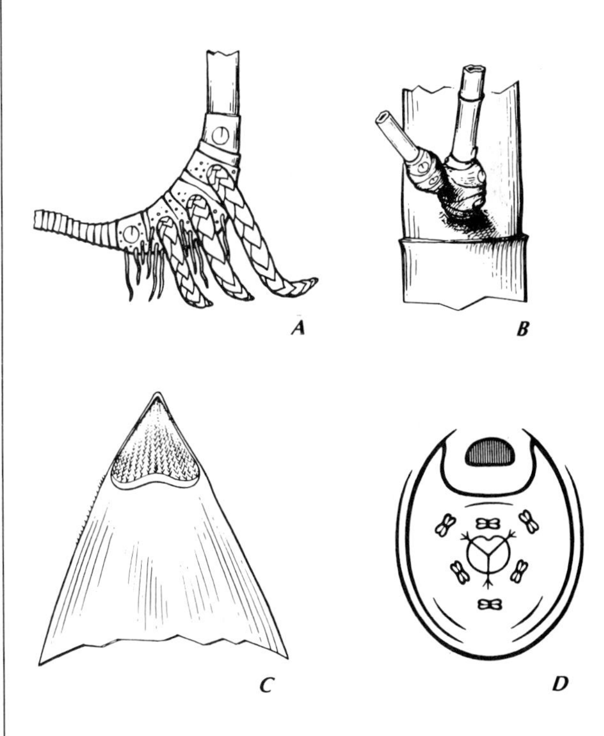

Guadua angustifolia: ***(A) Rhizome with necks of new rhizome branches that help support the mother culm. (B) Midculm branch complement. (C) Tip of culm sheath, inner (ventral) view. (D) Diagrammatic cross section of a floret.***

bamboo still remains serviceable. The original untreated siding, of boards of this bamboo, in a forty-year-old plantation house at Pichilingue in Department Los Rios, Ecuador, was still in a serviceable condition in 1945, long after the hardwood floors had had to be replaced."[190] For an extended treatment of this species, see Chapter 5.

Guadua capitata: 20 feet.
Native to Brazil, used for crafts.[191]

Guadua inermis: 50 feet by 5 inches; *tarro* (Central America). Native from Mexico through Central America. Thick walled, even completely solid at the base, the wood around the lumen is honeycombed with many small lumina. Broadly triangular culm sheaths with well-developed but very fragile auricles are persistent below, deciduous above. Branches usually single, but as the culm matures extra branches develop from buds basal to primary branch, especially in the upper reaches of the culm. Lower culm node buds usually do not develop, are spineless if developed or bear an occasional short spine. Leaves, coarse in size and texture, partially deciduous in dry season. Uses include house and fence construction.[192]

Guadua latifolia: 26 feet by 3 inches.
Native to Brazil and Venezuela; used in crafts.[193]

Guadua paniculata: 32 feet by 1½ inches.
Native to Brazil; used in crafts.[194]

Guadua superba: 75 feet by 5 inches; *marona* (Brazil). Native to territory of Acre in western Brazil, occurring in large stands along the Purus River, a tributary of the Amazon, with specimen plants in the Botanic Gardens at Rio de Janeiro and Belem. Straight culms with short, 3–5 inches, internodes at base of culm, up to 16 inches or more in midculm. Nodes somewhat inflated with branches at nearly all of them. Culm walls ⅝ inch at base. Its wood was recommended above all bamboos native to the West—with the possible exception of *Guadua angustifolia*—for the manufacture of ski pole shafts when the supply of Tonkin cane *(A. amabilis)* from China was cut off during World War II.[195]

Guadua tagoara: 33 feet by 4 inches.
Native to Brazil; used in crafts.[196]

Guadua tomentosa: 15 feet by 1½ inches.
Native to Peru and Brazil; used in house construction.[197]

Guadua virgata: 26 feet by 2½ inches.
Native to Brazil; used for crafts.[198]

Guadua werbertauneri: 32 feet.
Native to Brazil and Peru; used in construction of houses.[199]

Melocalamus.

Melocalamus compactiflorus: 32–114 feet by 1¼ inches.
Native to Thailand, Burma, India. A climbing species used in crafts.[200]

Melocanna.

**Melocanna baccifera:* 70 feet by 3 inches *(Melocanna bambusoides, Bambusa baccifera).* In eastern Bengal and Burma, this species is one of the most common and useful bamboos, growing in groves extending nearly 700 square miles with estimated annual yields of some 300,000 tons in one area of its distribution. Rhizomes with necks up to 3 feet long form the open clumps that characterize *muli* bamboo, as this species is known in Bengal.

Its erect culms, nodding slightly at the tip, are durable, straight grained, with nonprominent nodes and internodes up to 20 inches at midculm. The lower half or more of the culm is usually without branches; above, the usual complement consists of numerous slender, subequal branches that snap off easily with a blow of a stick.

Easy to propagate from rhizome cuttings that produce vigorous clumps recovering quickly even

Melocanna baccifera: ***(1) Flowering leafy branch with stigmas emerging. (2) Leafless flowering branch with some flowers in anthesis. (3) Culm sheath, outside (dorsal or abaxial) view. (4) Upper part of pseudospikelet (lacking prophyllum). (5) Floret. (6) Lodicule. (7) Stamen. (8) Fruit, much reduced. (9) Fruit in longitudinal section.***

after clear-cutting harvests, muli bamboo yields 6,000–10,500 culms or 24,000–33,000 pounds of air-dry culms per acre per year. Much used for house construction, mats, and basketry, it is a chief source of excellent paper pulp in its native area. In pulping studies of one hundred bamboo species available in the West, *M. baccifera* ranked among the top six. The main material for economical housing in its area of densest growth, flattened-out whole culms of muli are woven into prefabricated walls as large as 10 by 30 feet.

Its flowering cycle has been variously estimated, on the basis of conflicting reports, at thirty-five to sixty years. The fruit produced, thick walled and pear shaped, is the largest of any known bamboo—nearly 3 inches in diameter—and the seed germinates even while still hanging on the branch. This species is one of seven examined at some length by McClure.[201]

"It is said that the fruits of *M. baccifera* in northeast India and Burma are 'readily devoured by cattle, elephants, bison, rhinoceros, deer, pig, and other animals.' Admittedly, *Melocanna* has an unusually large fruit and (a quite exceptional feature in the grasses) it bears viviparous embryos which depend for early nutrition on the swollen starchy fruit wall or pericarp rather than upon the endosperm tissue within."[202]

Language reflects use: Eskimo peoples have twenty-eight words for snow. McClure lists thirteen different local names for this species: *terai* bamboo, *muli, metunga* (Bengali), *tarai* (Assam), *wati* (Cachari), *artem* (Mikir), *turiah* (Naga), *watrai* (Garo), *kayoungwa* (Magh), *kayinwa* (Burmese), *paia, taria, pagutulla.*[203]

Ochlandra travancorica: ***(1) Leafy flowering branch. (2) Culm sheath. (3) Pseudospikelet. (4) The one-flowered spikelet, with two bracts still attached, and the flower in anthesis. (5, 6) Bracts. (7) Lemma. (8) Palea. (9) Lodicules. (10) Stamen. (11) Gynoecium. (12, 13) Stigmas, much enlarged. (14) Fruit, still surrounded by lemma, palea, and bracts.***

Nastus.

Nastus elongatus: 65 feet by 1¼ inches.

Native to Madagascar; used for paper pulp and small crafts.[204]

Ochlandra.

Ochlandra capitata: 33 feet by 1–2 inches.

Native to Madagascar; used for houses, paper pulp, baskets, and musical instruments.[205]

**Ochlandra travancorica:* 6–20 feet by 1–2 inches.

"Elephant grass," as this species is commonly known, occurs in vast stands in the South Travancore and South Tinnevelly mountains (India) from 3,000–5,500 feet. It grows in impenetrable tracts that even elephants cannot enter, extending for many miles, often to the complete exclusion of all other vegetation. The culms are thin walled and straggling, sometimes supported by the dense clump; internodes are grey green and rough, nodes slightly inflated. The immense exuberance of its distribution and the maximum length of the cellulose fibers of the culm tissue—9 mm—(roughly ⅓ of an inch) are the most remarkable properties of the species in relation to possible human use. "Its culms . . . yield, by the sulfate process, a pulp containing 92.66 percent alpha cellulose, with an ash content as low as 0.2 percent, characteristics indicating . . . possibilities for rayon manufacture."[206] A mill equiped to produce 5 tons of rayon and 1¼ tons of transparent paper daily was established in Travancore in the mid 1940s to exploit the huge stands of *O. travancorica* there.

Field studies are needed, according to McClure, "to determine whether *O. travancorica* is a desirable silvicultural subject, in relation to the high- and sustained-yield requirements of a paper mill. The reputedly short flowering cycle and subsequent death of culms may, if true, be a disadvantage to its cultivation as a source of cellulose pulp. However, the plant apparently fruits freely, and there is the possibility that the development of a seedling progeny would be sufficiently rapid to restore the stand to a productive state by the time the dying flowered culms had been harvested—if the stands maintained are of sufficient extent to provide several years' supply from a single progressive-harvesting cycle. Flowered culms of

Dendrocalamus strictus are usable, and give a slightly enhanced cellulose yield of undiminished quality after standing in the field four years."[207] This species was successfully introduced in 1951 by the USDA at the Mayaguez, Puerto Rico, Experiment Station.[208]

Oxytenanthera.

Oxytenanthera abyssinica: 48 feet by 3 inches;
arkai, chommel. Native to Ethiopia, Angola, Gold Coast, Malawi. Used for house construction, farm equipment, basketry, and production of paper pulp.[209]

Oxytenanthera albociliata: 32 feet by 1½ inches.
Native to Thailand, Burma, Laos; used for basketry, farm equipment.[210]

Oxytenanthera nigrociliata: 50 feet by 5 inches;
poday (Andaman), *washut* (Garo), *bolantgi bans* (Orissa), *lengha* (Java). Native to Thailand, Burma, Indonesia, Malaysia, India. General uses for farm equipment and furniture construction.[211]

Pseudostachyum.

Pseudostachyum polymorphum: 50 feet by 1 inch;
filing (Nepal), *purphiok, paphok* (Lepcha), *wachall* (Garo), *bajal, tolli, ral* (Assam), *bawa* (Burmese). Native to India (eastern Himalayas, Assam), Sikkim, and upper Burma. Thin-walled culms with long internodes are used for lath, matting, withes for tying frames of houses, and small crafts.[212]

Schizostachyum.

Schizostachyum hainanense: 100 feet by 1 inch;
tang chuk (Chinese). Native to Hainan Island. Thin-walled culms with long internodes are used for lath, matting, crafts.[213]

Schizostachyum lima: 25–30 feet by 1 inch;
bolo, bagacay. Native to Philippines, Luzon, and Davao. Culms have very long internodes, thin walls. Uses include matting, shingles, and lathing.[214]

**Schizostachyum lumampao:* 60 feet by 3 inches;
lakap (Bosayan), *tamblang* (Bila-an). Native to the Philippines, Luzon. Culms are very straight, thin walled, with some 40 feet to first branch. Used to make boards and shingles.[215] This is an extremely useful species deserving wide distribution in Latin America. Its thin walls permit weaving of the entire culm after opening it out flat and removing nodal tissue. See p. 16, "airplane skins."

Sinarundinaria.

**Sinarundinaria nitida:* 20 feet by ¾ inch
(Arundinaria nitida).
Native to Szechuwan and Kansu Provinces of China, this clumping species is reportedly hardier in England than many of the running, temperate bamboos. In its native region it grows up to 10,000 feet on the northern slopes of mountains. The slender dark purple culms are there used locally in basketry, sieves, fencing, and light construction. Classed by some botanists as an *Arundinaria,* Lawson lavishes on this species his most raptured superlatives: "*A. nitida* is without doubt the most graceful of all the Arundinaria family . . . in some countries known as the 'Queen of the Arundinaria.' The leaves dislike direct sun, and the margins will bend towards midrib, forming a channel, at the first touch of brilliant sunshine . . . As soon as the clouds again cover the sun, the leaves will once more open out to their graceful normal shapes. . . . Grow this dainty bamboo where there is some partial overhead cover . . . The slender purple canes are densely packed into a closely circumscribed clump. Their lower halves are bare, and rise almost vertically, but at the higher levels, the wealth of foamy foliage bends them outwards in a filmy mass. The individual leafed branchlets are so fine that the foliage gives an impression of floating in the air. The thin whiplike new canes sway above the mass of foliage below."[216]

Named *nitida* ("shining" or "lustrous") by Freeman-Mitford, now commonly known as "fountain bamboo," the species was introduced into Russia, thence to England, by seed in the 1880s. Believed by the Chinese to flower once in a century, that rare moment may at present (1984) be returning. Lawson states that no flowering outside China has ever been recorded.

Culms, light to deep purple, are without branches the first year, then four to five per uninflated node, 6 inches apart, in the second season with more branches and twigs growing as the canes mature. The pale purple culm sheaths, hairy and often as long as the internode they cover, are thin textured and persistent. Small leaves, 3½ inches by ½ inch, are paper thin and bristled at the edges—a brilliant green above, matt green below. Delicate in appearance, the leaves weather severe winters with minor leaf scorch. Strongly recommended as an excellent garden or tub plant for northern locations because of its hardiness, the striking beauty of its culm and foliage, and its tendency to stay put where you put it.[217] More on this species under *Thamnocalamus spathaceus,* "umbrella bamboo."

Teinostachyum.

Teinostachyum dullooa: 20–30 feet by 1–3 inches;
dulooa (Assam), *paksalu, pogslo, wadroo, gyawa.* Native to India (Assam) and Vietnam;

thin walled with internodes up to 40 inches long. Used for lath, matting, and crafts.[218]

Thamnocalamus.

Thamnocalamus falconeri: 30 feet by 1¼ inches

(Arundinaria nobilis, Arundinaria falconeri, Bambusa floribunda).

Native to N.E. India, in the Himalayas, introduced in 1847 into England. The pliable, thin-walled canes are much used for fishing rod manufacture and basketry. Distinctive identifying characteristics to note include the tiny, paper-thin leaves, not visibly tessellated, and the nodes, which are stained a purple brown, especially noticeable on older canes. The species grows in a tight clump of stiffly erect canes, surrounded at the edge by culms curving gently outwards, with dense foliage beginning almost at the base on many delicate branches and twigs.

"Noble"—the specific given by Freeman-Mitford, who called this bamboo *Arundinaria nobilis*—describes the tall canes well. Olive green when young, dull yellow when old, the new culms appear from mid to late May on. The culm sheaths, pale crimson fading to a dull purple shade, are frail and deciduous, and truncated at the tip. The nodes, roughly 10 inches apart, are rather prominent, and stained about a half inch on either side with the characteristic brownish-purple ring, more pronounced in sunnier locations. The species is not invasive; it expands slowly in a clump. "Makes an excellent ornamental in shady spots in warmer gardens, and is especially recommended as a tub species for the conservatory."[219]

**Thamnocalamus spathaceus:* 14 feet by ½ inch

(Arundinaria murielae, Arundinaria sparsiflora, Fargesia spathacea, Sinarundinaria murielae).

A hardy, although clumping, Chinese bamboo from the mountains of western Hupeh Province, growing at some 10,000-feet elevation. First collected by European botanists in the late nineteenth century, live specimens arrived in Harvard's Arnold Arboretum in 1910 sent by E. H. ("China") Wilson, who called it the handsomest bamboo he'd ever seen. From Harvard, they were sent to Kew Gardens outside London in 1913 to become the most widespread of the ornamental bamboos in Europe, commonly known as "umbrella bamboo."

The culms are bright green during their first season, maturing gradually to a deeper green and finally a dull yellow. Deciduous pale green to cream color culm sheaths, bristled with fine hairs at the base, fade to straw. Late-shooting culms may keep their sheaths until the following spring. Uninflated nodes, some 6 inches apart, bear three to four branches the first year, which become more numerous as the culm matures. Leaves, 3–4 inches by ½–¾ inch, are light pea green above, duller pale green below. Long

Thamnocalamus spathaceus: ***rhizome system and culm tips.***

considered an *Arundinaria,* this species began flowering in Denmark in 1975, revealing its true identity as a *Thamnocalamus.*

In an interesting article detailing the century of botanical confusion that surrounded this popular species, Soderstrom permits the nonbotanist to savor something of the slow drama of taxonomy.[220] The related "fountain bamboo," named *Arundinaria nitida* by Freeman-Mitford in 1895 and reclassified as *Sinarundinaria*—a new genus erected in 1935 by the Japanese botanist Takenoshin Nakai—last flowered in 1886. According to oral tradition among the Chinese of its native region, the species has a flowering cycle of about a hundred years. Soderstrom suggests that when flowers become available for examination, *A. nitida,* the "queen of *Arundinaria*" will reveal itself to be *Thamnocalamus nitidus,* the fifth member of this limited Sino-Himalayan genus to be described since Munro established it in 1868. Both umbrella and fountain bamboo are sympodial, clumping species, unique in their hardiness as well as their acknowledged loveliness. This species is reported by Lawson to grow in the north of Scotland. Both fountain and umbrella bamboo are named for their similar dense clump formation, in which the thin canes, burdened with foliage, arch out at the edges, trailing branches groundward in a fountainlike manner.[221]

Thamnocalamus spathiflorus: 15 feet by 1 inch *(Arundinaria spathiflora).*
Native to the western Himalayas of India, Nepal, Sikkim, and Bhutan at elevations of 7,000–10,000 feet. First described by Munro in his 1868 monograph on the bamboos as the type species of a new genus that he then established. "He derived the generic name from *thamnos,* meaning 'shrub,' and *calamus,* meaning 'reed,' since the bamboo formed small, brushlike clumps of densely packed culms . . . [This species] occurs in the undergrowth of forests composed of deodar cedar, silver fir, Himalayan spruce and oak . . . *Thamnocalamus* can be characterized as clump forming with relatively narrow culms of low stature. A number of other features distinguish the genus botanically: more-or-less equal branches borne above the node line, rhizomes sympodial, spikelets many flowered and arranged in racemes enveloped in spathelike structures, and flowers with three stamens and three feathery stigmas."[222]

Introduced into England in 1886, this species is delicate in appearance and not terribly hardy in fact: to be grown in areas sheltered from cold winds. Like its near relative, the fountain bamboo *(Sinarundinaria nitida,* whose imminent flowering will probably reveal it to belong to the genus *Thamnocalamus),* its leaves curl in full sun, open again when shaded. "At one time in India this bamboo supplied the bulk of the raw material used in the manufacture of walking sticks, umbrella handles, pipe stems, and numerous other articles."[223]

Its erect culms, bright to pale green distinguished by a shade of purple pink, inclined to zigzag, are marked with a blue-white bloom the first season. Sheaths are deciduous, hairy at the base, some 7 inches long and quite glossy on the inner surface. Inflated nodes, marked by a white ring on their lower portion, brown above, bear two to three branches colored a pale pink purple. Leaves, paper thin and fringed with tiny bristles, are finely tessellated, up to 6 inches by ½ inch, pale green above, and a flat grey green below. Forms tight clumps; "extremely handsome and elegant . . . an excellent specimen plant for a sheltered spot and a good tub plant."[224]

Thyrsostachys.

Thyrsostachys oliverii: 50–80 feet by 2–2½ inches;
thanawa (Burmese), *maitong* (Kachin). Native to India and upper Burma to 2,000 feet. Culms "greatly in request" (Gamble) for general construction purposes. Also used for paper pulp and farm equipment.[225]

Thyrsostachys siamensis: 25–40 feet by 1½–3 inches;
kyaung-wa. Native to Thailand and Burma. Close clumped, unbranched for 15–20 feet, short, 8 inches, regular internodes on the culms supporting thick, feathery foliage. "One of the two most beautiful Burma bamboos, and by far the finest for planting at lower elevations. It is cultivated from end to end of the country, found almost always near Buddhist monasteries. The culms are much used for umbrella handles and make ideal vaulting poles."[226]

PSEUDO BAMBOOS

A number of plants, although not classified by botanists as bamboos, resemble them in appearance, serve similar uses, and are considered bamboo by the people who coinhabit their terrain.

**Arundo donax:* 20 feet by 1 inch.
One of the most widespread of the pseudobamboos is *Arundo donax.* Native to Europe, introduced by the Spanish in the New World in the sixteenth century, *carrizo* or *caña de Castilla*—as the plant is called in many areas —grows as far north as Texas and is broadly used throughout the American tropics. The pale green internodes are shiny and thin walled but durable. Uninflated nodes usually bear a single branch of sparse pale green leaves, leathery and broad. Frayed portions of the culm sheath remain on cured culms which mellow to a rich yellow brown, darkening with the years.

Still sound after over a century in walls of *bajareque* construction (see Chapter 9, pp. 259–260), carrizo can also be seen in ceilings forty to fifty years old, outlasting some of the wooden rafters with which it is alternated to support the tiles. Widely used for basketry and mats, occasionally as spools in textile industries, carrizo also forms the shell for skyrockets *(cohetes),* an integral part of any Latino fiesta. *Rondadores* (panpipes) and simple shepherd's flutes are other ancient uses of *Arundo donax*—which also provides reeds for saxophones. The erect or sometimes slightly curved culms of "Spanish cane" are cultivated as a garden ornamental as well.[227]

**Gynerium sagittatum:* 25 feet by 1¼ inches.
Grows usually at low elevations in tropical America. The plant is covered with persistent sheaths. Beneath these, the culm is sticky and pale green to cream colored when fresh, the internodes filled with pith that shrivels when dry. From the uninflated nodes at midculm the single branches, growing nearly parallel to the culm, bear long narrow leaves that cluster also at the culm's tip. From a distance, the plant resembles sugarcane. Its uses include basketry and construction of houses, in which it serves for partitions, horizontal lathing in the framework of

bajareque walls, and roof "sheathing," which supports the tiles and is supported by rafters. The name of the plant varies from country to country: *caña arava* in Cuba, *caña blanca* in Panama, *caña amarga* in Venezuela, *vara de tusa* in El Salvador.[228]

TAXONOMY

With some seventy-six genera and twelve to fifteen hundred species of bamboo worldwide, each possessing quite different tastes in climate and soil, quite distinct physical properties making different species suited to different tasks, a precise way of determining their classification is obviously imperative. Taxonomy—from the Greek *taktos,* meaning to order or arrange—the science of this classification of plants and animals according to biologically related groups was established in the West by Linnaeus (1707–1778). The problem with bamboos is that the flowering portions necessary for precise identification are not usually available. When bamboo flowers, vegetative production creeps or ceases so leaves and culm sheaths are generally not present. McClure's solution to this botanical impasse was to establish a living garden.

> Familiarity with the living plants is an essential preparation for sound taxonomic work on this group. To achieve this, one must be able to return again and again to the same plants to make notes, take specimens at different stages of development, assemble a complete array of structures essential to identification. This conviction led me to bring together living plants at Lignan University, and to date nearly six hundred introductions have been made. Every introduction is given a distinctive number, plants are tagged, their position in the garden plotted. A record is kept of the source of each plant, date of introduction, vernacular names, uses, and other pertinent observations. Herbarium specimens are made of all available structures, and a record kept. Usually neither culm sheaths nor flowers are available at first. As these structures appear, they are collected and recorded. As the picture of the plant becomes more complete, its identity becomes clearer. The identity between different numbers becomes apparent, and, finally, I have not only an adequate idea of the range of variation of each species, but I know its geographical distribution as well.[229]

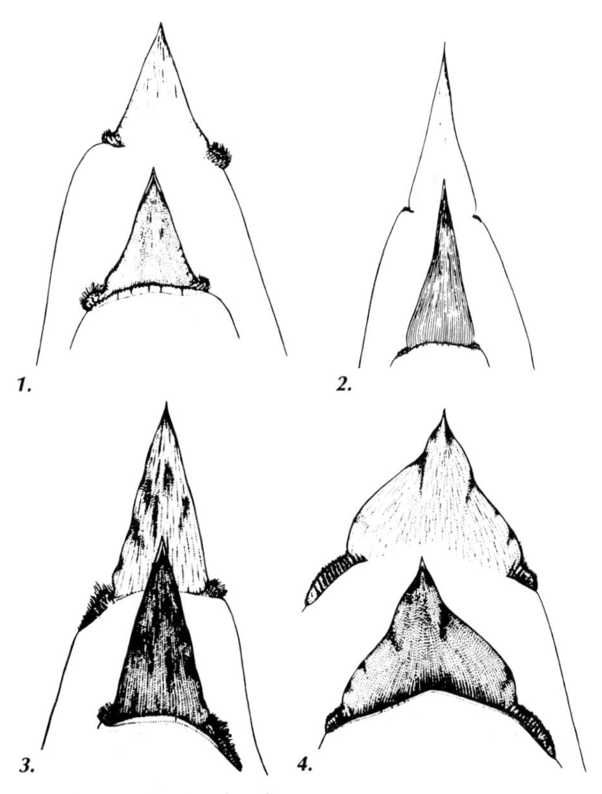

Bambusa midculm sheaths, the outer (dorsal) view above, the inner (ventral) view below (× ⅕).
(1) **B. tuldoides.** *(2)* **B. textilis.** *(3)* **B. pervariabilis.** *(4)* **B. tulda.**

Collecting specimens for identification.

For those who wish to dig into the matter up to their muddy elbows, collecting bamboo specimens becomes important at some step along the bamboo path. For the less active reader, also, curious what a botanist must go through in the field, we include the following suggestions. They were prepared by McClure for the Smithsonian Institution, which has the largest collection of dried bamboo specimens in the United States and one of the largest in the world.

Among the twelve to fifteen hundred species of bamboo, many that may look alike have vastly differing properties. The importance of precise species identification is therefore obvious; this is the procedure for obtaining it when confronted with an unknown bamboo in the field.

It is important to collect all bamboos, in flower or not. A properly selected, labeled, and preserved series of the vegetative structures listed below will be entirely adequate to identify a given bamboo. Reproductive structures are traditionally required for identifying bamboos and many other plants, but vegetative structures have proven to be a practical —and necessary—basis for field identification of bamboo owing to the infrequent flowering of many species.

Never mix material from two distinct plants under the same number, assuming they represent the same bamboo. Two or more different bamboos may grow so close together they appear as one plant. Exacting care is needed, as mixtures cause much confusion.

Make specimens for permanent preservation. Fragmentary specimens collected hastily, "just for identification," frequently turn out to represent new species or new records. Such specimens, often too fragmentary to be identified with confidence but too intriguing to discard, may be more a hinder than a help. So make specimens of each bamboo as complete as possible. Represent omitted or fragmented structures by sketches or photos and notes illustrating them whole. Seedlings and small plants may be collected entire. For all plants, the following structures are essential:

CULM SHEATHS. At least two complete and in good condition, preferably from midculm nodes of mature-sized culm. Mark with node number (counting from base) and collector number. Press flat. If too large, cut or fold to 10 by 15 inches or less. Keep all parts. If sheath will not spread without breaking, do not press. Let it roll up and tie paper over tip to protect fragile parts. Persistent sheaths may be left attached to culm section of sheath length and dried. Young shoots slender enough to dry readily by artificial heat should be sent in whole. The more complete the series of culm sheaths, the more reliable the identification. Represent the whole series as fully as circumstances permit.

LEAFY TWIGS. Include big and little leaves, young and old, healthy and diseased (if any). *Press promptly* before they curl in driers thick and soft enough to prevent crinkling. At first change of driers, arrange leaves so that some show upper surface, some lower.

BRANCH COMPLEMENT. At least one typical example from middle of series on culm of mature size with at least 12 inches of culm itself. Cut branches 2 inches from base. Mark node and collector number on culm. Split and discard culm half opposite branch if space is limited. Additional specimens from lower and upper culm are desirable but not necessary.

CULM NODES AND INTERNODES. Best represented by segment of mature culm including nodes four and five above ground level and the internode between, marked with node number and collector number. Cut any branches back to 2 inches. Segment may be split to save space or speed drying.

RHIZOME. At least one complete example; if space permits, two or more rhizomes attached together to show typical branching habit. Wash and trim roots. Mark or tag with collector number. A sketch or photo, showing proportion and branching habit, will serve in place of actual specimen if facilities are limited.

FLOWERING BRANCHES. If present, collect longest possible series to show range of variation in habit, leafiness, stages of development, and so forth. Seek fruits and put some in a small folded paper to call attention to them and prevent their loss. Mature fruits usually fall very promptly. *Distribute specimens thinly between thick driers; change often to dry promptly* and prevent breaking up of spikelets. Unless absolutely necessary to save paper, do not pile dried specimens together, but keep in original folders to avoid serious damage to spikelets. Put single example of long, fragile spikelets in separate envelope of folded paper to ensure floret count.

SEEDLINGS. Special search should be made under and near flowering bamboos. Seedlings should be given a separate collector number from that of the supposed parent plant, but cross-reference should be made in notes, setting forth evidence of their supposed relationship. The "seed" still attached to very small seedlings is sufficient for identification if kept intact.

HANDLING BULKY SPECIMENS. Node and internode, branch complement, rhizome, and culm sheaths that cannot be pressed flat—especially specimens that come from large plants—should always be marked properly with node number and collector number on clear surface, durable cloth tag, or tough paper. Do not put bulky specimens in press with leafy or flowering twigs. Dry separately and promptly in sun or over fire. Keep in open air as much as possible; don't wrap or store in closed container until thoroughly dry.

MINIMAL NOTES. These should include: clump and culm habit, maximum height and diameter of culms (at base), length and diameter of fifth internode, length and number above ground of longest internode. (Make measurements on largest available

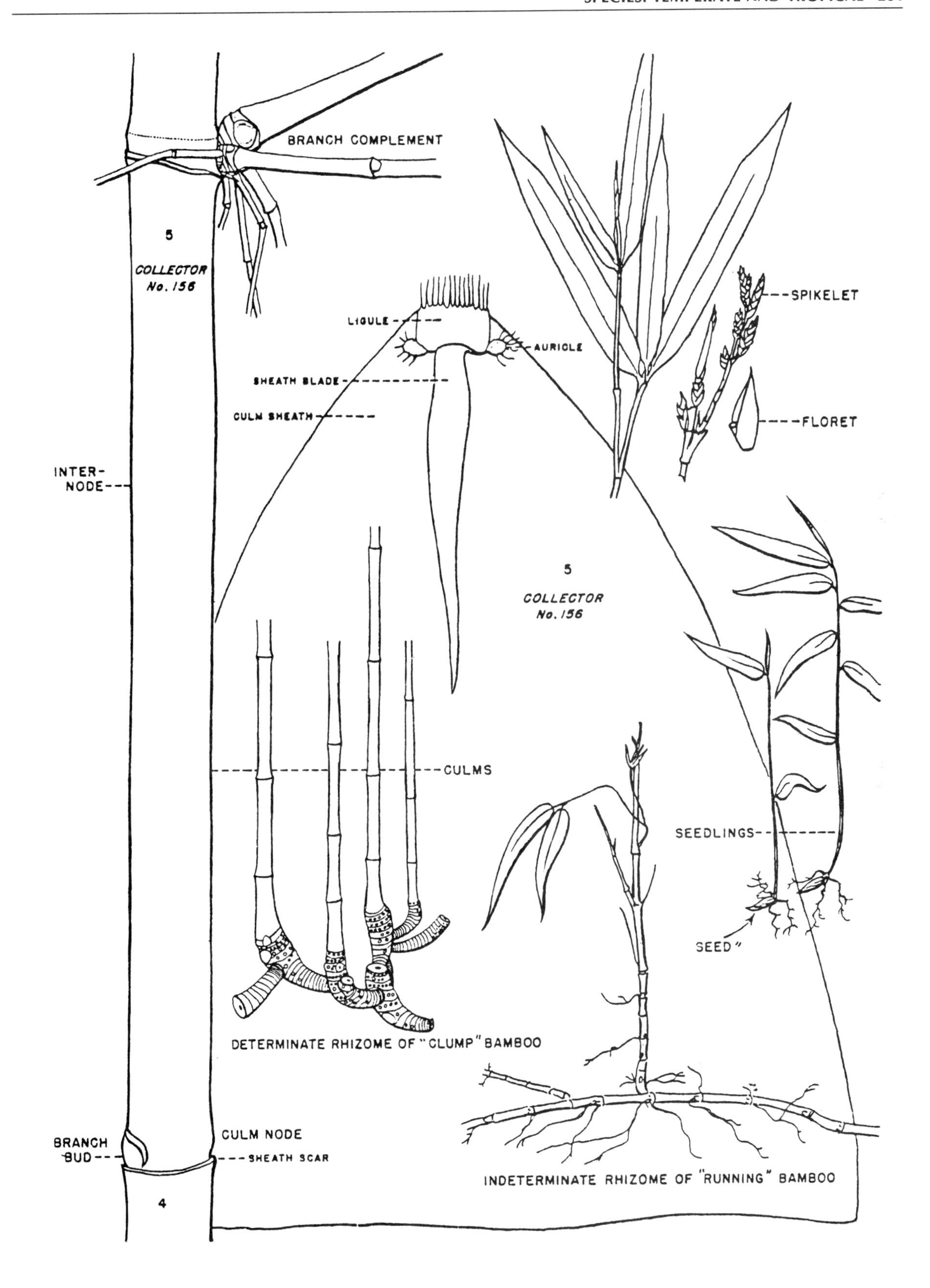

Collecting specimens.

specimen; estimated measurements should be designated as such.) Location should be described or sketched in relation to nearest inhabited place and/or point shown on available map so that it can be found again. Habitat, local names (dialect), local uses, date collected, collector's field number, and reference to photos, sketches, and so on, if added, are kept in a separate book.

PHOTOGRAPHS. These are useful to show habit and proportions of typical clump and the following features, particularly if time is insufficient for making specimens of them: branch complement, culm node and internode, and the rhizome habit and branching. Habitat pictures are useful if the names of the associated plants are recorded or if ecological features are clearly shown. Be sure to correlate photographic numbers with collector numbers at the time the picture is taken.

SKETCHES. The simplest diagrammatic sketch may be used to represent structures called for above but which, for whatever good reason, must be omitted from the specimen.[230]

PRIMARY GENERA OF BAMBOO

The name after the genus is that of the botanist who established it.

1. *Apoclada,* McClure. Some four species, native to Brazil.
2. *Arthrostylidium,* Ruprecht. Around thirty-five species, all native to tropical America; generally not important economically.
3. *Arundinaria,* Michaux. Over one hundred species, mainly southeast Asia and adjacent islands, Japan to Madagascar. In Africa, two possible species; two from the United States, including the genus "type," *A. gigantea.* Greatest concentration in Japan.
4. *Aulonemia,* Goudot. Some twenty-four species from Brazil to Mexico.
5. *Bambusa,* Retzius. Some eighty-five species, all native to Old World tropics, southeast Asia principal center of distribution. Closely related to *Dendrocalamus* and *Guadua.*
6. *Cephalostachyum,* Munro. Three species in India and Burma area.
7. *Chimonobambusa,* Makino. Some five species native to China and Japan. A hardy genus named for its rare habit of shooting in late fall or winter (Greek *cheimōn,* meaning "winter"), famous for "square-stem bamboo," *Ch. quadrangularis.*
8. *Chusquea,* Kunth. Some ninety species, all New World, native to mainland and islands, Mexico to Argentina and Chile. Grows to 12,000 feet, probably widest altitude range of any bamboo genus. Slender, pithy, sometimes solid culms have a constellation of branch buds in place of one bud characteristic of all other genera.
9. *Colanthelia,* McClure and E. W. Smith. Some seven species in Brazil.
10. *Dendrocalamus,* Nees. Some thirty species, all large tropical plants from southeast Asia, mainland and islands. Many locally important for construction and crafts. *D. strictus* is the world's most studied and most drought-resistant species, forming—with *Bambusa arundinacea*—the basis of India's bamboo paper production.
11. *Dinochloa,* Buse. Some four species form this small southeast Asia genus of bamboos equiped to climb other plants. *D. andamanica* is the longest world bamboo, with a culm 100 yards long.
12. *Elytrostachys,* McClure. Two species with distribution centered in Colombia and Venezuela. Likes riverbanks from 200 to 1,500 m in elevation (650 feet–5,000 feet).
13. *Gigantochloa,* Munro. Some thirty large tropical species from southeast Asia, mainland and islands, Burma, Indochina, to Malay Peninsula and Philippines. Many important where plentiful for building and daily use. Java species were introduced there.
14. *Guadua,* Kunth. Some thirty species, all native to Central and South America except *G. philippinensis. G. angustifolia* is probably the world's most durable bamboo.
15. *Melocalamus,* Bentham. Monotypic, that is, a genus with one species, *M. compactiflorus* is found from east Bangladesh to northwest Thailand. Related to *Dinochloa.* Fruits germinate while still on parent plant. Used in basketry.
16. *Melocanna,* Trinius. Some three species include "type" *M. baccifera,* occurring in vast (700 square miles) natural groves in Burma and India. Much used for building, woven ware, and paper pulp.
17. *Merostachys,* Sprengel. Over twenty species, mainly from Brazil, with one species each in Paraguay, Peru, Belize, and Guatemala. Internodes up to 36 inches in one Brazilian species (*M. argyronema,* 10–12 m by 5 cm) used for flutes, resonance tubes beneath marimba keys, and basketry.
18. *Myriocladus,* Swallen. Twenty species in

Venezuela between 1,025–2,500 m (3,400–8,350 feet).

19. *Neurolepis,* Meisner. Some nine species are found between 2,900–4,500 m (9,650–15,000 feet), in Colombia, Venezuela, Ecuador, Peru, and Trinidad. *N. elata,* from the Andes, is distinguished by leaves 5 m long by 0.5 m wide (16½ feet by 20 inches)—the largest known among bamboos.
20. *Ochlandra,* Thwaites. Some seven species in India, Sri Lanka, Madagascar, and southern India. *O. travancorica* (q.v.), "elephant grass," is the most important economically, for paper pulp.
21. *Oxytenanthera,* Munro. Some five species in Africa and Asia, from Angola to Java. *O. abyssinica* most used in Africa; *O. nigrociliata,* in Indonesia.
22. *Phyllostachys,* Siebold and Zuccarini. Some forty species, mostly native to central and southeast China, a few in areas to immediate south. Most thrive in moist, warm-temperate climate.
23. *Pleioblastus,* Nakai. Some twenty-six species, mainly native to Japan.
24. *Rhipidocladum,* McClure. Some eleven species found from Mexico to Bolivia and Brazil.
25. *Sasa,* Makino and Shibata. Some thirty species native to Japan and mainland of north Asia. *S. kurilensis* (from the Kuril Islands northeast from Japan) is the most northern bamboo, the only species native to Russia.
26. *Schizostachyum,* Nees. Some forty species, from tropical and subtropical south China, through Southeast Asia (mainland and islands including Hawaii) to Madagascar. Thin walls in all species, and "by the siliceous, whetstone-like surface you can tell a *Schizostachyum* in the dark."[231]
27. *Thamnocalamus,* Munro. Some four species of this Sino-Himalayan genus include hardy though clumping *Th. spathaceus,* "umbrella bamboo," most widespread European ornamental. West China through Nepal to India.
28. *Thrysostachys,* Gamble. Two species, from India, Thailand, and Burma.
29. *Yushania,* Keng. Two species, from Mexico to Honduras. Known locally as *otate* in many places and much used. This genus is also placed by Lessard (1980) in Taiwan and the Philippines.

This list of twenty-nine genera represents less than half those known—roughly seventy-six—but includes the main genera of economic importance.[232]

CHAPTER 7.

1. Freeman-Mitford 1896:186.
2. Freeman-Mitford 1899:271–2.
3. Lawson 19.
4. McClure 1953:32. Lin 1972:1.
5. McClure 1931 and 1966:150–7 offers extensive descriptions of culture, harvest, curing, and use of *A. amabilis.* Photos and text of *National Geographic* October 1980 bamboo cover article also focus on this species.
6. Jackson 1949.
7. Lawson 67.
8. Ibid.: 74–5.
9. Ibid.: 75–6.
10. Ibid.: 76–7.
11. McClure 1953:32. Lin 1972:1
12. McClure 1953:32.
13. Ibid. Lin 1972:2.
14. Young 1945a:194.
15. Lawson 85.
16. Young 1945a:194. Young 1961:32. Lawson 83–5.
17. Lawson 85.
18. Ibid.: 85–6.
19. *Southeast Forest Experiment Station Research News,* No 5, May 1949.
20. Quoted in West 1935.
21. West 1935:254.
22. Young 1945a:196.
23. Lawson 111. See also Hughes 1951.
24. McClure 1954c:19–20.
25. McClure 1963b:134–6.
26. Young 1945a:178.
27. Lawson 87–8. Young 1945a:178.
28. McClure 1953:32. Lin 1972:2.
29. Satow 49–50. Fuji 1963:15. *For* Han Shan, see *Cold Mountain,* tr. Burton Watson, Columbia U.P. 1970. Gary Snyder's translations of Han Shan have appeared in a number of anthologies after publication in *Riprap and Cold Mountain* (1957?).
30. Satow 78–80. Fuji 1963:35.
31. Lawson 89–90.
32. Ibid.: 90–1.
33. McClure 1953:32. Lin 1972:2.
34. Ibid.: 33. Ibid: 2.
35. Ibid. Ibid.: 3.
36. Lawson 107–9. Young 1961:19.
37. McClure 1954c:19.
38. Young 1961:19.
39. Ueda 1960:118–9.
40. Lawson 113.
41. Ibid.: 112–4.
42. Ibid.: 114.

43. Ibid.: 114–5.
44. Young 1945a:179; 1961:20.
45. McClure 1953:33. Lin 1972:3.
46. Lawson 96.
47. Ibid.: 94–6.
48. Ibid.: 147–8.
49. Freeman-Mitford 1899:276–7.
50. McClure 1957a:2.
51. Ibid.: 12–3.
52. Ibid.: 13–5. Glen 1954.
53. McClure 1957a:15–8. Young 1961:20–1.
54. Ibid.: 18–20. Ibid.: 21.
55. McClure 1966.
56. McClure 1944.
57. Lawson 120.
58. Young 1961:22–3. McClure 1957a:24. Lawson 120.
59. McClure 1957a:20–5 and Young 1961:22–3 describe *P. bambusoides* and variants.
60. McClure 1957a:25.
61. Ibid.: 25–7.
62. Ibid.: 27–9. Glen 1954.
63. McClure 1957a:29–30.
64. Young 1961:23–4.
65. McClure 1957a:32–4. Glen 1954.
66. Lawson 122–3. McClure 1957a:34–6. Young 1961:24.
67. McClure 1957a:36–8.
68. Ibid.: 38–9.
69. Ibid.: 39–42. Young 1961:24.
70. McClure 1957a:42–5.
71. Freeman-Mitford 1896.
72. McClure 1957a:45–6.
73. Young 1961:25.
74. McClure 1957a:46–7.
75. Freeman-Mitford 1896:151.
76. McClure 1957a:47–8.
77. Young 1961:25–6.
78. Ibid.: 26. McClure 1957a:48–9.
79. McClure 1957a:49–50. Glen 1954.
80. Lessard 1980:48,57.
81. Young 1961:27.
82. Satow 35.
83. Young 1961:27. McClure 1957a:52.
84. McClure 1957a:51–3.
85. Freeman-Mitford 1899.
86. Glen 1954.
87. Glen 1954. McClure 1957a:53–6.
88. Sturkie 1968.
89. McClure 1957a:56–60.
90. Ibid.: 61–2.
91. Lawson 139.
92. Young 1961:28.
93. McClure 1957a:62–5. Young 1961:28–9.
94. McClure 1957a:65–7.
95. Fuji 1963:15.
96. Young 1961:29.
97. Lawson 92.
98. Young 1945a:190; 1961:29–30.
99. Satow 43–5.
100. Lawson 79–80. Young 1961:30.
101. Lessard 56.
102. Young 1961:30–1.
103. Lawson 141–3. Young 1945a:185–7; 1961:30–1.
104. Lawson 103–4. Young 1945a:178.
105. Lawson 105.
106. Lawson 104–6. Young 1945a:180.
107. Lawson 144.
108. Ibid.: 143–4. Young 1946d:36.
109. Young 1946d:36.
110. Soderstrom 1979a:161.
111. Young 1961:20; 1945a:180.
112. Lawson 145.
113. Young 1945a:182; 1961:31.
114. Lawson 147.
115. Young 1961:33.
116. Lawson 151–2. Young 1961:32–3.
117. Huberman 1959:40.
118. Munro 1868:4.
119. McClure 1953:33.
120. Young 1961: 37. See also McClure 1954c:20.
121. McClure 1953:33. Lin 1972:5.
122. Young 1961:37–8; Lin 1972:4.
123. McClure 1953:33. Lin 1972:4.
124. Lin 1972:4–5.
125. McClure 1958b:No. 3, p. 40.
126. Lin 1972:5.
127. Ibid.
128. McClure 1948c: exhibit 12.
129. McClure 1944:48; 1954c:21–3. Young 1961:38–42.
130. McClure 1953:33.
131. McClure 1944:45.
132. McClure 1958b:No. 3, p. 41.
133. McClure 1953:34. Lin 1972:6.
134. Young 1961:42. Lin 1972:6.
135. Lin 1972:7.
136. Ibid.:6
137. Young 1961:43.
138. Dickason 1966:7.
139. Gamble 1896:37.
140. McClure 1944:49. Young 1961:43.
141. Lin 1972:7.

142. Young 1961:44.
143. White 1945:846.
144. Lee 1944:129. See also, Young 1961:44–5. McClure 1944:51; 1958b:No. 3, p. 42.
145. McClure 1944:55. See also McClure 1948c: exhibit 10; 1958b:No. 4, p. 53. White 1946a. Young 1961:45.
146. McClure 1948c: exhibit 14; 1958b:No. 4, p. 53. Young 1961:46.
147. Plank 1950. See Chapter 8, p. 219.
148. McClure, unpublished *Guadua* text, from Grass Library, Smithsonian Institution.
149. White 1945:843–4.
150. McClure 1954c:23. The Jinotepe bamboo cannery mentioned by McClure apparently no longer exists (1983).
151. McClure 1953:34.
152. Lin 1972:8. Simmonds 1963:334.
153. Lin 1972:8.
154. Watt, George: *Commercial Products of India,* London, 1908. McClure 1944:62.
155. Lin 1972:9.
156. Lawson 148–51.
157. Lin 1972:10.
158. Ibid.
159. Ibid.
160. Ibid.
161. Young 1961:47.
162. McClure 1953:35. Lin 1972:11.
163. Ibid.
164. McClure 1953:35.
165. Ibid. Lin 1972:11.
166. Lin 1972:12.
167. Watt, George: *Commercial Products of India,* London, 1908. McClure 1944:62.
168. McClure 1953:35.
169. Ibid. Lin 1972:12.
170. Ibid.
171. McClure 1953:35–6.
172. McClure 1966:169.
173. Deogun 1937:79.
174. Huberman 1959:40.
175. McClure 1966:171.
176. Ibid. Young 1961:48.
177. Lin 1972:13.
178. Dutch patent No. 53,471. 1942.
179. McClure 1955:140–1; 1958b:4,53–4.
180. McClure 1953:36. Lin 1972:13.
181. Lin 1972:13–4.
182. McClure 1953:36. Lin 1972:14.
183. Lin 1972:14.
184. Ibid.
185. McClure 1966:172–9.
186. Ibid.:176.
187. McClure 1955:142–3; 1958b:4,54.
188. McClure 1948c: exhibit 1.
189. McClure 1953:36.
190. Ibid.; 1966:179–87. Hidalgo 1974, 1978. Castro 1966. Lodoño 1970.
191. Lin 1972:15.
192. McClure 1948c:exhibit 2. Lin 1972:15.
193. Lin 1972:15.
194. Ibid.
195. McClure 1944:75.
196. Lin 1972:16.
197. Ibid.
198. Ibid.
199. Ibid.
200. Ibid.
201. McClure 1966:187–97. Chap. 4, "Selected Species," pp. 147–201, is the easiest part of McClure's masterwork for nonbotanists to enter.
202. Simmonds 1963:336. See also McClure 1958b:4,55.
203. McClure 1953:37.
204. Lin 1972:17.
205. Ibid.
206. McClure 1966:197.
207. Ibid. Deogun 1937:115.
208. McClure 1966:197–201; 1958b:4,55. Gamble 1896:121–8.
209. Lin 1972:17. McClure 1953:37.
210. Lin 1972:17.
211. Ibid.:17–8. McClure 1953:37.
212. Lin 1972:22. McClure 1953:38.
213. Ibid.
214. Ibid.
215. Ibid.
216. Lawson 1968:100–2.
217. Ibid.:100–3. Young 1961:49.
218. Lin 1972:24. McClure 1953:38.
219. Lawson 1968:81–3.
220. Soderstrom 1979b:22–7.
221. Lawson 1968:96–8.
222. Soderstrom 1979b:27.
223. Lawson 1968:110.
224. Ibid.:110–1. Soderstrom 1979b:27.
225. Lin 1972:24. McClure 1953:38.
226. Dickason 1966:4,7.
227. McClure 1948c: exhibit 17.
228. Ibid.: exhibit 8.
229. McClure 1952b.
230. McClure 1945a.
231. McClure 1952b.
232. Hidalgo 1974:6–21. McClure 1955, 1973.

8. CULTIVATION, HARVEST, CURING

"Every farm in the South should be supplied with a small forest of these valuable plants, in the same manner as it is now supplied with a wood lot. . . . I have growing in my experimental gardens in Louisiana sixty-four varieties entirely hardy, unhurt by frequent temperatures of 15 degrees F."

—E. A. McIlhenny

DIXIE BAMBOO

"It is impossible for the people of the United States, who have lived among the lush forests that covered North America, and which still cover much of our land, to look forward to the time, a few hundred years hence, when our land will be as bare of forests as China is today, and for that reason, and that reason alone, we cannot visualize the value of the rapid-growing and maturing bamboo capable of producing many times the tonnage of usable wood, many times as rapidly as anything else we know of that will grow from the ground, and with but little care after planting. When bamboo becomes established here, it will be as indispensable to our existence as it is now to the Chinese.

"The U.S. imports millions of dollars worth of bamboo yearly that should be grown at home. Now is the time for our state agricultural departments to see that plantings of bamboo are made wherever they will thrive so that the people of the southern states may be ready for the time which surely will come when our forests are no more, and we will be obliged to rely on the quick-growing timber bamboo to supply the wood for all necessities, from the making of paper to the making of houses, and the furniture for the houses."

McIlhenny was writing this in the mid 1940s. China is no longer bare of woods. Tree cover has doubled according to FAO estimates, and Saint Barbe Baker says it has increased from 7 to 27 percent of China's land area. Deforestation is *not* inevitable, as McIlhenny apparently presumed forty years ago. This is a momentous shift. For the first time in the cultural existence of our race, a significant part of it is proving we *can* regreen the globe. Tree planting has become a kind of permanent national picnic in China. Sometimes such basic sanity is contagious. It could happen in many places now that we know it can happen. After the first runner broke the four-minute mile, suddenly many more found their legs could do it, too.

Fortunately, affection for forests is not something we need to "awaken" in people. It is inherent to our eye, nose, and lung. All we have to do is stop

Bamboo energy is boldly expressed in the phallic power of a newly shooting culm. (*Gigantochloa *is the genus shown.)

Dixie, or Dixieland:
the southern states; probably from *dix,* dixie, a $10 note, widely current in Louisiana before the Civil War, with a large French *dix* (ten) in the center of the reverse. (Webster's New Collegiate Dictionary.)

stifling it, foster a love of leaves early in our homes, gardens, schools, and rituals. Then, perhaps, bamboo will be the cheerful friend of our forests, instead of the replacement that McIlhenny imagined. Broad cultivation and use of bamboo's versatile fiber can help us lean more lightly on the woods and thus encourage trees to come back to live with us again.

BASIC GARDEN GUIDELINES

Propagation.

Hardy *monopodial* bamboo is generally propagated from young rhizomes with or without an attached culm. The rhizome should be yellowish, 15–40 inches long, with at least ten good buds. Cut with a saw to avoid shock, and plant 1 foot deep. The attached culm should be young, of the same or preceding year. It is trimmed to about 5 feet, the lower branches left—or culm can be cut off to within 1 foot of ground level. A rhizome alone, some 20 inches long with ten to fifteen nodes and plenty of roots, can also be used. The soil should be washed off and the rhizome wrapped in damp moss or burlap and plastic if it is to be sent any distance. By this method, up to one hundred suitable rhizomes may weigh as little as 3 pounds so it is shipment efficient. Rhizomes without culms are preferably planted 8 inches deep in a nursery and transplanted the following spring when new shoots appear.

For *sympodial* species, a healthy culm is selected with its rhizome and similarly planted at the beginning of the rainy season in early spring. Divisions of clumping bamboos are obtained by cutting culms just above the second or third node. As many as three culm stumps may be included in a single division, which may weigh up to 30 pounds, or more, depending on the species. In an eight-hour day, a laborer may perhaps prepare only three such divisions, whose weight and bulk greatly increase transportation costs. The growth of the parent clump is also retarded by the division, and the amount of propagating material per clump is quite limited.

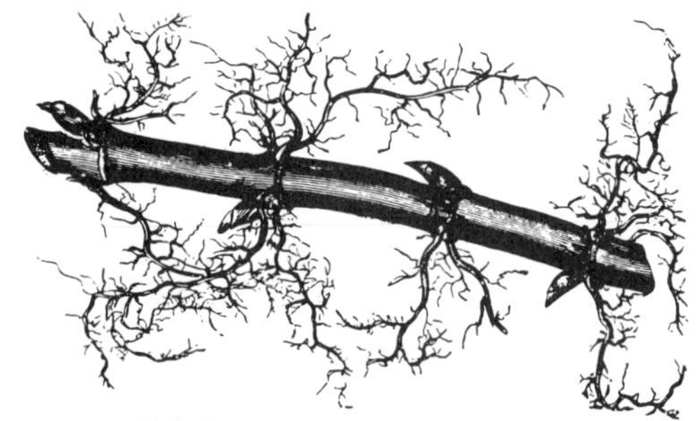

A monopodial rhizome **(Phyllostachys viridi-glaucescens)** *prepared to plant.*

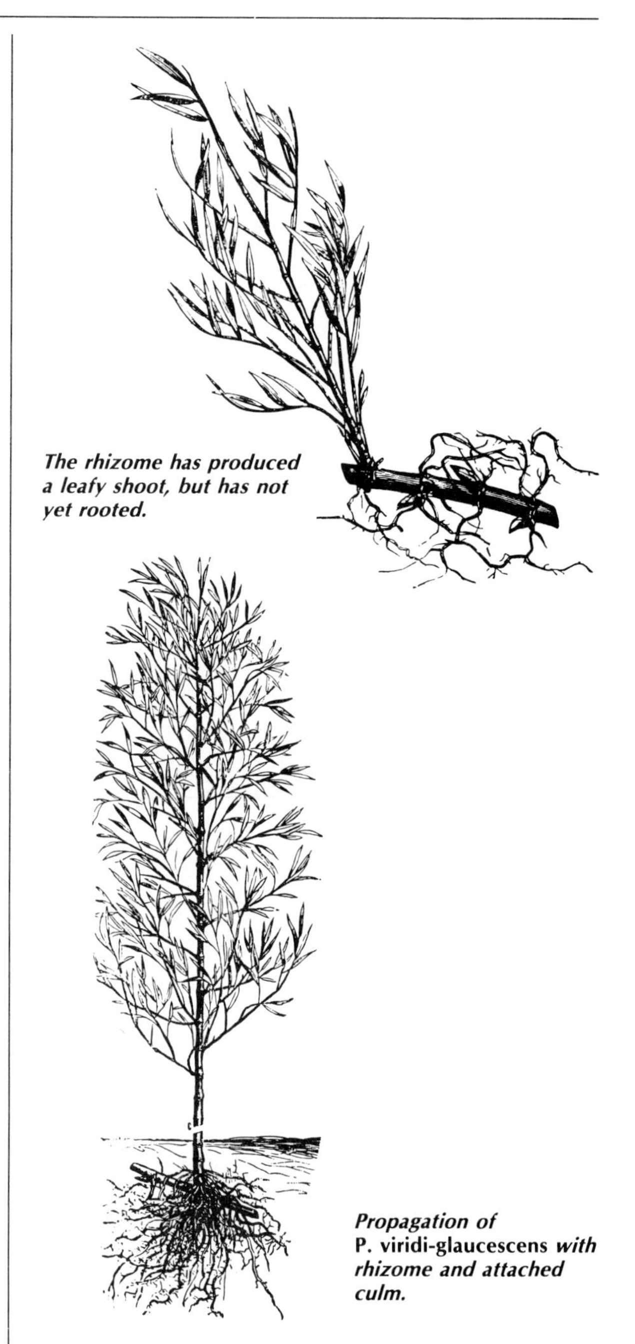

The rhizome has produced a leafy shoot, but has not yet rooted.

Propagation of **P. viridi-glaucescens** *with rhizome and attached culm.*

Whole-culm plantings are also practiced with tropical bamboos as is propagation by culm segments: a length of two-to-three-year-old culm, 20–40 inches long without branches, horizontally planted 1 foot deep will also root easily in many sympodial species. This method, however, is rarely productive with monopodial bamboo.

Transplanting by whatever method should be done as quickly as possible to prevent the culm and rhizome from drying out, and newly planted bamboo should be watered at least weekly.

> Unlike many woody plants, especially deciduous species in a dormant state, uprooted bamboo plants or cuttings dry up easily if exposed for any length of time to sun or wind. Disregard of this fact is probably responsible for more failures or partial failures in bamboo plantings than all other causes combined.[1]

SOIL AND SITE. Bamboo likes moderate water in a soil that is fertile, well drained, and mixed with gravel. It grows well on slopes where rhizomes and dense leaf fall both help retard erosion. Strong sun should be avoided, except for black bamboo which darkens well in direct sunlight. Sites facing west into a harsh afternoon heat are not ideal. In warm or mild climates, locations facing north are preferred; in colder regions, the south.

CUTTING. For the first three years of a new grove, all shoots should be allowed to grow. Thereafter, young bamboo should not be cut, but five-to-six-year-old culms should be removed each autumn to reduce insect damage and make room for new sprouts. Cutting thin culms at two years and leaving larger culms uncut favors the rhizomes producing sturdier culms and tends to increase the average culm diameter in the grove. Thinning is important. Greater density can mean fewer new culms each year and less harvest weight. One grove, reduced from 1,200 to 900 and then 750 culms, steadily increased the number of new culms produced annually from 165 to 179 to 218. The harvest weight rose from 1,650 to 1,920, to 2,180 pounds.

***Gigantochloa** species two years after propagation. Culm segments propagate most readily in species with branches swollen at the base. These repeat the anatomy of the bulbous sympodial rhizome at the base of the culm, and assume its functions when planted.*

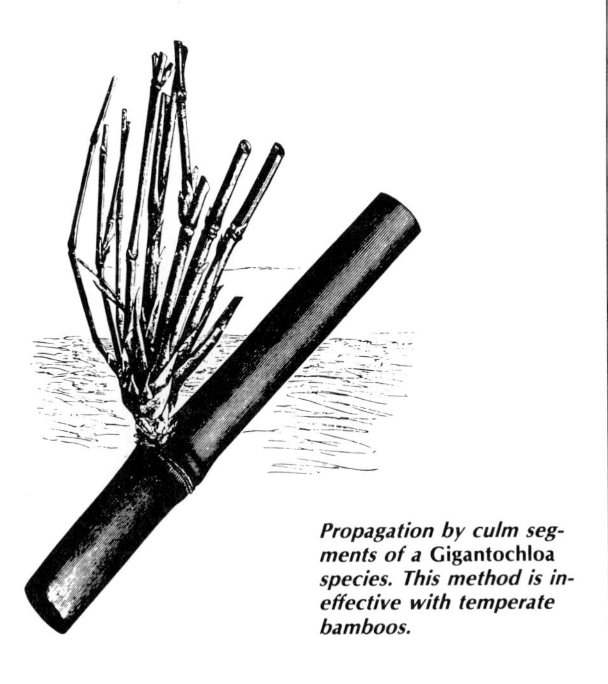

*Propagation by culm segments of a **Gigantochloa** species. This method is ineffective with temperate bamboos.*

OVERTHINNING. Overthinning can result in excessive sunlight, which yellows the culms and dries the ground. Enough culms should be left to provide good shade for the grove itself even in broad daylight. Harvest three-year culms, which implies keeping track of their age: a thin wire can be loosely placed each fall around the base of new culms. A code of twists in the wire serves to indicate the year of growth.

"Too much beauty can get out of hand."

Listen, bamboo, chief mischief of my yard, thou shalt not chew the concrete of my neighbor's drive. Feel these sheets of corrugated metal, two dozen inches deep? There ends your bed, my lovely. So by all means stay green, tall, and beautiful, but *stay put.*

The average American homeowner wants to construct a brief Eden in his yard for green relief from commuting to the job. He doesn't want to come home from battling his boss to battle his bamboo. Ruth McClure, brisk widow of F. A. McClure, is insistent in her warnings: "I'm afraid I'm not much use to you. I just keep warning people that bamboo's sneaky. My husband knew a man who *died* trying to get it out of his yard. Sure, it's beautiful, but too much beauty can get out of hand. So my husband always warned anyone he gave bamboo to: *it has to be contained* if you don't want it everywhere."

McClure issued his warning in print as well: "*Sinobambusa tootsik,* a Chinese bamboo once highly prized as a garden ornamental in Honolulu, has come to be regarded as a dangerous weed there because it escaped from cultivation and now dominates many acres of once pure native vegetation."[2]

FERTILIZING. Compost or, less recommended, commercial fertilizer, increases grove vitality enormously. The number of culms and harvest weight can be doubled, with culms both taller and greater in diameter. Ten parts nitrogen, six parts silicate, five parts each of potassium and phosphate are applied about a month before spring shoots appear and again before the onset of rhizome growth, which immediately follows completion of culms. For 1 acre, apply 80 pounds of nitrogen to 48 of silicate, 40 of potassium and 40 of phosphate. Bamboo leaves, roughly equivalent in weight to the annual harvest in some species, are left in the grove. A dense leaf fall of up to 4 inches annually provides fertilizer and excellent mulch. They are 6 percent silicate, and their decomposition increases the availability of this toughening component of bamboo. Theoretically, older groves should produce harder culm wood. (Fertilizing bamboo is not recommended where durability, not quantity, is sought: see below, Soil and site, pp. 215–216.)

CONTROL. If bamboo is to be confined to a limited area, rhizome growth in some monopodial species must be controlled. Galvanized metal sheets or 1–2 inch concrete slabs are sunk 20–24 inches deep, with about 2 inches above ground. Concrete made with peat moss replacing some or all of the sand provides a barrier porous enough to allow groundwater to pass. Pebbled paths around the grove will harden the soil, reduce the shoots, and also reduce rhizome development. Sprouts that emerge where unwanted should be eaten or kicked over while still young and brittle. Make sure you *need* to control your specific species before you go to the trouble and expense of confining it. See p. 215 for more on the bamboo invasion. Also, "Eat your lawn," p. 278, for direct action.[3] Basically, bamboo becomes a problem only if you choose the wrong species, and don't *use* it. If you can't use your harvest, call the crafts teacher at your local school and offer a copy of this book and free harvest in exchange for taking your bamboo abundance off your hands.

TRANSPORT

Selecting and shipping bamboo starts.

Go to the grove prepared with digging and pruning tools and enough waterproof material to wrap plants and keep them moist at all times. Find an area in the grove where small, strong, disease-free plants seem most plentiful. Make exploratory digs to get a clear picture of underground organization of rhizomes. At the young edge of a clump, choose three to four adjacent culms. One or more of them should be at least one year old with branches or visible buds at nodes below a height of 18 inches.

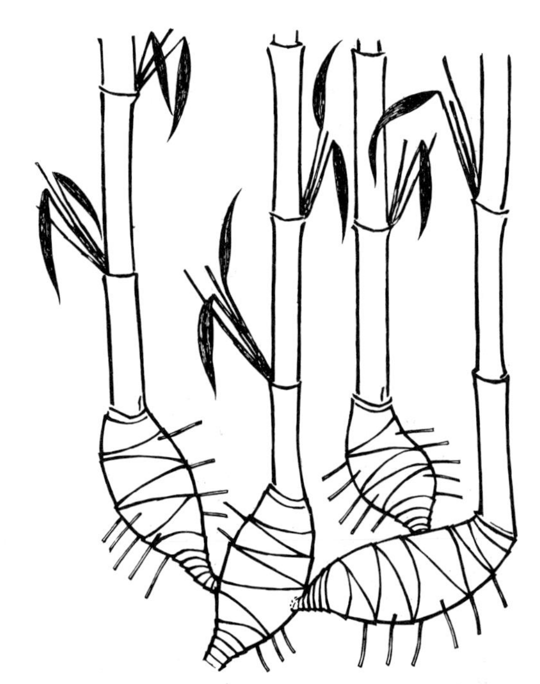

A group of sympodial rhizomes prepared for shipping and planting. Two or more connected rhizomes, two or more nodes of a culm segment, will survive more readily than a single propagule.

Dig out a part of the clump large enough to include a ring of culms beyond the selected portion, and remove the soil to expose where the rhizomes join. Cut the part wanted from the rest in such a way that all the culms of your propagule remain joined by their rhizomes. Prune culms to roughly 18 inches and branches to 4–5 inches. Don't cut branches close to culm. Remove most of remaining foliage, and prune roots to about 3 inches. Cover propagules with moist litter as soon as work on each is complete. To assure that at least one will survive, prepare, if possible, at least four propagules. Wrap them in waterproof material with moist litter (sphagnum if possible) for transport to base where packing for shipment is done.

Packing for shipment.
Prepare a batter of rock-free clay, about 1½ gallons, the consistency of mayonnaise or thick gravy. Wash off soil from propagules and dunk the rooted base of each into batter until roots and rhizomes are completely coated. Pack in moist (not wet) sphagnum, an equal number of bases at each end of the pile, then wrap in waterproof plastic or heavy wax paper. Wrap finally, and firmly, in heavy manila paper and tie well. Ship by air express or other rapid means.[4]

Receiving bare-root bamboo.
"Keep packages in a cool place and unpack in shade, protected from any wind. Don't try to remove all the packing material as this may damage young roots. Set plants in moist sphagnum or similar material in a windless, shady place until ready to plant. It is best to prepare ground before shipment arrives. A nursery is most convenient, efficient, and effective. Plants can be observed frequently and needed attention promptly given, till well established; they will be planted in permanent locations at the onset of the next rainy season.

"Set plants in rows 5 feet apart, 1 yard between plants. This is the minimal distance, good for no more than one year. Rows follow land contour to provide natural drainage. Avoid poor drainage, which bamboos abhor—except for some *Guadua* species. Water should be handy for irrigation or at least sprinkling during droughts until plants have dug in well.

"Dig holes of ample size. Spread roots out thoroughly while adding well-sifted soil. Tamp the soil firmly while adding it *before adding any water.* Then water generously, adding a thick mulch of loose organic matter. Plants set out in dry season require temporary shade—and windbreak for any strong, dry winds. Sketch a map of species position immediately in case plant markers get lost or confused. Keep weeds down, especially climbers."[5]

EXPANDING GROVE

Bamboo bonsai.
Chinese ornamental gardening developed largely as miniature idealizations of natural scenery. Travel was difficult, space limited, plants much loved by the Chinese people—so nearly every home had a tiny garden whose aesthetic traditions were extended, in time, to dish-gardening, *p'oon tsui,* or *bon-sai* in Japanese, the more familiar name for this type of microgarden in the West. Plants used are either naturally miniature or capable of thriving indoors while being subjected to the severe dwarfing techniques evolved in bonsai practice over the years.

Usually, bamboos coming from extreme north or south latitudes are natural dwarves. *Sasas* from northern Japan are the most common known in Western gardens. But all species can be shrunk in stature by reduced water and nutrients, small containers that confine roots, and the practice of removing culm sheaths from growing culms, which stops growth of the internode above it.[6] (See description, Chapter 6, "Seedling," pp. 151–152.)

Sympodial bamboos are used in pot culture by cutting them off to a short stump just before new growth would normally begin from the rhizomes. The new shoots are quite small compared to the ordinary size of the plant.

Monopodial bamboos are potted by digging a rhizome a foot or so long and planting it upright or at a slight angle with 3–4 inches in the soil. Buds underground produce short shoots, while those above send out branches and leaves from the rhi-

Bamboo bonsai. The bamboo path is endless. It can begin at a tiny window with a tiny grove.

zome, which turns green in response to light, thus resembling a tiny culm with short internodes.[7]

Some favorites for potted plants or small gardens include those with swollen internodes, like *Bambusa ventricosa* or *Phyllostachys aurea,* the square-stemmed *Chimonobambusa quadrangularis,* or the green-striped yellow culms of horticultural forms of *Bambusa vulgaris,* and *P. bambusoides.* White powdery bamboo *(Lingnania chungii)* is a popular species long honored in Chinese art and poetry, introduced in the West, but little known or distributed as yet.

Pseudosasa japonica is an indoor favorite for which propagating material is widely available. *B. glaucescens,* also, in its horticultural form of Chinese goddess bamboo, "does well as a potted plant, and will stand more shade than most bamboos. Chinese gardeners use it a great deal in their miniature gardens, where it assumes, under skillful treatment, a stature of a few inches, with all parts in proportion."[8] Extended advice on bamboo bonsai techniques is provided by Austin in the broadly available *Bamboo.* The text is by the eminent Japanese bamboo master Koichiro Ueda and is painstakingly detailed about some of the refinements of the art:

> When the shoots in the nursery have grown to about 4 inches in height, their sheaths should be peeled off. This is done by inserting scissors between the sheath and the culm—starting from the lowest node—and gently cutting the sheaths from the top down to the node into thin strips, which are then removed carefully and without damaging the culm. The sheaths and the strips must not be torn off by hand as this has a serious effect on the culm. After removal of the first sheath, several hours should be allowed before starting on the second, and so on. The treatment should also be followed through the evening, since bamboo continues its growth by night. Once the stripping operation has been completed, the bamboo bonsai should be given a little oil cake, or other compound fertilizer, in spring and in early summer to improve its color. The tender shoots without sheaths should initially be protected from the sun and must be watered if they get dry. Finally, they should be transplanted into the bonsai pot in the autumn after the new shoots have completed their growth. The rhizomes are cut to fit into the pot and a mixture of large and small stems planted to simulate the variety of a bamboo grove.[9]

Potted bamboo. Mini-Eden: the courtyard garden.

Large containers of potted bamboo are gallons of dirt beyond windowsill bonsai efforts at deliberate miniature but still represent considerably less effort and space than a grove. The Chinese and Japanese are both famous for moving their gardens where they go (see Morse 1886, Keswick 1978), and potted bamboo is the obvious option of modern, mobile, urban populations in the West. *Small* is no obstacle for the Chinese or Japanese gardener, long accustomed to population densities that people in other areas of the world are now only approaching. The infinite variety of oriental adaptations to density, their Tao of Crowd, is one main aspect of their culture that makes it seem so apropos in the now denser West.

Inspiration for bamboo use in a cramped place is lavishly presented by *The Japanese Courtyard Garden. Landscapes for Small Spaces* by Shigemori (1981). But offering a model for the population at large—to which we speak—is not an objective of Shigemori's book, whose price—$150—suggests its intended audience.

Chic critique.

Bank Zen, marble downtown Tokyo corporate garden architecture, is the book's unfortunate main focus. The spare effects, the bamboo chic it makes breathlessly available—on the page but not to the people—is actually at considerable variance from the attitudes of its design grandfather. The mini-Zen "grass hut tea," the humble shack of Rikyu and surrounding garden, which is the cultural bone of Shigemori's flawlessly beautiful places, was deliberately *a norm for all.* It demanded a certain attitude, not extravagant resources. Nevertheless, Shigemori is a source for striking and suggestive photographs. Look for it in your local arboretum, along with Austin (1970) *Bamboo,* and *The World of Bamboo* by Shinji Takama (1983), the latest and by far the most inspiring of these books.

All of these are full of beautiful photographs, and all suggest that bamboo is still basically the coffee-table preserve of the few. There is an irony —and a publishing gap—in a major global resource being limited to a handful of initiates, and of books focusing, again and again, more on the cosmetic possibilities of the plant for the landscapes of the overrich than on its useful beauties for everyone. Shigemori's book misses the more globally relevant mark of a book on small gardens for the nonelite. Takama's obvious reverence for bamboo puts his book in a class definitely apart, and classless, transclass in appeal.

For more on the central position of bamboo in oriental gardening, see Schaarschmidt-Richter

(1979) *Japanese Gardens;* Engel (1959) *Japanese Gardens for Today;* Sirén (1949) *Gardens of China;* Keswick (1978) *The Chinese Garden;* Davidson (1983) *The Art of Zen Gardens.* Bamboo was also a standard component of the small teahouse garden and path (see p. 91 and p. 110). The current interest in the teahouse and adjacent garden seems attested by a number of recent books we can only acknowledge in part and in passing: Hammitzsch (1980) *Zen in the Art of the Tea Ceremony;* Sen (1979) *Chado: The Japanese Way of Tea.*

Hedges.

> A hedge between
> keeps friendship green.
> —Mother Goose

Hardy bamboos tested for hedges at Savannah, Georgia, USDA groves include *Pseudosasa japonica, Arundinaria fastuosa* (narahira bamboo), and three species of *Phyllostachys: P. meyeri, P. nigra,* and *P. purpurata. P. aurea* and *P. aureosulcata* were also recommended. All these species branch close to the ground in plants of moderate height, producing a close hedge.

Into a trench 18 inches deep and 36 inches or more wide, excavated and filled with fertile soil high in humus content or well-rotted manure, transplant cuttings of rhizomes formed the previous year. These are distinguished from older rhizomes by the presence of sheaths or bud scales. Rhizomes with at least six to eight good buds are planted vertically in two to three rows on each side of trench with cuttings 6 inches apart. Vertical planting encourages quick culm growth and retards rhizome development. Horizontal planting in rows parallel to trench sides will yield greater rhizome development along trench axis. Plant bamboo with stems cut back to about 1–2 feet, covered with 5–6 inches of loose soil to the same height as the natural soil line on the culm, with attached rhizomes as parallel as possible to trench sides. Mulch with leaves, peat moss, or organic litter to help keep soil moist and loose. Plants that die should be replaced.

Plants or rhizome cuttings can also be established in a nursery for a few years, 3 feet apart in rows 5 feet apart. When these are transplanted to the desired site, move them with a ball of earth 18 inches square, 8–10 inches deep, with all their rhizomes carefully preserved and placed parallel to the trench. When desired hedge height is reached, prune its top and sides with slanting rather than right angle cuts, 1–2 inches above nodes of culms and

Hedge of living culms lashed to split bamboos of a larger grove.

branches. Always prune *after* new shoots have reached full height, extended branches, and opened leaves.

A hedge can be kept trimmer along the sides by stretching a strand of #12-gauge galvanized wire slightly above mid height on each side on posts close enough to maintain tension with cross wires running between them to keep hedge the desired width. Periodic use of organic or commercial fertilizers rich in nitrogen will assure dense and deep green foliage.[10]

Windbreak design: the harvested hedge.

"In 1956 a hedge of *Pseudosasa japonica,* a hundred yards long, was planted at Rosewarne Experimental Horticultural Station in Cornwall (England) and is now an effective barrier against the Cornish gales. Four years after planting, many individual canes stood 8 or 9 feet high. In another trial, four hundred species of hedging plants were carefully compared and assessed for possible agricultural use as shelter hedges, and *Pseudosasa japonica* was one of the eight finally selected as most suitable.

"Wind velocities are reduced by a solid barrier for as far as thirty times the height of the obstruction, but the degree of protection needed for ordinary garden purposes probably does not extend farther than ¼ to ⅓ this distance. It has been found that a hedge presents a more satisfactory barrier than close fencing, stone, or brickwork, and so on. A solid wall, ironically, gives rise to air turbulence on the leeward side, and the swish and eddies of the disturbed air do more damage than would have been done had the wall never been erected. Experimentally it is shown that the most satisfactory barrier is one with a 50–60 percent degree of permeability. A bamboo hedge meets these requirements."[11] Lawson also notes that, unlike brick walls or other inorganic barriers, a bamboo hedge also complements any garden by providing materials for stakes, arbors, trellis work, fences, tool sheds, and so forth from harvested culms.

BAMBOO FARM MANAGEMENT

Silvicultural summary.
Bamboo silviculture, or the scientific management of bamboo plantations, has been most developed in India, which probably has the world's largest reserves of bamboo. About one seventh of India's forest resources, some 25 million acres, are bamboo groves. The Forestry Research Institute at Dehra Dun has since 1878 been the principal world center for bamboo research.

A few rules of the big groves: "A rich, well-drained soil is desirable. The two other most important factors governing bamboo growth are temperature and moisture."[12] To establish a large grove of monopodial bamboo, rhizomes set out are 8–10 feet apart, requiring around 550 or so rhizomes per acre. Until the grove is well established, hand weeding will be necessary since even shallow cultivation would destroy the rhizome growth immediately below. Fertilize with up to 7,000 pounds of manure per acre.

Immature culms should be cut only if attacked by insects. Old or injured culms should be removed first. Remember that cutting around the edge of a sympodial grove decreases its growth. If practiced regularly, the rhizomes surrounding a grove die, forming a barrier and driving the remaining rhizomes back into the already overdense center of the grove.

Cuts, 6–12 inches above ground, are made

Bamboos in snow, by Kuo Pu (1280–1335). Culms can be flattened or snapped by severe snows, and commercial groves are often pruned at the top to prevent snow damage.

level with the node so no rainwater collects to rot first the culm, then the rhizome below. Culms cut higher not only waste the broadest section of the culm but make future work more difficult through increased congestion: branches often grow from nodes of these tall stumps, blocking movement more. Culms in flower should be cut after the fruit falls, not before. Culms which are badly formed, dead, sick, or useless in any way should be removed at each harvest. Cleanliness in a grove both reduces insect infection and clears way for new growth.

Environmental impact: bamboo invasion.
To play the devil's advocate, we collected published warnings on bamboo invasiveness from other lands. They weren't many. In northern Japan, *Sasa* species, one sixth of the forest biomass in Hokkaido, prevent erosion but disturb tree regeneration and hinder summer work of foresters. Use for flakeboard is feasible but harvest presents economic problems.

In Malaysia: "Forest departments regard bamboo as weeds interfering with timber growth and regeneration. Control options are available: starve clumps over a period of years by kicking over young culms in growing season or eliminate with various chemical sprays. Medway (1970), a zoologist, considered bamboos a nuisance for reducing species diversity of forests by suppressing other plants, reducing fauna to narrow selection of bamboo-loving species.

"In cleared or logged forests, such as in Sabah (northeast Malaysia), *Dinochloa* species [a climbing bamboo] can become a very serious weed problem, preventing regeneration of commercial timber." (Quotes from Lessard 1980: 56, 95, 129.)

Selective introduction: control by use.
Of the more than twelve hundred bamboo species, the more vigorously spreading varieties turn weed or wonder depending on *use*. Many useful bamboos are extinct or much reduced in former habitats: Human interest is an overly effective control. We recommend, obviously, an integrated approach to cultivation: Choose species whose use to people takes care of the question of harvest and control. If you choose a useful, unrampant bamboo and cultivate it well to grow culms of maximal size and encourage maximal vigor of rhizomes, you can, with little effort, distribute plants and cured culms locally. (See pp. 210, 248 on school use of bamboo in relation to these questions.)

Soil and site.
"A favorite site for a bamboo grove is the base of some range of hills or a broad valley where some mountain stream has brought down and deposited a mass of alluvium. These situations have the double advantage of suitable soil and shelter from strong winds. Wind shelter, all growers agree, is important: The young shoot thrashed about by winds with branches of older culms injures its growing tip, stopping growth and producing an imperfect culm. Windbreaks of conifers sometimes protect groves in an exposed position.

"*Soil quality* materially influences texture of bamboo. In Japan, a particular mountainside reputedly produces the hardest, flintiest bamboo in the country. The culms grown at Togeppo are cut up into the cylindrical ash boxes, or *haifuki,* upon the edge of which the smokers strike their metal-trimmed pipes in order to knock out the ashes. After years of use the edge of Togeppo ash boxes remain smooth while that made from a stem grown in the lowlands is splintered to pieces."[13]

"Relatively poor soils and dry sites are commonly used in preference to fertile ones among bamboo growers in China where strength and hardness of wood is more important than large size of individual culms or total annual yield per unit area. Culms of *Bambusa textilis* grown under this regime

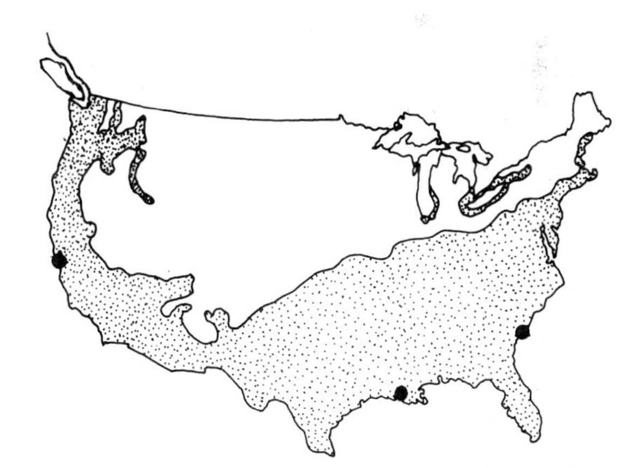

Bamboo cultivation extends north of Vancouver, B. C., on the Pacific Coast. Stars indicate suggested regional experimental centers for distribution of plants and training of workers. Savannah, Georgia, and Avery Island, Louisiana, are natural sites for eastern and central regions. The San Francisco Bay Area provides a climate friendly to both tropical and temperate species, and a large oriental population with deep affection for the plant. Early plans for extensive bamboo groves in Golden Gate Park are beginning to be realized. The arboretum and research library there provide an ideal base for an urban bamboo information center on the West Coast.

give years of service as bean and cucumber stakes in the Canton delta."[14]

"Canal embankments, pond borders, and riverbanks are suitable locations, especially in dry regions. Large clumps grow along canals in Egypt. Algiers has many varieties growing in her trial gardens watered only by irrigations. There are in California, Oregon, Texas, and throughout the Gulf and Southern states thousands of suitable locations. The banks of small streams, the deltas of rivers, low, irrigated islands, like those in the San Joaquin and Sacramento rivers, would produce big forests of these valuable plants, while the banks of irrigation canals, wherever such occur in mild climates, could be made beautiful by them."[15] Especially in areas with long dry spells, locate plantings along ponds, streams, and near springs, just before rains start. (If propagating material is available, try drought-resistant species such as *Bambusa ventricosa* or *Dendrocalamus strictus.*)

Nursery.
Anyone planting a large area with bamboo will consider whether to plant at once in the field or in a nursery first. A nursery recommends itself for several reasons. You can establish the same optimum conditions of soil preparation and care, watering, watching, weeding—whatever seems required. Cost is diminished, efficiency increased. Dead plants are more quickly noted, more easily replaced. More time is available to choose and prepare a permanent site.

Planting distances depend on time plants will remain in the nursery and also vary with the species. Leave 1½ to 5 feet between rows, 1 to 3 feet between plants, depending on species and size of propagule. As little as ½ to 1 inch of soil can cover rhizomes if mulch covers the soil: well-weathered sawdust, wood chips, cocopeat (meal from coconut husks) are some alternatives; others depend on your surroundings.[16]

Slope or level?
In general, sympodial bamboos thrive better on land not too steep; monopodial bamboos seem to develop best on steeper slopes. Sympodial rhizomes tend to arise from progressively higher levels as the clump develops and soil is gradually carried away by erosion. The rhizomes of monopodials, on the other hand, seem able to burrow as deeply as circumstances require. If land with a steep gradient must be planted to sympodial bamboos, it should be terraced, if possible.[17] Even on level land, sympodial bamboos thrive best when some fresh earth is thrown over the rhizomes each year. In the culture of this type of bamboo for shoots in southeastern China *(S. beecheyanus* and *S. latiflorus),* the earth is pulled away from the base of each clump every year in December or January and the deadwood of old rhizomes removed. Earth is then heaped afresh and systematic application of fertilizer, usually diluted urine, is begun. In addition to protecting the rhizomes and roots from undue exposure and drying, these heaps of earth protect young shoots from light until large enough to be harvested.

Monopodial bamboos are grown on both level lands and hillsides. Aside from fertility, which is usually higher in level land, hill land seems preferred in part, perhaps, because bamboos loath poor drainage. Maybe, also, the slope stimulates rhizome use of greater vertical range of soil, evident in hillside cultures. This postpones the competition between rhizomes common on level land.[18]

Field planting distances.
Sympodial (tropical clumping) bamboos are planted at distances that vary with species, climate, and soil. The rough rule of thumb in the accompanying table for equidistant triangular plantings of five common species was developed at the USDA Federal Experiment Station in Mayaguez, Puerto Rico, a major site of Western experiments with tropical bamboos.[19]

Monopodial bamboos: Large species of temperate, running bamboos are most happy in an area at least as long and broad as they are tall. This counsel isn't followed by many urban gardens be-

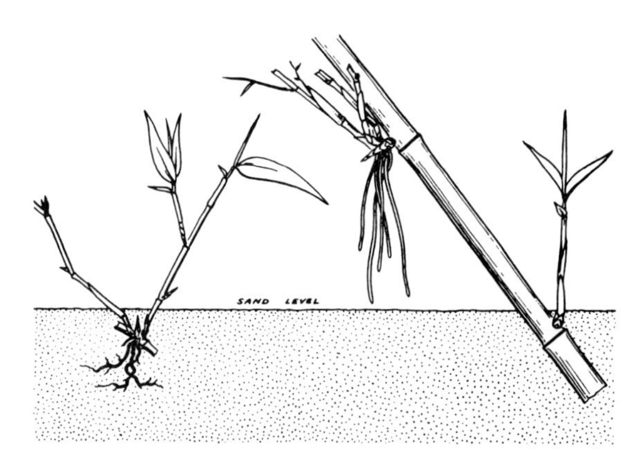

After 36 days in a humid propagating frame, culm segments with branches of **Bambusa vulgaris** *developed roots under the sand* **(left)** *and in the air as well* **(right).** *This species is the world's most widespread bamboo.*

Cultivation Table

SPECIES	POOR SOIL AND CLIMATE		GOOD SOIL AND CLIMATE	
	FEET APART	PLANTS PER ACRE	FEET APART	PLANTS PER ACRE
Bambusa longispiculata	15	223	25	80
B. textilis	12	348	20	125
B. tulda	18	158	25	80
B. tuldoides	15	223	25	80
Sinocalamus oldhami	18	158	25	80

cause they haven't enough room, but for commercial production of bamboo it is an applicable rule. Plant 8–10 feet apart, 550 rhizomes per acre, roughly—depending, as with tropical bamboos, on local climate and soil as well as rhizome vigor of the species in question.

Propagation notes.

For reasons of economy, culms are usually taken up singly, but a unit composed of a new culm and its "parent" gives much surer and quicker results. The earth is dug away from the rhizome, and the culm or the unit is severed from its parent by a transverse cut through the narrow "neck" at the base of the rhizome. Too great stress cannot be placed upon the importance of care in digging and in making the cut. One careless stroke of the hoe or knife can ruin a bud, and this damage to the tissues can cause the loss of the plant or serious impairment of its growth through the introduction of fungi.

Bamboo species that normally lack branches, and even branch buds at the lower nodes, present a special problem. Make sure that the culm is severed at a point high enough to ensure the presence of at least one, preferably several, strong branch buds or better still, a complement of fully developed branches. Foliage, when present, should be

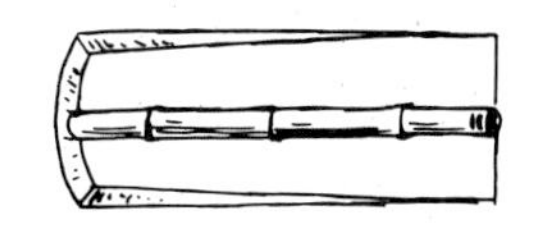

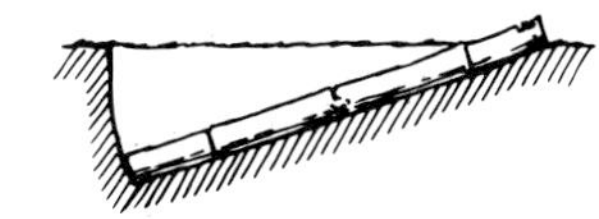

Age aids vitality:* *Meter-long basal cuts of* B. vulgaris, *var.* vittata *were used in an experiment to determine the effect of culm tissue age on the yield of rooted plants. Six age groups were tested, with a code indicating each group on the visible tip of culm segment above ground, as shown. Vitality increased steadily with age to five years, the oldest cuttings tried.

AGE GROUP (MO.)	NO. OF CUTTINGS	PRODUCED ROOTED SHOOTS (PERCENT)	PRODUCED UNROOTED SHOOTS (PERCENT)	DIED (PERCENT)
<2	110	20	22	58
ca. 6	256	19	18	63
12–18	283	26	19	55
24–30	228	30	19	51
36–40	98	51	29	20
48–60	44	50	34	16

reduced drastically, but a few leaves or parts of leaves should be kept on unless plants are likely to dry out. The aim of Chinese experts seems to be to keep the water in the plant moving upward, if possible, by keeping some foliage functioning continuously. It is claimed that chances of success are greatly enhanced if this condition is maintained.[20]

Whole-Culm Planting

SPECIES	NUMBER OF PLANTS PER 10 FEET OF CULM	AGE OF PROPAGATING MATERIAL IN YEARS
Gigantochloa apus	9.5	3
Guadua angustifolia	9.1	2
Sinocalamus oldhami	8.2	3
Bambusa ventricosa	7.4	2
Bambusa tulda	6.8	3
Cephalostachyum pergracile	4.1	2

Whole-culm planting.

Although unproductive with monopodial species, experiments have shown that whole-culm planting works well in tropical bamboos. Species, position on culm, and age of material are three important variables.

Ten species were tested, with material one, two, and three years old. Material two and three years old was superior consistently to one-year-old

A single node of sympodial culm (a) is planted with bud or branch on uppermost side. Larger sections of two or more nodes (b) are more effective, but use more propagating material. Buds or branches are placed to the side when two or more nodes are planted; the uppermost portion of internodes is opened with a V cut or drilled and the internodes filled with water (c).

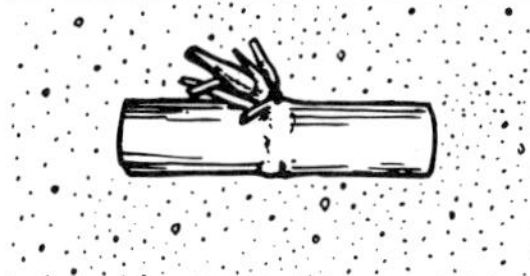

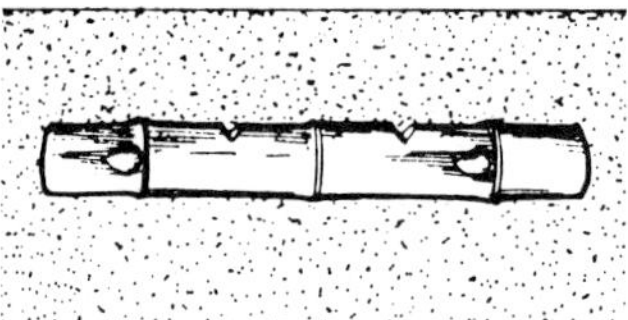

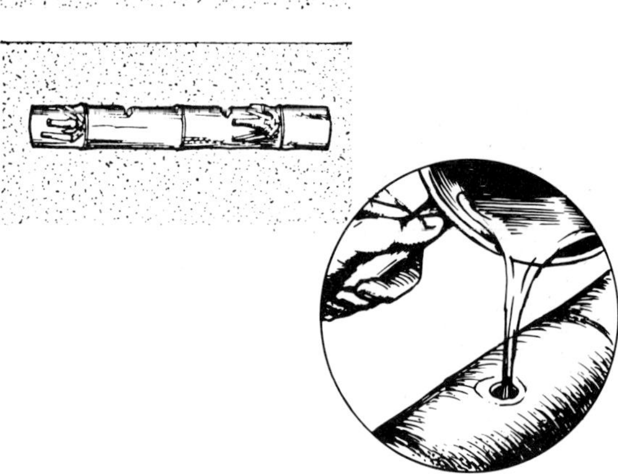

material. In six species, two-year-old material was better than three. In four species, three-year-old material was superior to two. The tip of the culm was most productive in thirteen cases; midculm, in nine cases; the base, in seven.[21]

Marcotting.

In the Philippines, *marcotting* was 68.9 percent successful with *Bambusa blumeana.* Bend a one-year-old culm so that all nodes are within easy reach—easier to do if an undercut is made at culm base. Branches are pruned to about 1 inch, carefully, so that no dormant buds are injured. A mix of garden soil and leaf mould around each node is then wrapped with coconut fiber, old burlap, or other suitable material. This method is not recommended for hardy bamboos.[22]

Seed storage, planting, and international exchange.

Bamboo seeds are notorious for rapidly losing vitality. Seeds of *B. arundinacea* were used in Puerto Rico USDA experiments to determine best storage methods to enhance seed longevity. Storage over calcium chloride at room temperature gave best results. Storage over hydrated lime or charcoal was also good if refrigerated. Drying to a moisture content of 12 percent increased longevity of seed stored under refrigeration over hydrated lime. Drying otherwise had little or no effect. Exposed seed lasted longer than seeds sealed airtight.[23] Seed sewn ½ inch deep, 1 inch apart in rows 3–4 inches apart germinated in about a week. Seedlings grew rapidly, were transplanted to gallon containers at 6–8 inches, and to the fields at 2½–3 feet. "Growing the plants from seeds is undoubtedly the most economical and convenient method of propagating large numbers of plants."[24]

An international network of experimental stations should be established to share seeds as they become available from flowering species. Only a bamboo fool can know the intense frustration of walking a flowering grove of a commercially valuable species, ankle deep in places with fallen seeds, and no mechanism yet established to distribute them. Creating such an international exchange service, making seeds available to nurseries and parks, would be a significant act towards making the flowering of groves a positive rather than negative event.[25] International gatherings of bamboo researchers routinely agree that international seed collection, storage, and distribution are among the

Square bamboo is formed by placing long forms over growing shoots. The area in cross section enclosed by forms should roughly equal the area in cross section of the emerging shoot's base when grown.

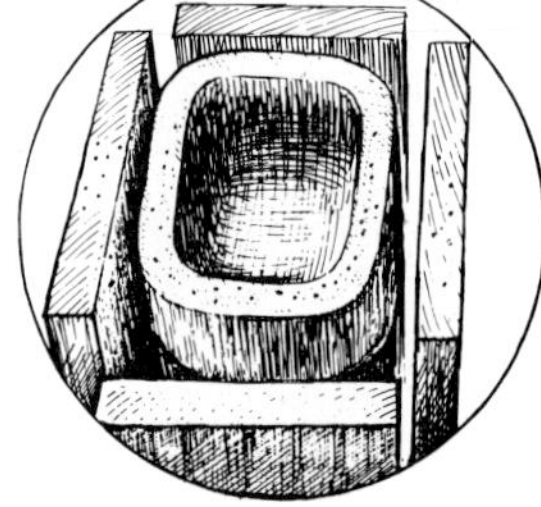

The forms are tied together to allow expansion to the dimension of the growing shoot. The finished culm is used in the Orient for interior decorating.

highest priorities facing the world community of friends of bamboo.

HARVEST METHODS

Dating culms.

Bamboo should be three years old before harvesting. This requires a convenient system of marking the shooting year. One way is to rub the thin waxy film from a small area of a new culm with a piece of coarse cloth and paint on the year with India ink. Oil paint on a ½-inch brush also works. These methods last three to four years under ordinary conditions. Years can be color coded as well. A quick method used successfully at Savannah USDA groves is to stamp new culms with a small hammer and steel die which gives a permanent mark that needn't be so deep as to injure the culm. Mark all culms at same level for ease later in finding the year.

Clear versus selective cutting.

"Clear cutting at determined intervals and selective cutting are two general methods to manage groves. At first thought, cutting selectively, at a rate determined by ecological conditions, removing only mature culms three years old, would be the most natural procedure for maintaining a grove in a condition of sustained yield. The functioning of the grove as an organism is minimally disturbed, and removal of mature culms that would die eventually in any case would provide a natural stimulus to regeneration, similar to that produced by the moderate pruning of any plant.

"But economics might limit this biologically sound preference for selective cutting, which is several times more costly than clear cutting. Comparative, long-term yields, per unit of area and unit of time under the two harvesting systems must be determined by actual trial.

"A combination of the two methods may be best: immature culms, a small percent of the total stand, would be allowed to stand, providing a sustained source of nutrition to the network of rhizomes from which new growth would be produced."[26]

Harvest rules from Dehra Dun.

1. Cut no culms younger than three years or in the rainy season or from a flowering grove.
2. No cuts lower than second node or higher than 30 cm above ground.
3. Remove branches, culm tips, and all harvest trash: Debris obstructs growth, encourages disease, and makes later harvests more difficult.
4. Leave leaves for mulch. Their 6 percent silica helps harden later culms.
5. A minimum of six mature culms are left uncut in each clump of tropical species, to sustain grove vitality and insure steady yield.

Horseshoe harvest.

Clumping, sympodial bamboo's growth form presents a harvesting problem. Centrifugal growth of new culms around the grove edge leaves oldest stems most fit for harvest surrounded by immature culms whose silica content, and consequent hardness, is still increasing. The solution is a horseshoe harvest. Cut into the grove from the direction that sacrifices the fewest younger culms. Store these

A horseshoe harvest is imposed by dense tropical clumps in order to cut a minimum of young culms on the periphery and reach the seasoned canes at the core. Dark circles indicate harvested culms.

separately for basketry or other use. Harvest third-year mature culms from within the grove, gradually extending the harvest year by year to ripening younger culms. Some feel it's better to cut smaller species with a saw to avoid leaving dangerous points to wound later harvesters.[27]

Battle beetles better with clump-cured culms.

Harvest bamboo at the beginning of the dry season. Leave culms standing four to eight weeks in the groves, propped on stakes or rocks, with branches and leaves uncut to increase evaporation surface and diminish insect entry points offered by freshly cut skin. This clump cure not only reduces starch content, which the bamboo beetles seek, it also greatly decreases the tendency to crack while producing a pleasing uniform color on the culms. *Bambusa vulgaris* cured this way was 91.6 percent less attacked by beetles than untreated culms in USDA Puerto Rico experiments at Mayaguez.[28]

Removing branches.

Although bamboos harvested for construction obviously require less care than culms cut to make flutes

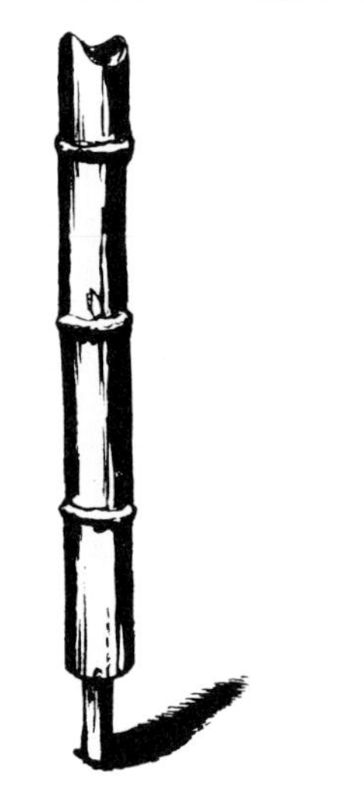

A stake in the basal internode keeps the culm off the ground during a clump cure.

or furniture, scratching a culm diminishes beauty and value. One way to avoid scratching the culm when removing branches is to cut a third of the branch thickness from underneath with a hacksaw. (A molybdenum blade is best with eighteen to twenty-four teeth per inch.) Then snap down on branch. If cutting with a machete, always cut up-culm, that is, towards tip. Swinging down-culm tends to scalp the internode below the branch removed.

Making the crooked straight.
Practicing any of the following cures for the bamboo bends will quickly communicate the commercial importance of ramrod erect species like *A. amabilis* or *B. textilis,* which don't require the touch of human correction to find the shortest distance from toe to tip.

Low tech, long term: hang a curved pole, when freshly cut, by the tip in the shade, tying a stone or other weight to the base. Two to four months will straighten it, depending on the curve, the species, the weather, and the weight used.

Drive a line of nails at 16-inch intervals along a board or floor. Bind crooked culm to them where necessary, weight where it curves up from floor surface. Presoaked dry poles also respond to this treatment.

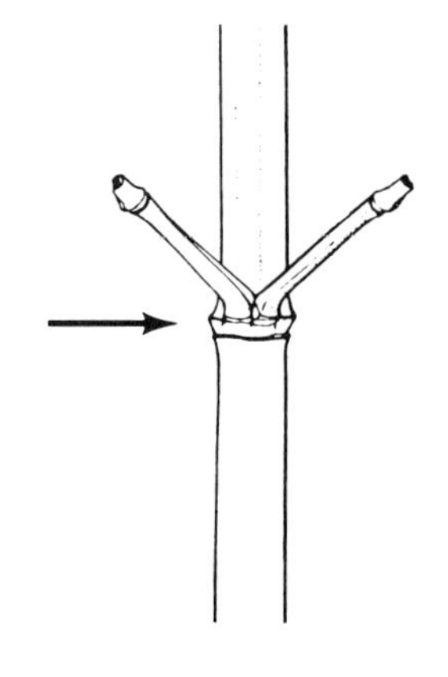

To remove branches, cut at arrow and snap down. Leave branch stubs or entire branches to support tomato, bean, and other garden crops without tying. (A **Phyllostachys elegans** *midculm node is shown.)*

Heat is the most common, fast, and effective treatment to straighten crooked culms: with wood or charcoal fire, with blowtorch, Bunsen burner, or gas stove. Steam heat, where available, is best. But all these alternatives, enacted on a large scale, are energy dear; so long-term bamboo development implies growing erect species, not burning up the local forest to straighten the local groves. Ridiculous waste levels, unadvisable for the individual, impossible for a planet norm, can no longer be so casually included in our cultural designs. But straightening a few culms over campfire is neither costly nor difficult. You can adapt the following method of professionals in India to your particular hearth and need.

Market preparations in India.

> Few forest operations in India are more interesting to watch than the preparations of raw bamboos for the market and the conversion of the rough, crooked, dirty-looking stem as it comes from the forest into the highly polished, rich brown lance stave or tent pole. Extremely primitive methods effect this remarkable change, as efficient as more up-to-date appliances. First, stems are cut to length and nodes are cleaned by a gang of small boys with sharp adzes. A broken skin means a ruined stem, so knot cleaning requires a certain skill. After an apprenticeship on cheaper grade bamboos, the boys become remarkably good, rarely making a bad shot . . . the *kammaggars* then warm each stem over a hot fire of two large logs. This supples the culm, imparts the fine brown color, and by melting the dirty waxy covering gives the stem a clean and polished appearance. Once the stem is sufficiently pliable, curve and kinks are taken out by handing with some force in an opposite direction. Two tools are used: for heavy stems, an upright pole, bored with slanting holes, is planted in the ground. The kammaggars put the warm bamboo in a hole and press with considerable force to bend it in the right direction. For lance staves and other delicate work, the kammaggar holds a stout stick with a groove in it in his right hand, in his left the stem to be straightened, which he works carefully along its entire length. Several firings may be required for perfect straightness. Considerable skill is necessary to warm without scorching the bamboo, then bend without breaking the fibers.[29]

Straightening *Arundinaria amabilis* in China.
"Each worker has a thick-walled earthenware firepot, without a chimney, in which a kind of smoke-

less coal is burned, a very hot bluish flame playing above the incandescent coals. Two large bricks are laid across the top of the firepot with a space of about 1½ inches between them, through which most of the flames pass. The bamboos to be straightened are stacked for a time, in their original bundles, on the racks which are about a foot above the firepots. Thus they are gradually warmed up. The worker sits on a low stool before his firepot, from which position he can reach and pull down the bamboos from the rack without rising. The culms are thrust, one at a time, into the glowing channel between the two bricks, kept there in motion for the brief space of two or three seconds, then withdrawn and subjected to a vigorous straightening process by means of a wooden tool. The various crooks in each bamboo are straightened separately, the heatings and bendings following each other in quick succession. In straightening the larger culms, some of which are 2 inches or more in diameter, a slightly different technique is used to get the necessary leverage, but the fundamental process is the same. They must be held rigidly in the desired position until the tissues are cool for a permanent set.

"Essentially the same technique is used to straighten green bamboo and bend semicured stems for furniture. The pectic compounds which cement plant cells together are soluble in hot water. Although the culms are comparatively dry after repeated sunnings, perhaps the water present is sufficient, when heated, to cause a softening of this material which makes possible the slight adjustment between the tissues necessary in the straightening process. When the tissues cool, the pectic layer then hardens again to hold the tissues rigidly in their new relation."[30]

Storage and shipping culm bundles.
After removing branches, cut culms to desired length and store by species in curing sheds that are rainproof but breezy. Store largest diameter culms lowest, on horizontal racks, with supports at intervals of a few feet to avoid sag and consequent bend in culms. If cost or time prohibits ideal storage, at least keep stems shaded, ventilated, and dry. Avoid leaning without turning as this creates bends.

To ship culms, tie in bundles roughly a foot in diameter tightly so that they don't scratch one another. Then wrap in burlap or similar material and wire the bundle securely.

A wooden tool used in China to straighten smaller culms after heating (× ⅓).

BUGS LOVE BAMBOO: PROBLEMS AND SOLUTIONS

Resistant Eastern species move west.
"After a number of years in Oriental countries, the writer in 1934 came to the Western Hemisphere tropics and was impressed by the absence of articles of bamboo in farm and household life . . . It quickly developed that there was one outstanding factor, in the West Indies at least, preventing much use of bamboo. *Bambusa vulgaris* was the most common species there, with culms comparatively soft and susceptible to boring insects. Articles made from this bamboo, often within a few months, were riddled by the small powder post beetle, *Dinoderus minutus.* Species of *Dinoderus* were widespread in the Orient, so we concluded that bamboo used for farm and industrial purposes in the Far East must be resistant to this borer . . . Numerous Far Eastern species were immediately secured to determine resistance to borer infestation.

"At Mayaguez, Puerto Rico, standardized tests were made, using *Dinoderus minutus* as the test insect. Bamboo samples of different species were placed in small cages of about 1½ to 2 feet of cubic space. Into each cage, five hundred powder post beetles were turned loose for thirty days. Counts of borer channels on samples then gave quantitative measures of susceptibility. Howard Plank continued and improved these tests. He found *Bambusa tuldoides, B. tulda, B. textilis, B. longispiculata* and *Dendrocalamus giganteus* notably resistant, and a number of still untested bamboos can be expected to be resistant. Age and maturity of the bamboo greatly alters resistance. Even *B. vulgaris,* if cut at five to six years and left standing four to eight weeks in the clump with branches untrimmed, was much less susceptible to borers than one- or two-year-old culms. Plank also established a relationship of carbohydrate content to susceptibility. The beetle is seeking starches and sugars, which explains the function of exhausting the carbohydrates by soaking in water, cutting five- to six-year-old culms, and hardening them in the clump after cutting."[31]

Puerto Rican tests and conclusions.
Clump curing culms with branches uncut, leaving them to stand four to eight weeks propped in the grove, controlled up to 90 percent infestation. Best results if harvested in hot, moist weather; the culms remain alive a month or more and lose most of their starch. Shed curing at least eight weeks after clump cure reduced susceptibility even more. Infestation was reduced 94 percent by eight-week water cure of freshly harvested culms—but culms became stained, light, and brittle. Complete submergence for twelve weeks was ineffective. Clump curing before the water cure was better than water alone, and best was clump-shed-water.

Starch content largely determines susceptibility and is measured by the iodine spot test. Species, age, time of harvest, and physical properties of the wood are determining factors. *B. vulgaris,* with the most starch of eleven tested species, was most attacked. Relative susceptibility of one-year culms in eleven species varied from 44.2 percent in *B. vulgaris* to 0.3 percent in *B. textilis.* With minor variation, susceptibility diminished in ascending internodes of culms. Harvesting by phases of the moon, contrary to popular belief, does not influence durability. Kirkpatrick also found that traditional bamboo moonlore is not confirmed by tests, and claims ripe bamboo (five- to ten-year-old culms) is actually more subject to borer attack.

Bamboo cut with moon on wane
Will ensure financial gain,
But beetles bore it very soon
If cut upon the waxing moon.
Moreover it's a well-known fact
That ripe bamboo is less attacked.
So say the chaps who ought to know—
Alas! It really is not so.
For Science reared its ugly head
And knocked these superstitions dead:
The lunar myth is utter tripe
And borers like *their bamboo ripe.*[32]

CHEMICAL METHODS. Injecting poisoned water solution into sap stream was explored. Copper sulfate (by the stepping method) gave 93 percent control in two tests and was found more effective in dry weather for rapid absorption. Hydrochloric acid injected into sap stream failed to hydrolyze starch in culms and also weakened them. Other methods tested were also found unfeasible. Many other chemicals, including those used for lyctus control in lumber, were tried and found ineffective—and also stained culms or reduced their strength or quality. DDT, 5 percent in kerosene, brushed on freshly harvested culms reduced internodal infestation 94 percent at 2½ months after application. DDT, 5 percent in diesel fuel oil, a ten-minute dip, reduced infestation 98 percent at 3 months, 91 percent at 12 months, and was still toxic to beetles three years after application. Previous clump curing doesn't improve control in DDT treated culms, but splitting culms or breaking out nodes increases absorption and effectiveness of dip. Minimal handling of treated culms in storage is best—to leave residual coating of DDT intact; make only necessary inspections. At three months, a second dip is recommended. Water suspensions of DDT, preferably micronized forms, are better than oil solutions for culms to be glued.

Beetle populations in sheds can be reduced by spraying DDT on curing racks and inside curing sheds, workrooms, and wherever finished or in-process bamboo articles are stored.[33] *We report without recommending:* DDT is very foul stuff.

India: love, life, and hungry habits of the bamboo beetle.
In India, the estimated life of a bamboo pole after cutting is a year to a year and a half, owing to the ravages of *Dinoderus minutus,* the bamboo powder post beetle—a creature roughly a quarter the size of an *o* on this page—whose total jaws are at this moment reducing tons of bamboo to dust around the world. In a heavily infested piece, you can hear them chewing. They bore holes a bit larger than their own diameter, tunnels made en route to laying eggs inside the culm wall.

Each female beetle lays twenty eggs. A few days later, tiny white dots—grubs—emerge and begin four weeks of burrowing in the culm wall, reducing it to powder and beautiful ruins. These larvae then enlarge the ends of their burrows to become pupae or nymphs. Some eight days later, they emerge from their cocoons as mature beetles and fly off to destroy a different culm or remain to further devastate their native pole. The few holes visible on the culm outside no more gauge the damage within than the number of doors on a building tell you how many people are inside using it. Successive generations use the same entrance and exit, so a culm appearing mildly attacked may be a labyrinth of dust.

In warmer areas, *D. minutus* can pass through five or more generations in a single year. Females usually outnumber males, but even figuring 50 percent females, beetles laying twenty eggs and passing through five to six generations between April and October–November can multiply from 2 to 200,000 or even 2 million in a season.[34]

> A small number of holes may be a sign of resistance to beetle attack since they sample poles with ¼-inch holes and abandon those found least suited to their taste. The Timber Testing Section at the Forest Research Institute, Dehra Dun, has run strength tests which indicate strength varies greatly, pole to pole, independent of the number of holes. A distinct loss of strength does not appear until the number of holes approaches two hundred per pole. Storage methods can increase or diminish beetle attacks. Creosoted racks without contact between poles is best since beetles enter where bamboos touch surfaces.[35]

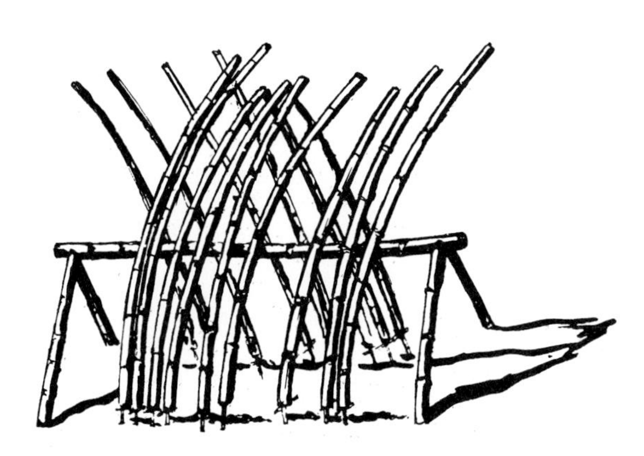

Curing in the sun is one way to preserve bamboo, but culms must be turned periodically to prevent the bending shown here. ***Culms are often bundled for ease of handling when sunning, and the bundle maintains straightness of individual culms.***

Treatments.

SUN AND SMOKE.

> Exposure to the sun can afford very effective protection. In fact, this is the principal device used in the Far East, where bamboo is so extensively used in making basketry, matting, and withes. In rural China, baskets and small bamboos for home use are often stored in the loft above the kitchen exposed to heat and smoke from fires, which provide effective protection.[36]

SOAKING. Preservation by soaking is the cheapest and simplest chemical treatment method. The culms, preferably green, have only to be kept immersed in a preservative solution for a period of five weeks or more, depending on species, age, thickness, and degree of absorption desired. Longer soaking is required for bamboo that will be in contact with the ground. Adequate absorption can be obtained by soaking: The disadvantage is the time required.

In split bamboo, the soaking period can be reduced 33 to 50 percent with 100 percent penetration of inner and outer wall. Rupturing the outer skin and use of high temperature hastens penetration. In full culms, knocking out nodes speeds and improves treatment. Soaking can be advised for the treatment of all bamboo for all purposes. It requires little equipment and technical knowledge, but the type and concentration of preservative and soaking time must be carefully worked out.

Make a simple tank: cut top and bottom off several oil drums, depending on length of bamboo to be treated. Weld together to form cylinder closed at both ends. Divide lengthwise.

1. Soak poles five days in water. A thick, shiny gelatinous substance exudes and is wiped off. This increases oil absorption.
2. Dry poles completely.
3. Soak forty-eight hours in Rangoon oil (a thick, heavy petroleum from Rangoon, Burma). Though stored in a beetle-infested area, fifteen thousand bamboos so treated were reported perfectly sound five years later.[37]

LEACHING. The most common preservative treatment for bamboo in the Orient is to leach out the starch, sugars, and other water-soluble materials sought by insects. This is done by weighting down and completely submerging freshly cut culms for three days to three months, preferably in running water since standing water can stain bamboo. Sea water is also an alternative if marine borers aren't present. However, water leaching at Mayaguez, Puerto Rico, was reported to result in excessive stains and brittleness of wood.

Bamboo grows well along rivers, so commercial or experimental groves are often planted at riverside locations for ease of transport. Water is therefore readily available to leach out the starch that attracts bamboo beetles.

Chemical preservatives.

The master disaster, the chief planet polluter, indirect cause of and carrier to distribute the New Poisons all over the maps, catastrophe par excellence of modern times, the car (and its extensions) is a recent inhabitant of earth that roams like a roaring lion devouring more civilized space than any old-fashioned species alive including its drivers, and certainly makes more impact per hour on the style and quality of our existence than any other single product of our culture.

By whatever name, the car looks as pretty approaching in all cultures, and smells as rank in its wake. Industrial countries are "advanced" in direct ratio to their pollution: superpolluted, superpower. Poorer countries, apparently, are determined to reenact the chemical disaster of modern settlement, transport, production, and agriculture as rapidly as possible. Each seems set to relive in turn the tale of the prodigal son, squandering limited resources and living like a hog before returning contrite, to the lap of nature mother's house to behave with ecological sanity *imposed,* not chosen, at last.

WICK ACTION, CULM-CUP CURE. Freshly cut bamboo with branches and leaves left on can be stood in preservatives to a depth of 1 to 2 feet (30–60 cm) for one to two weeks, allowing wick action to suck up the solution as the leaves and wood dry. Preservative can also be simply poured into the basal internode of an inclined bamboo or into a bike or car innertube firmly clamped or tied to the base of the stem with the butt end elevated enough to let the solution soak down into the culm.

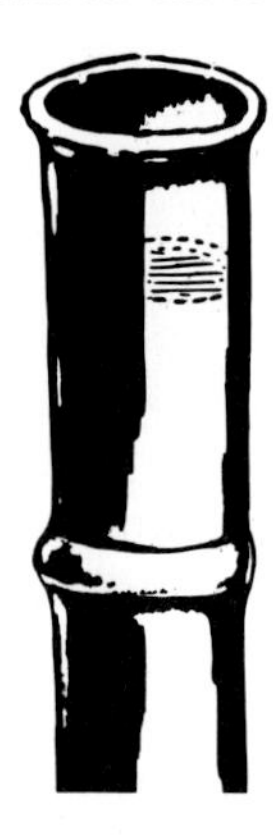

Culm-cup cure.

BOUCHERIE PROCESS. For freshly cut round bamboo with nodes intact, the "Boucherie process" is the most effective treatment. The preservative is gravity fed through pipes and rubber tubes clamped to stems from an elevated tank some 30 feet above ground, or pressured into culms at ground level by a simple hand pump, which considerably shortens the treatment period.

COATING. Whitewash, tar sprinkled with sand, and such coatings are effective only if they give a continuous coating at cut surfaces. Inner walls made accessible by splits in the culm cannot be effectively protected.[38]

Chemical cures.

Runner-up in the Dirty Derby of modern pollution, choking close to the exhaust of our National Car, is our National Chemistry Set. The chemical nightmare of industrialized cultures is an intimate co-function of the car. In addition to large-scale tactics to clean up our air and water, we each have to drive less and purchase fewer poisons if we expect to reverse present trends. Chemical preservatives have been much explored to make bamboo unappetizing to bugs, but they are all too expensive for broad scale use among the people, quite apart from their environmental contamination. Selection of proper species, proper harvest and curing proved sufficient for centuries of marvelous bamboo use in the Orient before the appearance of modern chemical techniques. Bamboo strips used for military archives and grocery lists have lasted over two thousand years in China. We have mentioned the bamboo organ in the Philippines still sound since its construction in 1821, cured simply by a six-month burial in sand.

So, in light of the above, we have decided to exile advice on chemical cures to a tiny footnote. For those interested, Narayanamurti's United Nations publication on bamboo use provides extensive lists of recommended pollutants.[39]

Culm anatomy affects treatment.

"The tissue of the bamboo culm is built up by parenchyma cells and vascular bundles consisting of vessels, thick-walled fibers and sieve tubes. The movement of water in the culm takes place through the vessels. The fibers are responsible for bamboo's strength. Nutrients such as starch granules are stored in the parenchyma cells which compose up to 70 percent of the tissue. The vascular bundles become progressively smaller in size and denser towards the periphery. The orientation of all the cells is in the vertical direction. The culm is covered inside and out by hard waxy cuticles which offer considerable resistance to the absorption of water. (This characteristic is important when impregnation of preservative chemicals is required.) *Fibers* constitute 60 to 70 percent by weight of the wood of bamboo. Fiber predominates in the periphery; parenchyma, inside. Fibers are most dense in internodes at ¼ to ½ culm height, where longest, most mature and thick-walled fibers are found. Fibers gradually decrease in length, maturity, and cell wall thickness towards culm tip. Bamboo fibers vary considerably in shape, size, and wall thickness. They are usually long and straight with tapering ends, the length roughly one hundred times the diameter.

"*Parenchyma tissue* abounds in bottom internodes and inner wall, descreases towards the culm tip and periphery. The *vessels* occupy only about 15 percent of the culm. In internodes, vessels are parallel to stem axis, without branching or contact. Intensive branching inside nodes makes horizontal transportation of liquids possible, and pits connect vessels. The vessels, through nodes, connect culm walls and some go into branches.

"The distribution of vessels affects preservative treatment, which occurs not only at culm upper and lower ends, but also through cut branches at nodes, from where preservatives can penetrate culm in both directions. The number of vessels available for treatment generally decreases with height. When bamboo is dried, the sap in vessels dries up and the vessels fill with air. Pits inside the nodes close; pit openings of parenchyma cells are closed by their own dry cell sap. These factors influence preservative treatment of dry bamboos. If a preservative is to penetrate into the vessels, it has to overcome surface tension and friction forces. In order to enter parenchyma cells, it has to dissolve out dried sap closing their pit pores or diffuse through the cell walls."[40]

Storage: function and description.
At the storage yard, bamboo is air seasoned under cover six to twelve weeks to increase strength and avoid cracking. Kiln seasoning can do the same job in two to three weeks, though at risk of splitting the outer membrane of several species if the seasoning is too rapid. To reduce fungal attack, guard bamboo against wetting by rain or contact with soil. Good ventilation and frequent inspection are important. The storage ground should be thoroughly inspected

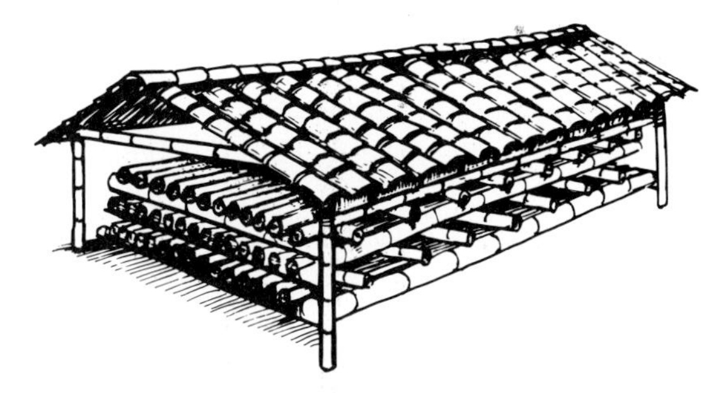

Curing and storage shed.

and cleaned before laying out, all refuse and useless timber and bamboo removed. Termite-infested areas should be sprayed with 4 percent emulsion of DDT or 0.2 percent emulsion of BHC or other suitable insecticide. Destroy termite colonies by breaking mounds open and pouring in insecticide. The ground should have good drainage facilities.

Bamboo should be stacked on high skids or raised platforms, at least a foot above ground for prevention of termite attack, stored so all sides may be readily and regularly inspected. Remove or treat attacked culms.[41]

Pests, smuts, and rusts.
Rats, porcupines, squirrels, hare, deer, and monkeys are reported to gnaw rhizomes, eat shoots and seeds. Goats and cattle trample rhizomes and devour foliage. Monkeys can also damage young brittle culms by leaping onto them, and elephants trample and feed on groves. To this list of bamboo enemies, some Indian authorities add wild pigs, thieving villagers, and mischievous boys—but the animals causing the most damage are the executives, agronomists, and field hands of transnational corporations, now landscraping with machines of immense dimensions but without the moral and ecological consciousness needed to manage safely a team of mules, as Wendell Berry remarks of the strip miners grunting after coal in Kentucky. As usual, the human race has been the chief nuisance to bamboo, exterminating vast tracts for cattle, cotton, or other crops. In other areas, deforestation occurs when growth cannot keep pace with indescriminate and untrained harvest.

Living bamboo holds its own very well against all pests but us, and most of the following are infrequent and minor visitors of its vitality. *Bamboo smut (Ustilagoshlraiana)* is a parasitic fungus related to the smuts of corn, wheat, oats, and other grasses. Its known occurrence is limited to the genus *Phyllostachys,* and it attacks only new twigs and leaves. Twigs swell, growth is checked, and later the sootlike spore masses break through. All infested culms must be cut down and destroyed, and their rhizomes—where the smut mycelium apparently lives perennially—must be dug and burned. There has been no bamboo smut reported in the United States since 1909 when an outbreak accidentally introduced from Japan was arrested.

BAMBOO RUST. Of the dozen or so bamboo rust fungi reported from around the world, only *Puccinia phyllostachydis* Kusano is known to reside in the United States. It only attacks *Phyllostachys* bamboo and, although widespread, does not cause serious damage. The rust attacks new leaves, with brown powdery spots appearing on the underside, yellow spots above. Rust-infected leaves should be removed before shipping or planting propagules. Locate any contemplated bamboo nursery a goodly distance from any infected plants in your area.

Myers lists fifteen scale insects, nine miscellaneous bugs, four mites, and one thrip thriving in the United States, together with their locations, the species they attack, their occurrence elsewhere in the world, and when they were first apprehended by the USDA.[42] Stanford University campus enjoys the distinction of having hosted the first recorded bamboo pest in continental United States (1899). Young describes various bamboo diseases and remedies in a broadly available USDA handbook: fungi, which attack new growth, including new twigs and leaves, and nematodes or eelworms, which infect roots, are mentioned, but scales are more important.[43] "Bamboo scales (sucking insects that feed by extracting juices from the leaves and stems) assume serious importance only when because of neglect infestation becomes heavy." Heavily attacked canes should be cut and burned. "More lightly infested ones should be sprayed in the spring when the young crawlers are present. A white-oil emulsion diluted to 1 to 2 percent oil and fortified with 1½ pints of 50 percent malathion emulsifiable solution per 100 gallons of spray should be used for spraying. It should be repeated weekly about three times to kill all the crawlers as they hatch."[44]

Young's descriptions of the physical appearance of pests and the symptoms they cause in bamboo are detailed and useful for bug identification. Some of the cycles are complex. For example, the *bamboo louse,* or aphid, found under leaves of *Pseudosasa japonica* and other *Sasa* and *Bambusa* species—often in large colonies—is pale yellow, ⅒ inch long, with some light brown and dusky markings. It excretes sugar water that in turn becomes a medium for a mold that gives the plants a sooty, ugly appearance. It is controlled like scales.

A *roundheaded beetle* or grub that damages *Phyllostachys* rhizomes reveals its presence by wilting young culms. The infested rhizomes must be dug up and destroyed. Various *mites* are noticed by small white webs on leaves. Control in rhizomes chosen for propagation is helped by dipping them in hot water—122°F—for ten minutes.

A *fungus (Melanconium bambusae)* occasionally attacks medium to large culms of large species of *Phyllostachys.* Its presence is often first noted when new culms reach almost full height: Basal internodes turn purple black or brown, beginning at their base and extending upward. Eventually the culm dies, its walls full of mycelial threads, the vegetative parts of the fungus, which is thought to attack only culms previously cut, bruised, or injured in some way.

Bugs versus bugs: biological pest control.

> Two destructive insect pests attack growing bamboos in the West Indies. Both are scale insects—*Asterolecanium miliaris* and *A. bambusae.* Spraying a crop such as bamboo to kill insects is not feasible in tropical, heavy rainfall countries and is impractical for small farmers of limited means. An effective remedy, however, has been developed. The writer observed lady beetles which were feeding upon and destroying these scale insects in Venezuela, Panama, and Jamaica. . . . The director of the Mayaguez, Puerto Rico, experiment station imported from Brazil, Trinidad, and Cuba some six or eight *Coccinellid* species which prey upon these scale insects. Harold Plank brought in another from Texas, and the writer still another good predator of this type from Martinique. They are now effectively controlling these two scale-insect species in Puerto Rico. This procedure should be possible in other countries.[45]

GROWING A GRACEFUL GIANT: MOSO CULTIVATION IN JAPAN

Hairy bamboo: *Phyllostachys pubescens.*

Moso bamboo was introduced into Japan from China by Zen monks, according to some traditions, early in the eighteenth century.[46] "Hairy bamboo," as it is called in China, reached Europe by 1880, the United States West Coast about 1890, and—by way of Rufus Fant's groves in Anderson, South Carolina—the USDA Savannah groves in 1926.

Softer tissued than *P. bambusiodes (madake),*

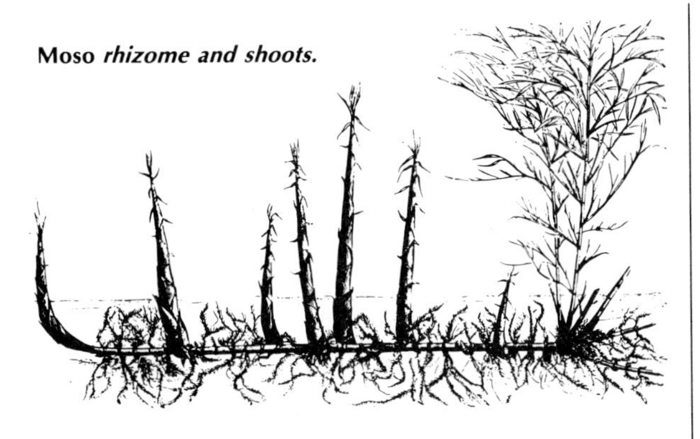

Moso *rhizome and shoots.*

moso culms are less valued industrially. But larger ones are prized as alcove posts in Japanese homes (the *tokonoma*), and many are used as floats for giant fish nets, each net requiring as many as a thousand culms. In Miyazaki-ken alone, some twenty thousand moso culms are used each year for these yellowtail nets, and many more are grown for export to Korean fishermen. Drainage pipes for ships, New Year's ornaments, flower vases, teapots, kettle pads, slop basins, trays, carved brush stands, dippers, cigarette cases, tea boxes, basins for wine cups, sake containers, tobacco cases, cups, rice bowls, soup cups, and confectionary packages are all made from moso culms of different types and from specific parts of the culm.

Some thousand boxcars of bamboo, each carrying about nine hundred moso culms 20 feet long with branches left on move from the coast to Tokyo each year at seaweed season. Young seaweeds (used widely in Japanese cooking) get tangled in the culms and branches and continue growing there. Because of its numerous branchlets, moso is also the preferred material for bamboo brooms. Culms are used according to their character: elastic ones with thick walls are used for lantern holders, and the framework for cages and fans. Durable culms become cages, wicker trunk frames, and winnowing baskets. Culms with toughest walls make small shovels and packing cases. Pliable poles make dippers and coat hangers; the most workable culms become combs, forks, knives, spoons, silkworm net frameworks, and varnished chopsticks.

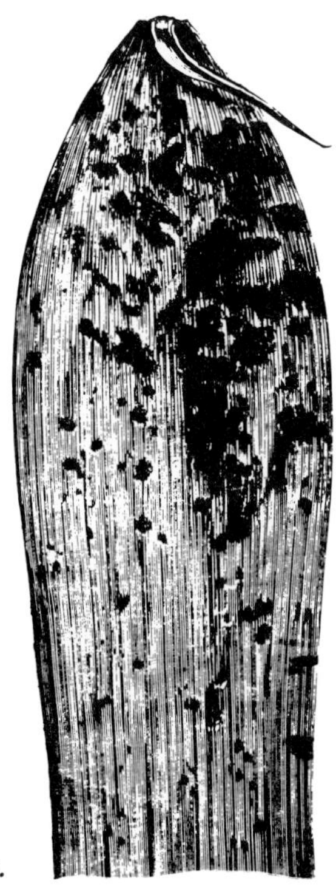

Moso *sheath tip and blade.*

The sheaths of moso are less used than henon *(P. nigra* var. *henonis)* or madake sheaths, but large ones are used for wrapping. Thick ones stuff slipper soles, and charred sheaths are used as a remedy for bleeding, stomachache, and wounds. Moso rhizomes, like those of henon and madake, are used to make canes, whips, and umbrellas. These same rhizomes, living, are effective for erosion control on riverbanks and the 7–15-degree slopes where moso often thrives.

Edible shoots are the principal product from moso in Japan. Roughly 40 percent of Japan's 123,000 hectares (307,500 acres) of bamboo—about 50,000 hectares (125,000 acres)—are devoted to moso cultivation. The export of shoots has grown to be a major industry all over the country, particularly as Japanese cuisine, requiring bamboo in many recipes, has become popular in America and other areas of the world. Production staggers in the wake of demand, and crops are sold before they're canned.

Rhizome.

Since moso is of such economic importance, its habits of growth, soil preferences, and propagation methods have all been subject to scrutiny for many years. The moso rhizome is where the story properly begins. Sometimes as much as 6 inches in circumference, growing 12 feet or more in a single season, the rhizome grows about 29 percent of this in June and July, 50 percent in August and September, and 30 percent in October before coming to a sudden and full stop, lasting from early November to late May. Beginning with 10 percent growth per month, the rhizome gradually triples its speed until its abrupt halt in the fall. The life expectancy of a moso rhizome is a decade or more, but it is most active from three to six years of age. When the tip of a growing rhizome is injured, a branch rhizome angles out from a bud behind it for 3–5 inches at an angle of 7–15 degrees and then angles back to run parallel to the original direction. In the eighth or ninth year, decay sets in and by the twelfth to thirteenth year it becomes complete.

Culms.

Moso culms complete their growth some forty days after the shoots appear above ground. They grow faster by day than by night and more in afternoon than morning hours. Maximum growth occurs some ten days before completion, when the culm grows more than 3 feet a day, eventually reaching 75–80 feet, making moso the largest—some claim the handsomest—of the *Phyllostachys.*

Cultivation for lumber is called a "grove"; edible shoot cultivation, a "garden." Moso garden practices include annual tillage to reduce weeds, relocating rhizomes more densely to increase shoot production in a small area, and dressing with mulch. Unlike madake or henon, moso grows fairly well in poor soil. Compared with other agricultural crops, bamboos have come through little artificial selection, so moso—like other grasses—is fairly resistant to disease and insects. Moso is, however, easily wind damaged owing to its luxuriant foliage, broad spacing in a grove, and shallow roots. Most rhizomes are in the top 1½ feet of soil, and few go lower than 3–4 feet. This means only 2–3 percent of its altitude is anchoring the moso culm, and high winds turn culms into 75-foot levers to uproot themselves.

Moso *shoot.* *New* moso *culm.*

Hillside culture.

Hillside lands with a slope not greater than 15 degrees are suitable for moso. They should slope towards sunlight, southeast, south, or southwest. Drainage is improved, and the slant catches more direct sunlight, hastening decomposition of soil nutrients and promoting growth of shoots in spring: A 10-degree rise in temperature can increase chemical processes in plants by 50 percent.

Soil types.

A clayey loam soil, which produces good sweet potatoes and taro, also yields excellent moso with bright color, soft texture, good fragrance and taste. In Fukuoka market, shoots with black soil on the sheaths sold cheaper than shoots with red soil. Crafty merchants would wash off the dark soil and rub on red, well known to produce top-quality shoots. Soil too heavy with clay (60 percent) should be avoided, however. Since clay is quite cohesive and a poor conductor of heat, there's less air and breath in the soil, which is always cool, thus slowing decomposition of organic matter and retarding moso growth. Humus soil (over 20 percent decayed vegetation) is easy to work but becomes acid easily, and the quality of moso shoots is not optimal. Marly soil (over 15 percent calcium carbonate finely divided) is one of the best for moso. Loamy soil (30–60 percent loam, 40 percent sand) is also satisfactory, but clayey loam is the soil most sought by Japanese moso growers.

Preparing land.

Shrubs and weeds are dug out, and soil is cultivated to a depth of 1½ feet. Moso rhizomes eventually go as deep as 3–4 feet, but most are in the top 1½ feet of soil so deeper cultivation is unnecessary and uneconomical, although clayey soil can be worked to 2 feet. One hundred workdays per acre are figured for hand clearing and cultivation. Fertilizer—manure, compost, hay, whatever—is worked in the summer before actual planting. Spot cultivation of scattered circular areas about 5 feet in radius is practiced when larger plantings cannot be afforded in a single year. Belt plantings 12 feet wide can also be staggered across a field. Contour belts on river-

Moso *branch complement.*

banks or sloping lands should be at right angles to the slope.

Drainage is very important. On flat lands with standing water, provide ditches to remove it. Soil acidity can be corrected with about 1,000 pounds of lime per acre, which will not only neutralize the acid but will also aid in making available otherwise insoluble nutritive elements and hasten decomposition of organic materials. Cedar or other trees should be planted as windbreaks to protect grove from cold winter winds in the area. Since moso uses liquid fertilizer, some water storage is necessary where running water is not available.

Choosing stock plants.
Vigorous young bamboo culms one to two years old are selected for stock plants, with a preferred circumference of about 5–6 inches. Older culms may initially produce more shoots around them—but it is rhizome development that's most desired in a new grove, and younger plants produce more rhizomes. Plants grown in sun are preferred for stock to shaded bamboos, which tend to be softer textured in culm, branch, and leaf. Early shoots are a recommended propagating material: They can be dug out easier since their rhizomes tend to be near the surface. They are marked for transplanting in the fall or the following spring. Clear cutting a grove in February produces a large number of slender shoots that are ideal for propagating material on a large scale. Rhizomes alone can also be used, cut into lengths 12–18 inches long with a minimum of five good buds on each length. Place immediately in a nursery bed, taking care they don't dry out. The bed should face south, with moderately wet sandy loam soil that is well tilled and fertilized with soybean cake or rapeseed oil cake. Make trenches about 2 feet wide; put rhizomes every 6 inches, and cover with a little soil and a straw mulch to preserve moisture and check weeds. Fertilize twice before September. In the first year, these rhizomes may produce some thin shoots resembling *Sasa* bamboos. In the second year, shoots may measure 2–3 inches in circumference. If the leaves of young plants turn yellow, use ammonium sulfate and soybean cake from September to October.

Transplant timing.
Plant bamboos any time except in extreme heat of summer or cold of winter, but best is March or October-November. Winter transplants should have a larger protective ball of earth; in summer, wet straw should wrap the roots and sometimes the culms as well. Transplanting is best when rhizomes have stored a sufficiency of food, in fall when growth is about over or in spring before it begins. One disadvantage to spring digging in a grove is the possibility of injuring emerging shoots. Nevertheless spring is preferred to autumn, especially in colder climates. The best day is a cloudy one with no wind, with a rainy day following. If possible, avoid transplanting on a rainy day.

Uprooting rhizomes.
Satow (1899) cites the Japanese tradition that a large bamboo clump requiring ten men to lift will establish itself in one year, and a bamboo propagule one man can manage will take ten years to reach mature growth. Choice of size depends on convenience, distance of transport, and other factors specific to each case. A trench is dug with a radius extending about 1½ feet around the bamboo to be transplanted, cutting off rhizomes at that point. Usually, rhizomes run in the direction of culm branches. In a tangle of rhizomes, be sure to cut only those belonging to the culm you're extracting. Strike the rhizomes with a piece of metal—a spoon will do—and listen closely at the culm for the most pronounced vibrations. Culms close together—6–10 inches—should be transplanted together.

Pry the clump up out of the hole with a stout pole, being careful not to hold culm when removing. Any pressure on culm can, increased by leverage, injure its delicate connection with the rhizome. Cut off branches with a fine-toothed saw to minimize drying and later wind damage: Tissue on cut branch or culm dies back locally to node but no farther.

Remove the soil from the rhizome and study

the buds: If buds are missing, damaged in digging, or not obviously alive, discard the rhizome. Vitality, not bud size, is the issue: It can be tiny as a soybean or already swelling. If possible, move bamboo with original earth around it; 120–200 pounds is recommended as optimal for a 2–3 inch culm. Wrap securely with rope, with burlap, then rope again.

Shipping plants.
Take care they don't dry out. Cover with mats or straw and keep shaded as much as possible. Avoid injury (by shaking and so forth) to delicate area where culm and rhizome join. Avoid injurying rhizome buds. If little or no soil is left on roots, wrap rhizomes in wet straw, then burlap, then rope and tie securely so that roots maintain snug contact with soil or straw. Ship by most rapid and economically feasible means. In one sizable moso shipment, there was 20 percent mortality among 135 rhizomes in twenty-seven days between digging and arrival at new location—even with these precautions. (See McClure's mud mayonnaise, pp. 210–211.)

Spacing.
Around 180 plants per acre—120 minimum, 240 maximum—is the rough consensus of moso experts, with 12–25 feet between plants. White (1945) recommends hexagonal plantings for maximal erosion control on slopes.

Planting.
The planting hole should be shaped according to the individual plant, larger than the ball of soil, and about 1½ feet deep. Rhizomes should run at right angles to the slope on inclined land. The old ground line is usually clearly visible on culms—green above, yellowish below. Replant at the same ground line, a little deeper on very windy land, a little more shallow where it's moist.

Fertilizer should be applied at transplant, under and around but not touching the rhizome—circling the plant on flat land, in a half circle on the uphill side on slopes. After transplanting, later fertilizing can be done in a 6-inch deep groove, then covered with soil to prevent runoff in heavy rain.

Use topsoil to fill planting hole, pushing and packing it with a thin stick so soil is snug around rhizome. Step lightly on ground around buried culms. Water to settle soil around rhizomes and give them a good drink. "Water culture" is practiced with plants that have drying roots and those stored a time after digging or transplanted in a dry season: In the bottom of a planting hole, water and loose soil is stirred to mud soup. The plant is placed in this hole which is filled with dirt, then more water is added. High success is reported with this method, although in clay or moist soils it can rot the rhizome.

Mulch a 5–6-foot diameter circle around culms with straw or cut grass. Tie three supports to hold bamboo in place if the culm is tall enough and with enough foliage to catch the wind. Tie supports tightly to culm. Loose supports can do more harm than good.

Planting of root stumps—base of culm only with rhizome attached—requires more time for a grove to establish itself, but severing the culm makes transport much easier. Root stumps are best transplanted just before emergence of new shoots. Select culms free from disease, two to three years old that have vigorous rhizomes with at least five new buds. Plant 70–80 in a quarter acre.

Extending grove area.
Bamboo rhizomes naturally extend towards nutrients and friable soil. Bamboo can cross a stone bridge on the soil falling from hooves of passing horses an old proverb says of the plant's readiness to move. So the easiest way to increase bamboo land is to fertilize and cultivate immediately around

Turtleshell bamboo, a new shoot of a moso *mutant.*

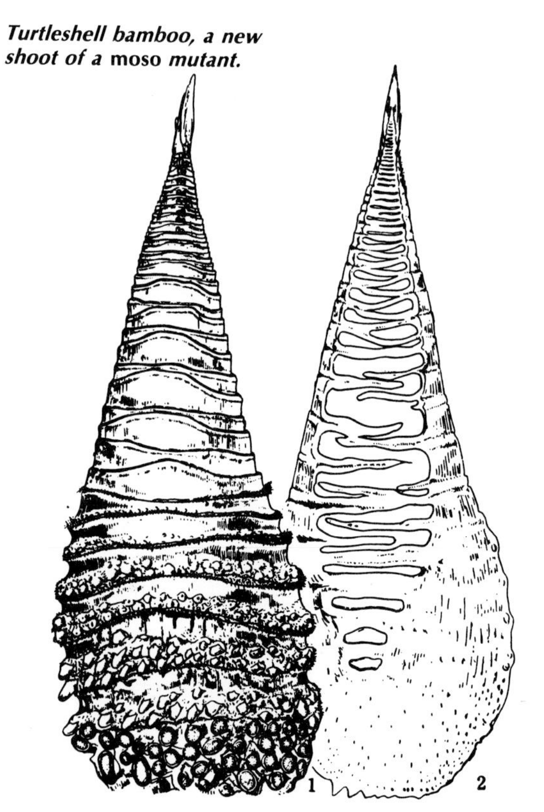

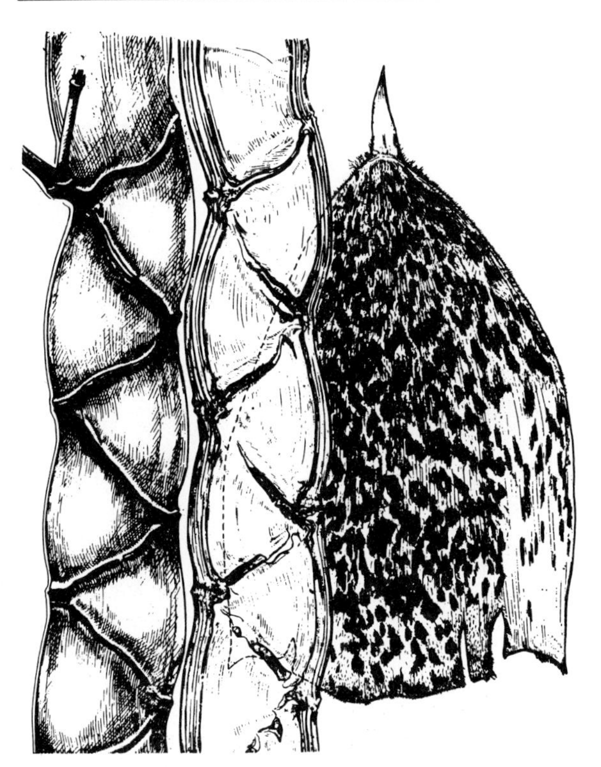

Culm of turtleshell bamboo.

a grove, preferably immediately after shoots emerge in spring. Remove earth 1½ feet deep, in a 12-foot belt around the moso grove. Place compost or stable manure in the bottom, refill with soil cleaned of roots and rocks, and mulch 2 inches or more deep with straw or other cover—or sow a cover crop, especially a leguminous nitrogen fixer that grows fertilizer on the spot instead of having to be lugged there. Fertilize the moso grove well to encourage rhizome development; moso rhizomes grow some 12 feet annually, so each year groves can be extended in belts of this dimension.

A ravine or ditch along a grove can sometimes be filled with soil to extend a grove area. Sometimes bridges of large culms up to 6 feet long are constructed across narrow but deep ditches. Nodes are knocked out, and culms are filled with soil. Rhizomes then grow through them to establish the grove on the other side.[47]

YIELDS

Yield figures for bamboo vary greatly, depending on the species, length of time since previous cutting, location, and grove management. Figures are sometimes for green weight of entire plant or culms only, sometimes for air-dry weight. Weight sometimes refers to clear cutting, sometimes to selective cutting of only mature culms. The following accounts are sample yields of a number of species from several locations around the whole wide world.

Louisiana.

"To test yield in the Gulf Coast area, cuttings were made at the Avery Island bamboo forests, where over 80 acres are devoted to sixty-four varieties. Clear cutting 1 acre of fourteen-year-old *Phyllostachys bambusoides* yielded 22,491 canes, 146 tons to the acre. One acre of fourteen-year-old *P. edulis* yielded 15,876 canes, a gross weight of 143 tons. Neither of these two plantings had before been disturbed by cutting." Twelve years later, in 1944, half the culms from a half acre were cut from these same groves, a total of 1,200 bamboos, up to 62 feet long and weighing up to 76 pounds. The half-acre yield was 39.6 tons. "Clear cutting 2.44 acres of *Bambusa argentea* and *B. Alphonse Karri* from a fourteen-year-old grove without previous harvest yielded 111,941 canes, weighing 349 tons gross. These two varieties of *B. glaucescens* are bunch-growing types, whose canes grow so closely together that even a dog cannot pass between them. Twelve years later, in July 1944, the same area was cut, removing two-thirds of the canes, yielding 24,600 canes, weighing an average 6 pounds each or 73.8 tons."[48]

Other figures for McIlhenny's groves for plantings five to six years old were summarized by

Louisiana Yields

SPECIES	HEIGHT (FEET)	CANES PER ACRE	TONS PER ACRE	
			GREEN	DRY
**Sasa japonica*	17	37,144	55	37
Arundinaria gigantea	11	61,299	45	22
***Bambusa multiplex* (argentea)	29	65,195	244	138
***Bambusa multiplex* (Alphonse Karr)	23	46,746	105	58

**Pseudosasa japonica* ***Bambusa glaucescens*

Sineath from an unpublished communication of Polly McIlhenny Simmons:[49]

Alabama bamboo experiments.
Bamboo experiments were begun in Auburn, Alabama, at the agricultural experiment station there in 1933. Fishing poles for local markets were the initial focus of interest. In the 1950s, search for new crops to replace those in surplus production and an increased interest in bamboo for paper pulp prompted extension of experiments to Camden in 1959. Results of thirty-five years experience with bamboo cultivation in the area were published in 1968 on the twenty-eight species included in the plantings, mainly *Arundinaria* and *Phyllostachys* varieties.[50]

P. rubromarginata, in addition to being among the three most hardy of the species tested, produced the largest tonnage of dry wood per acre and showed the highest survival in propagation by rhizomes (86 percent).[51] Clear cutting of seven species showed a yield of poles 15 feet or longer from 11,000 to 39,000 poles per acre; selective cuttings, 2,000 to 10,000 poles per acre per year, depending on the species, with larger species such as *P. bambusoides* yielding about 1,000 per acre per year. By weight, the yield of dry wood varied from 17 to 54 tons per acre excluding leaves but including stems and branches, depending on the species, from groves fifteen to twenty years old that were clear cut.

Bamboo cut in 10-foot strips every five years produced yields of 18–45 tons of dry wood per acre, with a four-year average of 28 tons per acre. Strip cutting was found much more productive than clear cutting since grove recovery was much more rapid from rhizomes bordering the cut. Green leaves were found to compose 10–16 percent of plant weight; limbs, 11–21 percent. In one experiment comparing loblolly pine *(Pinus taeda)* and *P. bambusoides,* the pine produced 8 tons of dry wood per acre eight years after planting; the bamboo, 14 tons. The numerous tables of Sturkie (1968) provide more detailed figures for the tested species.

Alabama.
Oven-dry yields (per acre per year)[52]
Phyllostachys aurea: 1.31 tons
P. aureosulcata: .98 ton
P. bambusoides: 1.83–3.16 tons
P. meyeri: 2.57 tons
P. rubromarginata: 3.43–5.70 tons (highest yielding species)
P. viridis: 1.67 tons

Georgia.
According to the unpublished correspondence of R. A. Young in 1935, studies of a well-established grove of *Phyllostachys bambusoides* at Savannah show a per-acre per-year yield of some 8–9 tons. Yield experiments with six *Phyllostachys* species at the USDA Savannah groves from 1956–1965 showed harvests ranging from an average 1.32 to 4 tons per acre per year.[53]

India.
In Chittagong Hills, east Bengal, figures of 1,200–6,000 culms per acre have been reported on a three-year rotation (Ahmed 1954). One thousand culms of *Melocanna baccifera* weighted 4.4 tons green, 2.5 tons air dry. These counts would mean air-dry weights of 3 to 15 tons per acre. In India, dry internodes for paper production gave 4.6 tons of *B. polymorpha* and 4.1 tons for *Cephalostachyum pergracile.* In Burma, *Melocanna baccifera* yielded 8.3 tons of dry culms per acre. *B. arundinacea* yields varied greatly from 1 to 14.5 tons of dry internodes per acre, with average of larger areas said to approach the higher figure.[54]

Japan.
Fresh-weight yields of culms (per acre per year)[55]
Phyllostachys bambusoides: 2–5.6 tons
P. nigra (henon): 1.2–2.8 tons
P. pubescens: 2.4–7.6 tons

China. See p. 290.

Taiwan. See Chapter 10, pp. 304–305.

CHAPTER 8.

1. McClure 1952b:6.
2. McClure 1961a:18.
3. See also Lessard 1980: 56, 95, 129, for the question of bamboo control in forest management.
4. McClure 1963a.
5. McClure 1948b.
6. Austin 1970:205–6.
7. McClure 1933.
8. McClure, "Bamboo Notes for El Salvador," unpublished manuscript.
9. Austin 1970:206.
10. Hodge 1955:123–7.
11. Lawson 1968:21.
12. McIlhenny 1945:124–5.
13. Fairchild 1903.
14. McClure 1958:56.
15. Fairchild 1903:20.
16. McClure: 1938c:14.
17. Ibid.: 15.

18. McClure 1957b:307–8.
19. White 1948:16.
20. McClure 1938c:5,6.
21. McClure 1966:232.
22. Lessard 1980:105.
23. White 1947.
24. White 1948:13.
25. Lessard 1980:30.
26. McClure, unpublished *Guadua* notes, abridged. Available at Grass Library, Smithsonian Institution, Washington, D.C. (see p. 305).
27. Varmah 1980.
28. White 1946a:268.
29. Wright 1921.
30. McClure 1931: 302–3.
31. Lee 1944:128.
32. Kirkpatrick 1958.
33. Plank 1950. He and colleagues published some fourteen articles on bamboo beetles from 1937 to 1950.
34. Stebbing 1910.
35. Trotter 1933.
36. McClure 1952b:5.
37. Stebbing 1910.
38. Narayanamurti 1972:21–6.
39. Ibid.: 91–2.
40. Ibid.: 17.
41. Ibid.: 6–7.
42. Myers 1947.
43. Young 1961:58–62.
44. Ibid.: 60.
45. Lee 1944:129. Lessard (1980) briefly reports on biological pest control in China bamboo work.
46. Oshima (1931) forms the basis of this section on moso, roughly 10 percent original length of USDA translation, available from the Grass Library, Smithsonian Institution (see p. 305). Cf. also Oshima, bibliography.
47. For moso cf. also McClure 1957a:51–3. Lawson 1968:127–8. Young 1961. These are abridged in the treatment of *P. pubescens,* pp. xx–xx.
48. McIlhenny 1945b:122–3.
49. Sineath 1953:11.
50. Sturkie 1968.
51. The other two most hardy were *A. fastuosa* and *P. viridis.* Three species suffered severe damage during the three hardest winters of the twenty years' observation: *P. aurea, P. bambusoides,* and *P. meyeri.*
52. Sturkie 1968. USDA 1978:3–4.
53. USDA 1978:3–7.
54. Huberman 1959: 40, abridged.
55. Ueda 1960. USDA 1978:3–4.

9. USES OF BAMBOO

Bamboo is both road and map
where use and beauty overlap
with learning in a roomless school
for the wisest or the fool,
for ancients creeping back to earth
or infants dripping fresh with birth.

Bridges, baskets, paper, flutes
in summer, shade, at dinner, shoots-
all from groves whose rhizomes will
mantle an eroded hill.
Count its uses? count instead
fingers of the thankful dead.

CRAFTS

Creating creation.

If people were created in the image of their creator, as it says in an old black book, we were created creative. A fluid creativity is what chiefly distinguishes our species from the many millions of others here with us. Cultures giving maximum outlet to the creative energies of the people will experience themselves as "expanding," independent of any economic index. And a culture that somehow, by accident or design, smothers or reduces creativity among its members will never be able to distract its people, by a show of material prosperity, from a certain contraction they feel in the womb of their spirit, which never bursts the water and comes to term.

We are not encouraged in our culture to go to the root of a dis-ease or dis-comfort, but to apply palliatives. But in the present case, all the mattresses in the kingdom will not rock us to sleep, because the pea is in the princess: A deep defeat of creativity is built into the very design of our culture. "Everything that rises must converge." Creation is always an uprising, and a great impulse towards union is the very bone of its blood. This impulse towards creative union in the United States now confronts a production system based on fragmentation and schools preparing their "products" to fit into it.

Any philosophy of economics or style of education that declares competition the base of its design rather than creation and that cheerful collaboration that creation requires of us in all its fuller forms, is doomed to a shrivelled existence. The implied objective of capitalist psychology is not to relate and create but to get the best of the bargain; not to meet, but to dominate. But "friend" means "free," according to its German root.

Perhaps bamboo became the Friend to the people of China and other countries in the Orient not so much for the things and shelter it provided, but for a richer gift: it created creation, made it more possible for millions for centuries to discover the endlessly inventive cunning, the nimble intelligence of human fingers.

"Shared bliss," a popular symbol in Chinese iconography.

BAMBOO COMMUNITY WORKSHOPS. It's good to gather together to do anything, but especially learning, meeting the new, is best practiced with others. Harvesting bamboo is easier in a group, buying it by

Bamboo community workshop.

the bale is cheaper, sharing discoveries and projects is more inspiring than everyone off in private corners, pursuing personal happiness. So in these reflections on bamboo use, we have imagined not only private individuals, but bamboo neighborhood workshops, bamboo introduction in schools, bamboo camps and travel for training. In its simplest beginnings, all this can simply mean two or three gathered together to work with bamboo or dig and transport and replant rhizomes in pots on their windowsills. When you come across or hear of someone tearing up part or all of a grove, it helps to have several hands ready to plant, and a place waiting to store and cure culms. Work in the local community prepares for work between communities, and between countries: the dimension finally implied by bamboo work is international.

FRIENDS OF BAMBOO. The Japanese teahouse seems an appropriate point of departure in the development of bamboo community workshops in the urban West. In its early "grass hut tea" format with Rikyu (1521–1591), it was intended as a popular form whose skillful embodiment did not imply pedigree blood or inherited fortunes, but a certain constellation of attitudes available to anyone. It was the tiny Temple of Friendship, in which the harmony of the participants rather than the grandeur of the architecture measured the significance of the ceremony.

The depth of a lake
is not in its fathoms
but in its dragons.

The height of a mountain
is not in its peaks
but in its people.

—Chinese Proverb

DO-IT-OURSELF. We have always kept in mind the *bamboo group.* No plant or animal flourishes alone. In traditional cultures, the family is the basic unit, and bamboo was in many areas the symbol of it: a group of apparently isolated individuals secretly single beneath the soil. In leaving the family and village world, modern people have come by certain new sets of freedoms—but we have lost touch with many central human joys, junked as slow-fashioned and out-moded en route to the Supermarket and Freeway.

We have traded a certain sure-durable hum of traditional ways for staccato bursts of sensation and insight, great leaps forward followed by brand new forms of catastrophe, all the dizzy rhythms of modern urban existence.

URBAN EXILE. Our schooling, mainly in books, and our living, mainly in towns, don't encourage us to fully inhabit our bodies. We stand in them less as owner and operator than as tenants who have rented a corner of their total range of possibilities. Psychologists announce that we use some 2–3 percent of our intelligence, and the accomplishments of Eastern yogis make clear that we have reduced ourselves to a similar shallow percentage of possibilities in our bodies as well. We automatically distinguish between the brain and the body, mental and physical, as if that distinction could ever survive except in our own minds. This assumption is but another example of our alienation from wholeness, our exile from Eden. Crafts, building, gardening—all help us reenter the body and earth. Plugging into bamboo could be an appropriate fad for do-it-yourselfers in the United States and other highly industrialized cultures partly because the simple hand tools required are a welcome relief from the high-energy norm in the home workshop bristling with equipment that has to plug in to function.

Tools.

The equipment binge of many people entering a new fascination is a way of avoiding coming too close to creation. Like the best hardwoods, the best tool box expands slowly, with reticence and restraint. Working with bamboo should be scaled down to tools you can carry with ease. A permanent homebase and workshop is a proper center of this edge. "Ponder 10,000 volumes. Wander 10,000 roads." The ancient Chinese prescription for an art braided of tradition and experience, firsthand, of the land, the people, and the materials of

Bamboo Tools

TOOL	USE	RECOMMENDED SPECIFICATIONS
Machete	Miscellaneous: felling and trimming culms and cutting to length; removing fragments of diaphragms from bamboo boards.	Preference of the user decides blade selected; long, fairly heavy blade recommended.
Hacksaw	Felling culms, removing branches, cutting to length.	Large size; ample supply of molybdenum steel blades, with eighteen to twenty-four teeth per inch.
Tripod or trestle	Elevating culms and holding them firm for sawing to length, cracking nodes.	May be made locally, following the pattern locally preferred.
Ax	Cracking the nodes of large culms to make boards.	Lightweight ax with a narrow yet thick, strongly wedge-shaped bit.
Hatchet or small ax	Cracking the nodes of smaller culms for making boards.	Similar to the ax, but smaller in size and fitted with a short handle.
Whetstone	Sharpening tool edges.	Carborundum: coarse-grained on one side, fine on the other.
Spud	Removing diaphragm fragments and excess soft wood at the basal end of bamboo boards.	Long handle, broad blade set at an angle to operate parallel with surface of board.
Adz	Same use as spud, which is more convenient, but the adz is more generally available.	Standard design, best quality steel.
Gouge	Removing diaphragms to make troughs and drainpipes from split or opened culms.	Curved (front bent). 1 inch and 1½ inch bits.
Chisel	Making holes in culms to accommodate lashings for end ties.	Best steel (molybdenum if available) ¾-inch bit.
Drill	Making holes to accommodate bamboo pins or dowels.	Hand or power-driven drill; *metal drilling* bits, best steel, assorted sizes, ⅛ inch–½ inch.
Wood rasps	Leveling prominent culm nodes.	Large size, with one flat side, one convex; coarse, medium, and fine teeth.
Splitting jig	Facilitating the splitting of whole culms or sections into several strips at once.	
Splitting knives	(a) For splitting small culms. (b) For making bamboo withes.	(a) Short handle, broad blade. (b) Long handle; blade beveled on one side only; to be specially made.
Rods of reinforcing steel	Breaking out the diaphragms of unsplit culms.	Suggested minimum: one each of ¾ inch by 10 feet and ½ inch by 10 feet. Other dimensions to meet special needs. Hardwood or bamboo pole may be substituted.
Wire pincers	For handling wire used for lashings.	Conventional type with long, narrow jaws and wire-cutting feature.

an art in all four corners of its kingdom—this design for the incubation of artists should be kept in mind in times ever more centralized and homogenized in experience. The diversity of modern living is mainly in the *content,* which is funneled through formats that abstract it from the sweaty, balmy, or freezing physical world into the comfortable uniform temperature of tv, radio, newspapers, schools of all sorts. Actual physical first-hand experience of the earth is becoming more rare as it becomes more reported and available as information you don't have to alter your life in any serious way to reach. Instead of program or book A, you just select B.

But the issue, finally, of what we make and build is quality rather than quantity, the maker rather than the made. We need tools that make us more human rather than less. In general, the more we *have* beyond the strictly needed, the less space for our necessary existence to *be,* with the necessary physical and cultural mobility *being* implies in its most ample state. Crude tools in master hands outwork the molybdenum of dabblers. The junkmen of China built the tools that built their junks from cast-off horseshoes from England.

We live surrounded by such an abundance of tools that the advantage of a material that requires few tools, and those hand powered and even handmade in many cases, is not so apparent in industrial centers as in the hinterlands where bamboo is most abundant.[1]

Bamboo's high silica content is famous for dulling tools. Tool effectiveness will be increased, time spent sharpening reduced, and work in general cheered by using molybdenum steel or an equally hard alloy. Many a bamboo house has been machete built, but more tools are demanded for more refined work, some peculiar to processing bamboo and therefore unavailable at standard tool sources. The Chinese bamboo tub and bucket maker requires some thirty different tools and gadgets to measure, cut, fit, and assemble his wares. Buy tools as the project requires. Begin simply. Make your own. The list on page 237 provides a fairly complete bag for building purposes.[2]

Techniques.

CUTTING. Bamboo is most easily cut into desired lengths with a hacksaw, eighteen to twenty-four teeth per inch. Molybdenum blades are best. (See pp. 219–220 on removing branches.)

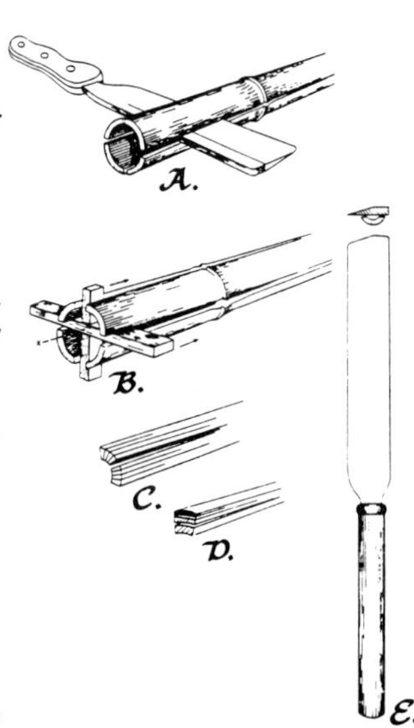

Splitting moderate-size culms to make withes for weaving and lashing. Quartering a culm: (A) Starting four breaches at upper end. (B) Driving a hardwood cross along the breaches to complete the splitting. (C) Dividing quarters radially, making center splits first. (D) Splitting radial divisions tangentially; the hard outer strip is best, and the soft, pithy inner strip is usually discarded. (E) Long-handled knife used for dividing and splitting; some workers hold a strip of bamboo on the blade to add to its effective thickness when they wish to speed up the work.

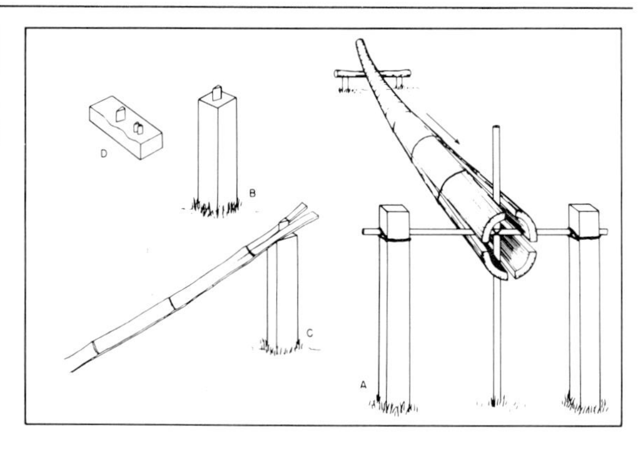

Devices for splitting heavy culms. (A) Cross of iron or hardwood bars (about 1 inch thick) supported by posts (about 4 inches thick and 3 feet high) firmly set in the ground; with an ax, two pairs of splits are opened at right angles to each other at the top end of the culm; these are held open with wedges until the culm is placed in position on the cross; the culm is then pushed and pulled, by hand, in the direction indicated by the arrow. (B) and (C) Steel wedge for splitting quartered culms. (D) Block with single and paired steel wedges for mounting on a heavy bench; adjacent faces of the paired wedges should be slightly closer together at the cutting edge than at the back.

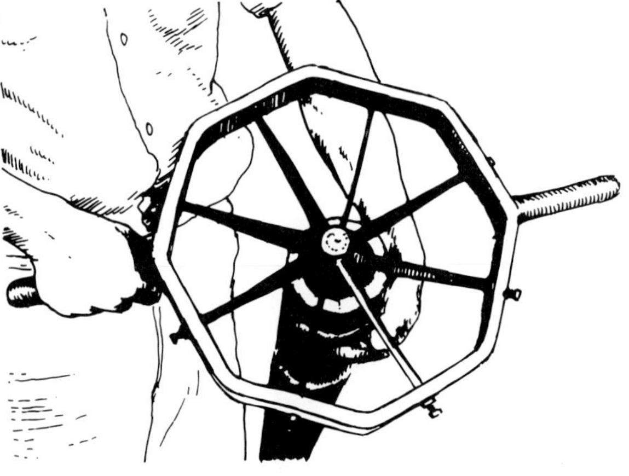

A splitter with changeable blades.

SPLITTING BAMBOO. This is ordinarily done with a good sharp machete. In larger bamboos, the blade is pounded along the culm like a wedge used on a piece of firewood. Ways have been devised to accelerate splitting when dealing with large quantities of bamboo. See illustration. For smaller pieces, for whittling kite frames or small models, use a knife with a replaceable blade. (See Kitebones, p. 253.)

BENDING. For large bamboos, see p. 220 on commercial methods used in India for straightening bamboo. A hot coal campfire is the best way to bend bamboo for furniture or comparable projects.

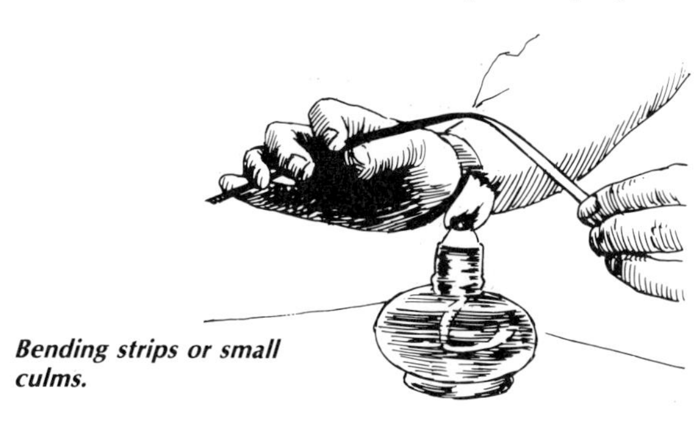

Bending strips or small culms.

For kites or other miniworks, soak the pieces to be bent overnight in water with a dash of ammonia, then tie in desired shape around a mold to dry. You can heat small pieces in a candle flame. The mortar holding the fibers in place becomes flexible with heat and permits bending to chosen shape, which is retained after cooling. Take care not to scorch or burn bamboo by leaving candle too long in one position. Try using a bucket of hot sand to shape small pieces, as eyeglass doctors do to shape plastic frames. Don't force the bamboo's pace, nor try to bend it too far, or you'll crack it.

JOINERY. Bamboo, hollow and round, demands different systems of construction from those used in building with wood. Nails split bamboo unless gently hammered into holes drilled with a diameter slightly less than their diameter. For lashings, galvanized wire, strong rope, nylon cord, vines, or bamboo strips from the tough culm surface are used.

COLUMNS AND BEAMS. Angled "ears," easy to cut with a machete, are used in walls, footbridges, or wherever a vertical bamboo supports a horizontal member of lesser diameter (1). One-eared supports in wall construction should have the ear outside, to brake the shove of the roof's weight (2). Two ears, cut neatly with a hacksaw, are an option to (1) above (3). For larger horizontal members, one or two ears can be added with small pieces of bamboo (4,5). The ear should be roughly ¼ the full length of the piece. Double columns to support a large horizontal beam are cut as shown, with a strip of bamboo or wood added beneath to distribute the load (6). Double beams can be used for large spans or heavy weights (7). Four horizontal beams can be used for especially heavy construction, or when larger bamboos are not available (8,9).

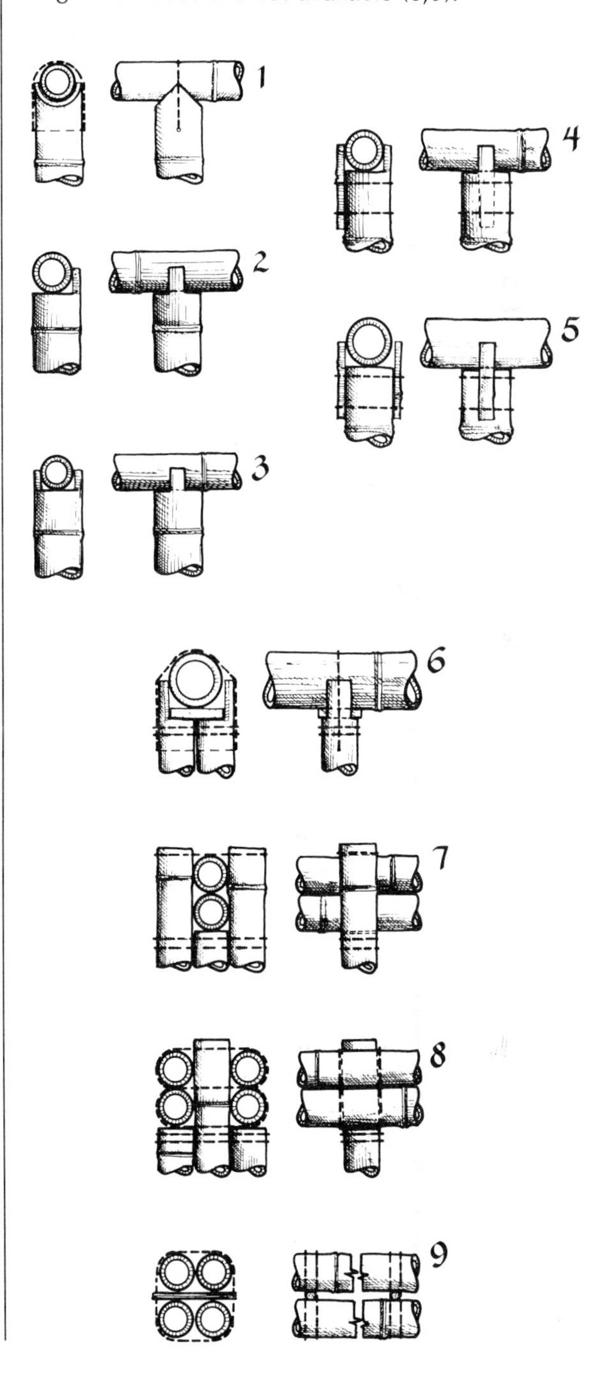

T JOINTS. For use in fences and walls (1). Lateral hooks can be used (2), or a pin permitting easy removal of the horizontal bamboo in gates or other constructions (3). T joints in corners (4). The horizontal members should not support much weight. Pins, of wood or bamboo, can be used when the beam is not under much weight (5).

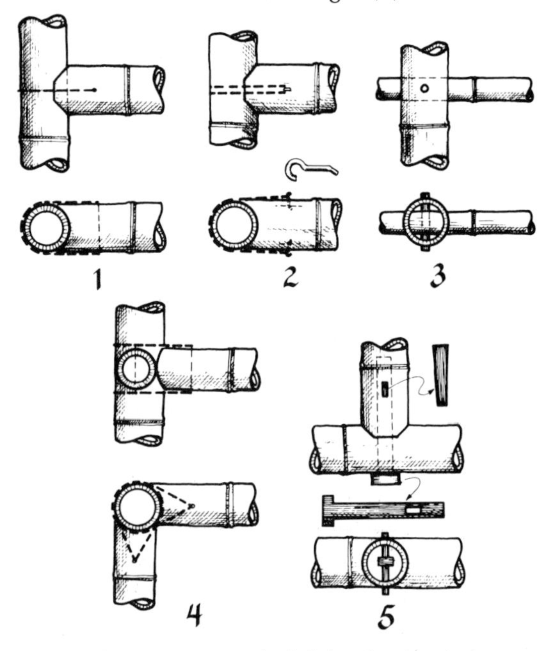

RIGHT ANGLE UNIONS. In joining horizontal to vertical members, the load-bearing capacity of horizontals can be increased by lashing on an external support, doubling the horizontal member, or both combined (1,2,3,4). Unions of joists and rafters in roofing permits various possibilities (5,6,7).

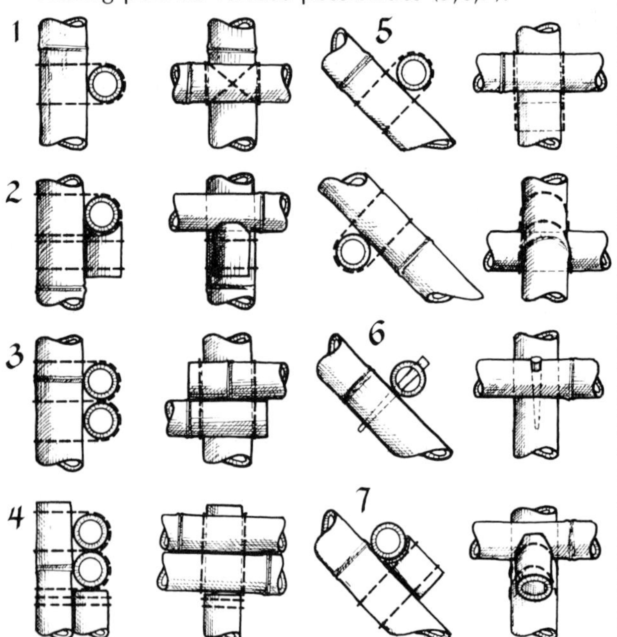

JOINING TWO HORIZONTAL MEMBERS. Of the seven alternatives, 4 is especially useful in furniture making, 5, 6, and 7 for bamboo pipes.

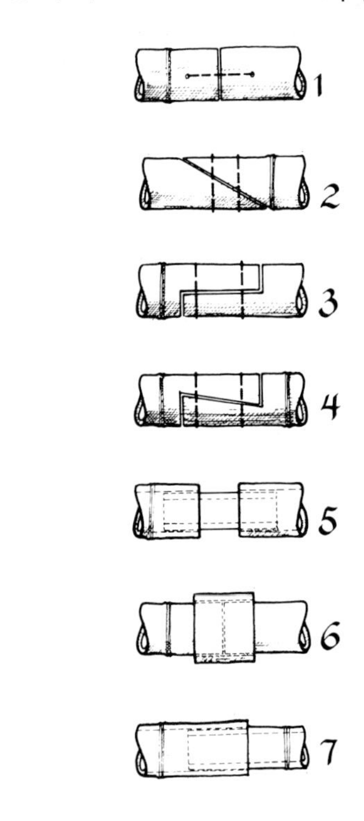

FURNITURE. Glue is required for 4 and 5, a plug of wood or bamboo for the latter.

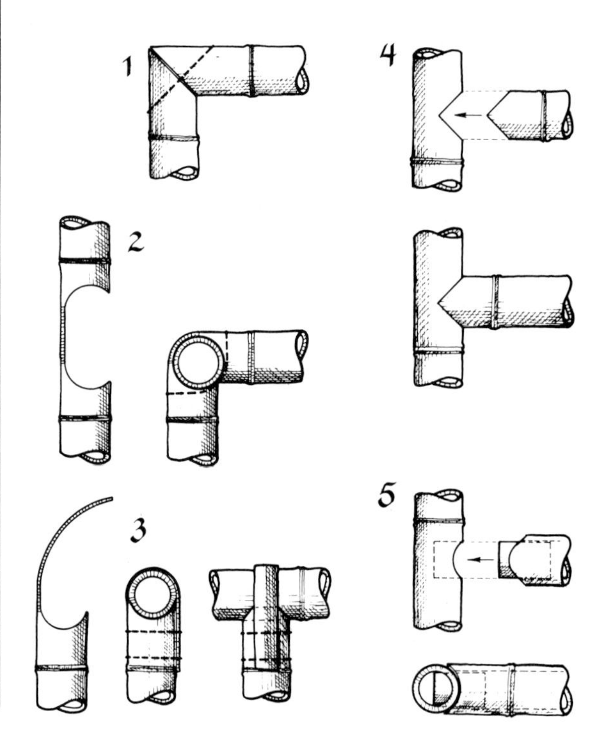

Knots and lashing.

The stability of a bamboo construction depends on the strength of the material itself; the right tight knots; and the rope, wire, vine, or bamboo used for lashing. The following are useful for all types of construction, the scaffolding, and vertical lift of construction materials. Hidalgo, the most complete source in this matter, provides 15 pages from which these have been selected.[3]

RIGHT ANGLE LASHING. This begins and ends with a clove hitch (see below).

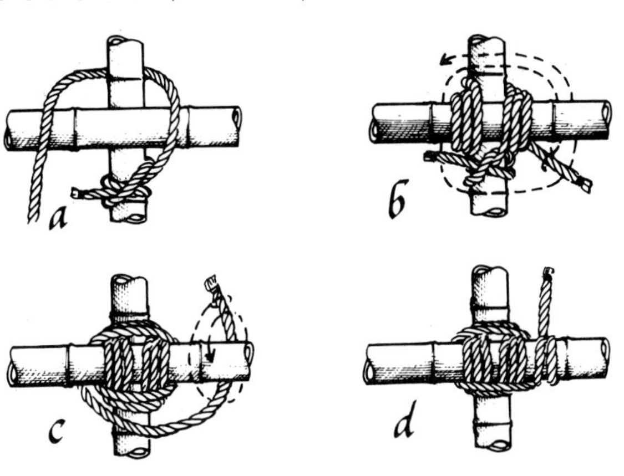

DIAGONAL LASHING. Used for diagonals of bridges or scaffolding. Begin as indicated and conclude with a clove hitch.

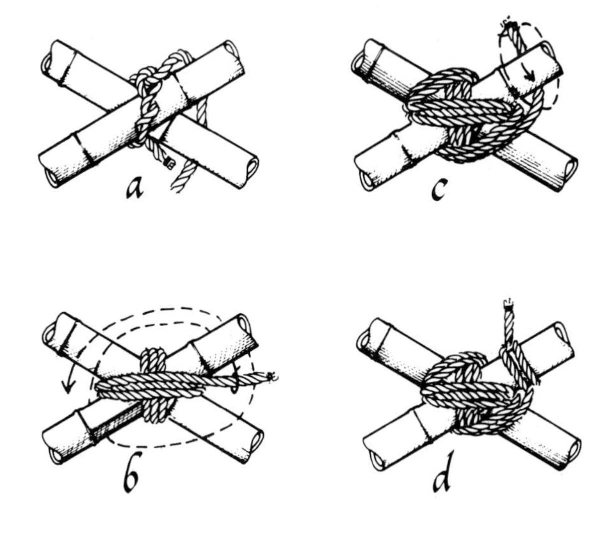

SCISSORS. To tie two bamboos together to hoist weights, the poles are placed parallel, a knot tied around #1; 7–8 turns of rope tied horizontally around both (a) are followed by 2 vertical turns (b). The rope is tied off around #2. Lashing must be loose enough so that legs open to required distance.

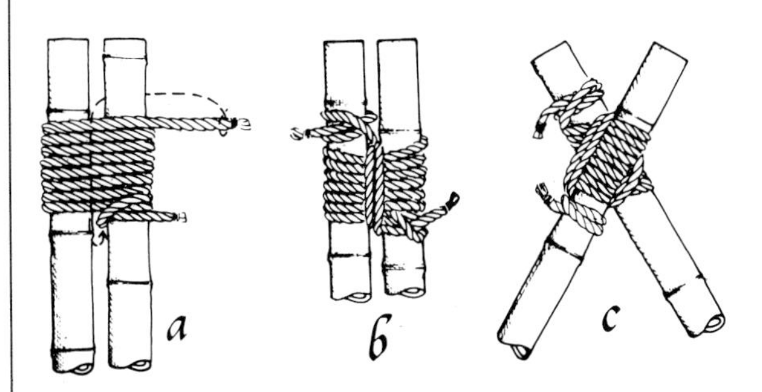

EXTENDING LENGTH. To unite poles at their ends to form a longer structural member, the same lashing is used, but tight and further tightened by inserting wedges of wood or bamboo. The final knot is tied around both poles.

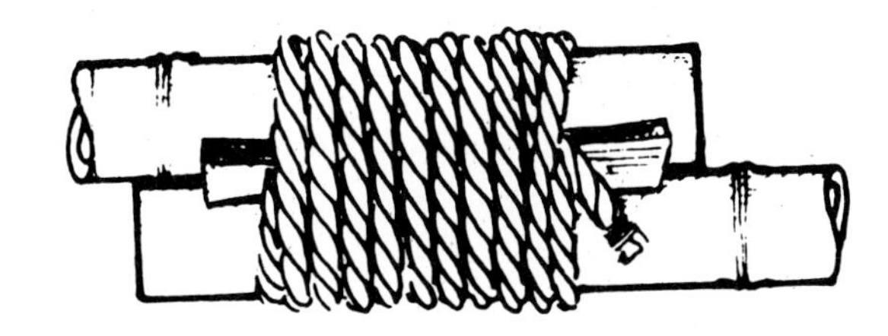

TRIPOD. Rope is tied to one pole, then woven horizontally around all three, as shown; then, before or after opening, it is woven or wound vertically through or around horizontal lashing and tied off. Lash loose enough and tight enough so legs just open to required distance.

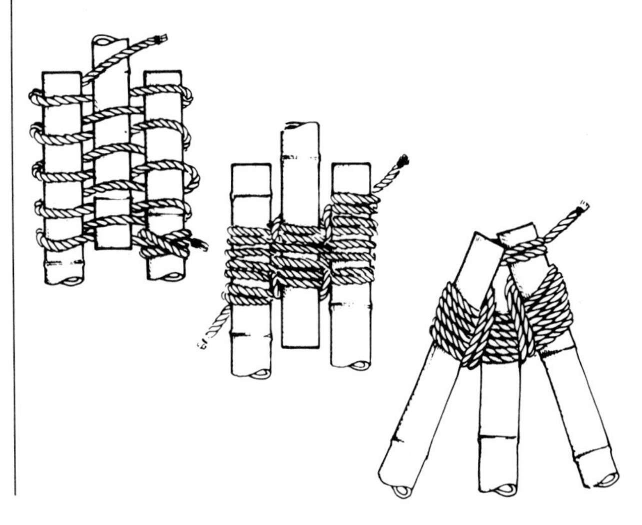

SCAFFOLDING. For hanging scaffolds, the rope is lashed as shown (a–d) and tied off (e). Other options to extend—but far from exhaust—your possibilities: the hangman's knot (1); a figure eight running knot (2); a running knot used also for vertical lifts in two versions (3); a knot hard to untie (4).

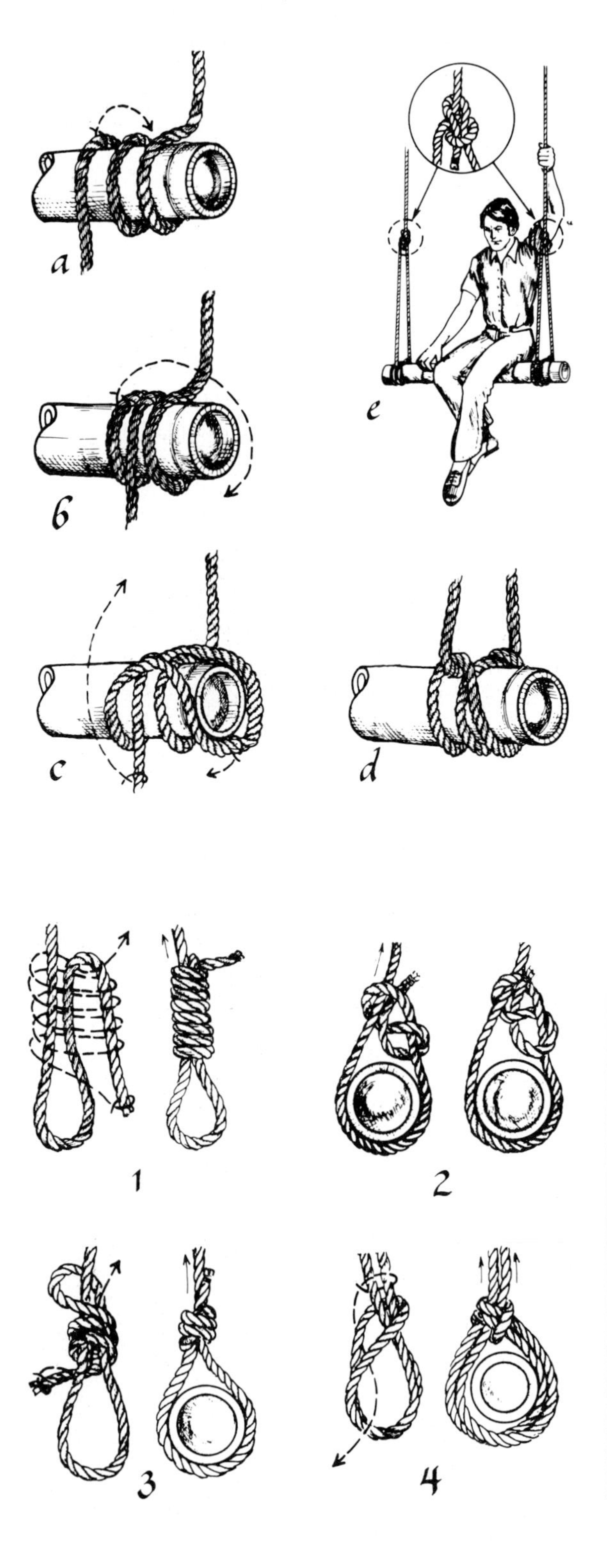

LIFTING MATERIALS. Three ways are shown, for use with or without a hook. Knot (3) under scaffolding can also be used.

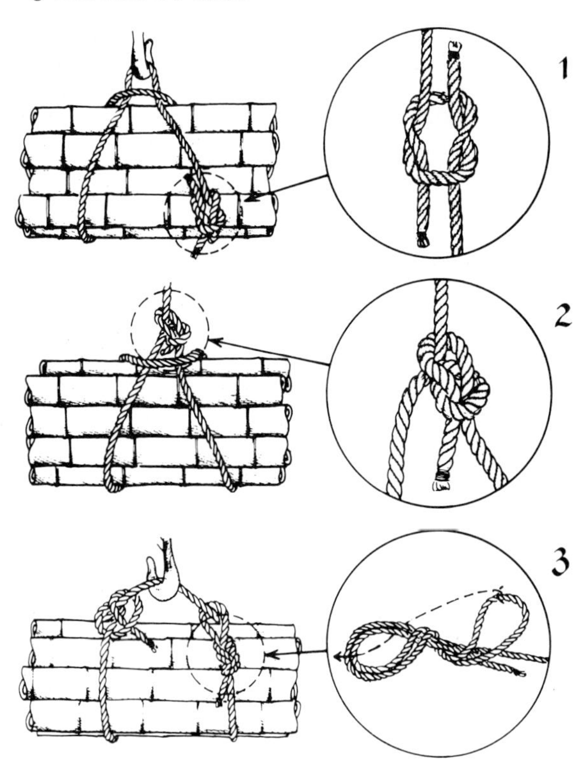

REDUCING ROPE LENGTH. Used to increase tension.

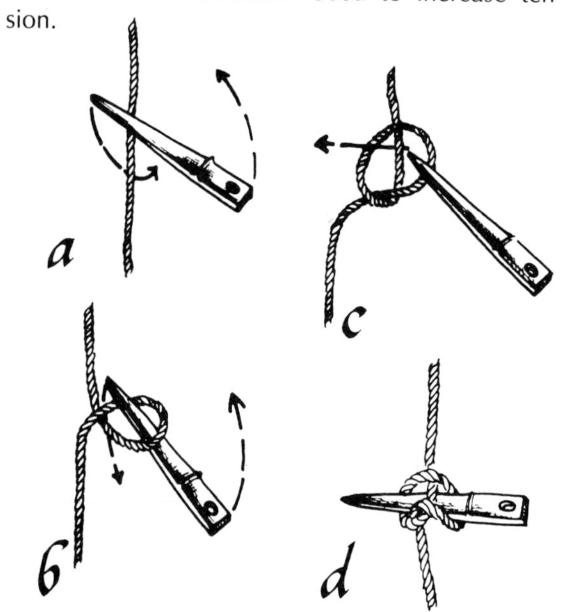

CLOVE HITCH. A basic knot with many uses in bridges, scaffolding, and turnbuckles; also useful for starting or concluding other knots (1). Another method is used when the top end of the pole is too high to reach (2). The knot can also be made first, and the pole inserted (3).

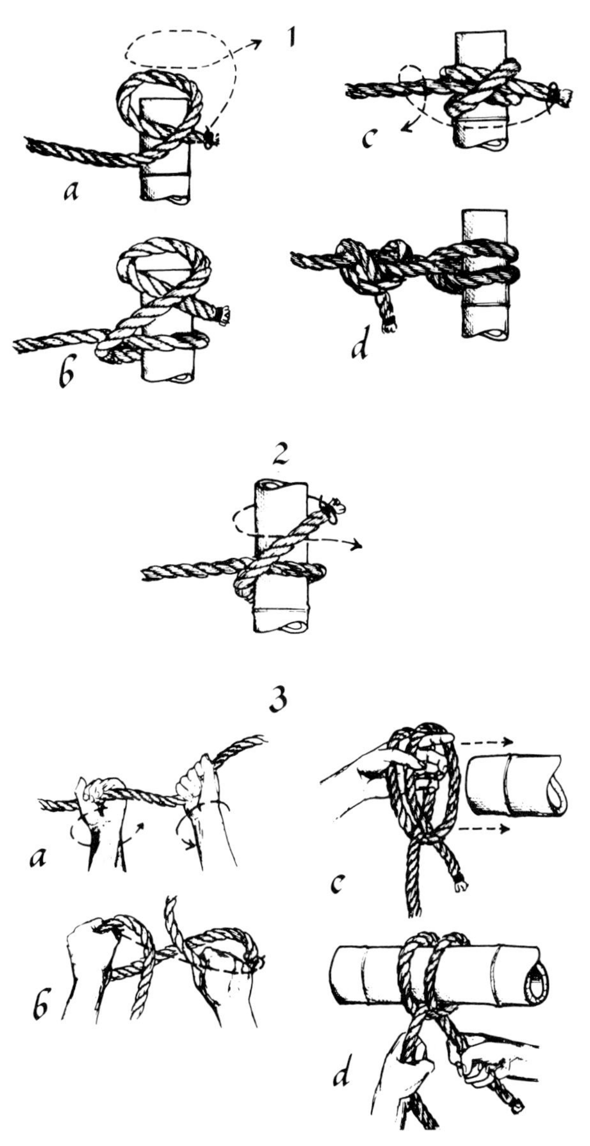

The cracking problem.
"In China and Japan where bamboo is an everyday commodity, humid climatic conditions both indoors and out are such that splitting of bamboo culms is a very minor problem. Not so in the United States, where in our dry overheated homes particularly, bamboo articles, when made from whole culms at least, are regularly subject to the danger of splitting."[4]

"Splitting of bamboo furniture results from uneven moisture absorption and loss in the tissues of different density in the bamboo. *Bambusa vulgaris,* for example, is well known for its tendency to split; it has soft pithy tissues on the interior while the outer tissues are much more compact and absorb much less moisture from the air. Fortunately, species with more uniform culm tissues are less susceptible to splitting. The problem is lessened by use of these species, or types of construction which call for split bamboo. Designs can sometimes be carved in the hard outer surface which permit expansion or contraction due to moisture absorption and loss."[5]

Seasoning behavior of nine species was tested in the round by the Wood Seasoning Branch.[6]* The study on seasoning behavior in air, kiln, and over open fire revealed that bamboos in the round are generally liable to considerable degrade in the course of drying in the form of surface cracking, splitting, collapse, and sometimes even fungal decay and borer attack. Air drying is the most practical and satisfactory procedure provided care is taken against too rapid drying in hot and dry winds, which aggravate cracking, and sluggish drying, which causes fungal decay. Air drying is completed within about three months in clear weather. Of the nine species, *B. arundinacea, D. hamiltonii, D. membranaceus,* and *D. strictus* dry with relatively little cracking or collapse in mature culms. Immature culms are liable to collapse. The rest of the species suffer severe collapse and excessive shrinkage in drying besides cracking.

"Sharma in 1972 successfully conducted trials on chemical seasoning of round bamboo for handicraft items without any cracking or splitting. The species used was *D. giganteus.* The antishrink chemical polyethylene glycol-1000 in a 50 percent solution was used for a soaking (diffusion) pretreatment of the bamboo wall in the green condition, after which no cracking occurred in air drying, which is a problematic proposition in round bamboo."[7]

To reduce the likelihood of splitting, keep bamboo objects from direct sun. Don't keep valuable bamboo objects in the open in a heated room. Put them in a glass case with an open container of water or store in tight containers, preferably in a room with 50 percent humidity. Lutz (1975) notes the propensity for brush pots to crack on the uncarved side, which suggests that carving breaks the tension, gives room for fiber expansion and play, thus reducing the tendency to crack.

Splits that occur can be filled with lacquer, glue, paint, wax, or epoxy. In case of a wide crack:

**Editor's note: B. nutans, B. polymorpha, B. tulda, D. calostachyus, D. longispathus,* and the four species resistant to cracking listed below.

put brush pot, flute, drum, or whatever bamboo object in a covered bucket with water in the bottom. Rest bamboo on brick above water level overnight or longer. The crack may narrow enough to glue and bind. Tendency to split depends in part on age of the culm at harvest. "Unripe" bamboo—under five years at harvest time—was four times more likely to split than "ripe" five-to-ten-year-old culms in Trinidad experiments.[8]

In South Carolina, Clemson College tests on bamboo studied the absorption of water-and-oil-soluble wood preservatives by bamboo culms to prevent splitting, among other problems.[9] Freeman-Mitford suggests soaking bamboo generously in linseed oil to avoid splitting.[10]

Kids love bamboo.

THE EFFORT/EFFECT RATIO. A two-year-old swings a door on its hinges, open and shut, again and again—amazed to move so easily something so much bigger than myself! Astonished such little effort yields such great effect.

The effort/effect ratio is also high in bamboo, which may be why the reaction of children to the plant is immediate and intense: a six-month-old baby can wave an 8-foot pole. A two-year-old shouldering a 20-foot culm grins with a sense of accomplishment. With a bundle of a hundred 6-foot by 2-inch *B. textilis* poles, five-year-old preschoolers can autoarchitect their own playground.

In Java, treats are hung on strings around nose level for children on bamboo stilts. Fruits and candies must be bitten off, without use of hands.

BROKE IS BEAUTIFUL. Schools everywhere are passing through a severe economic crisis. But in institutions intended to make people smart, broke is beautiful. Indigence is the great grandmother of invention. "Guerrilla economics" means inventing projects that pay their own way so the ticket is built into the trip. Educational systems have to increase designs for productive homework that then earns the cost of the course. Bamboo propagation or construction can diminish expenses and even create income—or useful home furnishings, as in bamboo school projects in the Philippines. Bamboo plants are particularly in demand just now, if we can trust the reports of plant sales at botanical gardens, so a bamboo nursery might at this moment make both dollars and sense, neatly combining earning and learning.

LEARNING DESIGN: NEW SCHOOLS AND STUDENT-CREATED CURRICULUMS. The explosion of numbers on the planet demands that we learn, rapidly, a planetary culture. This involves inventing new ways to pass culture on as efficiently as possible.

Bamboo childhood.

"I was born and raised in a house surrounded by bamboo. All my earliest memories are enmeshed with its sights and sounds: the changing seasonal colors; the whispering of leaves tossed by wind; the sharp cracking of huge culms, weighed down by heavy blankets of snow. In the narrow valleys, there wasn't a village that wasn't surrounded by groves. Even the poorest child could not help but become intimately aware of bamboo. The children played in the groves, they watched the harvesting of shoots; they saw brokers who came to railway station towns to bid high prices for the coveted sheaths of the madake bamboo *(Phyllostachys bambusoides)* once they had fallen. Many children would sneak into the groves, bundle sheaths, and sell them to their landlord, who in turn might sell them to the local butcher to use as wrapping for his meats.

"Making bamboo dragonflies, insect cages, and flutes were winter pastimes, and almost every household had children who became skilled in these crafts. My father was a carpenter, and his winter hobby was making shakuhachi from madake bamboo. After selecting the proper culm and getting permission to cut, he would have to age the piece. Then came the painstaking labor of carving, cutting, chiseling and polishing. From beginning to end the entire process took almost three years to finish one shakuhachi. I observed my father's great love and reverence for each piece of bamboo he handled; I can still remember it clearly today."

Tsutomu Minakami (Takama 1983: Introduction.)

The hoop is a toy that girdles the earth. Laminated bamboo strips can make one with plenty of spring for optimal hop in the hoop.

The old educational models are hierarchic: one teacher in front of a roomful of students. You don't have to work it out on the blackboard to notice that 1 faucet + 100 buckets = wait in line. The old design can serve as much to restrict information flow as to encourage it, as many of us have experienced in conventional schooling. So we are now all students of the homework presently assigned our species by our planet: Design an educational system or non-system that teaches people to teach themselves; phase out the hub-spoke model of the teacher's individual talent or data heap, and phase in the creative collective imagination of the taught.

More projects must be designed that turn the students loose with their own capacities to interact with their fellows in a cooperative rather than competitive manner. This frees up the teacher to wander among the students as a gentle guide, dealing with them one by one as a more experienced equal rather than directing them en masse from a distant height. Bamboo fits this decentralized, more active and interactive model admirably.

The diablo, an ancient Chinese humming top, was a smashing success when introduced in late 18th century England, and has remained sporadically popular in the West ever since.

A French print (1812) demonstrates the diablo dance.

THE PROPER STUDY OF PEOPLE IS PLANTS. Bamboo also recommends itself as an agent of revegetation in a time of ecocrisis on the earth when the homework of all children should be to plant and recover the planet with its ancient green blanket, snatched from Mother Earth in the dark night of industrialization. Now more than any time in history the proper study of people is plants. Maria Montessori calls the students of her dream-school *Erdkinder*—Earth children—because their education is based on agriculture, the basis of all culture and culture's earliest job.[11]

LOW-TECH CREATION COMPLETES HIGH-TECH CONSUMERS. Surrounding customs and personal taste confine us to routine mirrors of who we are already, but we are really completed, complemented, by our opposite. A society of high-tech consumers needs low-tech doing and creation to balance itself out. We don't need weekend amusement innovations that require more power tools screaming in the bored basements of a million home workshops. Untamed, unmanufactured, nonelectric, nonpollutant, lots of oxygen in abundant leaves—bamboo is an upright natural high that, broadly embraced in industrial cultures, could help unplug people from current, ease neon-withdrawal, reduce useless motion, and maybe even diminish driven driving in Americar, more full of distance than destinations. Genuine growth in a hypermobile world sometimes comes most readily without movement. Loaf and

Rattles of bamboo, for the newly born and for percussion instruments; try beans, grains, pebbles, screws. Where thick- and thin-walled bamboos are available, let children experiment with different sounds of different species, and with sound progression in internodes of varying lengths.

invite your most intimate creation. Carve a small object for neither applause nor profit. "Have the innocence to enjoy yourself."[12] Stop lashing your life to increased achievement and start lashing bamboo.

Prayer of a bamboo bum.

Preserve me, Buddha, from distress
of riches, wisdom, or success.
Grant me, rather, an immense
and creative negligence,
while my dry detractors flee
about with ulcered industry
till, plagued by Profit, they confess
the opulence of Idleness.
But let them wake too late to mend
their anxious, honored ways, Amen.

FRIEND TO CHILD FINGERS. "As iron sharpens iron, friend sharpens friend," according to an Old Testament proverb. Bamboo, the "friend of the people" in Chinese folklore, has indeed sharpened the skills of many oriental cultures. For centuries in Japan it honed capacities for patient repetition of detailed processes that demanded rigorous exactitude. The deft hands and minds that matured making finely woven baskets and minutely ribbed fans, were, in the twentieth century, designing electronic circuit boards.

Although one of the main teachers of the Orient, this friend is largely unknown in Western education. Through the influence of educators like Montessori, who demonstrated the imbalance of excessively bookish schooling in an Italian slum, *things* are finding their way back into the educational process to counterweight the dominance of words and print. "Words indeed are excellent, but things are still to be preferred," as William Penn opinioned in his seventeenth century reflections on education in the New World. If things are missing from schools, let the first thing be . . . providing them. Planting bamboo and other crops that can be incorporated into our curriculums would seem self-reliant homework to encourage.

Chasen. Powdered tea is stirred in Japan with a tea whisk made from a culm about an inch in diameter split just above a node into as many as 120 fine tines, delicately arched. The chasen is witness to the trim fitness of both the bamboo fiber and the fingers shaping it.

Some experience in gardening becomes increasingly vital for populations increasingly urban. The more our feet are exiled from earth on cement, the more urgent it becomes for health and sanity to spend at least some time with our hands in the dirt. Modern city children are starved for this experience. Neither their need for nature nor their need for work is adequately nurtured by the usual schooling in most industrialized countries.

We have mentioned a Western botanist's observation that a "Bamboo Age" preceded the ages of stone or bronze. Primary grades can reenact this cultural evolution for the child by providing bamboo as one of the earliest materials that unskilled but eager young hands can learn to work, after paper and before wood. Wood is a harder material than bamboo, requires heavier tools, and implies a greater age in the craftsperson. If you take a stroll around the Global Village, you'll notice that children of basketmakers are helping their parents sooner than the children of carpenters. There are years of schooling before children begin "shop" when they could be tooling up their craft capacities with bamboo. And the groves supplying culms for their crafts will also offer summer shade and winter windbreak, as well as a mellow green softening the sharp corners of whatever buildings they surround.

"Bamboo is one of those providential developments in nature which, like the horse, the cow, wheat and cotton, have been indirectly responsible for man's own evolution" (Porterfield 1933). Like the embryo in the womb, reenacting the physical descent of our species, growing and shedding gills and tail, the child in school relives the cultural evolution of the race. The complete absence of any knowledge of bamboo in our schools is an unfortunate gap in this theater of our collective past: unfortunate for our culture as well as for our children, because the educational process is perhaps *the* key point where the useful beauties of bamboo could be introduced to the West with maximal long-range impact.

Megaschool.

Only quite recently have people gotten the idea of educating everybody in a culture. They've gotten it,

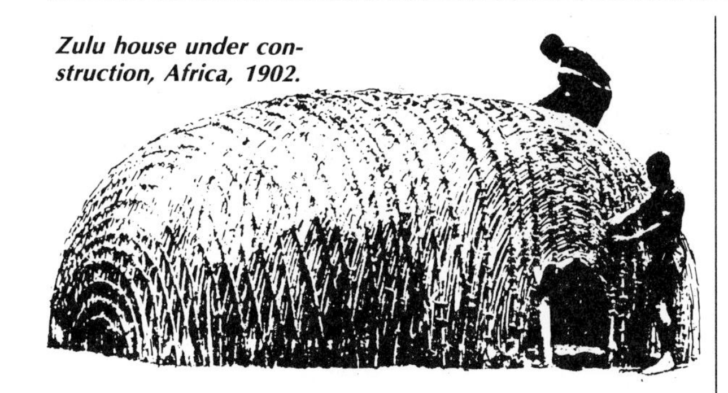

Zulu house under construction, Africa, 1902.

in all countries of the world, at a time without precedent for the closeness of human packing on the planet. This morning more people got up and went to school than at any time in the whole past history of the world. We need roofs over the heads of these millions, materials for these hands to work. What shall it, what can it be if we want to design for everybody? What is a feasible physical embodiment for school ideas? As wood and metal become more scarce, bamboo becomes more apparent as an alternative for many school systems going broke: to grow their own . . . or buy from a more southern school.

We suggest beginning with models because they require little material of good quality. Scrap or broken pieces of bamboo shades and screen and other artifacts found free on the street in large cities are one source; thrift stores are another. In models, any size of any bamboo, even some common dwarf bamboos fit for little else, can be used. This cheapness encourages a relaxed exploration of the material, a period of getting to feel and know bamboo in the hands, working with it in a small space like the kitchen table or, better, in oriental fashion, on the floor in some corner. Best, if you have a yard and space available, is a tiny shack of bamboo where you can keep the few tools you need to work it. By sitting on a mat on the ground, oriental people acquire an extra pair of hands, their feet, to grip bamboo when they're dealing with larger pieces.

Miniature magic.

ARCHITECTURAL LITERACY THROUGH DWARF DWELLINGS. We're no longer satisfied with a few professional readers in a village, and the rest listening. We should encourage, in our schools and lives, the practice of architectural literacy. Standing under dripping roofs, behind windy walls, crowded into dehumanized spaces, we're waiting for architects, construction workers, city planners, and real estate agents to take care of our shelter needs. Since housing is a primary need, architecture should be a primary and universal study—from preschool through adult education.

Making dwarf dwellings, models of schools, playgrounds, towers, and bridges, on a scale of 1 inch to 1 meter, we discovered how people are drawn to the diminutive. Stopping at the workshop window to study the little bamboo knobs on the bamboo doors swinging on coconut leaf hinges, people of all ages broke into appreciative smiles. There was much interest in nearly everything the workshop produced, but the miniature obviously had an extra special magic for everyone.

Working with small models requires little bamboo per project—of significance when working in

Nyakusa chief's house, southwestern Tanzania, c. 1900. The wall slope—outward here, inward in the Kinga house following—completely alters the external appearance and the inward feeling of space.

groups, schools, or community workshops with bamboo. Errors waste less. The careful measuring and cutting required are excellent preparation for later, larger projects with bamboo or other materials, for precision and attention to detail in general. Tools and supplies are also simpler and cheaper: a machete to split, a hacksaw to cut pieces to length, a file or sandpaper to refine cuts and smooth edges, occasional use of a drill and a few small bits, a ruler,

Kinga house, southern Tanzania, c. 1900. The trimmed bamboo rhizomes at the tip of the roof presumably represent ostrich heads. "Ostrich eggs were widely used to decorate apexes of roofs and were sometimes reputed to act as lightning insulators." (Denyer)

shoemaker's glue, some oil or varnish for finish, some strong thread.

Old bamboo curtains or broken baskets are frequently out on the streets in large U.S. cities. They are an excellent lumberyard for making small models. Keep your eyes out, pick up any bamboo you see, prop in a corner until inspiration strikes. In yards or parks, offer to clear out dead poles. You will soon have a regular harvest route. Get to know the gardeners in local parks and ask them to alert you by phone or card when a clearing of the groves is scheduled. Bamboo is often just thrown away if nobody's asking for it. If you yourself have excessive riches of bamboo, don't call the dump, call your local school. Children can harvest and carry it off—along with a few rhizomes.

A workshop in model making with the *Instituto*—a grade school–high school—in Rama, Zelaya, Nicaragua, led to the students preparing an exhibit on possible uses of bamboo in Nicaraguan education, both as material for classroom construction and for use in the curriculum in a variety of ways. Among twenty-five hundred exhibits prepared throughout the country by some eighteen thousand participating students in an appropriate technology science fair, the bamboo project came in third.

The "house" is a primordial toy, naturally fascinating to the child. How can we wed this fascination to bamboo, a primordial material for houses? From scale models to playground structures to child- and adult-sized schools or houses: This would seem a good teaching progression. California high school students are now building suburban houses. There is no reason students can't learn to build their own schools.

Pre-school Building Project.

Birds beat us to it in song, flight, and shelter. The youngest children can experience the pleasures of building in structures that are most nestlike. Here's one from Denyer's (1978) densely, wonderfully illustrated *African Traditional Architecture.* Cattail and any thin, pliable bamboo—*Pseudosasa japonica* is a common species that would do well—could serve as thatch and frame. A woven bamboo structure such as this could also support the cement-gunny sack construction mentioned in Chapter Four. Begin with a small model, about a foot in diameter, then one perhaps 4 feet in diameter and 3 feet high, just large enough for one or two kids to creep into. Work up to maybe a 9-foot diameter, building miniature bamboo scaffolding as necessary. Alexander's (1977) *A Pattern Language* is to be recommended, perhaps more than any other single volume, for inspiration in teaching the bones of architecture to children young enough for them to grow up with a sense of the possibilities of humanized space in their bones.

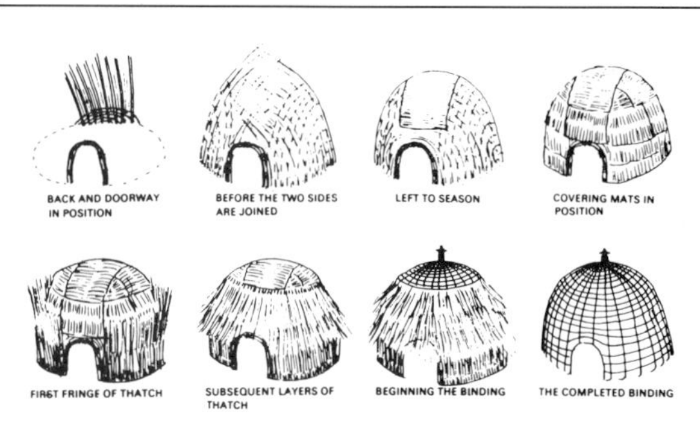

Swazi house construction, Swaziland, c. 1945.

Fulani roof interior, Guinea, 1956. "The apex was filled with a basketwork panel." (Denyer)

SHELTER STUDY: AGRO-ARCHITECTURE. Teaching architecture in schools implies some attitude toward the traditions of architecture. What will we set out to do, exhibit slides of the imperishable beauties of the Parthenon to a restless audience of sixth graders? Bernard Rudofsky, in *Architecture without Architects* (1964), voices an articulate departure from academic consensus. Instead of focusing on the familiar set of world wonders, he examines the virtually unexamined: popular anonymous architecture around the globe. His defense of the wisdom of this perspective, condensed above (pp. 115–116), contains hints for the design of a curriculum for architecture in grade schools and high schools that is quite different from the standard fare of university architecture departments. Architectural beginners should be taught the beginnings of architecture.

In the context of a renaissance of popular architecture, bamboo is especially relevant. As the earliest component of popular architecture in Asia, bamboo development is an appropriate tactic for agro-architecture in a broad temperate-tropical belt around the world. Putting housing back into the hands of the people implies and depends on first putting the proper plants back into their consciousness, gardens, forests and fields, for hardwood foundations, bamboo framing, and the best local species for thatch.

THE HOMEMADE HOUSE: MAIN PLANET TOY. "We have to cross the frontiers of our civilization into what until recently we used to call the world of savages to encounter children untouched by industrial toys and mechanized entertainment; whose curiosity has not been deadened by programmed activities; who never felt the breath of the child psychologist on their neck . . . With Nature as their inexhaustible toy shop, children in primitive societies get far better ideas of how to have a good time than the most zealous kindergarten teacher. 'The children in Uganda are very well behaved,' noted an explorer; 'I was struck by the way in which they amused themselves. Instead of making senseless mud pies, they made miniature villages which were almost exact copies of the dwellings around them. They would be thus employed for hours together, day after day, and would persevere until their models were completed.' . . . Kpelle children build houses which they furnish with household things made by themselves . . . Their power of observation is uncanny, their devotion to what they are doing total. Transition from work to play is imperceptible; when the children accompany their elders to the fields, they are given a small hoe and sickle to play with. It was Plato who counseled parents to 'give real miniature tools to those three-year-olds.' In Africa and Oceania, wherever the native life style is still intact, children make their own tools. On Borneo and the Gilbert Islands they build their own houses and canoes. 'The natural imitative skill of the primitive child knows no bounds,' says Miller. Ethiopian children playfully set up small-scale huts and corrals and stock them with snails, shells, and pebbles that are their cattle, goats, and sheep. The children of the Warega, a Congolese tribe, do not only build entire model villages but act out an accelerated version of a full day's community life. Preparations begin on the eve. In the morning the children proceed to a river, catch fish, and roast them on fires in front of their own little houses. This done, they go about their work much like their parents. At a certain moment, one of them calls out, 'It is night.' Whereupon they retire and pretend to sleep. After a while another one imitates the crowing of a cock, they awake and recommence their work."[13]

Ethiopian children at play. A school is a womb, and schooling in essence provides an accelerated reliving of the cultural life of the race.

"Japanese and other northern children build snow houses. The snow is cut into slabs, 4 inches thick and 15–20 inches wide, and set up in an ascending spiral to form a domed roof. The chinks between are filled with soft snow which hardens in ten minutes. Three skilled kids can build a snowhouse 9 feet in diameter and 6 feet high in about 45 minutes."[14]

PLAY MAKES COMPLETE PEOPLE. "The privileged children of Europe, by way of contrast, have been presented since the sixteenth century with the doll's house, complete with parquets and stuccoed ceilings, wallpaper and embroidered rugs, veneered tables and chests of drawers, chairs upholstered in silks and brocades, quilted bedspreads, glass chandeliers, marble fireplaces, and gilt-framed mirrors of Lilliputian proportions. Too precious to be touched by clumsy little paws, its function was reduced to a showpiece."[15] The German poet Schiller maintained that "play alone makes man complete. Man plays only when he is in the full sense of the word a man, and he is wholly man when he is playing." But prefabrications of the international toy industry miss the inventive spontaneity at the heart of true play. Schiller's later compatriot, Rilke,

Early bamboo shelter and baskets.

Early house history reinforces interest in the building process, as well as increasing confidence that we can do it ourselves. A glance at early shelters provides hints on how a bamboo camp could auto-construct itself with materials found or grown on location and shaped by the active energies of the sheltered kids.

POTS BEGAT HOUSES. As early as 7000 B.C., prehistoric peoples living in caves and in excavated holes crudely roofed with branches were already competent potters. The size of early pottery dwarfs later ceramics; large pots were used for storing oils, wine, grains and dried fruit, and as coffins. In time, people saw that they could not only cook their soup and keep their dead in their crockery, but live in it as well; the crock became the cottage. Wattle and daub construction evolved from basic basketry techniques combined with pottery. Uprights are warp to a weft of withes lashed with vines and coated on one or both sides with mud, the most abundant building resource in the world.

BASKETS BEGAT POTS. Baskets, even more ancient than pots, provided one early method of making pots. They were coated with clay and fired, consuming the basket and leaving the pot.

Hidalgo (1974) gives the most detailed guidance for applying this ancient technology to cement over a basket for water storage and to the use of bamboo mesh as reinforcement in building panels.

lamented that playing with dolls and doll houses provided an "initiation into the rigid passivity and emptiness of life." With ready-made toys, Rudofsky concludes, "the child picks up ready-made notions and acquires mental deformities that later put him in good standing with society."[16]

KID-BUILDABLE PLAYGROUNDS. Superlight, kid-liftable bamboo can be used as mobile components around wood or metal constructions in a playground. Instead of a fixed space, imposed once and forever by adults only, the playground could become more fluid, with children's own creative imagination at last welcomed as an ingredient in the design.

Tipis, pyramids, bridges, ladders, towers, mini-houses, climbers of all sorts can easily be made with bamboo. Climbing is a need deep in the feet and hands of children—a denied need or inadequately met in our urban architecture with its tragic absence

Bamboo observation perch.

Try building a chair with the seat about the height of your neck, rungs on the sides a little higher than your ankles, and others in the front and back a little higher than your waist. Climbing up into it whenever you want to sit down will give some notion of how uncomfortably we design our houses for our children.

Our architecture could be adjusted in many minor ways with major impact on a child's learning. One important example: children are fascinated by food preparation and many other forms of adult domestic work. But we provide no elevated space for them to watch it happen. An observation perch can readily be constructed of bamboo or whatever available material. It can be, in fact, the giant chair we began describing above, but built to facilitate ascent rather than frustrate it, a seat-tower light and small enough to move easily about the house to witness all action occurring at levels not conventionally available for inspection to those with eyes at the height of our knees. Ordinary high chairs are good substitutes for a really giant chair, but they aren't usually designed to be climbed, and lack the dramatic extra foot or two we have in mind: they are for sitting with adults, not for a panoramic vision from which to survey what's going on in the room. Infants in them are frequently arching their neck for a taller vision. All this may seem a farfetched suggestion, but if you have children in your home, make height available at will to them—and watch them use it constantly to watch. There is almost no simple, cheap piece of furniture we could introduce that would more transform the home for children than a giant, towering chair. Children of most traditional cultures are spared this vertical exile. Food preparation and most else still happens on the floor, readily accessible to be explored. Maybe the "generation gap" has something to do with our "norm" of chairs and tables that, through all the earliest years of their most intense capacity to learn, leave our children wondering on floor level what's going on. If we want to remain close to them when we are old and grey, we must let them come close when they're young. The house, the space we share, is the main tool we have to accelerate their learning in their early years, but domestic architecture rarely escapes the ghetto of adult-mind. Alexander *(A Pattern Language)* is, lamentably, *un*characteristic of modern builders in his attention to these themes.

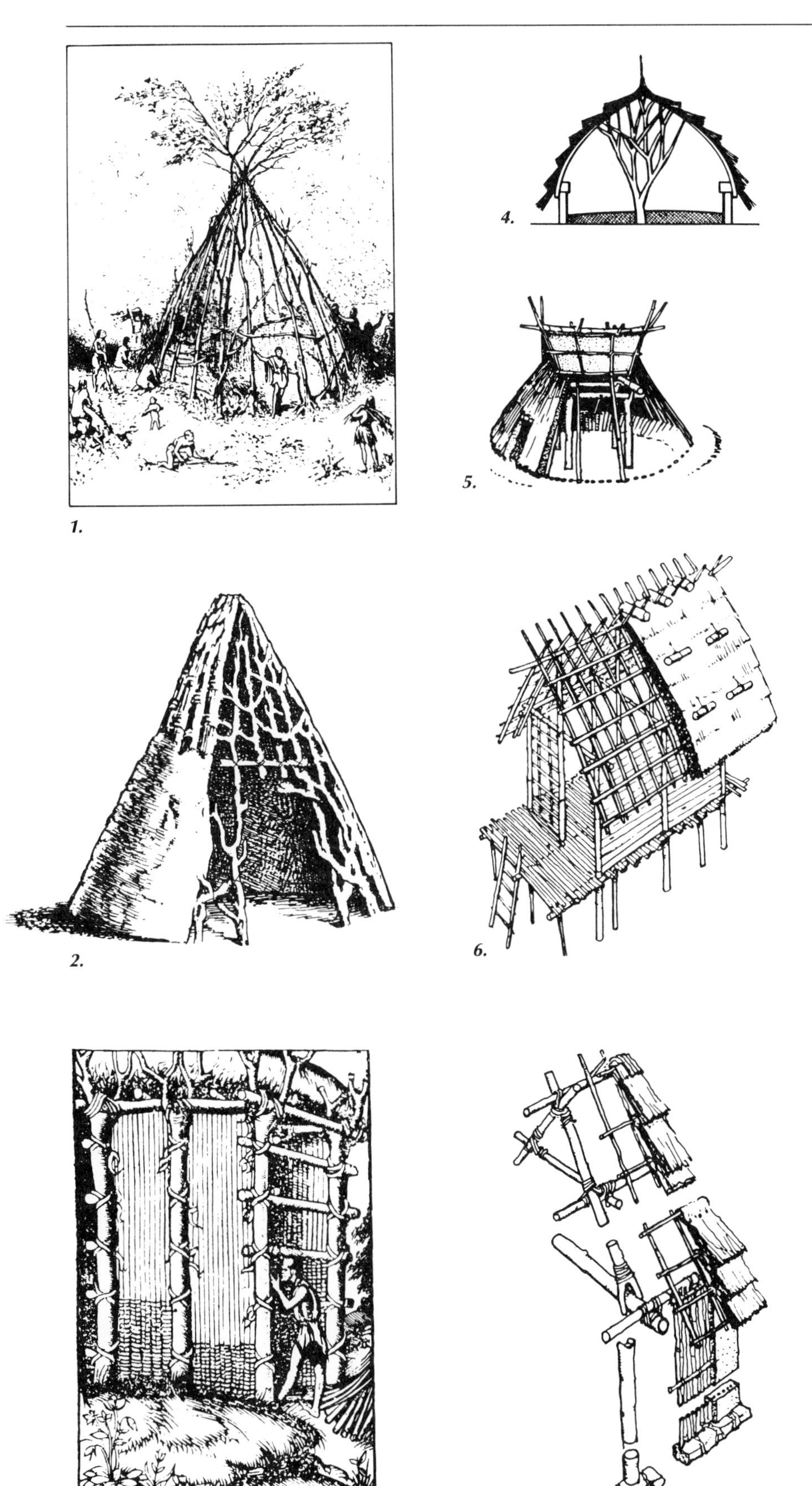

1. 2. 3. 4. 5. 6. 7.

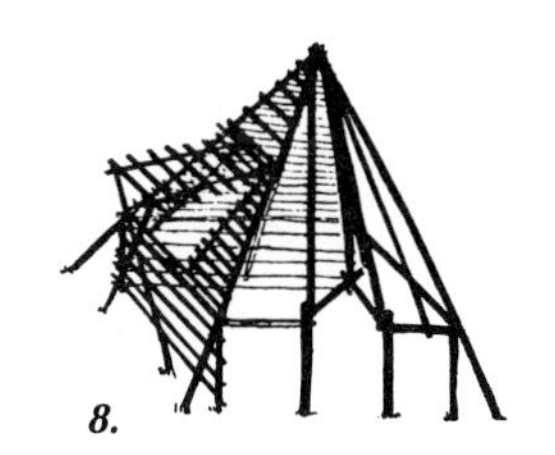

8.

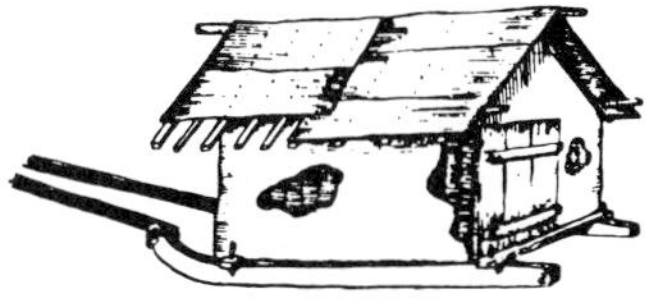

9.

Shelter genesis, shelter games. Kindershelter, *a curriculum of shelter games for pre-school or post-school architects. (1) "The First Building," after Viollet-le-Duc. (2) These late visions of early architecture are reproduced from Rykwert's* Adam's House in Paradise, *an intriguing study of the idea of the hut in architectural history. (3) Use of living trees is a constant in early shelter. (4) Sectional view of a Dinka hut about 40 feet in diameter discloses an umbrella construction, with the tree's branches supplying the spokes. The roof consists of layers of cut straw, Upper Nile. (Rudofsky 1977:132.) (5) Neolithic shelter, northern Japan. (6) Pole and bamboo stilt house, southern Japan. (7) Framing and thatch details. (8) Pyramid hut framing. (9) "Sleigh huts of Bulgarian nomads house entire families." (Rudofsky [1974:142ff] surveys sleigh houses from 3500 B.C. pictographs.)*

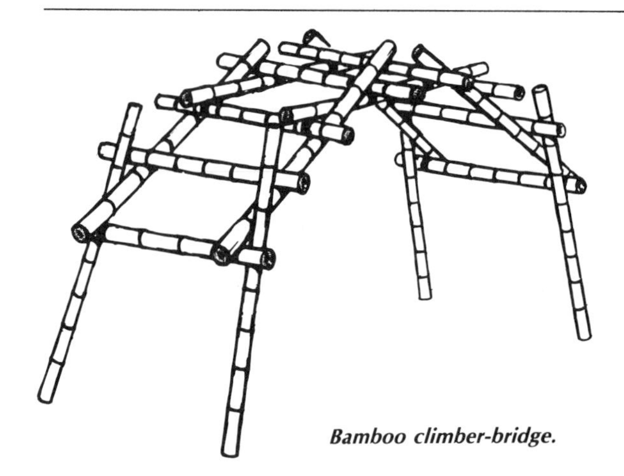

Bamboo climber-bridge.

of trees. Granted its height, bamboo is an excellent material for constructions whose main purpose is up. Up is beautiful—anything that puts children awhile bigger than big people, lets them look *down* for a change on the adult world instead of forever up at it.

Chime-mobiles and a whole percussion orchestra for kids to knock and drum on as well as the *booloo perindoo* or giant flutes doubling as ladders —the possible parts of a bambooed playground are virtually endless, and a bamboo craft shop on location would make it practical for older children to go on experimenting and building components. Playgrounds, a major hangout for children, deserve more conscious study by adults. In Japan, park designers wait to construct paths where people wear

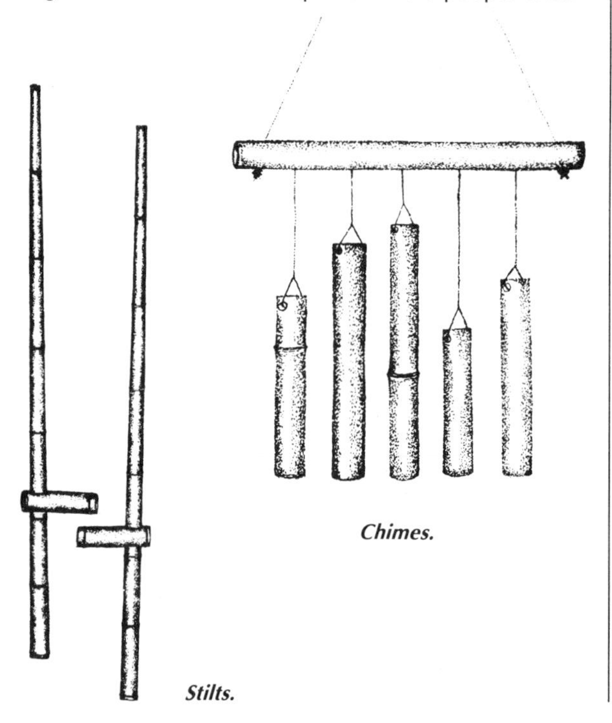

Chimes.

Stilts.

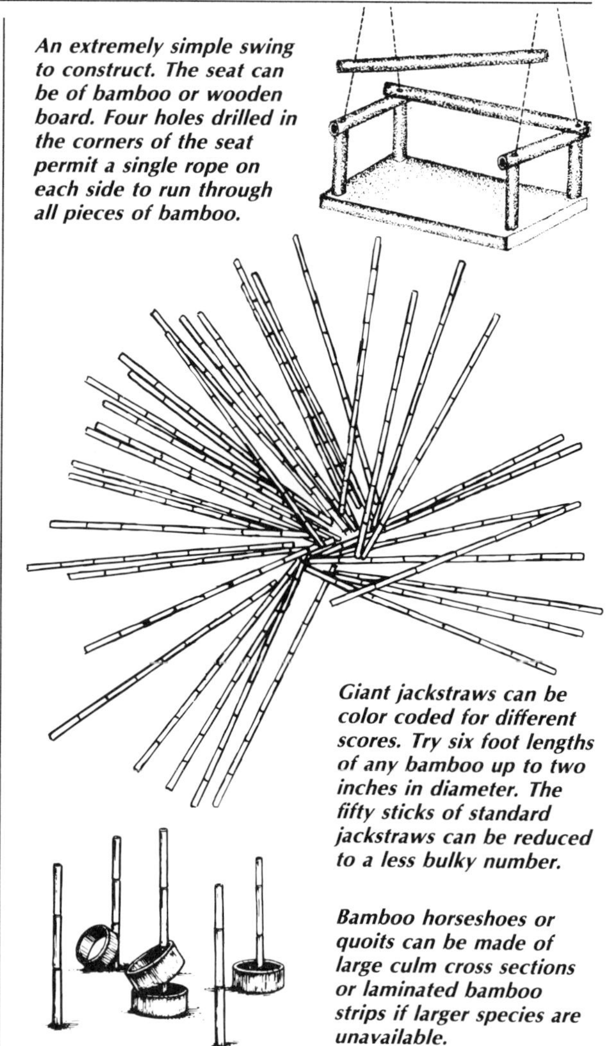

An extremely simple swing to construct. The seat can be of bamboo or wooden board. Four holes drilled in the corners of the seat permit a single rope on each side to run through all pieces of bamboo.

Giant jackstraws can be color coded for different scores. Try six foot lengths of any bamboo up to two inches in diameter. The fifty sticks of standard jackstraws can be reduced to a less bulky number.

Bamboo horseshoes or quoits can be made of large culm cross sections or laminated bamboo strips if larger species are unavailable.

them. The space is shaped partly by the spontaneous impact of those who use it. How can we extend that principle to playgrounds and invent processes that gradually phase children in as architects of their own kid-designed, kid-made environments?

KITES LIFT MORE PLEASURE PER POUND. It is time for academic culture to adopt—seriously and playfully —this tallest child of popular imagination around the world. Interest in kites, which dipped briefly after the invention of airplanes, is soaring again. You can learn more about flying by building a kite than by boarding a jet. Making them is an excellent project for schools alert to teaching the principles of aerodynamics in a format more vital than print. Kite crafting is an inexpensive introduction also to working with bamboo, requiring little material, producing more conscious fun per culm than almost any other bamboo project you could choose. Bamboo forms the bones of kite history, the spine of the art.

"The kite began not as the reasoned invention of a scientific mind, but as a wondrous and even magical link with the heavens. Early kitemakers felt a very special relationship with this toy or tool which allowed them to become a part of such a distant and powerful force."[17]

"Nature must have intended bamboo for kite frames. It is modular, each module structured like a bird's bone, probably stronger than steel for weight, and very easily worked."[18]

Sometimes bamboo whistles add an audio dimension. Kite lutes and wind zithers were also explored by the curious and nimble Chinese fingers, making these airy instruments for the winds to play. Ancient toy and stratagem of battle, modern instrument for experimental research, the kite is an ideal early learning device that unites botany and agricultural studies in growing the plant, the physics of flight, the industrial art of microdelicate handcrafting a material, experiments in papermaking, graphic design, drawing and painting. All are found in the simple, cheap, ancient, profound homework of kite construction.

Cost per pupil can be reduced to near zero by cultivating your own supply. Be alert to groves in your neighborhood producing an excess: bamboo needn't be a "problem"; integrate it with the local school.

A whole candy store of modern supertech from Mylar to fiberglass is available in kite stores in New York, San Francisco, and many other cities and towns. Newman lists forty-two kite organizations, shops, materials sources: forty United States, one English, one Canadian.[19] These refinements are not available in poorer countries, but the lightweight strength of the most old and widespread material for kite frames, bamboo, is broadly available already and could easily be made sufficiently available for all children everywhere. For the Global Schoolhouse of the Global Village, kites are a cost-possible planetary norm.

KITEBONES. Traditional framing materials for oriental kites were cypress and bamboo, both still favorites for their lightweight strength. Weight critically affects flight potential, especially in a light wind. The most versatile kite-framing material is bamboo—extremely light, flexible, and easy to work. A machete—or large kitchen knife and hammer—can be used to split large pieces. Once rough size is reached, use a sharp carving knife and sandpaper for details. Shave finely with blade perpendicular to surface: scraping off minute pieces gives more control, necessary for Indian fighter kites, for some oriental designs using varying thicknesses of bamboo in their construction, and for any highly refined flier. Chinese designs employ sizes ranging from solid, small-diameter shafts for the spine, un-

Birds, dragons, fish, insects of all sorts—the styles of Chinese kites are virtually endless.

tapered strips for standard secondary spars and spines, broader and thicker slats of bamboo as center spines or as spars near the leading edge, and especially thin reedlike strips to serve as guide strings along the outside edges.

For the bow of a fighter, a ¼-inch piece will be fine. Trim rough edges, taper ends slightly, bow strip to sight irregularities, and pare nodes and other thick sections with a knife. Avoid overcutting. Moving knife blade back and forth quickly, perpendicular to strip, shaves small slivers, further refining the bow. Constantly check curve symmetry while cutting and shaving. The bow is the key to a good fighter kite. One of the last steps: sand the bamboo. "The finished bow is a work of art."[20]

Kite frames are a good introduction to the art of bending bamboo. Remember that overheating makes a brittle, fragile flyer. Bend to slightly less than exact shape desired: This maintains the frame's structural tension, the kite's shape and spring, which help it adjust to minor wind changes.

KITE SKINS. Kite skins have been made of bamboo twigs and plaited leaves of various plants, of plaited straw by Caroline Islanders, of a wide range of vegetable fiber. But the first kite cover was silk (prohibitively expensive), then paper of many sorts in many countries: tissue paper in India and Bermuda, fine bark paper in Japan. Making the paper teaches more per square inch than buying it. Once mastered, papermaking is an art with polyapplications: lampshades, room dividers (Japanese *shoji*), window blinds. (For paper, see p. 277. For more on kites, see p. 48.)

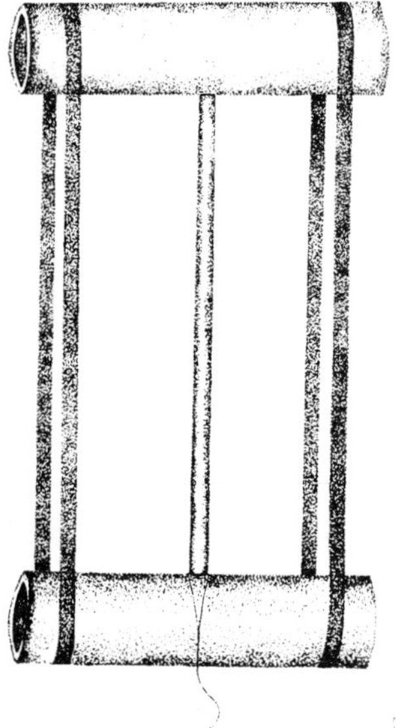

A bamboo hummer. A thin branch section running between the middle of larger sections is fitted snugly in burnt holes and braced against the inside of the culm wall. Thick rubber bands at the ends hold the hummer in tension and provide the hum.

Make a note of it.

Bamboo music in the Global Village school has not yet arrived. No bamboo musical instruments are presently available in the general school systems of many countries where bamboo is plentiful, for the usual musical direction of these countries is away from their own roots. Their ear is listening to the new sounds of the international music industry.

Bamboo wind and percussion instruments form an important element in the musical traditions of many oriental and Latin American countries, but they are often virtually unknown in schools, even when present in the surrounding (and vanishing) popular culture. In Nicaragua, plastic flutes from Hong Kong are available in the Masaya market, but no wind instrument more sophisticated than a toy whistle is produced locally, and cheap musical instruments for school use do not exist.

The massive introduction of easy homemade instruments seems indicated in schools as appropriate homework at this point to keep time with present-day harsh economic realities and dehomogenize our musical education. Many global sounds will soon be silenced unless honored in their region's curriculum. Participation is the keynote of appropriate music: making music, not merely listening, and ideally making the instruments as well. We need a highly graphic inventory, detailed enough so skilled craftspeople can pick up subtle aspects of construction of the most successful bamboo instruments throughout the world.[21] Until the happy day when we have that resource, here are a few notes on the global bamboo band. (See also, musical instruments, pp. 52–54, 82–87.)

BAMBOO BAND. A number of simple instruments can be easily constructed of bamboo. *Rhythm sticks* are among the simplest: Tape or tie together six or so bamboo plant stakes that florists and nurseries stock. Two-foot by ⅛-inch stakes work fine. Two-foot lengths of an old split bamboo curtain serve the same purpose, or use a piece of whole bamboo of a comfortable length split to about 4 inches from one end left intact as a handle. Philippine *tinakling poles* are pairs of 8–10-foot canes 1–2 inches in diameter played like this: Two people sit at the ends, a pole in each hand, and a small log or wood block maybe 2 feet long in front of them, on which the poles are tapped while a third player dances in and out of the poles, in imitation, it is said, of the long-legged crane pacing the reeds by the shore of a windy lake. Left foot, then right hopping between the bamboo poles as they tap, tap the logs: Then the poles clap together with the dancer-crane hopping outside them.

Finger castanets are a small percussion instrument of bamboo requiring only two nodes cut from a pole about 2 inches in diameter. From the internodes between, bamboo *wind chimes* or panpipes can be cut. For *panpipes,* experiment with lengths of bamboo roughly ½ inch in diameter. For a C scale, try the following lengths, which are a little long to permit tuning: C–6¾ inches; D–6 inches; E–5¼ inches; F–4¾ inches; G–4¼ inches; A–3¾ inches; B–3½ inches; C–3¼ inches; D–3 inches. Lash the nine pieces together with two strips of bamboo on each side. These can be glued first for easier lashing. Tune pipes by putting clay in the lower ends.

A bamboo *fiddle?* From the base of a pole, cut a section leaving a node at each end. Cut a groove under two strips in such a way that they remain attached at both ends and lift them from the surface of the culm with bridges cut from another piece of your bamboo. For the bow, cut a piece from the tapering end of the culm, leaving a node at one end, with some 4 inches beyond the next node at the other end for a handle. Hollow out a string, leaving it attached also at both ends and raising it with

Hu ch'in, a Chinese bamboo string instrument.

bridges. Rub violin-bow rosin on the strings of fiddle and bow. In Nigeria, some eight to ten of these segments are laced together to form a zither.

Shepherd's pipe: In a 12-inch piece of bamboo with at least ⅝-inch inside diameter, cut a slanting ½-inch air hole 1 inch from the blowing end. Cut a 1-inch cork or wood plug with an air passage along the top slanting from 3/16 to ⅛ inch. Try the plug in the bamboo and enlarge the air passage slightly if the sound is weak. Tune pipe to lowest note by sawing off the end, ⅛ inch at a time. These measurements should give a C scale going up from middle C. Draw a line from the middle of the air hole to the opposite end, marking the finger holes beginning 2 inches from the end of the pipe.

Drill a 3/16-inch hole at C, blowing to test the pitch, which is raised by enlarging the hole slightly on the side towards the mouthpiece or lowered by enlarging on the side towards the other end. Tune each note before drilling the next.[22]

The *didgeridoo* of the Australian aborigines is an impressive bamboo wind instrument made from a hollowed tube of bamboo 5–6 feet long, 2–4 inches in diameter, with a mouthpiece of beeswax. For an extended account of the complex subleties of the didgeridoo within the music of an amazing people, see "Traditional Music of the Australian Aborigines."[23]

Experiment freely. In some islands of the Pacific, bamboo bands with up to thirty different types of instruments are found. Rampant innovation is a bamboo norm.

FLUTE-MAKING IS EASY: A TRAVELER'S ART.[24] A friend came through a few years ago, an itinerant flute-maker. Before we knew it, we were making flutes, and bamboo became central to our lives. Making a flute is simple. Get a re-bar red hot over a fire and push out the membrane at the nodes. Any thin piece of iron will do, you end up inventing your own tools. I like to clean out the inside as well as possible and sand down the nodes inside flush with the inner walls of bamboo. In some traditions, they're very casual; pieces of the membrane are even left in. In others, the bore is well sanded and given several layers of lacquer, so there are extremes either way. Just do it. There's no "right way." You experiment, pick up on what people say, and feel your way, finger by finger.

After you clean the inside, you make an edge for the breath to flow over for end-blown flutes or a hole for side-blown flutes a few centimeters from one end. Traditionally in Japan the edge is sawed, but I don't like the feeling, so I file the edges. Making an edge isn't that complicated, and when you've got your edge, you've got your tone.

I look at a piece of bamboo and decide whether I want to make an end-blown or side-blown flute and how long it's going to be. I like to

put both ends on a node, and the flutes I make are 54–70 cm (21½–28 inches). My own preference, for tone, is 60–65 cm (24–26 inches). I decide what scale I want to put the flute in, and starting from the low tone of the flute, I feel it through. There's the traditional shakuhachi scale, Arabic scales of various sorts, and I use another scale I made up. I went to bed one night, heard a series of tones I thought would be fine in a flute, so next morning I made them and it worked out well. Now I make a lot of flutes in what I call the "river scale."

The end-blown flutes I make are narrower in diameter, the tone somewhat dark. The side flutes have a larger diameter and much brighter tone, both in the shakuhachi scale and in a kind of Hasidic scale somewhat similar to the blues scale. I've heard that the traditional shakuhachi scale was derived from where the fingers feel comfortable, and I made a flute like that once with the little girl next door. We just marked the points where her fingers felt good and punched out the holes. When we played it, I couldn't believe how close it was to a pentatonic scale.

Generally, though, I measure very carefully. For the traditional shakuhachi formula, for example, the first hole is put in $\frac{2}{10}$ the length of the entire flute. The second, third, and fourth holes are spaced $\frac{1}{10}$ the length apart, and the thumb hole underneath is $\frac{2}{3}$ of $\frac{1}{10}$ the length behind the fourth hole. I measure the flute and vary the formula according to wall thickness and bore diameter. I put the holes in small and get a flat tone. Taking into account at each step what I've learned from the hole before, I put in all the holes and then enlarge them carefully to bring it into correct tune.

After the holes comes the finish. The bamboo is so beautiful—sometimes you want to touch it, sometimes you don't. The skin isn't absorbent unless it's sanded, so sometimes I sand the outside and use Haines walnut oil liberally, pulling a piece of towel on a string through the flute to oil the inside. The Japanese don't use oil. If you don't sand or oil them, your fingers will rub down the skin if you play a good deal, and the oil from your hands gets into the flute so they're changing constantly if you play a lot, especially if the walls are thin.

Naturally, you want to protect a flute. I thought about lacquering them and preserving them in various ways and finally decided, nothing's permanent. They're going to crack eventually—or they're not. I do keep them well oiled so they don't get too dry or too damp. The problem is the skin: if the fibers dry up, they shrink and separate. If they're too wet, they swell and crack the rigid skin. Binding helps prevent cracking.

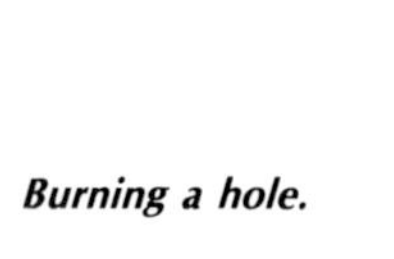

Burning a hole.

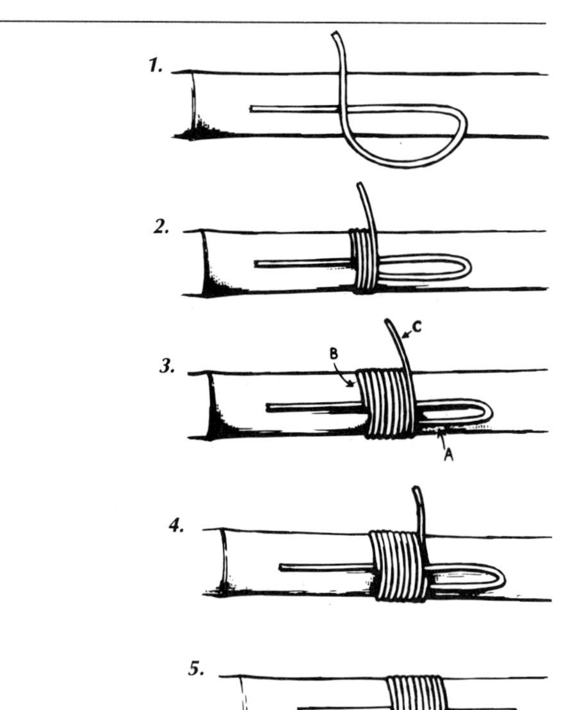

Flute knot: (1) Make a loop. (2) Wind cord around the bamboo and the loop, pulling tight at each turn. (3) Pull tightly at A until B is tight. Pull cord tight at C and cut off. (4) Insert newly cut end through the loop and pull both ends of cord. (5) Loop is pulled under the wrapping where friction holds it tight. Cut ends off flush with binding.

A day or two workshop in flute-making would be easy in schools at a minimal cost. Even buying by the piece, you can get a 9-foot, $\frac{1}{2}$-inch diameter pole for about $2–3. This is good for a number of flutes and leaves end pieces to play with. When we sold at fairs, we always had a box of little flutes made from the scrap ends of poles, free to those who got a tone out of them—and nearly every kid who really tried could play them. I started making scrap wind chimes too, out of cracked or end pieces, or flutes with an inaccurate scale. We spent a lot of time on the beach so we collected shells and combined them with bamboo.

Materials: Metal rods to burn holes. Hacksaw

to cut length. File: "Dragonskin"—a rough metallic sandpaper to wrap around dowels.

Inward Oz.

If I can talk, I can sing.
If I can walk, I can dance.

—African Proverb

Making a flute, like weaving a basket yourself or turning out your own first piece of paper, is one of those simple, wonderful things you could just this afternoon stop and do for the first time in your life. Even if your fingers are ten idiot thumbs, they can heat a poker and burn a hole. No matter how you botch or muff details, you will receive a shock of pleasure, discover an inward Oz through holding a rare modern artifact: Made in Me.

Though the casualness of this approach may horrify experienced flute-makers, what of the mute, musicless millions with nothing but an AM dial for their fine fingers? Elite artists, in gifted isolation from the norm, must add to their gifts the art of sharing them with others, the rest of us here in the culture. Honor beginner's mind. We must escape professional performance as the norm: as is is OK. Otherwise we stifle people's willingness to explore their ignorance without nervousness. Making one's own first clumsy flute can completely change one's relation to music. This music is light and easy to carry anywhere, requires no batteries, invites you to take it out anytime and visit a note. Soon enough you'll want to make a better one.

Wonders of mountains and forests travelled by people become a market. Triumphs of art and music bought and sold become merchandise.
—Ming dynasty epigram

FLUTE-MAKING IS HARD. Support the fugitive flautist and small press: consult Shepard (1978, 1979) and Levenson (1974) for loving instructions on how to shape and toot your future flutes. Flute-making is an endlessly complex art and science that will always recede before the rippling centrifugal circumference of your competence, leading you on in delight to kiss the flute as it flies before you into unknown areas of skill.

CONSTRUCTIONS

Fiber grass.

Bamboo is an abundant resource for shelter that known land-management principles globally applied could make superabundant. Contemporary scarcities of tensile fiber through deforestation and high energy costs for steel manufacture coincide with a new respect for tensile structures as Western culture turns away from a long history of heavy monumentalism, stone piled on stone. Bamboo fibers embedded in pith make for a lightweight and flexible strength, useful to the plant in wind and to people in scaffolding, quake-zone shelter, and basketry. Architecture has imitated this composite structure: Bamboo-adobe dwellings of tensile culm and compressive earth, the yin-yang of popular architecture, are found worldwide.

Much of contemporary Western technology also imitates bamboo's shrewdly composite design. Combining tensile fiber and compressive filler is the structural design of "two-phase" materials such as fiberglass and reinforced concrete. Contemporary skeletons of steel are also heirs to millennia of bamboo engineering. The bamboo ancestor of the Brooklyn Bridge spans 750 feet over the River Min in Szechuan, China. So bamboo anatomy is reflected in new materials, and traditional bamboo engineering in modern structures historically descended from it. The following notes on bamboo construction techniques complement earlier con-

Wonder Lumber. Few life forms in nature come so ready to use as bamboo. Imagine a tree without bark, with few or no branches, growing in sizes fit for immediate use, splitting easily into equal parts, easily bent to shapes it will hold after heating over a simple fire, straight as an arrow and light as the feathers on it, with a natural finish that requires no planing or sanding or any preparation apart from simple scouring with wet sand. It would be a wonder lumber. And is.

siderations of the subject (see pp. 97–102, 104–107, 126–129).

FOUNDATIONS. Bamboo is not generally used in direct contact with the ground in house foundations. Stone or hardwood is preferable. If used, treat bamboo with a preservative or put in gravel postholes to accelerate drainage. Where large bamboos are not available for main pillars and beams, smaller stems can be bound together to form large structural members.

FRAMES. See pp. 99–100, 104–107, 127.

FLOORS. On ground level, grade earth floor slightly to provide natural drainage. Place bamboo boards (see below) on loose soil, then compact. A raised floor of bamboo boards offers a sheltered spot below the house for adults to sit, children to play, chickens to nest, tools or farm equipment to be stored—however you'd like to use the area. A first floor raised 6 feet or more off the ground is more hygienic, more breezy in hot climates, and doubles a one-story dwelling space. The bamboo boards are easy to sweep clean—many cracks let the dirt fall through. They are resilient, so require more supports than wood floors, but are more pleasant to walk on or sleep on. Floors can be built of large whole opened culms, many small whole culms or main branches of large species, or strips cut from the culms of the larger bamboos.

WALLS. If using whole culms, half culms, or strips, remember they drain better vertically than horizontally. Mat walls woven of whole opened and flattened culms of thin-walled bamboos (for example, *Schizostachyum* species in the Philippines) are one of the most efficient and attractive low-cost housing alternatives in areas of bamboo abundance. They can be prefabricated in 10-foot-by-30-foot panels, rolled up and delivered by ox cart, and moved easily to another location when necessary. A design success for millions over centuries of low-cost bamboo technology throughout the Orient, this type of wall has a possible future in Latin America (lacking

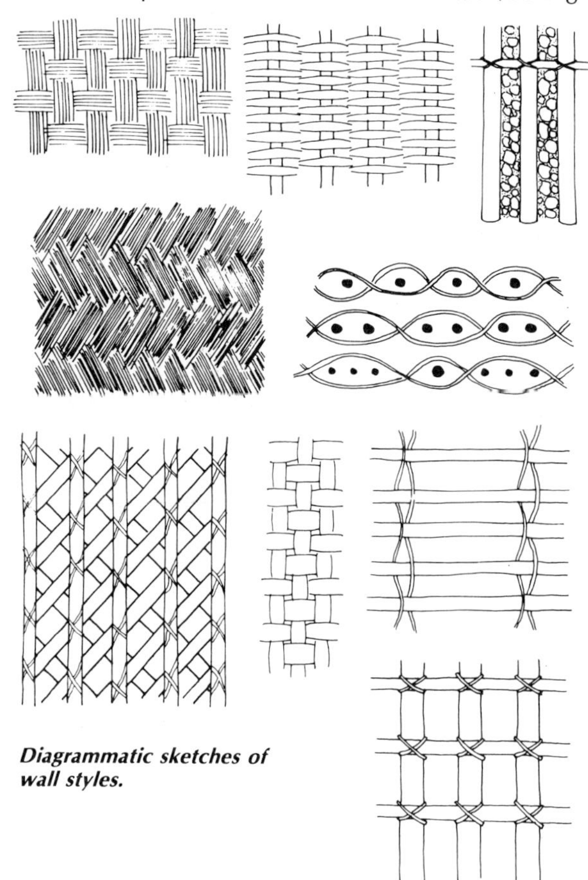

Diagrammatic sketches of wall styles.

20 million homes), which has not yet been seriously considered.

BAMBOO BOARDS: ESTERILLA. Where large bamboos are available, boards are made by splitting and opening the whole culm, minus the tip and basal parts, of species with thick walls. The process in Colombia, Ecuador, and other parts of Latin America is described by McClure: "The picador holds the bamboo culm in position on the ground with his foot and strikes the blade of an ax* into each node at intervals of an inch or so right around the culm. The incisions in the different nodes are short and entirely independent so that the wall of the culm clings together as a fabric in spite of the great num-

Editor's note: Sometimes this is done with a well-greased bit. The ax may also be used to remove diaphragms at nodes, or else a machete, adze, or curved spud can be used.

Side framing in Japanese construction.

ber of splits with which it is rent. When every node through the length of the culm has been cracked in this way, the picador makes a single continuous split from one end of the culm to the other. The culm opens out and may be pressed flat. Debris at each node is removed. The boards are stacked with first inner then outer surface uppermost, alternately. The stack is weighted with stones to prevent curling, and the boards dry out flat."[25]

BAMBOO-CLAY WALLS IN JAPAN. "Even in developing countries, where modern construction techniques have been introduced, cement and synthetic materials have been replacing, little by little, the use of clay as a covering for walls, to the point that it is now used only in isolated rural buildings or houses of people of quite limited means. Nevertheless, in spite of the fact that Japan is the most industrialized country in Asia using the most modern construction technology, the Japanese continue to use the traditional clay wall, not only in their houses but also in contemporary buildings of steel and concrete, in which they use the same materials and techniques of construction employed in building the most refined houses of the Middle Ages in Japan. The reason for the persistence of this tradition lies in the tone and beauty of its texture as well as its low cost and stability in weather changes, although it is not extremely resistant to daily wear and tear.

"The clay wall is generally constructed in a wood frame formed by structural elements of the house such as beams and columns, to which is fixed, as reinforcement, a grill or framework of thin canes or strips of bamboo. This framework is given greater strength by interlacing it with crosspieces of wood, horizontally and/or vertically. Once this is made, three to five coats of specially prepared clay are applied to both sides.

"The bamboo framework is made of thin canes 8 to 13 mm (⅓–½ inch) in diameter and/or strips of the same dimensions split from larger poles. These canes or strips are divided into primary and secondary members. The primaries are 30 cm apart, vertically and horizontally, with their ends tied into the wooden structural elements forming the frame and to wooden crosspieces by means of mortises or notches cut into these. Secondary members are placed between primaries at a distance of 2.5 to 4 cm, tied to these with straw cord having an average diameter of 5 cm. (2.5 cm= 1 inch.)

"The wall is made of three to five coats of clay of different mixtures applied to both sides of this bamboo framework. Only three coats are used generally, the composition of the final coat depending on the climate and traditions of the site. For the initial coats, clays of different colors and consistencies are used, mixed with fibers, glues, lime, and sand, all of which give distinct qualities to the clay. Fibers are used to prevent cracks in the clay produced by contraction when drying, Among the most common materials used are: fibers of raveled hemp 1.5 cm long; fibers of raveled manila rope in lengths 2.5 cm to 5 cm; separated, raveled, and chopped fibers of jute; manes of horses and hair of other animals which has little use for other purposes in Japan; cut up Japanese paper; chopped straw of various sorts. Straw 2.5–10 cm long is used for the first coat. Straw from the waste of mats and ropes in 2.5–5 cm lengths is mixed with the second coat. Fine straw 2.5 to 3 cm is mixed with the final coat."[26]

MUD-STUFFED BAMBOO WALLS. On the floor or cement footing, a wooden or bamboo footing is laid on which bamboos of equal dimensions are placed upright no more than a foot apart. A horizontal of bamboo or wood holds them in place above. Diagonal lathing of bamboo should be added to reinforce the whole structure. On this are nailed horizontal lathes at least 4 cm wide, 8 cm apart—(1½, 3¼ inches) far enough to let you get your hand inside. The *cascara* or outside wall of bamboo should be nailed against the structure, with the inside (*corazon* or "heart") facing towards the worker. Mud is then packed in well between the two bamboo walls of the wall, leaving it level with the outside surface of lathing. Then it dries four weeks before the surface coat is applied.[27]

Standard mud recipes include mixing earth, straw, dung—often cactus boiled and added for adhesive assistance—and a liter of the local lightning for seducing neighborhood cronies to lend a hand in the work.

ROOF. "Because of their high strength–weight ratio, bamboos are used to excellent advantage for structural elements in roof construction. In designing the roof, account must be taken of the nature and weight of the roof covering to be used, whether it be grass or palmleaf thatch, halved bamboo culms, bamboo shingles, corrugated sheet metal, or tiles. The dimensions, spacing, and framing of the individual structural units that support the roof covering are varied to conform to the requirements of the case."[28]

ROOF NOTE. Around 50 miles north of Lima, Peru, big groves are visible from the highway in Chancay, startling enough after many miles of desert to make one investigate how they came there. In November 1980, we found Don Guillermo, already ancient, who fifteen years earlier—around 1965—had planted twelve bamboo windbreaks, each about 1,000 feet long, to protect his large orchards of oranges, mandarins, and tangerines. He got them originally from the *Escuela Experimental de la Molina* in Lima, and showed us, with some pride, the roof of the attractive, even sumptuous, home he had designed. Bamboo rafters were treated with 4 large tablespoons of oil and insecticide poured into a ⅜-inch hole drilled in each internode and then sealed with plaster. Nylon fertilizer sacks treated with a waterproof material lie under a roof of earth with 50 percent bamboo leaves mixed into the soil to lighten it. "Earthquakes don't affect the construction because the roof is so light. With a heavy cement roof, when the earth moves, the walls move too. And a conventional roof would be four to five times as expensive as this one I've designed of bamboo."

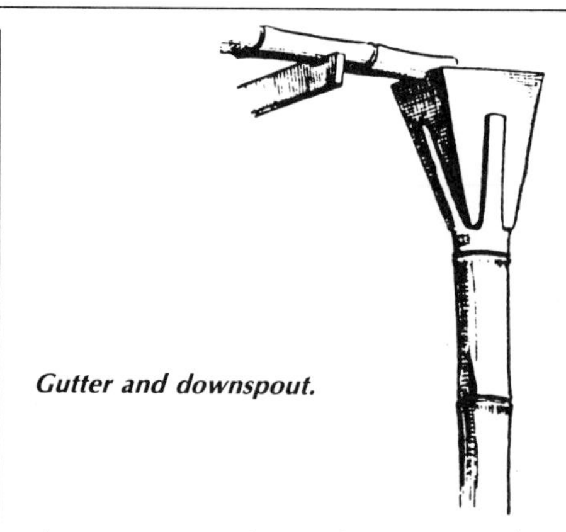
Gutter and downspout.

Bamboo, the "wood of the poor" in India, remains the experiment mainly of the well-to-do in much of the West. How much he saved on his roof doesn't really matter to Don Guillermo personally. He lives in Miraflores—a posh section of Lima; one son is the local lawyer for Braniff, another a doctor somewhere else, and so forth. The class of people who could most profit by his inventive uses of bamboo are his neighbors, the impoverished villagers — who don't respect or seek bamboo and don't have the economic margin to be culturally curious.

GUTTERS. "The usual form of gutter for conveying water from the roof consists of a large bamboo split lengthwise, with nodes removed. This is held to the eaves by iron hooks, or by long pieces of wood nailed to the rafters, and notched in the upper edges to rest the bamboo. This leads to a bamboo downspout with the upper end cut to leave four long spurs, whose elasticity holds in place a square and tapering funnel of thin wood forced down between them."[29]

Thatch.

If you have ever lived and lounged in a huge, cool cavern of high-pitched and softly rustling thatch

Moving day in Vietnam.

roof on bamboo framing, you will know thatch has no equal for vitality: eaves alert to the least wind; intricate innards and shelter to various communities of insects complete with lizards in pursuit; birds borrowing a tiny portion of your mega-nest to tuck into theirs. No machine-stomped material feels quite so amiable and fresh to flesh as this dry fiber, unprocessed, earth color that, from a distance, gently stitches shelter and land into a seamless garment, the human artifact of housing smoothly continuous in shade and contour with the hill.

If you have climbed tall palms to cut thatch, high above most surrounding jungle, with a tossing sea of green leaves below you; if you have dived down under a lake to cut the cattails where they bunch at the base some 8–10 feet below the surface and then slowly shoved a raft of them down shore to patch your hut—if you have ever been even briefly involved in the technology of thatch, you know there is no more deeply and simply satisfying task than to cover a space with fronds and grasses —the warp to a woof of bamboo or other lightweight raftering—and then to sit dry and satisfied as toast while the afternoon storm breaks loose in all its rambunctious tropical glory. The same lavish rains that produced your thatch will test your hand at using it.

"It is indeed a matter of wonder that someone in building a house in this country [the United States] does not revert to a thatched roof. Our architectural history shows an infinite number of reversions, and if a thatched roof were again brought into vogue, a new charm would be added to our landscape. The thatched roof is picturesque and warm, and makes a good rain shed. In Japan an ordinary thatched roof will remain in good condition from fifteen to twenty years, and I have been told the best endure for fifty . . . When properly constructed, they shed water very promptly, and do not get water soaked, as one might suppose.

"It is customary in the better class of houses having thatched roofs to pave the ground with small cobblestones, for a breadth of 2 feet or more immediately below the eaves, to catch the drip, as in a thatched roof it is difficult to adjust any sort of a gutter."[30]

"The thatched roof is by far the commonest form of roof in Japan outside the cities. The slopes of the roof vary little; but in the design and structure of the ridge, the greatest variety of treatment is seen, each province with its own peculiar style, probably due to isolated life in feudal times . . . Straw, grass, reed, and rush are various materials

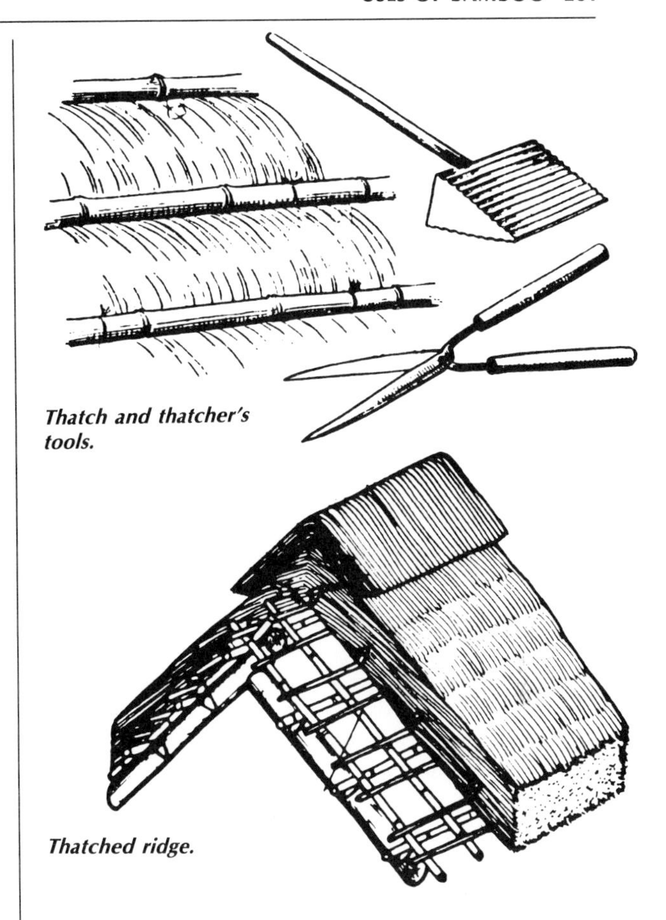

Thatch and thatcher's tools.

Thatched ridge.

employed for thatch. Rafters and framework should be close enough together properly to secure and support it. A bamboo framework is sufficient for a small roof.

"The thatch is formed in suitable masses, combed with the fingers, secured to the rafters and bound down to the roof by bamboo poles which are afterwards removed. While the thatch is bound down in this way, it is beaten into place by a wooden mallet of peculiar shape. The thatch is then trimmed into shape by a pair of long-handled shears . . . When a roof is finished, it presents a clean, trim, and symmetrical appearance, which seems surprising, considering the nature of the material.

"North of Tokyo, in many cases the ridge is flat, and supports a luxuriant growth of iris, or the red lily."[31]

FENCES. Fences consume more raw material than any human artifact except buildings. Bamboo has probably made as many miles of them as any other earth resource.

FENCE SHACK. Once a fence is built, there's only three walls and one roof to go before you have a

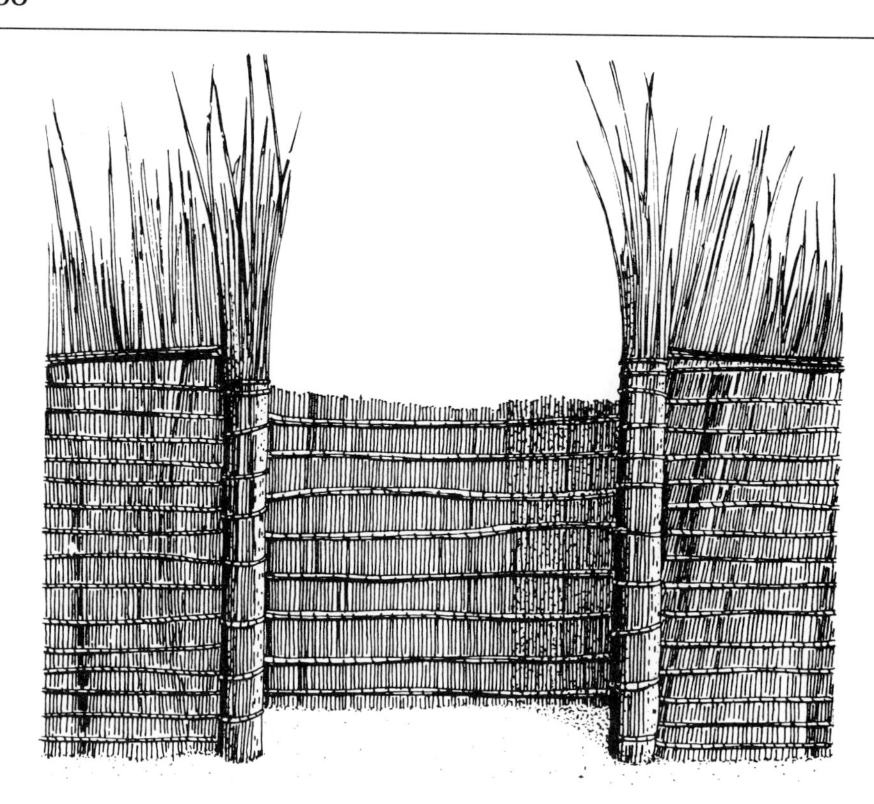

Gate of Ganda Palace, Uganda, c. 1910. "The palace was surrounded by a tall woven fence about 3.5 meters high made of elephant-grass and supported at intervals by stout posts cut from wild fig trees, which eventually took root. In the palace there were about 450 buildings for the Kabaka and his wives, and hundreds of smaller houses and cooking huts. The main buildings were built in a similar way to ordinary houses, but of much larger proportions and with much finer and thicker thatch, sometimes as much as 30 cm [1 foot]. To build just one, two hundred men would be at work for at least two months." (Denyer 1978)

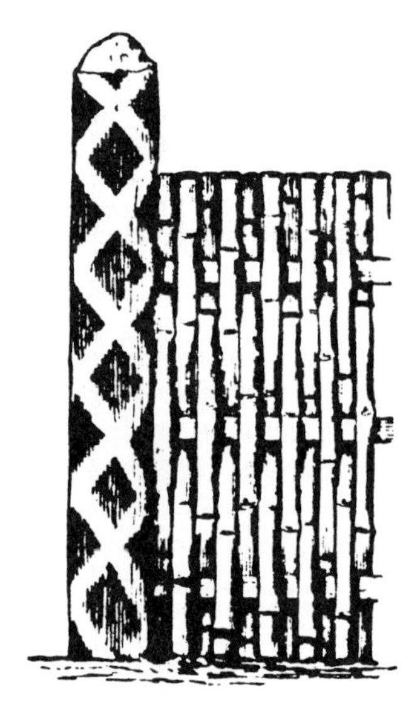

A simple and effective construction method using strips or whole culms of bamboo wedged in an alternate weave between three horizontals. For low-tech decoration, the hardwood post is stripped of bark, wrapped in wet straw or rope in any desired pattern, and the spaces between burned lightly or deeply over hot coals. (Japan c. 1880.)

Basket-fence. The walls of the basket are woven with bamboo or any available materials, and periodic reinforcements hold them in place. The fence shown is of Guadua angustifolia, *in a design from Colombia. This is an especially durable species, but bamboos are not generally recommended for use in direct contact with the soil.*

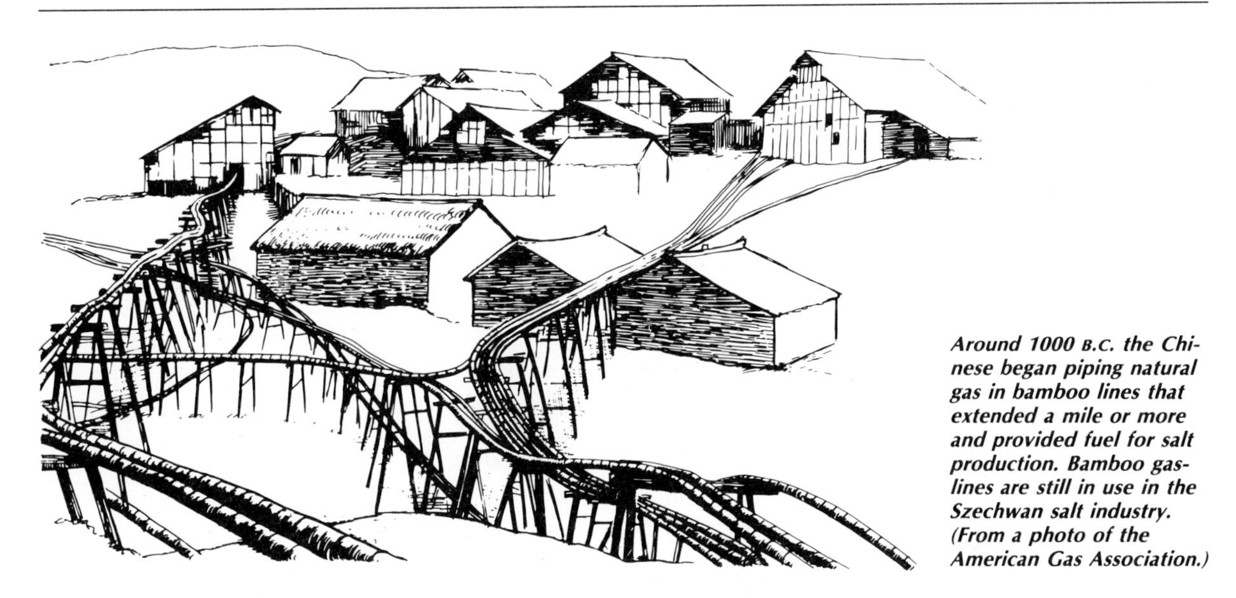

Around 1000 B.C. the Chinese began piping natural gas in bamboo lines that extended a mile or more and provided fuel for salt production. Bamboo gaslines are still in use in the Szechwan salt industry. (From a photo of the American Gas Association.)

tiny hut, big enough for the smallest twig-sized stove, a few tools, a book or two, a shelf for potting plants—or else make just a 3-foot thatch roof, without walls or with waist walls woven of wasted branches from your grove, room enough for you and a flute or poem to rest together a moment, half-sheltered, half-exposed.

PIPES. "Bamboo pipe, easy to make with unskilled labor and local materials, is extensively used in Indonesia to transport water to villages. In many rural areas of Taiwan, bamboo is commonly used in place of galvanized iron for deep wells up to 150 meters [500 feet]. Bamboos of 50-mm [2-inch] diameter are straightened by means of heat, and the inside nodes knocked out. The screen is made by punching holes in the bamboo and wrapping that section with a fibrous matlike material from a palm tree, *Chamaerops humilis.* In fact, such fibrous screens are also used in many galvanized iron tube wells. *Tools and materials:* chisels, nail, cotter pin or linchpin, caulking materials, tar, rope.

"Bamboo piping can hold pressure up to 2 atmospheres (about 2.1 kilograms per square centimeter or 30 pounds per square inch). It cannot, therefore, be used as pressure piping. It is most suitable in areas where the source of supply is higher than the area to be served, and the flow is under gravity.

"*Health aspects:* With piping for drinking water, the preservative treatment recommended is to immerse green bamboo completely in a solution of 95 percent water and 5 percent boric acid: borax. After a bamboo pipe is put into operation, it gives an undesirable odor to the water. This, however, disappears after about three weeks. If chlorination is done before discharge to the pipe, a reservoir giving sufficient contact time for effective disinfection is required since bamboo pipe removes chlorine compounds, and no residual chlorine will be maintained in the pipe. To avoid possible contamination by groundwater, an ever-present danger, it is desirable to maintain the internal pressure within the pipe at a higher level than any external water pressure outside the pipe. Any leakage will then be from the pipe, and contaminated water will not enter it.

"*Design and construction:* Bamboo pipe is made of lengths of bamboo of the desired diameter by boring out the dividing membrane at the joints. A circular chisel for this purpose is shown. One end of a short length of steel pipe is belled out to increase the diameter, and the edge is sharpened. A length of bamboo narrow enough to slide into the pipe is used as a boring bar and secured to the steel pipe by drilling a small hole through the assembly and driving a nail through the hole. This nail is also

Support systems for bamboo pipe.

Bamboo aqueduct depicted in a carved wood ramma (Japan, c. 1880). The ramma is a decorative panel that gives expression to the exacting tool control of Japanese craftsmen.

known as a cotter pin or linchpin. Three or more chisels ranging from the smallest to the maximum desired diameter are required. At each joint the membrane is removed by first boring a hole with the smallest diameter chisel, then progressively enlarging the hole with the larger chisels.

"Bamboo pipe lengths are joined in a number of ways. Joints are made watertight by caulking with cotton wool mixed with tar, then tightly binding with rope soaked in hot tar. Bamboo pipe is preserved by laying the pipe below ground level and ensuring a continuous flow in the pipe. Where the pipe is laid above ground level, it is protected by wrapping it with layers of palm fiber with soil between the layers. This treatment will give a life expectancy of about three to four years; some bamboo will last five to six years. Deterioration and failure usually occur at the natural joints, which are the weakest parts. Where the depth of the pipe below the water source is such that the maximum pressure will be exceeded, pressure relief chambers must be installed. These chambers are also installed as reservoirs for branch supply lines to villages en route."[32]

Efforts to apply the information above are reported in a leaflet from Africa: "Bamboo Piping: An Experiment in Mezzan-Teferi, Ethiopia." (See Christian Relief and Development Association in Bibliography.) "The authors did not have much success with a circular punch tool for knocking out the inner dividing walls of the bamboo. They developed a simple drilling bit which can be easily made by a blacksmith [drawings and photos are provided]. With this tool three workers could easily bore out twelve 7-meter bamboo poles in one hour. The experimenters also developed a unique joint sealing system in which soaked cowhide is wrapped around the joint twice and sealed tight with two pieces of galvanized wire . . . Pipes carried water for irrigation and domestic use."[33]

"The tested crushing strength of bamboo used for water pipes has been determined to be approximately 766 pounds per lineal foot for 3½-inch-diameter canes. Two-inch-diameter canes will withstand a pressure of 374 pounds to the lineal foot. Clay tile measuring 3½ inches in diameter has a crushing strength of only 276 pounds."[34] See also Lippert (1976) and Herrera's feasibility study on bamboo *(G. angustifolia)* and plastic pipe in Colombia (1974).

Hen house: appropriate cages.
"A bamboo poultry house with thatch roof and slat walls provides good insulation. The elevated slat floor keeps chickens clean and healthy while the egg catch and feed troughs simplify maintenance. Costing mainly labor to build, housing healthier hens, such coops are used successfully in the Philip-

Bamboo animal cages and feeding troughs reduce expensive importation of metal cages, are more amiable to the caged, and can be repaired from locally available material. Their construction lies well within the tool supply, economic means, and skills of any villager. These same qualities make them a good school project for early architects. Light cages are easily elevated on poles or hung from trees. Multiple stories economize space.

pines and Liberia. The house is built on a frame of small poles, with floor poles raised about 1 m [3 feet] from the ground. The floor poles are covered with large bamboo stalks, split into strips 38 mm [1½ inches] wide, spaced 38 mm apart. Floors so constructed have several advantages: better ventilation, no problem of wet mouldy litter during rainy season or dry dusty litter during dry season; droppings fall between split bamboo to ground, away from chickens. This eliminates parasites and diseases normally passed from hen to hen through droppings remaining warm and moist in litter. Wide spacing of floor and wall slats might invite marauders such as weasels and snakes. Tin shields on the support poles will keep rats and other pests from climbing.

"Walls are made from vertical strips of bamboo 38 mm [1½ inches] wide spaced 6 cm to 8 cm [2½–3 inches] apart. This also allows ample ventilation, needed to furnish oxygen to the chickens and to allow evaporation of excess moisture produced in the droppings. In the tropics, the problem is to keep chickens cool, not warm. Using a closed or tight-walled poultry house with a solid floor would keep them too warm and result in lowered production and increased respiratory problems. Shade over and around these houses is very important. If the ground around the houses is not shaded, heat will bounce into them.

"In Liberia, thatch roofing keeps the birds cool, is cheap and readily available to the small farmer or rural family, and so is most likely to be used. The roof must have an overhang of 1 m [3 feet] on all sides to prevent rain from blowing inside the house. It may be desirable to slope the overhang toward the ground.

"Feeders and waterers are made from 10-to-12.5-cm-diameter bamboo [4–5 inches] of the desired length, with a node or joint left intact in each end of the bamboo section to keep the feed or water in. A section 7.5 cm to 10 cm [3–4 inches] wide around half the circumference of the bamboo, except for 7.5-cm sections [3 inches] on the ends, is removed to make a trough. All nodes between the ends are removed. These feeders must be fastened at the base to keep them from rolling. The feeders are fastened to the outside of the walls about 15 cm [6 inches] above floor level. The hens place their heads through the bamboo strips to feed or drink, thus conserving floor space for additional chickens.

"Nests: The demonstration nests are 38 cm [15 inches] long, 30 cm [12 inches] wide and 35.5 cm [14 inches] high. The strips used on the floor of the nest are about 13 mm [½ inch] wide, spaced 13 mm apart, and must be very smooth. The floor slopes 13 mm from front to back, so that when the eggs are laid they will roll to the back of the nest. An opening 5 cm [2 inches] high at the back of the nest allows the eggs to roll out into an egg catch, resulting in cleaner and fewer broken eggs of better quality because they begin to cool as soon as they roll out of the nest. The eggs are also out of reach of egg-eating hens and can be conveniently gathered from outside. Placing the nests 1 m [3 feet] above the floor conserves floor space and permits more laying hens to be placed in the laying house. One nest is put in for every five hens.

"Split bamboo nests give unobstructed ventilation. Conventional lumber nests are hotter; this may cause hens to lay eggs on the floor instead of in the nests. This means more dirty eggs, more broken eggs, and more likelihood of the hens eating the broken eggs. The only way to cure a hen of eating eggs once the habit is formed is to kill her. Without an 'egg roll' in the nest, the hens sit on eggs laid previously, keeping them warm. The quality of eggs deteriorates very fast under these conditions."[35]

Bamboo and chickens go well together. Groves provide shade, roosts, protection from predators and chilling winds, a food supplement recommended for chicks (prepared from chopped leaves rich in vitamin A), and a leaf fall that shelters lots of edible insects. The chickens return these favors with excellent fertilizer for the groves.

Bamboo camp.

BOY SCOUT BAMBOO. Backpacks, fishing poles, frames to stretch a tarp on—camp sometime by a bamboo grove, and you'll find that the instant handiness of the living plant has a true chance to emerge, given an abundance to play with. Such a bamboo camp for people to visit, to cut from and maintain, and to experience the full moon through the culms, take tea by a fire on one side with a flute on the other—such groves should be available near any big city in climates mild enough for bamboo cultivation. We playfully address ourselves to Scouts because they—girls and boys together—are the main massive group of young townspeople trying to maintain some touch with the untown, the terra firma beyond cement.

But a bamboo camp is a serious, low-cost, high-consciousness design for schools, churches, and other collectivities to consider for young and old. Inexpensive ways to exit from town must be

Appropriate travel for urban children is a high priority of contemporary cultural design. The bamboo camp is a cost-feasible norm.

designed for the town's poorer people, especially families. Since shelter and food are two main travel costs, we should design this as a small hostel-garden where guests are also hosts, responsible for building and planting and all the work of being alive directly on the earth. The youth hostel system of Europe is a crude analogy, but a popular bamboo camp would be participatory and, in fact, a school in rural survival. The high cost of deprivation in such organizations as Outward Bound (which specialize in wilderness experiences for children of well-to-do urban parents who value bare feet) can't process sufficient numbers of city children to make a real impact on cultural consciousness in town. We provide below a sampling of possible projects for camp, school, or home.

FURNITURE. *The Philippine Craftsman* ran a series of articles from 1912–1914 on furniture and basketry native to the islands, intended to introduce profitable industrial arts work into the school system and directly improve the comfort and convenience of the Philippine home. The 7 articles comprise a guide with 154 pages of text and 133 illustrations showing 195 different basket styles and sizes with 43 exercises for making them. The furniture section includes 64 photos and designs of chairs, desks, settees, couches, rockers, clothes hangers, foot stools, tables, hat racks, washstands, armchairs, beds, screens, and wardrobes, together with joint designs used in their construction and simple machinery used in the process. The group of articles is somewhat stiff and dated, but it represents the most extensive consideration in print of

The Chairman's chair. Mao's chair, of the 19th century or earlier, is standing in the cave museum of Yen-an. The splay legs are high, typical of the northern style in China.

Oil lamp stand, 19th century, Kuang-tung. A long piece of bamboo, shaved thin at the middle, is bent for the handle, and holes are drilled for joining the legs. The design can be expanded for a stool, high chair, or ladder.

The short legs and high back are of the southern Chinese style, 19th century.

The hidden curriculum of the bleak cube.

Blocks are such a staple toy in industrialized countries that it's surprising to learn they're only about a century and a half old. "The bleak cube as plaything was devised by a mild-mannered, humorless, early nineteenth century German schoolteacher, Friedrich Fröbel, the originator of kindergarten." Before becoming a schoolmaster, Fröbel had been a forester, architectural apprentice, museum curator, and soldier in the Prussian army. "Wartime duty taught him the merits of discipline, yet it was probably his bout with architecture that led to his infatuation with the cube.

"Fröbel's choice of blocks was prophetic. At the time he received his call to mold the minds of the little ones, the better part of the world was still tidy and fragrant, abounding in forests and clear streams, with the sea's plangent waves unsullied by oily squish. However, in the course of time it was to undergo a fateful change that led to a new attitude toward life. It saw the destruction of the woods and the darkening of the waters, the growth of gigantic factories and block after block of equally gigantic housing projects. The infant had a preview of things to come right in the nursery, for hard-edge building blocks were his first encounter with the hard-bitten reality of life. At home and in schools on shelves within easy reach, the cubes are stacked in rows, allegedly for the child's amusement, actually for imprinting on his mind the principle icon of our architectural creed."

(Rudofsky 1977:355–6)

The circle and the square.

"You have noticed that everything an Indian does is in a circle, and that is because the Power of the World always works in circles, and everything tries to be round. In the old days when we were a strong and happy people, all our power came to us from the sacred hoop of the nation, and so long as the hoop was unbroken, the people flourished. The flowering tree was the living center of the hoop, and the circle of the four quarters nourished it . . . Everything the Power of the World does is done in a circle. The sky is round, and I have heard that the earth is round like a ball, and so are all the stars. The wind, in its greatest power, whirls. Birds make their nests in circles, for theirs is the same religion as ours. The sun comes forth and goes down again in a circle. The moon does the same, and both are round. Even the seasons form a great circle in their changing, and always come back again to where they were. The life of a man is a circle from childhood to childhood, and so it is in everything where power moves. Our teepees were round like the nests of birds, and always set in a circle, the nation's hoop, a nest of many nests, where the Great Spirit meant for us to hatch our children.

But the Wasichus (the whitemen) have put us in these square boxes. Our power is gone and we are dying . . . You can look at our boys and see how it is with us. When we were living by the power of the circle, in the way we should, boys were men at twelve or thirteen. Now it takes much longer to mature . . . It is a bad way to live. There can be no power in a square."

—Black Elk (Neihardt 1961:198–200)

Bamboo Puppetry

No camp is complete without a fire. No fireside is complete without a tale. No tale is told more fully than by puppets, an ancient and supremely popular method of storytelling around the world.

Puppetry should be regarded as an art as valid as any other, and much more demanding than most, because it combines so many arts in its fully successful embodiments. Traveling with a traditional family of puppeteers should be included in the schooling of any serious student of the art.

Living on the road with puppeteers, the student sleeps on stage below the limp figures that only a few hours before entertained an entire village. The priest, the young lovers, the one-eyed thief, and all the rest are motionless now, but their shadows quiver in the frail light of a kerosene lamp, and the student lies there remembering the stories and the dance of *the dolls who come alive . . .*

Bamboo can furnish tent poles, scenery, houses, and armature for puppets up to giant dimensions.

The Bamboo Cutter and the Shining Maiden

Bamboo legends abound wherever the plant is plentiful. Bamboo wombs are an element in many stories and often, as in this thousand-year-old tale from Japan, the moon is somehow intimately related to the plant. The tale is told at greater length, with 18th century Japanese illustrations of sumptuous elegance, by Fisher (1980).

A poor and childless bamboo cutter was pursuing his solitary trade as usual when early one evening, by the soft light of the rising moon he noticed a stalk of bamboo glowing at the base as though lit by a candle inside. Carefully cutting the internode open, he found a tiny girl, some three inches tall and beautiful beyond belief. He carried her home cupped in his hand like a baby bird, and built a tiny basket for her to sleep in, which she quickly outgrew. In fact, she grew as quickly as bamboo itself, and within three months she was a young woman of normal size, but of a beauty not only far beyond all local norm but almost, it seemed, beyond mortality. If anything, her beauty had increased with growth. Something in her flesh still glowed with the faint, unearthly light the bamboo cutter had first noticed in the grove. Rumor of her rippled to the four corners of the kingdom. Her downcast eyes were the torment forever after of all who caught a furtive glance of them. Her dark lashes lay at the restless center of a thousand dreams. Suitors flocked to her gate like bees swarming to the world's last rose, but the shining maiden would have nothing to do with them. As months dragged on into years, her reputation ripened to terrible but established fact, and all except five persistent idiots finally went home.

This small group remained loyal with a tenacity that was either bestial or divine; human behavior offers no parallel. In summer, their jackets stuck to their sweaty flesh; in winter, snow muffled their lutes. Every evening they presented a fresh sonnet or performance of some sort, sometimes even a small skit in which all took parts. Long residence before the same frustration had reduced them to unity; they had grown brothers rather than rivals, five fingers of a single courting hand. If any had been accepted at this point, the four others would have chipped in for the ring, both for the sake of a familiar friend and to be relieved at last of an ordeal that had already claimed too much of them to be abandoned, but which was devouring the rest of their sanity and fortune at a pace too rapid to last for long.

At last the poor bamboo cutter decided to discuss the matter, and suggested to the shining maiden that she marry one of these poor men. At last she agreed—provided each run an assigned errand first. The old bamboo cutter hurried out to tell the suitors, who were just finishing their evening serenade. One was playing mournful music on a bamboo flute while a second sang a song declaring he could live without food if he could kiss, daily, a spot freshly shadowed by her hand, while the others whistled and beat time with their fans. "Listen," he called to them from the balcony, interrupting the performance, "the shining maiden has agreed to marry whoever returns from a quest with what she wants." They were overwhelmed; and when the shining maiden herself came out, they were even more overwhelmed to actually see her after all these completely unencouraging years. They were most overwhelmed when she told them what they had to get. One had to fetch the Buddha's begging bowl; another, a jewelled branch from the golden tree with silver roots growing on Paradise Mountain in the Eastern Sea; another, a nonflammable robe sewn from a hundred hides of a certain magical species of Chinese rat with habitat of fire; another, a jewel with five colors from a dragon's head; the last, a special charm for easy childbirth, plucked from the womb of a nesting swallow.

The suitors were completely flabbergasted. They felt hopeless, angry, and tricked. "Why don't you just say, 'Go home and stay home.' None of these things can be bought in Japan!" The shining maiden just turned aside to gaze at some bamboo growing by the cutter's gate. "I don't see what's so difficult in any of this," she said quietly.

Some of the suitors tried to trick the shining maiden, faking the begging bowl, hiring teams of jewellers to fabricate a jewelled branch . . . all failed and either went home filled with shame or embarrassment, prudently abandoned the search in mid-quest and never returned to the maiden, or died. All of this naturally found its way to the ear of the Emperor and made the shining maiden appear a fitting conquest for the Imperial Charm. But the elaborate strategems of the Emperor met with little more success than the rest, and when, in time, the moon-men came to take the shining maiden home again, 2000 Imperial troops sent to prevent them became rag dolls, sprawled helpless before the radiance of these creatures from another plane. . . .

The shining maiden left a last fond letter to the Emperor before departure and he went to burn it on the tallest mountain, where it smoulders still, so people call it Fuji, the Never Dead.

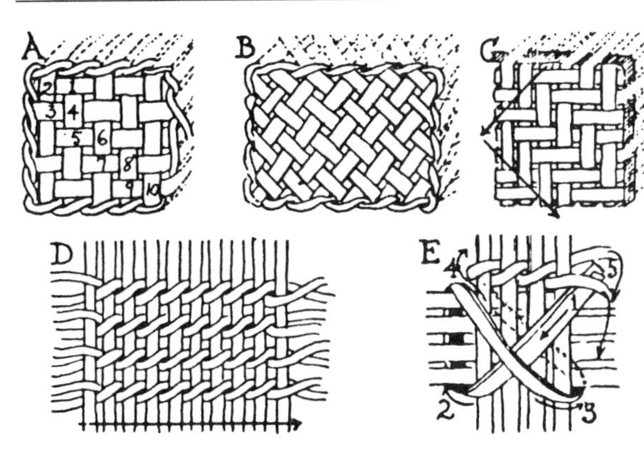

Square-bottom bamboo baskets: basic base weaves. Plaited base with checked weave (A); diagonal checks (B); twilled stitches (C); twined base (D); cross-tie center (E).

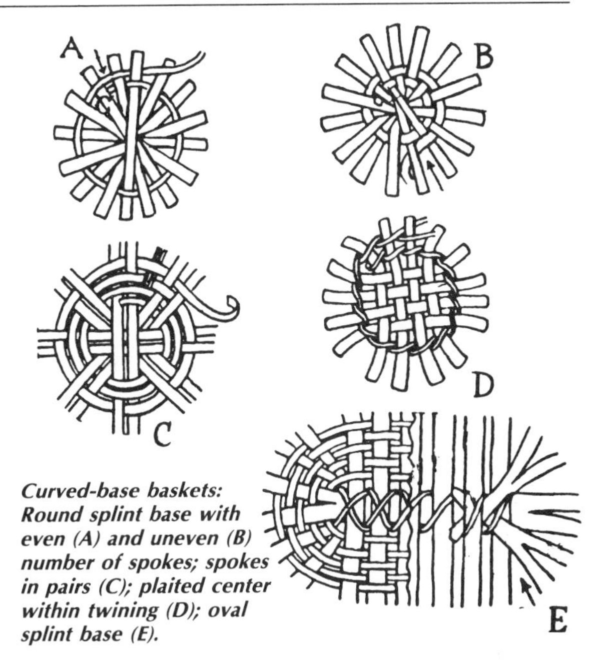

Curved-base baskets: Round splint base with even (A) and uneven (B) number of spokes; spokes in pairs (C); plaited center within twining (D); oval splint base (E).

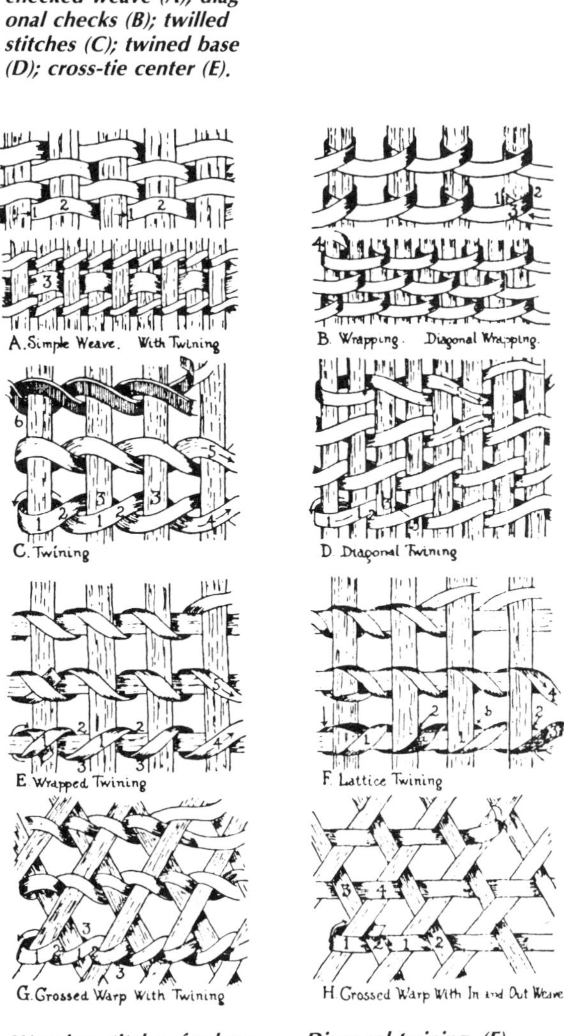

Weaving stitches for bamboo and other flat materials. (A) Simple weave (above), with twining (below). (B) Wrapping (above), diagonal wrapping (below). (C) Twining. (D) Diagonal twining. (E) Wrapped twining. (F) Lattice twining. (G) Crossed warp with twining. (H) Crossed warp with in-and-out weave.

Basket borders for bamboo or other flat materials. (A–G) Simple locked border. (H–M) Double locked border. (N–O) Border with twined edge. (P–Q) Simple wrapped border. (R–T) False braid border. (U,V) Twined border over groups. (W,X) Interlaced border. (Y,Z) Openwork braid border.

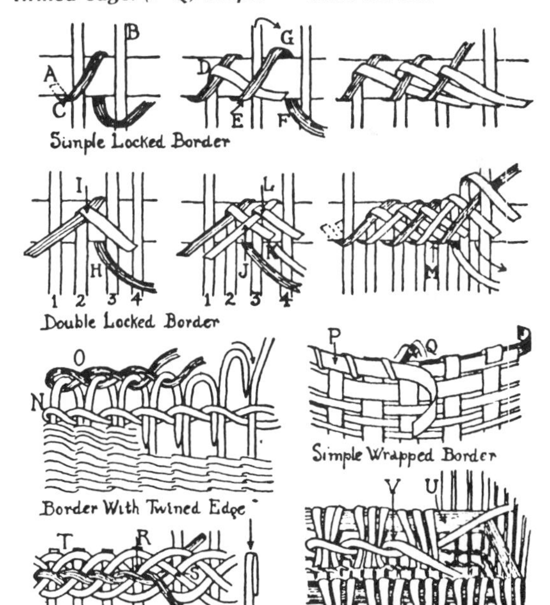

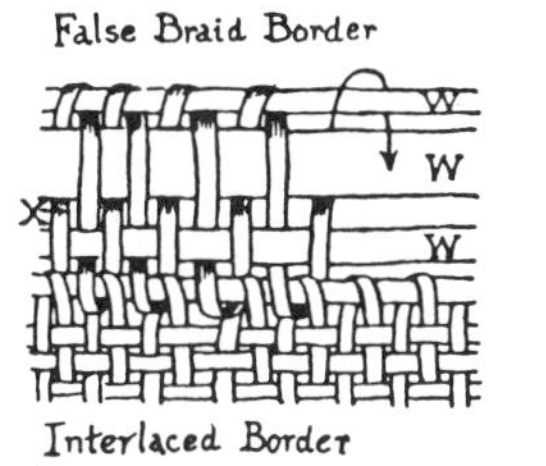

Basket birdhouse of bamboo with cornhusk thatch. Bamboo's rapid foliage can attract birds to cities in greater numbers and provide them with nest-building twigs, leaves, and sheath fragments.

practical works with bamboo in education. A condensed version should be made widely available, supplemented by other world bamboo designs for toys, furniture, musical instruments, playground equipment, kites—the whole range of small projects possible with bamboo, graphically and suggestfully presented.

Dramatic architecture. Bamboo's light weight, strength, and floatability have inspired shelter solutions distant from our present norms and provide inspiration for exciting structures. Tree shelters, water sites, and rolling road homes are amply and philosophically considered by Rudofsky's Prodigious Builders, *from which the following illustrations and most of these thoughts derive.*

Nest dwellings with bird's-eye views and basket elevators are an architectural option explored in the South Pacific. The basket elevator occurs often in legend and history. Saint Paul escaped from jail in one, and at Mount Athos, the Greek monastery, entering monks were supposedly lifted in them to the dizzying heights of the famous retreat.

Moroccan picnic hut, open, bright, and breezy, is lightweight enough to be hung from a tree in a smaller version for playground or camp.

BAMBOO TIPI. A bamboo tipi is easy to make and unmake. Laubin, in *The Indian Tipi,* notes how children among the Sioux imitated the shelters of their tribe: "Little girls had small tipis made under their mother's direction, in which they played house. They even made little tipis for their favorite dogs. Children also made tiny tipis and villages from the larger leaves of the cottonwood, pinning them together with splinters or thorns. Following the Custer battle, the paper money found on the soldiers was turned over to the youngsters, who made play tipis of it. A small tipi for children in the corner of the yard needs only nine poles in the frame, 11 to 12 feet long, and two for the smoke flaps . . ."

Tipis and other tribal shelter forms suggest themselves as an excellent introduction for children to building principles. A few forms are given here, but many more can be easily found in *Shelter* (1973) and other publications on buildings by the people.

Tree houses require less land for foundations than any other form of domestic architecture, an interesting feature in a crowded world. In the Orinoco River (Venezuela), air and water provide sites for woven and thatched treetop nests.

Bamboo island villages of up to 200 families on the Yellow River astonished 17th century European travelers in China. "The best artists in Europe would but coldly be able to make the like of the same stuff, a common reed which the Portuguese call bamboo, on which they set up their huts with wives and children. They keep and feed aboard their island all manner of tame cattle especially hogs . . ." Another traveler reported flowers, vegetable gardens, and orange trees as well. (Rudofsky 1974:146.) Anyone who has ever lived on a houseboat knows the lure of a floating home, which should be indulged wherever children, water, and bamboo of sufficient dimensions occur together.

PAPER

The place of paper.

Rags make paper
paper makes money
money makes banks
banks make loans
loans make beggars
beggars make rags.

—Anonymous English Verse

Many centuries before rags dropped off the legs of beggars for the task, bamboo was busy making paper in the East. That role is being examined with increasing interest by the Food and Agriculture Organization (U.N.) and other planet planners in the context of a dramatic rise in demand for paper around the globe. World paper use increases 5 percent annually, roughly twice the population growth. In some countries, it rises as much as 30 percent annually. About 85 percent of the world's paper comes from six countries, and most of it goes to the United States, Western Europe, Australia, and New Zealand, where per capita paper consumption shows a "spectacular discrepancy" (FAO) when compared with rates elsewhere in the world. Newsprint, which accounts for over a third of world paper use, is one example of this imbalance: Asia, with over half the world's population, uses 3 percent of the world's newsprint.

The 75 cubic meters of wood now standing in the world per person will be reduced to half that by the year 2000 at current deforestation rates—an acre a minute in tropical forests, according to some estimates. "Though developing countries contain three fourths the world's peoples and more than half its forest, they account for just 13 percent of global consumption of industrial wood. The average American consumes annually about as much wood—1 cubic meter—in the form of paper as the average resident in many Third World countries burns as cooking fuel."[36]

Paper plenty in densely industrialized societies forms an important part of their dominance over areas that are paper poor. Wealth depends in large measure on information storage and retrieval. The ease of information flow determines the school style and training capacity of a culture. In a number of ways, paper is the root of money and helps to broaden the gap between those with and those without. Developing countries trying to crawl from the deep pit of economic colonialism could profitably invest energy in being paper free, independent of imports. Many, like India, are in the hot and hungry middle of the world where bamboo happens most abundantly.

PAPER HISTORY.

> Man's high state of civilization can be easily construed to be more directly dependent on the inventions of paper making and printing than to any other development: these permitted wide dissemination of knowledge in time and space.[37]

Many surfaces have been used by many cultures as a message package. Papyrus (a plyreed mat from which our word *paper* comes) and woven cloth were used in Egypt. Parchment—from the inner hides of sheep—dates from about 1500 B.C. in Asia Minor. Vellum, from the whole hides of goats, lambs, and calves, was also used, but vegetable fibers were a more common message base. In North America, birch bark was a natural choice; in Asia, palm leaves. Sometimes, as in Ceylon, these were strung together on a string for "leaves" of books. Hemp and mulberry were employed in the South Pacific and in the Aztec and Mayan cultures of the Americas. The latter boiled the inner bark of these plants in a solution of wood ash and mashed the resulting mush on a flat surface, forming scrolls and accordion-fold books, few of which survived the Spanish conquistadors zeal to sow a monoculture and monocreed. The art, however, survived its artifacts: The technique persists in southern Mexico among Aztec and Mayan descendants.

In China, records before paper were kept on thin strips of bamboo, which proved durable enough to survive over two thousand years in fresh and legible condition. Military grocery lists and funeral inventories have survived (by chance) longest, and the classic volumes of Chinese history and philosophy were also long preserved by this means: "The Great Learning is the core of Confucius, the gate of sanity. If we today can see how the people of old went about their study, it is due solely to the conservation of these strips of bamboo."[38]

China's cultural longevity is in large part owing to its respect for the past, and part of the early affection for bamboo derived from its help in erecting an extensive cultural memory. The bamboo shavings were nearly a foot long, one character wide, punctured at one end, and tied together with silk or leather. But the thin strips proved a fat storage device, with a low index of data per unit of space: a 300 B.C. emperor required a number of carts to carry the books needed on a journey. A light but strong paper can weigh less than one third of an ounce per square yard (10 grams per square meter). "Miniaturization," increased performance per pound, a chief feature of all evolving technologies, is dramatically embodied in the invention of paper.

It is hard for us to reenter the wonder feeling excited by the world's first paper for those who actually witnessed its arrival in the midst of the bulky obsolescence it replaced. Making your own paper for the first time is perhaps the most direct route to experiencing that historical moment when true paper joined the human race, around A.D. 105, fashioned through the cunning of a Chinese eunuch using boiled fishnets. The essential process of papermaking has remained unaltered for nearly two thousand years since its invention; though mechanization, begun some two hundred years ago in Europe, has vastly accelerated the procedure.

True paper, as distinguished from the vegetable and animal writing surfaces described above, is made by "cooking" rags, straw, bark, bamboo, wood, or other fiber materials until the fiber can be separated from its mortar in the living plant. This pulp floats in a weak consistency—about 1 percent —of water, its fibers completely separated from one another in a large vat. A screen is passed through the vat collecting a mat of fibers. As this screen is removed from the water, the surface tension of the water draws together neighboring fibers, and tiny fibrils on their surface lock them together,

A world paper shortage is inspiring a closer look at bamboo. This is pulp of* Guadua angustifolia, *×75.

European papermaker, 1558. Arabs founded the first paper mill in Europe, at Xativa in Spain, A.D. 1150. Monks—the only scribes in the society—initially opposed paper because it came from the Arabs and was thought less permanent than vellum (calfskin) or parchment (sheepskin). Handmade paper in the West has become a craft of the few, but mechanization needn't eliminate knowledge and appreciation of the homemade process. Gandhi encouraged village papermaking for rural self-sufficiency in India.

producing a uniform sheet some four to five times as thick as the final, dried piece of paper.

This art took a thousand years to diffuse to Europe, traveling fewer miles per year than a person could walk in a few hours. It was introduced by the Moors to Spain around A.D. 1150 and spread quickly north within the next century. Combining paper technology with movable type in the first printing of the Bible (A.D. 1440) opened up the "Gutenberg Galaxy" of Western cultural evolution, some thirteen hundred years after the Chinese had first made paper and four hundred after the Chinese invented type (A.D. 1041).

An important early technical advance in the industry was the invention of the screen or sieve on which the wet fibers of pulp were lifted from the vat. Thin strips of bamboo—as many as twenty to forty to the inch—provided a surface from which the wet sheets could be "couched" or removed rapidly, permitting quick reuse of the screen. Metal replaced bamboo as the screen material in the West. It releases paper very wet and permits a finer weave, producing a smoother surface, with about fifty-two strands per inch. In modern screens, metal has been replaced by polyester plastic.

First in France, then in England, as the eighteenth century moved to the nineteenth, machines began to be invented with an endless loop of metal screen. The basic method of making paper was not changed but mechanized, with innumerable refinements continuing to the present mammoth papermaking monsters pouring out bands of paper 30 feet wide at roughly 30 miles per hour, to be wound on immense rolls 4 feet in diameter. One machine can produce 600,000 pounds of paper in one 24-hour period, roughly equivalent to ten years' production—at around 200 pounds per day—for one vat man working by hand.[39]

HANDMADE IN CHINA. For nearly 2000 years, bamboo paper has been handmade in the East. For seventeen of the nineteen centuries since its birth, all paper was made *only* by hand. All the variants of the process as practiced in a number of oriental countries with bamboo cannot be examined here, but two brief accounts of some phases of the trade as hand done in China, paper's native land, will suggest the main method. They come from observations by Westerners written in the early decades of this dwindling century.

"The stems are cut into lengths, made into bundles, and immersed in concrete pits, being weighted down under water by heavy stones. After three months they are removed, opened up, and thoroughly washed. Next they are restacked in layers, each layer being well sprinkled with lime and water, holding potash salts in solution. After two months, they are well rotted. The fibrous mass is then washed to remove the lime, steamed for fifteen days, removed, thoroughly washed, and again placed in concrete tanks. The mass is next reduced to a fine pulp with wooden rakes and is then ready for conversion into paper. A quantity of the pulp is put into troughs with cold water, and mucilage prepared from the roots of *Hibiscus abelmoschus.* An oblong bamboo frame, the size of the desired sheet of paper, having a fine mesh, is held at the two ends by a workman and drawn down endways and diagonally into the liquid contents, which are kept constantly stirred in the trough. It is then gently raised to the surface, and the film which has collected on the top is deposited as a sheet of

A

B

C

D

Paper production by hand in China: illustrations from the* T'ien Kung K'ai Wu*, a work of the Ming dynasty (A.D. 1368–1644) dealing with the arts and industries of ancient China: (A) Preparing culms for retting in a vat with lime to free the fibers from the starchy parenchyma tissue that mortars them in place. (B) Cooking the pulp. (C) Vatman forming sheets with a bamboo screen, with tubs of sizing to the left of the vat. (D) Pressing the finished sheets.

moist paper when the frame is turned over. After the surplus water has drained from the mass of moist sheets of paper, the whole is submitted to pressure. It is then dried; the cheaper papers in sunlight, the superior quality in kilns. Much water is necessary in papermaking, so mills are always erected alongside streams."[40]

"A paper mill depends on sufficient pulp in easy reach, a steady clear water supply, and cheap digesting materials, such as quick lime, soda ash, or potash. Primitive methods employed in old mills where paper is handmade are inadequate for refining the highly lignified tissues of mature bamboo culms. Better grades of paper are therefore made only from young and still leafless culms. Requirements are less exacting for cheap papers, and a wider range of species is employed. Probably any abundant local species, reasonably priced, may be used. Mature stems are acceptable for the very coarse dark papers of common use for filters, wrapping, and so forth—for which mature culm tips, a by-product of the split-bamboo industry, are used in Southeast Asia. Digestion time is very long, often a full year. Pulping methods are not highly refined.

"In construction of the common mold, on which the finest paper is still made by hand in the Orient, bamboo is always used: a flexible screen of slender wirelike units fastened together in parallel order with hair, silk, or ramie. Peripheral wood of large moso or madake culms make the best screens. After preliminary splitting, strips are reduced to size and cylindrical form by pulling through holes in a piece of steel, which produces wirelike strips of marvelous uniformity and fineness. Finished screens, treated with lacquer, are objects of great beauty and unbelievable durability. The binding fibers, the 'warp,' wear out before the strips. When these can no longer be repaired, the bamboo strips are salvaged and worked into a new screen.

"Bamboo finds numerous incidental uses in the average Oriental handmade papermill. Half-stuff is carried from the digesting vat to bamboo treading troughs in bamboo baskets suspended from a bamboo pole. Finished pulp is 'combed' with bamboo loops to remove coarse fibers—'shives'—which have escaped reduction by digesting and treading. In the dipping vat, water is added and pulp agitated with bamboo stirring rods to disperse fibers evenly. Vat man and drier work by bamboo lamplight at night. Bamboo rope is used on the windlass for applying force to the press. Bamboo forceps are used to pick up the corners of the wet sheets from the block as it comes from the press. Old bamboo culms too highly lignified for handmade pulp are commonly used as fuel for drying the paper. Bales of finished paper are often covered with bamboo culm sheaths and bound with bamboo bands. A bamboo tool, combining the functions of a gauge and an awl, is used to space the bands on the bales and tuck in the twisted ends.

"The principal technical problems blocking bamboo use for pulp in modern mills have been solved. Many variants of the process have been patented in those countries where paper is made on a large scale. At least one of the several modern papermills established in pre-Communist China used bamboo pulp, making ninety types and grades of paper, ranging from wrapping paper and tissues to bond and ledger."[41]

MODERN METHODS. It is not China but India that has taken the lead in the Orient and the world in the production of bamboo paper. With some 20 genera, 113 species of bamboo covering almost 3 percent of her land—an estimated 9.57 million hectares or nearly 24 million acres—India is blessed with the largest bamboo reserves of any country in the world. Of roughly 3.23 million tons of air-dry bamboo—20 percent of India's total annual wood production—about 2 million tons is consumed by the paper and rayon industries.[42]

The extent of bamboo reserves in the country has had an interesting influence on bamboo research. In spite of the fact that the United States has only two native species and England none, English is a primary language for bamboo research. In the mid 1800s, England's empire—far-flung, literate, and addicted to print, connected and ruled by messages on paper—needed a backyard bigger than Sherwood Forest to grow paper pulp. Queen Victoria declared herself Empress of India in 1877 and became legal owner of vast resources of bamboo, which Englishmen like Routledge (1875) and Raitt commented on with sufficient persuasion to inspire the opening of bamboo paper work in India in 1910, soon making India the world's largest user of bamboo.[43] The stream of publications in *The In-*

dian Forester on bamboo since 1875 and the research center at Dehra Dun—north of New Delhi—have helped make English the main Western language for bamboo studies.

As surely as the foot of the ox is followed by the cart's wheel, increased bamboo study has been followed by increased bamboo use. The 6,000 tons of bamboo used for pulp in India in 1925, by 1953 had increased to 80,000; by 1959 to 450,000; by 1970, to 800,000; by 1980, to 2.2 million tons. Projections are for 3.1 million tons to be used by 1985, and 3.5 million, by the year 2000.[44]

Meanwhile, other countries have followed India's lead. Pakistan, Burma, Indonesia, Taiwan, China, the Philippines, Kenya, and Brazil—among other nations—now use bamboo paper pulp on a large scale. Increased need is a principal reason for increased interest in bamboo. As modern packaging and literacy grow worldwide, so also does the consumption of paper. In Thailand during the 1960s, for example, paper needs tripled, but 75 percent of the paper was imported. Properly managed, large bamboo forests in northern Thailand that presently provide only 15 percent of Thailand's paper could supply all of the country's domestic need and leave excess for export as well.

WOOD AND BAMBOO PULP. Use of bamboo for paper pulp on an industrial scale is only sixty to seventy years old, but papermaking already accounts for a significant proportion of the world's bamboo consumption. Improved grove maintenance, the world paper shortage, and diffusion of techniques and machinery are expected to increase dramatically the quantity of bamboo pulp production in the years ahead; but the principal reason for that increase will be the natural advantages bamboo possesses over wood as a raw material for papermaking.

Trees such as pine must be individually planted, require fifteen to thirty years to mature, and yield only one cutting. Reforestation is necessary after each harvest. Bamboo, on the other hand, spreads unaided after planting, and a single rhizome and its descendants yield hundreds of culms in the total life of a grove. In China, twenty-four bamboo stems of moso *(P. pubescens)* were planted, tended, and left unharvested. In twelve years they increased to 3,200 stems.[45] Bamboo matures in three to six years, but one-to-four-year-old culms are preferred for pulp, and the grove flourishes through regular thinning in cutting cycles of one to four years. It should be stressed how much harvesting increases the vitality of a grove. In one experiment lasting eight years, reduction of the number of culms by 40 percent actually increased the total harvest weight over one third.

In the thirty years one pine harvest is maturing, an established grove of bamboo may be harvested ten to twenty times, depending on the species and other factors determining the cutting cycle. "Bamboo has no bark to be removed and its high specific gravity ($\pm$ 0.6) makes possible 10 to 20 percent increase in the effective pulp capacity of the digester. It is estimated that six or seven times as much cellulosic material can be obtained per hectare from a bamboo forest as compared to a coniferous or other broad-leaved forest. Bamboo fibers are longer than those of hardwoods but are shorter than those of most coniferous woods. Length to width ratios are higher than those of wood fibers; the bamboo fibers are strong and flexible rather than stiff and brittle—features that are better for most papers, especially high-quality ones like facial tissues, bond papers, and stationery."[46]

One constant of the history of paper manufacture has been a chronic shortage of raw resource. Around 1850, when cotton and linen rags fell drastically short of paper demands, cereal straws and then esparto grass supplied sufficient pulp for some twenty years. Around 1870, acute shortage was relieved with new chemical processes to convert coniferous woods into cellulose. But increased use increased demand for paper. A half century after conifer pulp rescued the papermaker, the world was consuming 20 million tons of paper products annually, and cellulose industries of many sorts were evolving to further devour the forests. At Dehra Dun, the Forest Research Institute was concluding, with increasing experience to firm their feeling, that bamboo was the only feasible alternative available in sufficient quantities to meet the global demand for paper.

Drawbacks in bamboo paper production include the large amount of rootstock necessary for propagation, the adaptations in pulping machinery necessary for processing bamboo, the periodic flowering and death of groves, and the high cost of shipping. The solutions presently researched, proposed, or enacted include tissue culture of bamboo to start thousands of plants from the tip of one shoot, reliance on a number of species so that flowering cycles do not coincide, and processing at the groves as far as the pulp stage to save on transport.

Bamboo pulp can also be a by-product of other industries. In Taiwan, annual waste from bamboo-shoot canning plants of 35,000 metric tons of culm sheaths has inspired an effort at creative garbage disposal. The Taiwan Forestry Research Institute began experimental runs for printing paper and the corrugated guts of kraft paper board with satisfactory results: The product was comparable in quality to the printing paper made from 80 percent rice straw and 20 percent long wood fiber. Paper yield was about 30 percent, comparable to rice straw. Strength and brightness of the culm sheath papers were rated better than rice straw papers. For the corrugated innards of kraft paper boards, yield was 45.8 percent, lower than *bagasse** (50–65 percent) and conifers (60–75 percent). The fiber dimension of bamboo sheath was inferior to bamboo culm and conifer fibers, but better than some hardwoods. Dehydration of the wet culm sheaths from the canning factories, in a simple and economical form, appears to be a critical problem.[47]

PAPER MAKES SCHOOLS. The paper beneath the words you're reading may be the first you've seen in your life inviting you to think about paper. We haven't been taught to look at the paper, but at the words—which are curiously silent in our schools about the paper they're sitting on. Paper is the invisible foundation and environment of the educational method, but its manufacture and world distribution and the ways to make it from scrap paper or fibers of many locally available plants are not discussed or demonstrated. Simple methods of paper production could profitably be added to school curriculums in an effort to increase awareness of this material so central to the cultural evolution of any person or nation in our times. Making one's first sheets of paper is an astonishing experience for anyone who has dealt with paper for years without making it. It is an experience easily incorporated into the school system of any country—with or without bamboo. Lampshades, screens, curtains, paper partitions in houses after the manner of the Japanese, packages, kites—these are a few immediate uses for handmade papers. Here, briefly, is how: You don't have to wait for a bamboo harvest. Junk mail will do.

PAPERMAKING IN YOUR KITCHEN SINK. Save scrap paper. Cook a while until soupy. Squeeze a handful of pulp damp dry and put it in 3 cups of fresh water

Pulping by hand. Use just enough water for slushy consistency in pulp. A bottle base serves as a beater. When well pulped, test a little with twice as much water in a jar. There should be no recognizable particles of paper left.

in a blender. A fine screen on a wooden frame built the size of paper you want to make is fitted semi-snugly into a wooden box with four sides but no bottom or top. Leave just enough room around the frame for your fingers. The box is then placed in the kitchen sink or tank of like size, which is filled with water to within a few inches of the top of box. The 3 cups of water with dissolved pulp is then poured into the water above the screen and gently agitated to distribute the fibers uniformly through the water.

Slowly lift the screen, and the fibers will become evenly distributed over its surface as it is raised from the sink. Turn the screen over on a paper towel or other absorbent surface, and soak up any excess moisture through the screen carefully with a sponge. Then lift the screen off the latest sheet of handmade paper in the world, and let it dry in the sun.

The texture and color can be infinitely varied by small additions of vegetable fiber such as grass clippings, greens of vegetables such as carrot tops, beets, leaves, corn silk, shredded corn husks, or onion greens. Dozens of alternatives await your experiments in your backyard, groceries, or nearest vacant lot. For a school project, Chinese handmade papermaking methods could be imitated on a small scale. Handmaking paper in schools will not solve the world's paper shortage. But it will make for a

Editor's note: Bagasse: the fibrous residue of sugar beets or sugar cane after the juice has been extracted.

more paper-conscious populace, give some paper geniuses an early start with their obsession, and perhaps even inspire governments—especially in developing countries—to acknowledge the central seriousness of designing paper self-sufficiency as a precondition for many cultural processes. Ministries of education, particularly, should assign the national student body the homework of designing and creating national paper autonomy. Any countries poor in paper but rich in bamboo reserves or blessed with a climate friendly to bamboo cultivation will be knocking on the door at Dehra Dun to collect advice.[48]

FOOD

Eat your lawn: bamboo shoots.

> The best way to control bamboo is to eat it.
> —David Fairchild

In Taiwan, a single factory turns out 150 tons of bamboo shoots daily:[49] roughly 8,000 tons a year are consumed there and roughly as much in Japan. This amounts to 22 tons each day and indicates how common a part of the oriental diet sprouts are.

"Those who savor the roots of wild vegetables know the meaning of life."—Wang Hsin-min (Sung dynasty)

Arundinaria ***bamboo shoots.***

But in the West, except for those with their own grove, shoots are available only as a canned delicacy. In a few locations, such as Chinatown in San Francisco, shoots are sold out of great tubs of water. The price in 1980 was $1 per pound.

When bamboo is cultivated for shoots, workers walk the fields barefoot, heaping soil around those spots where their toes feel a slight rise in the ground. The bulge heralds an emerging culm, which is kept covered as long as possible, sometimes with a wooden box, because when exposed it becomes fibrous in a short time. The shoot is harvested when about 1 foot tall, by digging down and carefully severing it at the point where it joins the rhizome bearing it. Shoots are boiled a half hour or more, and the outer sheaths removed. They are white in color, with the look of a raw potato, crisp in texture. In taste they are like young field corn, slightly sweet.

Fairchild's advice—to control bamboo's spread by eating it—has probably never occurred to most people who have a problem with excessive growth in a small urban garden, but shoots of almost any bamboo can be eaten. The principal genera used for sprouts include *Arundinaria, Bambusa, Dendrocalamus,* and *Phyllostachys.* Although size is one limiting factor, even diminutive *Sasas* in Japan provide some 600 tons of shoots annually.[50]

In Taiwan, per capita consumption is six times that of Japan, though total volume of shoots—8,000 tons annually—is about the same. *P. edulis* and *D. latiflorus* use has reached a high degree of specialization, including processing, canning, and export. Shoot export value in 1977 reached almost 25 million in U.S. dollars. Exported bamboo products the same year valued at $40.8 million U.S. was double the $20 million or so of 1973.

In Japan, *Phyllostachys* managed for shoots—mostly moso, *P. edulis* (or *pubescens*)—are topped off (pollarded) at 30–40 feet so that sunlight and warmth help prevent snow damage. Minimum temperature for shoot production is 20°C; shooting seasons are April–May and November. Yield of about 10 tons per hectare per year is valued at 1 million yen (yen = $0.00455 U.S.) or $4,550 U.S. per hectare. Costs invested are generally 10–30 percent harvest value. Soft, quality shoots require yearly soil dressing, straw litter, and farmyard manure.

In Korea, shoots are collected April to mid May, nearly 9,000 lbs. per acre (10,000 kg per hectare) from an intensively managed grove. At 280 won each, shoots value about $0.47 (won = U.S. $0.001671). Taste research is conducted, and the most vigorous shoots are preserved as mother culms.[51]

COOKBOOKLET

The bitterness of some bamboo sprouts is removed by boiling in two or even three changes of water:

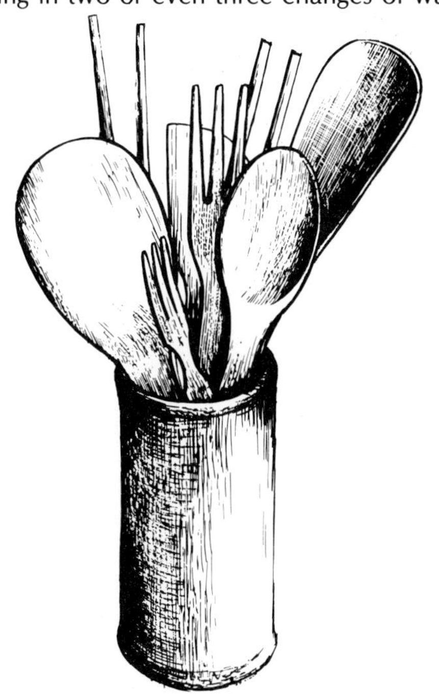

"A sprout properly dug will have a rooty, woody, basal portion, increasing sharply in diameter upward for a short distance from the very slender part which was attached to its rhizome, and then tapering to a point. With a sharp paring knife cut lengthwise through the sheath only, from tip to base of the sprout. Beginning with the lower sheaths, remove all except the tender ones at the tip. If there is a grayish layer (owing to age) next to the lower nodes, pare this off. Remove the tough basal part. Cut the core portion diagonally or crosswise into rather thin slices. The lower, firmer portion should be cut across the grain, not thicker than ⅛ inch, but the more tender middle and upper parts may be sliced thicker or cut into various shapes, according to the recipe. Sprouts may be served alone, drained, with butter melted over them, after boiling for about twenty minutes. Salt is added near the end of the boiling period. If the fresh sprouts are unpleasantly bitter to the taste, there should be a change of water after the first ten minutes of cooking. The more pronounced bitterness of some of the tropical edible bamboos (of the genus *Bambusa* and others) may be removed by a third change of water if the bamboo is not sliced too thick and a good volume of water is used."[52]

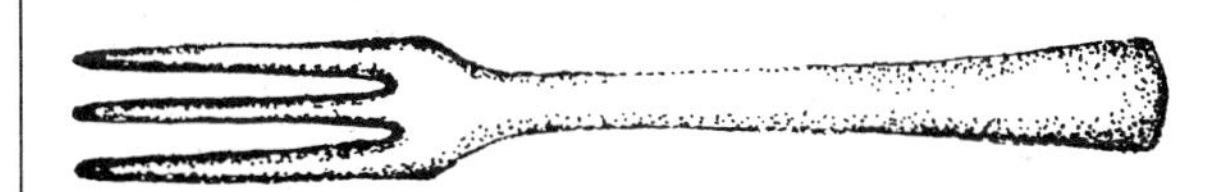

Scalloped bamboo.

3 cups parboiled sliced bamboo
4 tablespoons butter
4 tablespoons flour
1 teaspoon salt
4 tablespoons grated cheese
paprika

Place the bamboo in a greased shallow baking dish. Prepare a sauce of the butter, flour, milk, and salt; then blend in the cheese. Pour this over the bamboo and bake in an oven at about 350°F for 30 minutes. Serve with paprika sprinkled over the top.

Sautéed bamboo.

2 tablespoons butter or other fat
3 cups sliced parboiled bamboo
1 teaspoon salt
pepper

Heat the fat in a frying pan, add the bamboo, sauté for about 5 minutes until slightly brown, and stir occasionally. Add the seasoning and serve on hot cooked rice with a cheese sauce.

Bamboo potato salad.

2 cups diced cooked potatoes	1 cup finely cut parboiled bamboo
1½ teaspoons salt	2 teaspoons chopped onion
½ cup French dressing	1 cup salad dressing
1 cup finely cut celery	½ cup diced cucumber

Add the salt and French dressing to the potatoes and chill. Add the celery, bamboo, onion, and cooked salad dressing. Mix together carefully to avoid breaking the potato. Add the cucumber and serve on crisp lettuce. Bamboo may also replace the celery, wholly or in part, if desired.

SPROUT FOOD VALUE. The food value of bamboo sprouts is roughly equal to that of an onion, with the following composition per 100 grams in the case of *P. edulis.* (One ounce = twenty-eight grams.) Values for fresh sprouts are followed by figures for the canned product: crude protein, 2.5 g, 1.9 g; crude fat, 0.2 g, 0.1 g; carbohydrates: sugar, 2.9 g, 2.9 g; crude fiber, 1.0 g, 1.8 g; water content, 92.5 percent, 92.8 percent; calories 23, 20; ash, 0.7 g, 0.4 g; lime, 1 mg, 1 mg; phosphorus, 43 mg, 26 mg; iron, 7 mg, 1 mg; vitamins: A, 50 iu, 50 iu. B-1, 0.10 mg, 0.5 mg. B-2, 0.08 mg, 0.05 mg. C, 10 mg, 0 mg.[53] Slightly different figures are given by Leung for unspecified *Bambusa* and *Phyllostachys* species.[54]

HAY AND FORAGE. The horses of the Chinese imperial stables were fed on bamboo leaves, which—half a world away—Jamaican exports regard as a "hard" forage conferring superior physical tone and stamina on horses compared with animals raised exclusively on succulent grasses. A favorite food of many wild animals—elephants and buffalo in India, pandas in China, and gorillas in Zaire, to name a few—bamboo provides relished feed for cattle, sheep, and horses to farmers all over the world where the plant abounds. *A. gigantea,* one of two bamboos native to the continental United States, is the highest yielding native range in its land of birth. Other species make significant contributions to animal food needs elsewhere. Among fifty-two species discussed by McClure, some of the more significant include the following.[55]

B. arundinacea is listed by Indian government publications as one of the fourteen most important shrub and trees for fodder in India and Ceylon.[56] It is unfortunately poisonous to livestock when new shoots are small and contain lethal amounts of hydrocyanic acid.

B. ventricosa has shown a remarkable tolerance for drought during severe dry seasons when it produces "an astonishing mass of succulent foliage." As such, McClure recommends it for trial as a forage in areas with prolonged dry seasons.[57]

B. vulgaris leaves are a valuable feed, very rich in nitrogenous material, eaten with zest by cattle and horses. Cut to the ground, clumps spring up again rapidly, providing abundant forage. Cattle fed on it during long dry periods equaled the condition of those fed ordinary fodder, and bamboo leaves of this species are claimed to make for more resistant horses. Dried leaves of *B. vulgaris* and *B. ventricosa* have been found an excellent source of vitamin A in chick rations.[58] In spite of hydrocyanic acid found in young shoots of *B. vulgaris,* there are no reports of cattle poisoning. Various species of *Chusquea* have local importance in many areas for forage in South America. In India a three-year plantation of *D. strictus* yielded 19½ tons of fodder per acre, which proper management could increase to 40–60 tons. The spear grass commonly found on equivalent soil was producing 1½–2 tons of green fodder per acre. The green leaves of *D. strictus,* supposedly curative for certain animal ailments, are fed to cattle and ponies with broken wind or hoof and mouth disease.

Of the *Phyllostachys* species, trimmings from all those at the Savannah, Georgia, groves of the USDA were routinely fed to cattle and mules. *Sasa paniculata* is used as pasture for horses and sheep in Japan.

The chemical composition of bamboo leaves varies through the year. Their nitrogen content diminishes from May to October, for example, from as much as 25.8 percent to 6.6 percent.

Analyses of *A. tecta* in North Carolina revealed great changes in amounts of crude protein, calcium, and phosphorus. Digestibility of bamboo leaves diminishes with age.

Bamboo shoots are routinely protected from cattle, for the sake of both the animals and the grove, since several species of bamboo have sufficiently high concentration of hydrocyanic acid for 5–10 ounces of their shoots to kill a full-grown cow within two hours. Boiling quickly rids sliced shoots of cyanogens; there is no fear of poisoning people from cooked shoots.[59]

"Daniel Boone, the deer-skin man
needing land or fuel for free,
took his gun and axe in hand
felled an Indian or a tree . . ."
—Popular ballad.

ECOPROTECTION

Revegetation: heal thy planet.

By nature man is a forest dweller. He was cradled in the tropics. His food was the fruit of trees. In his forest setting man was conscious of his unity with all living things. The memory of that Golden Age has come down through the ancient Egyptians, the earlier Greeks, the Aztecs, and is told in the folk-lore of African peoples of today. The paradise or garden was a clearing in the forest where gourds and other vegetables were grown. The folk-lore and legends of the Golden Age echo the scriptures of many religions and show that trouble first came when man forsook the garden culture and became a herdsman. The traditional "field" or "felled" was a clearing in the forest. In the garden people lived in harmony with Nature and learned to understand the seasons for sowing and harvest, as well as the best ways of storing seed and food; and from the art of horticulture other forms of culture grew, with seasonal dances.

—St. Barbe Baker

Bamboo is managed by the forestry service in most countries, and in its natural state is a recurrent companion of trees. Any consideration of bamboo use on a significant scale to people now implies a careful look at this larger context of the world's forest use and abuse. As coda to these notes on bamboo, we add a chorus of warnings on the central seriousness of devegetation in our time and the importance of tree and bamboo farming as part of a worldwide effort to revegetate—especially among the young in schools. The uses of bamboo for people are complemented by its uses for the planet, as weather buffer for ecologically vulnerable terrain such as hillsides and riverbanks; as bandage for lands already bruised by human enterprise; as friend of forests and ally of appropriate wilderness.

These pages grew, nourished by a huge hope that we find our way towards a balanced and intimate agriculture, humane and permanently possible, in which the best crop of the kingdom would be not bamboo or any other plant, but we ourselves, the gentle stewards of the earth's sufficiency, complete people. As the Chinese proverb says,

Complete people grow neither fast nor slow,
like a tree on a mountain—
root by root without rest
gripping the earth more firmly,
leaf by leaf without haste
climbing sky. —I Ching, Tree on a Mountain

Earthskin.

Early peoples gradually improved the flowering plants and developed edible grains such as wheat, barley and rye; stores and granaries were erected; fields were enlarged; more and more forest was removed and frequently the water cycle was broken. It took time to recognize the connection between tree cover and the growth of food crops. In the course of centuries people discovered it was easier to raise and graze domesticated animals than to hunt the wild ones. Soon flocks and herds spread over the land. The animals, especially goats, denuded ever-widening areas of trees. In order to protect their flocks the owners hunted for wolves and other predators. When the grass grew sparse they felled more forest, cutting and burning to make new grazing lands. The pressure on the land encouraged hamlets to swell into towns, and towns into cities, where the inhabitants developed trades and professions, demand for meat increased and crops were diverted to feed animals and walls were built around settlements to keep the animals out.

The Ancients believed that the earth was a sentient being and responded to the behavior of man upon it. As we have no scientific proof to the contrary, should we not accept this point of view and behave accordingly? If a man loses one-third of his skin he dies; the plastic surgeons say, "He's had it." If a tree loses one-third of its bark it dies. This has been proved scientifically by botanists and dendrologists. Would it not be reasonable to suggest that if the earth loses more than a third of its green mantle and tree cover, it will assuredly die? The water table will sink beyond recall and life will become impossible.

—St. Barbe Baker

ONCE UPON A TREE.

> Forest-field-plow-desert: that is the cycle of hills under most plow agriculture. Field wash, in the United States, Latin America, Africa, and many other parts of the world, is the greatest and most menacing of all resource wastes. We are destroying our soil faster and in greater quantity than has ever been done by any group of people at any time in the history of the world.[60]

So many problems are pounding on the door of human history, it hangs lose on the hinges; but among the more insistent are food, fuel, and water shortages, all related intimately to soil loss. We have spoken of firewood's relation to the present ecological disasters on the earth.[61] "Eckholm [*Losing Ground,* 1976] sees the shortage of firewood as a central feature of this crisis. The uncontrolled clearing of remaining or replanted trees has its severe ramifications: Precious topsoil erodes not in centuries as in the past, but practically overnight; a disastrous increase in flooding occurs on lowland plains; and new deserts and grass wildernesses are created in drier zones by inhabitants removing the remaining ground cover. Economic development projects in Africa, Asia, and Latin America are now dimmed by accelerating destruction of the land's productivity. Slash and burn shifting cultivation has in many places increased beyond the ability of the sensitive forest to recover. Massive forest-clearing operations by governments and corporations in places such as the Amazon river basin and Borneo are often followed by heavy grazing, which completes the land's destruction."[62]

HAMBURGER HUNTERS: MEAT YOUR MAKER.

> For Africa, estimated meat production fell to one-sixtieth its natural level when we cleared, fenced, ploughed, sowed pasture and introduced exotic cattle. In Australia, we find one grazier barely existing on lands that supported two to three hundred Aborigines.[63]

World woods are being lost for a number of reasons—inefficient cook stove design, increasing populations increasingly "civilized"—that is, requiring more packages, Kleenex, and newspapers. Each Sunday edition of the *New York Times* knocks down 140 acres of forest. But a major reason in many areas for forest loss—and bamboo grove destruction—is the hamburger hunters. Humanity's addiction to dead cows, the idea that being made of meat we need it to survive helps create drastic changes over large areas of the earth's surface. These changes effectively reduce protein production, dramatically alter weather, augment erosion, extinguish habitats, accelerate extinction of species . . . and human tribes. Many effects more messy than cow pies fall in the wake of the herd. Meat, in the modern world, is not an "everybody idea," not a feasible planetary norm.

Four of these are bacteria-burgers, four real hamburgers. Can you tell the difference? Correct answers will be found in a fortune cookie under the plate.

"The energy content of the food on the McDonald's menu is just one tenth the energy expended to get it from farm to consumer. The reverse is true in a primitive society of subsistence farmers in New Guinea. They get sixteen times more energy from the vegetables they raise than the human energy used in farming and cooking them. Their farm produce goes almost directly into their mouths, without ever being machine-stamped into 1.6-ounce hamburger patties 3.875 inches wide, quick frozen, shipped great distances, grilled and neatly packaged in a paper box. Only in America and other affluent countries does food take such a long, energy-wasting route to the consumer."[64]

Even in revolutionary countries committed to feeding all, governments often find it hard to revolutionize the mass diet, adjust it to necessities imposed by present population densities and to the possibilities offered by technologies that are known but unenacted. It's interesting, for example, that the biggest mammals eat the smallest lunch. Whales, the size of a house and a half, feed on plankton as small as the words on this page. The human race, a quarrelsome giant on the planet now, could take a hint from this fact of nature. The most bulky thing we eat, a steer, is the least efficient crop we have

Faunavores and Floravores.

How strange it is that communities fail to realize the importance of preserving tree cover on tree slopes. Man has a bad record as a forest destroyer, cutting and burning greedily and recklessly, destroying the built-up fertility that has accumulated through the centuries. He has been skinning the earth alive in his greed and folly and to satisfy his unnatural appetite for the flesh of his fellow creatures.

In some countries, such as the U.S.A., up to three-quarters of the land has been degraded to the use of growing crops to feed animals which they kill to feed themselves. Surely a round-about way of getting food, when it is possible to feed ourselves directly from the earth through fresh vegetables, fruit and nut-bearing trees. It is possible to produce meat and milk direct from the plant kingdom. These foods are not only of better taste, but richer and more wholesome—easy to prepare and cheaper to use. Above all, everybody who has his own piece of land can easily raise his own raw materials for these foods on his own land.

Village communities of the future, living in valleys protected by sheltering trees on the high ground, will have fruit and nut orchards, live free from disease and enjoy leisure, liberty and justice for all, with a sense of their one-ness with the earth and all living things.

Under existing systems food looms large and there is a constant threat of famine over wide areas, but if we treat reforestation as seriously as we do national defence, and turn from an animal economy to a sylvan one, we shall be able to look forward confidently to the time when food will worry us as little as the air we breathe.

—St. Barbe Baker

for protein production. A "diet for a small planet" with a big family might sensibly eat more bugs and fewer cows. Half the world's people have nothing to eat tonight but their fingers. None of them are ministers of agriculture. If they were, they'd already be acting on this information, which is already old:

> The full potential of protein bacteria is easier to grasp if it is compared to a properly fed 1,000 pound steer. The steer stores up just 1 pound of protein a day. In the same 24 hours, a half ton of selected microorganisms, feeding on oil, increase in size and weight by five times, and half this gain is useful protein. While the steer is making 1 pound of protein, the bacteria-in-oil produces 2,500. The line of meatlike, meatless foods includes ham, sausage, frankfurters, chicken, steaks, meat loaf, chipped beef, and luncheon loaves. They have no bones, skin, or excess fat. Surprisingly, most of these rated high both in taste and appearance.[65]

MONSTER MAKERS: INAPPROPRIATE TECHNOLOGY. In the United States and other hyperdeveloped countries, herbicides are another major source of devegetation. The area under transmission lines of utility companies in the United States, for example, adds up to some 32,000 square miles, equal in size to the combined areas of the five states of Massachusetts, New Hampshire, Vermont, Connecticut, and Rhode Island. One tall tree beneath the wires can interrupt the power cables, so right-of-way managers in the past have dumped on the ground beneath them each year 4 million pounds of 2,4,5-T, a herbicide that was the major component of Agent Orange, the dreadful defoliant used in Vietnam. This herbicide contains dioxin, one of the most toxic of all manmade substances, identified as a cause of liver cancer. It is also a "teratogen" or "monster maker": Its use was temporarily suspended by the Environmental Protection Agency on rights-of-way, pastures, and forest lands in 1979 as a result of a sudden rise in miscarriages and birth defects in Oregon, where it was sprayed on forests before planting Douglas fir and other evergreen seedlings. Oddly enough, the use of 2,4,5-T on croplands and rangelands was not prohibited by this partial ban.

STABLE LOW-SHRUB COMMUNITIES. Highways, railroads, and pipelines are other open spaces controlled by herbicides in the United States—a total area of some 200,000 square miles or 34 million acres. An alternative to this chemical control is described by Tillman (1979): the establishment of stable low-shrub communities that are nonpollutant and aesthetically pleasing. They encourage wildlife by providing food and cover not available in forests or grasslands and offer greater erosion control with roots deeper than grasses. "Many plants have been identified as being good components of a stable low-growth community. Sumac, various dogwoods, viburnums, small willows, sweet fern, snowberry, blueberry, blackberry, and bear oak are examples of desirable shrubs. Goldenrod, meadow fescue, bluestem grasses and bracken fern are favored herbaceous species."[66] Allelopathy, the production by plants of chemicals that inhibit growth of other plants around them, is one mechanism responsible for creation of a stable, low-growing

plant community that is currently being researched more at centers such as Cary Arboretum, in Millbrook, New York.

"With so many benefits to shrub communities, one may wonder why all rights-of-way have not already been converted. Unfortunately, it is more a problem of converting right-of-way managers than converting the plant communities. Most of the present generation of right-of-way managers have been trained at forestry schools that emphasize techniques of growing tall trees quickly, not the low and slow-growing vegetation needed on rights-of-way. Most managers are not eager to have to learn about a whole new discipline, vegetation science."[67]

The possible role of bamboo in that science is, of course, completely unexplored. But shrinking land per person will eventually impose more productive planning on these millions of presently wasted acres. Groves dividing dual-lane highways,

for example, could provide a visual shield for headlights and a soft crash zone where firm-yielding culms would gently brake a speeding vehicle, while providing as well a crop for construction, paper pulp, or schools.

AGROFORESTRY. Some think that we shall never see crops more productive than a tree. "Nothing gives more yet asks less than a tree," remarked Jonathan Chapman, known in his shady wake as Johnny Appleseed. A number of modern agronomists agree: "The 'tool' with greatest potential for feeding people and animals, for regenerating the soil, for restoring water systems, for controlling floods and droughts, for creating more benevolent microclimates and more comfortable and stimulating living conditions for humanity is the tree. In food productivity alone, tree crops can produce ten to fifteen times as much food per acre as field crops."[68]

> If much of the tropic forest is to be preserved, we must make use of tree crops. Tree crops will safeguard fertility while producing food for man. In most cases there can be an undergrowth of leguminous nurse crops of small tree and bush to catch nitrogen, hold the soil, make humus and feed the crop trees—nuts, oils, fruits, gums, fibers, even choice weeds (Smith, *Tree Crops: A Permanent Agriculture,* 1950).

> It has been predicted that within the next twenty-five to thirty years, most of the humid tropical forest as we know it will be transformed into unproductive land . . . Agroforestry is a sustainable management system for land that increases overall production, combines agricultural crops, tree crops, and forest plants and/or animals simultaneously or sequentially, and applies management practices that are compatible with the cultural patterns of the local population. Trees are the dominant natural vegetation in most of the tropics, and with few exceptions must remain so if the land is to be used for the greatest benefit of man (Bene, *Trees, Food and People: Land Management in the Tropics,* 1977).

But we mustn't forget the forest in the trees. Any natural forest is a community of many species creating a symbiotic blanket of life, woven of many plants and animals. A broader notion than "reforestation" is required.

> *Revegetation* is a concept that needs wider circulation. The planting of many different suitable types and sizes of vegetation makes the widest possible use of the capacity of a landscape to support life. A wide range of plant life also further expands the amount and diversity of animal life that the area can support, and this includes the human animal. A community of plants composed almost solely of trees ignores the potential that shrubs and ground covers can contribute to the productivity of a landscape: animal life, both wild (deer and fowl, for instance) and domestic (such as pigs, cattle, and geese) will not do as well when there are only trees. *Reforestation* usually creates a place for humans to come and get lumber and firewood and little else. *Revegetation* creates a place for a greater variety of plants and a larger number of animals, including humans, to live . . . and provides lumber and firewood too (Weber, *Reforestation in Arid Lands,* 1977).[69] Reforestation programs are part of larger conservation efforts. Increasingly they are being conducted with the realization that it is very difficult to separate reforestation from other revegetation efforts—range management, sand stabilization and similar activities.[70]

ECOSCAB. Recent studies of the ecology of bamboo in the East have cast an interesting light on its possible future in South America.[71] Research suggests that bamboo functions as a healing scab for gashes on the planet. It has already swarmed tall and green over gutted areas of Vietnam defoliated during the war there and appears to have flourished, historically, in the wake of man's shifting agriculture throughout the East. In Thailand, for example, bamboo replaces the cleared teak forests. Bamboo is less prevalent in South America, according to this view, fundamentally because of the ancient harmony of the native population with the forest, which did not suffer large-scale clearings until after the arrival of Europeans. In the present destruction of the original forests, particularly in the Amazon, men may unwittingly be clearing for bamboo, in the long run. Its rhizomes are more tenacious and insistent than our shifting purposes.

In their excellent survey of bamboo, Smithsonian Institution botanists Thomas Soderstrom and Cleofe Calderon suggest that we can look to the oriental past for the role of bamboo in South America's future: "In the Old World, man found that as his primary forests diminished, one plant above all filled his many needs, the native bamboo."[72]

EROSION CONTROL. Flood tamer, soil builder and saver, windbreak and earthquake refuge, companion of forests—bamboo's effects form a litany of positive ecological roles. "Bamboo forests prevent landslides and washouts. They protect riverbanks. On July 5 of this year, 1972, typhoons hit Shikoku's Kochi Prefecture [Japan] and in the resultant landslides some sixty-four people were buried alive in their houses. No bamboos were growing in the disaster areas, and no damage occurred in places planted with bamboo . . . The vitality of the plant is well attested to by the popular injunction to flee into a bamboo grove in an emergency such as an earthquake or typhoon."[73]

"Bamboos, growing thick, standing single— put all your roots together and all is well in the mountains and rivers."[74]
—Sengai, 19th century Japanese Zen Master.

The emergency has become permanent in the ecological decadence—the "decology"—of our time. Crisis is our habitat. The earth under our mismanagement is trembling like a wounded animal, as a survivor of the 1972 Managua earthquake said of the havoc there that claimed 100,000 lives. Bamboo recommends itself to dress the wound because of its rapid growth; its vigor on hillsides where ecological deterioration is most severe; the density of its rhizomes binding soil; its massive foliage braking rain, gentling sun, preserving moisture, and building humus; and for the multiplicity of its use.

A roaming mass of roots and interlocking rhizomes—up to 85 percent in the critical top foot of earth—hold firm even in earthquakes. Living, the bamboo leaves form a leaky umbrella that shields the ground from pounding rain. Dead, a 4-inch annual leaf fall equal in weight to the new growth of culms in some species, clogs the small ditches where erosion starts. The 2–4-inch mulch bamboo leaves create makes it easier for earth to drink and hold water while increasing its organic content—the depletion of which is a major cause of man-made erosion. Finally, the harvest of bamboo does not disturb the soil surface, a distinct advantage for a hillside crop. Hexagonal plantings 15 feet or less apart are recommended rather than square patterns on inclined land since they break the flow of water more.

Researchers in Puerto Rico found bamboo *(B. vulgaris)* as good or better than any other plant for controlling landslides above and below the winding roads that wander her mountainous interior. A river there periodically devastated chunks of trial fields at the Mayaguez Federal Experiment Station. Bamboo revetments at critically sharp curves, backed by plantings of living clumps, effectively solved the problem. "The fibrous mass of roots binds the soft banks, and the thick culms arrest strong currents during flood periods."[75]

The constant moisture of stream banks fortifies bamboo's growth and consequent capacity to hold the ground against the current. Streamside locations for bamboo are particularly recommended in areas with a pronounced dry season.

B. vulgaris is effective in erosion control and is among the easiest bamboos to propagate in frost-free areas. But other species, more resistant to ravages of powder post beetles in harvested culms, are preferred as a more useful crop. These include *B. textilis, B. tulda, B. longispiculata,* and *B. tuldoides* among clumping, tropical bamboos. Among monopodial species, *A. humilis, A vagans,* and *S.*

pumila are especially recommended to resist erosion. See descriptions of separate species for other good soil-stitching species.

FRIEND OF THE WOODS. We have noted that bamboo likes the neighborhood of trees. Trees also find bamboo a companionable plant in the forest. The friend of the people in China, in India, the wood of the poor, bamboo is also recognized in both countries as the friend of the woods and watersheds. In Colombia, species of guadua growing on steep lands keep many million tons of mountains from becoming river bottom, delta, and ocean floor. (See p. 124.) This report of positive ecological impact comes from Dehra Dun: "Understory plants like bamboo, particularly in present deteriorating ecological conditions, can help a lot to maintain continuity of nutrient cycle for sustained fertility in extensive areas of Indian forests. Bamboo leaf fall, particularly in fire-free season, builds humus. Moderate shade of bamboo suppresses dense bush and shrub shaders, enemies of timber species seedlings. Their high canopy of foliage also helps conserve moisture so bamboos have performed a remarkable role in the regeneration of important timber species like teak in the semimoist and moist teak forests of Madhya Pradesh. In these areas, under patronage of the bamboos, teak is slowly but steadily increasing. Sal seedlings in Balaghat forests of M.P. also enjoy higher survival where bamboos are prevalent.

"The notion that shallow rooted and self-thinned bamboos can mar the regeneration of timber species is not backed by the concept of origin and succession of vegetation. Especially in present-day fast changing environmental conditions in forests—with heavy logging, frequent fires, grazing, and recent intensive use of bamboos—the bamboos offer no threat to the environment. On the contrary, they deserve to be preserved and propagated with great care and vigor."[76] There are reports from several countries of local species of bamboo in competition with trees. Others, like *Sasa kurilensis* in northern Japan, hinder the work of foresters.[77] Such reports are few in bamboo literature, and the problems described are avoidable in future Western plantings, in which species will be chosen rather than simply presented by the landscape.

THE CLOUDY BOUGH: TREE WORSHIP AND WEATHER CONTROL.

> Then they cut down our lemon trees
> And the Spring fell from our eyes!
> —Mahmud Darwish[78]

"In the religious history of the Aryan race in Europe, the worship of trees has played an important part. Nothing could be more natural. For at the dawn of history Europe was covered with immense primeval forests. . . . In the forest of Arden [England] it was said that down to modern times a squirrel might leap from tree to tree for nearly the whole length of Warwickshire. . . . Amongst the Germans the oldest sanctuaries were natural woods. Sacred groves were common and tree-worship is hardly extinct amongst their descendants at the present day. How serious that worship was in former times may be gathered from the ferocious penalty appointed by the old German laws for such as dared to peel the bark of a standing tree. The culprit's navel was to be cut out and nailed to the part of the tree which he had peeled, and he was to be driven round and round the tree till all his guts were wound about its trunk. It was a life for a life . . . Worship of trees was prominent among the Lithuanians, at the time of their conversion to Christianity towards the close of the fourteenth century. Some of them revered remarkable oaks and other great shady trees, from which they received oracular responses . . . In the forum, the busy center of Roman life, the sacred fig tree of Romulus was worshipped down to the days of the empire, and the withering of its trunk was enough to spread consternation through the city."[79]

The physical tree was regarded first as the body, later as the dwelling of the tree spirit. Trees were widely believed to give rain, control weather, encourage crops, increase herds and the fertility of women. Our Christmas trees are a remote vestige of these feelings. The practice of bringing an evergreen into the house at the deepest, darkest valley of the year, the winter solstice, is a custom with many variants depending on climate and available flora. The constant in all cultures is the belief that the tree, the tallest and most massive life form in the experience of all peoples, is intimately linked with the regenerative powers of life. Modern findings in meteorology and agroforestry confirm ancient feelings about the central role of trees in biosphere maintenance and the health of human cultures. Conclusions of the ancient tree-worshippers and contemporary ecologists differ not in essence but in expression. In myth or statistic, both acknowledge the crucial place of trees in the fabric of planet life and atmosphere. The Greeks made Zeus the god of both oak and rain. The Chinese, in the *I Ching,* established a family of images, a periodic table of the elements of change in people and the natural world: Father Sky, Mother Earth, and their six sons and daughters: Thunder, Water, and Mountain the sons; Wood-Wind, Fire, and Lake the daughters. Yoking wood and wind in the eldest daughter is an example of the intuition of all early cultures that trees had a good deal to do with weather control. FAO's 1978 study of the agricultural benefits of

ECOLOGICAL LULLABY

Rock-a-bye raincloud
on a woods' back,
where the oak grows
the thunder will crack.
Where the bough creaks
the weathers will fall,
and up will leap flower
and forest and all.

Rock-a-bye hillside
gnawed to the bone,
where the herd grazes
grasses will groan.
Where the plow plunders
gullies will crawl,
and off will run tillage
and village and all.

Rock-a-bye rainbow
on a cloud cap,
where the rose blows,
the lightnings will clap.
Where the root wanders
waters will stall—
and soon'll shoot daisies
and crazies and all.

China's present intense effort to retree herself confirms the primitive belief that fecundity of life on all levels, including human reproduction, is intimately linked to the holy groves around the village.

BIRTHDAY PLANTINGS. One of the most persistent and widespread rituals of the cult of trees is the practice of planting them to mark human birth and death. The Maoris bury the umbilical cord in a sacred place and plant a young sapling over it. The child will dwindle or flourish in sympathetic response to the tree's vitality and die with it as well. In Aargau, Switzerland, an apple tree is planted for a boy, a pear tree for a girl. In Mecklenburg, the afterbirth is placed at the foot of a young tree, with which the child then grows.[80] For a wealth of tree tales, see Frazer, *The Golden Bough* (chapters 9, 10, 15, 67.2), on tree worship, its remnants in modern Europe, the particular worship of the oak, the "external soul" of people in plants—many a curious legend on the divine rights and powers of trees.

The practice of planting a tree on the birth of each child should be expanded to keep pace with current rates of deforestation per capita. We are losing 1 acre per minute of tropical forests. A tree could be routinely planted by all earth peoples each birthday. Large cities should plan reforestation zones upwind of their bad air, birthday-tree picnic areas prepared to receive the planting energies of parties of friends as a standard ceremony in the culture. Bamboo and other valuable plants could be included as an option to more familiar tree species.

GREEN CHINA. China and South Korea are unique among nations in having reversed deforestation, demonstrating to the rest of us that it can be done. The 1978 publication of FAO, *China: Forestry Support for Agriculture,* reports on a study tour in 1977 analyzing the Chinese integration of forestry and farming: "Striken by a series of natural calamities throughout history, China appears determined to tame rivers, regulate water systems, reverse soil erosion, establish a favorable climatological balance and thus banish the feeling of helplessness against natural disasters. Forestry has played a major role in achieving these objectives.

"The *participation of the people* has been a central concept in China's forestry efforts. Research concentrates on practical problems and includes commune members; much is learned from the practical experiences of field workers. Education of the people is seen as a requirement of successful

Ecological activists.

The angel cried with a loud voice, saying, Hurt not the earth, neither the sea, nor the trees.

—Revelation 7:3

At long last I am beginning to see a tree sense coming into being all over the world. The foresters of China have made a shelter belt 680 miles long to stop the sands of the Gobi Desert spreading into China proper, and the Old Wall of China is becoming a tree wall stretching for 2,000 miles. Millions of little trees like a green mist stretch as far as the eye can see on either side of that great rugged wall. Latest reports indicate that as many as thirty-two million people are employed permanently in afforestation and that during the planting season about half of the white-collar brigade in the cities don their oldest clothes and go out into the country to help the peasants plant trees on the mountain-sides.

Peking, which was once treeless, has now become a city of trees with avenues comprised of up to ten rows of trees. The latest reports show that tree cover has been increased from 7 percent to 27 percent.

—St. Barbe Baker

tree planting programs.[81] As a result, the average Chinese is much more knowledgeable about forestry than the average person in any other country, and protection of reforested areas is not a problem. Forested lands have doubled since 1949. Tree planting has had direct economic benefits in the form of timber, fuel wood, livestock fodder, fruit, and other products. In some areas, shelter-belt forestry is considered the primary factor in dramatic agricultural gains, ahead of irrigation, fertilization, and improved seeds."[82]

The FAO group also concluded that countries with only a professional forest service on wages could never match the committed work force of the people of China, where tree planting, everywhere, is everyone's labor of conscious love.

Two is not too early to begin training ecological activists. Chinese children's stories are an interesting study in how she shapes her citizens at an impressionable age, providing positive social models for little children before they can even read. Main themes include: clever analysis of everyday problems, often through homespun, shoestring

Paper tractor. Contemporary Chinese papercuts feature tractors, and power transmission lines stepping over mountains in the misty distance of landscapes. In China, a deep and long cultural respect for nature and for traditions has prompted a judicious grafting of modern methods onto the tao of native farming as practiced for centuries. The Chinese are feeding their people, re-leafing their cities, restoring their land, and doubling rather than diminishing their forests.

Green Guerrillas

Tree-planting is closely associated with the love of one's country and it is interesting to see that in People's China the care and planting of trees has become the main pastime in the schools and recreation over the weekends for perhaps the greater part of the population. To inspire this there must be widespread education in trees' biological contribution to life, a sympathetic feeling for the earth, and a desire to restore its green mantle. I have indications that there is a growing desire throughout the world to co-operate in a gigantic tree-planting programme not only in the Sahara but in a green belt to encircle the globe.

Of the earth's 30 billion acres, already more than 9 billion acres are desert. Land is being lost to agriculture and forestry much faster than it is being reclaimed. At the same time the world population is exploding. Already half the human family is on the verge of starvation, for man breeds and lives beyond the limits of the land. Yet if the armies of the world, now numbering twenty-two million, could be re-deployed in planting in the desert, in eight years a hundred million people could be rehabilitated and supplied with protein-rich food grown from virgin sand.

—St. Barbe Baker

scientific experiments; friendship rather than competition in sports; self-reliance in personal cleanliness; study before play; alert concern for the comforts and convenience of others; a brisk readiness to learn from old people and share care of children younger than oneself; offering alternatives instead of merely criticizing; returning lost objects promptly and repairing old toys; sharing resources and working cheerfully—"everyone wants the heavy load."[83] It is instructive to compare Chinese child models—in which cooperation is the constant moral—with the violent justice and clenched face of U.S. comic superheroes, in which hyperachievers display miraculous and intensely personal powers to an astonished multitude of helpless spectators. Which town has the tantrum most deeply programmed into its early consciousness? What harvest will such a social psychology reap—or rape —from the earth around us?

Along with all the civic virtues, mentioned above, Chinese children's books include revegetation. A wordless series of drawings for prereaders pictures a collaborative tree planting by grade-school green guerillas, who complete their task, wave goodbye to one another, and head their separate ways home. En route, the child we're following has his hat blown off in a wind—and remembers the saplings, left in the wind without supports. He arrives back at the planting site just as the rest are returning to stake the young saplings—with bamboo. The story provides a tidy parable, for just as forestry is integrated with agriculture and supports it in China, bamboo complements and supports their intensive agroforestry. As a part of reforestation, China is committed to rebamboo.

FAO REPORT.

> Bamboo is one of the major cash crops planned for planting widely. An indication of the potential of bamboo cultivation in found in Chairman Mao's instruction to research stations to carry out research on the acclimatization of southern bamboo species to the ecological conditions of northern China.[84]

A sample production brigade of 750 hectares (1,650 acres) grows 500 hectares (1,100 acres) of an unmentioned species—the context suggests moso, *P. pubescens*—with a density of 2,400 stems per hectare (1,090 per acre). Around 75 tons of organic manure are applied per acre. Exploitation begins seven years after planting in selective cutting limited to culms five years old, with two thirds of the grove usually left unharvested. The average shooting/stem ratio is about 0.8 shoot per culm after five years. One commune performed yield experiments: 24 culms of *P. pubescens* planted and tended twelve years increased to 3,200, a shooting ratio calculated at 0.6 shoot per culm up to nine years, when it dropped to 0.5. At this rate, 2,400 culms per hectare would produce 1,440 shoots. Selective cutting of one third the grove would yield 800 culms yearly. (FAO 1978.)

EPILOGUE

The Great Disorder.

> Shu ruled the South Ocean, Hu the North. They often met to do business in the Center, where Chaos ran a traveler's tavern called The Great Disorder. He fed and sheltered them so well that Shu said to Hu: "We have seven holes to see, hear, breathe, and eat. This poor fellow has none. One good turn deserves another. Let's repay his kindness by helping him out." They set busily about and drilled one hole each day. At week's end, Chaos died.
>
> —Chuang Tzu[85]

Not doing—*wu wei*—is a Taoist term for that spontaneous and organic doing of the natural order that brought us here and holds our bones when we die. It is outside as well as inside: It whirls galaxies, pumps your lungs, and knits the eyes of embryos in the midnight womb. Human efforts to "conquer nature," bully the breasts of Mother Earth into more quarts of yield, have consistently produced short-term profit for a few, followed by a long-term mess for the many to correct or endure.

> *A great disorder is an order.*
> *A violent order is a great disorder.*[86]

Alert to the sad havoc wreaked by the violent orders of the chemical tractor in half a century of

industrialized farming, appalled at the accelerating extinction of species and peoples, the multiple fouling of lake and stream, the shocking rate of topsoil depletion—many concerned citizens of the biosphere are reevolving the design of an agriculture permanently viable, a sane and stable "permaculture" in harmony with the orders of light and leaf encompassing our human disorders. The principle harvest of permaculture is enlightened farmers capable of sustaining it, delicately tuned to climate and soil, selecting the sanest traditions of the dead, feeding the living with a healthy sufficiency for all, and leaving for the unborn down history from our brief stewardship a still more fertile and handsome planet than we found.

Natural rhythms, life patterns of other species, are not eternal; but they are at least much older than the human race and can provide models of duration for those trying to glimpse the shape of a possible permaculture. "Tempeculture," however, is ruled by ticking time and imagined temporal advantage rather than eternal laws. Its short-range goals remain ecologically reckless of consequences downstream from its ecofolly. "After us the deluge . . . or dustbowl." People and planet are plundered with the same ruthlessness. In contrast to the respect for time's rhythms, which characterizes permaculture, tempeculture impatiently sacrifices steady, prolonged yield for instant profit. The fairy tale of the man who killed the goose laying the daily egg of gold in hopes of harvesting all dozens at once is a fit parable for modern farming. Its advocates have had time to prove that, in agriculture as in art, rush is the mother of ugly.

> The most important question in assessing agricultural systems is whether yields are sustainable over long periods and do not overtax the land's capacity to continue producing for succeeding generations. Traditional manual and organic horticultural methods in the Orient have for over forty centuries shown dramatic results, managing to feed 1.5 to 2 times more people per hectare than industrialized farming now does using mechanized chemical techniques.[87]

Permaculture lasts as long as we will last. It is a "whole-systems" approach to agriculture allowing nature to do most of the work and people to do most of the harvesting: working with rather than against nature, cyclic renewal replacing linear exploitation, long and thoughtful observation replacing long and thoughtless toil; looking at plants and animals in all their functions rather than treating any as a single product system; meeting human needs with respect for the finite resources of the earth.

Permaculture stores rainfall, accumulates energy reserves, and creates new microclimates that add useful species. Tree crops, edible landscaping, biological pest control, organic waste recycling, the agricultural forest, and revegetation—these are themes related to the configuration of acts and attitudes that make up permaculture. Like a good move in chess, permaculture designs maximal functional connections. Multiuse species create nutrient cycles, food chains, and successional trends that work together to give *sustainable yields.*[88]

The place of bamboo in such a permanent balance of earth resource and human need has been the meditation of these pages. It is, perhaps, significant that some of the most mature and permanent agricultures on the earth's surface have been those most closely woven with bamboo cultivation and use.

THE CROP OF PERMACULTURE IS COMPLETE PEOPLE. "The bigger the job, the greater the challenge, the

more wonderful they think it is. It would be good to give up that way of thinking and live an easy, comfortable life with plenty of free time . . . I think that the way the animals live in the tropics, stepping outside in the morning and evening to see if there's something to eat, and taking a long nap in the afternoon, must be a wonderful life.

"For human beings, a life of such simplicity would be possible if we worked to produce directly our daily necessities. In such a life, work is not work as people generally think of it, but simply doing what needs to be done. Pure natural farming goes nowhere and seeks no victory. Gandhi's way, a methodless method, acting with a nonwinning, nonopposing state of mind, is akin to natural farming. When it is understood that one loses joy and happiness in the attempt to possess them, the essence of natural farming will be realized. The ultimate goal of farming is not the growing of crops, but the cultivation and perfection of human beings."[89]

CHAPTER NINE

1. See Levenson 1974:22 for tool sources.
2. McClure 1953:44–5.
3. Hidalgo 1974: 108–32.
4. Hodge 1957:128.
5. Lee 1944:129.
6. Rehman and Ishaq 1947.
7. Varmah 1980:19.
8. Kirkpatrick 1958.
9. Glenn 1954, 1956.
10. Freeman-Mitford 1899:269.
11. Montessori 1948.
12. Christian Morgenstern (1871–1914).
13. Rudofsky 1977:358–60, abridged. Queen Victoria gave wheelbarrows and gardening tools to her children.
14. Rudofsky 1977:362.
15. Ibid.: 364. See also Wilckens 1980.
16. Rudofsky 1977:354.
17. Newman 1974:1.
18. Aero 1980:149.
19. Newman 1974:209–11.
20. Ibid.: 46–7, 84–7.
21. Eric Sloane's Americana books provide a useful model for graceful clarity.
22. Hunter 1977:149.
23. May 1983:154–71; didgeridoo, 158–9.
24. Ellen Green, itinerant flute maker. Personal communication 1981.
25. McClure 1953:18–9.
26. Hidalgo 1974:222–4.
27. Ibid.: 182–3.
28. McClure 1953:5.
29. Morse 1886:83.
30. Ibid.: 105–6.
31. Ibid.: 59.
32. VITA 1963:86–8. Hidalgo 1974:273–9 provides this and more information on bamboo piping. The VITA article derives from "Water Supply Using Bamboo Pipe," #3 of "Water Supply and Sanitation in Developing Countries," U. of North Carolina at Chapel Hill, October 1966, International Program in Sanitary Engineering Design.
33. Darrow 1981:213.
34. McIlhenny 1945b:121.
35. VITA 1981:245–7.
36. Eckholm 1979, in Darrow 1981:496.
37. Sargeant 1976:4.
38. Chu Hsi's preface to "The Great Learning" in Pound 1969:25.
39. Sargeant 1976 is the main source of this history.
40. Porterfield 1927:48, quoting from E. H. Wilson: *A Naturalist in Western China* (1913).
41. McClure 1958c:393–5; savagely abridged.
42. Varmah 1980:1. A highly recommended report is available on request from Dehra Dun. (For address, see p. 305.)
43. William Raitt published some twenty-four articles on pulping processes in the first forty years of the twentieth century, laying the basis for modern methods of bamboo papermaking. He was far from alone. Because of its industrial importance, paper is the subject of the greater part of world publications on bamboo, as can be verified by a trudge through the 1,034 books and articles listed by Sineath (1953:157–230). By an unfortunate fatality that is not the lot of bamboo alone, the most critical information industrially is also consistently the most dull for the common reader. Unless you are a paper chemist or planning to open a mill, extensive tables on such matters as the length/width ratios of bamboo fibers, the world species most suitable for pulp, distinctions among soda,

sulfate, and sulfite pulping processes, or the intricate mysteries of comminution, bleaching, and digesting are likely to be an excess of riches. Fortunately, none of this is necessary to understand how and why bamboo has assumed increasing importance in the industry of papermaking. Sineath 1953:24–54 provides an excellent summary of technical matters on paper production for those equipped or concerned to dive deeper into that data vat.

44. Sineath 1953:27. Austin 1970:201. Varmah 1980:13.
45. FAO 1978.
46. Varmah 1980:21–2.
47. Kiang 1973:18–21.
48. The directions for making paper are mainly from Sargeant 1976, a recommended how-to. Directions for papermaking are included in many general crafts books now. This is a hopeful trend: perhaps we can soon expect papermaking to be a regular component of early education in the West.
49. Soderstrom 1979a:166. His figure should perhaps be 15, however; it conflicts with Lessard 1980:56.
50. Lessard 1980:56.
51. Ibid.: 109–10.
52. USDA pamphlet, "Edible Bamboo Sprouts," undated.
53. Ueda 1960:56.
54. Leung 1952:18.
55. McClure 1958:609–44.
56. Semple 1952:134.
57. McClure 1958a:614.
58. Squibb 1953:2.
59. McClure 1958b:57.
60. Smith 1950.
61. See "firewood," pp. 37–40.
62. Darrow 1981:76.
63. Mollison 1979:1.
64. *San Francisco Examiner,* 12 November 1972. Via *Shelter.*
65. Papanek 1971: 270–1.
66. Tillman 1979:13.
67. Ibid.: 14.
68. Darrow 1981:77, quoting Douglas, *Forest Farming,* 1976. See Darrow for reviews of most forestry books mentioned.
69. Quoted in Darrow 1981:500.
70. Darrow 1981:501.
71. Drew 1974.
72. Soderstrom 1979a:164.
73. Junka 1972:22.
74. Suzuki 1971.
75. White 1945:845.
76. Singh 1969:497.
77. Lessard 1980:56, 95, 129. See p. 215 for warnings on need of careful selection of species when introducing bamboo in new terrain.
78. *Palestine,* vol. 9, no. 3, 1983. Back cover photo caption.
79. Frazer 1922:126–8, condensed vigorously.
80. Ibid.: 791.
81. This is in keeping with dialectical farming. "Theory without practice is empty. Practice without theory is blind," Karl Marx.
82. Darrow 1981:498–9.
83. *Good Children* and *Little Red Guards* are two picture books from Peking Foreign Language Press that U.S. children seem to like a lot.
84. FAO 1978:74.
85. Chuang Tzu. *Field Guides to Enlightenment* (vol. 3), Random Road, Box 666, Bolinas, CA 94924.
86. Wallace Stevens: slightly altered from "Connoisseur of Chaos," *Poems by Wallace Stevens* (ed. Morse), Vintage, 1959:97.
87. Darrow 1981:74.
88. Mollison 1979 provides the name *permaculture* and nearly all of this description.
89. Fukuoka 1978:115, 119.

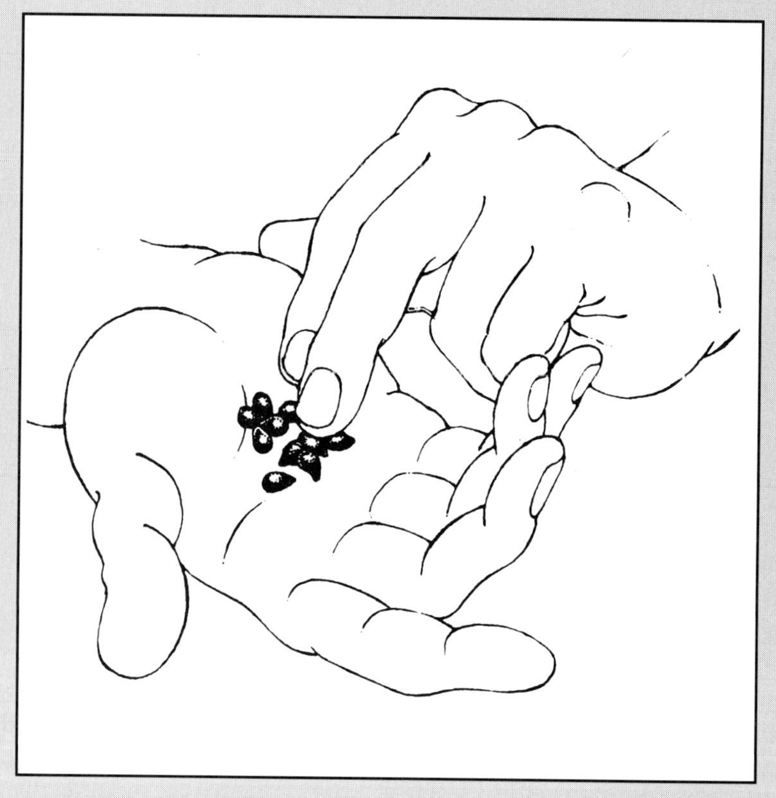

10. SOURCES AND DIRECTIONS

"Only the sun on his shoulder
knows where he has come from,
where he is going,
how he'll return,
well laden with wandering,
another morning by another road.[0]

BOTANICAL ACTIVISTS

Johnny's Appleseed.

Staring at the seeds in his hand left over from an apple he'd just finished one early morning in Massachusetts, John Chapman realized that if we all planted the seeds of every apple, there would soon grow a durable Eden, enough for everyone. Barefoot, with a sackful of dry apple seeds on his back, cook pot for hat on his head full of dreaming, Johnny—as were so many then around him—moved west.

He brought no gun, he sought no fortune: He planted apple seeds, gave small bagfuls to the scattered settlers and stayed weeks on end with little groups of pioneers, helping them to chink the cabin or clear the field as well as plant the orchard . . . and moved on to munch an apple in another wilderness village and to gaze, with an amazement never lost, at the dense wisdom of nature that let him hold the future of a whole healthy hillside in the palm of his hand.

The wet little seeds never lost his respect, however many he handled over how many years. Nobody kept track of either the decades or seeds scattered through them, up and down a thousand valleys, least of all the children who crowded round him, insistent for another and another tale from this magical stranger who kept weaving back through their lives to check his orchards at an ample rhythm.

He romped also with bear cubs in his beloved forest, they say, with the mother watching and enjoying it, gentled in tooth and claw by the subtle alchemy of friendship. Socially, too, he was a bridge uniting what custom drove apart: In a vast countryside where white and Indian feuds and full-scale warfare were the norm, he tried to plant human harmony along with his orchards. Privately, at least, he succeeded: He was loved and welcomed by the Indians everywhere on his wanderings. One bad winter, they found him half frozen in the anxious inspection of some threatened and remote planting of apples and nursed him back to life.

His creative respect for a tiny, unhonored, habitually wasted resource accomplished more with meager means over a wider area and larger population in his time and long after than many a later largely funded-and-staffed government blitz to

"beautify" America—usually mere cosmetic apologies for bad behavior.

His gentle, gardening spirit was antipode to the national habit of belligerence, in family, state, international, and ecological affairs. His economics were closer to Jesus, Francis of Assisi, or the Buddha than to the harsh ring of our national cash register. He played lifetime hookey from success scaled by any common norm. In a nation embracing commerce as fondly as the drunk hugs his jug, he was no more commercial than the wind and weather wrapped around his wanderings or the soil under them.

Chapman, from the Anglo-Saxon *cēap-man,* means "trademan" (cēap-trade). Johnny was a chapman with a unique, composite trade of sorts, it's true, and his love for apples was grafted onto a skill at wandering. He was shy of traffic, but was himself a fluid road. So he had the trade and tread to make a merchant—what he lacked was the profit motive. He was a dealer who neither bought nor sold, a peddler of gifts whose wares and person were not for hire for wages, but free for a neighborly asking in any mile of his meander, like a river you can toss a bucket in anywhere and find water.

Bamboo messenger: in green memory of McClure.

Bamboo, make us an instrument of your peace.
Where you find exhausted asphalt,
let there be . . . bamboo leaves,
drinking weather and sunlight,
sinking back to soil.

Let rhizomes wrap the troubled plumbing of our time,
and may your friends inspire in others
that green excitement (flecked with lemon)
that you stir, alert and trembling in the least breeze.

Let children, learning early your handy magic,
grow old enjoying it.
Let luckiest spirits die in a grove,
nothing between last eyes and April sunlight
but bamboo leaves.

He left among leaves: McClure died with his obsession on, in his bamboo grove, digging a plant in springtime for a child. He split clean and even, like the plant on which he'd lavished a lifetime of affection, dirt on his hands and knees, busy with a giving errand, April 15, 1970.

Born in 1897, his father a farmer-schoolteacher in Ohio, McClure was raised surrounded by living plants, the thousand chores of preindustrial farm life, and neighbors whose fields and beasts rarely left them time to put on airs. After a B.S. in agriculture at Ohio State in 1919, he did not return to the farm as intended. Instead, a spirit rambling as a rhizome led him to China as a lecturer in horticulture in Canton. He loved languages and learned Cantonese well enough to pass as Chinese if heard unseen. Plant collecting in areas with various dialects, he could dismiss his local interpreter within a week. He had been raised among country people, and he moved easily among the peasants of China whom he grew to love as much as he began loving the most omnivisible assistant to their way of life—bamboo.

McClure was accompanied on his plant collecting trips by a tough peasant named Kang Peng (1877–1926), a reformed drinker, gambler, and street fighter who had killed "more than one of his antagonists." Kang Peng was a sturdy assistant, sufficiently seasoned and risk-addicted to wander the mountains of 1920s China, ripe with revolutionaries and thieves, "on the flimsy excuse of collecting and drying plant specimens between papers." He died some six years after McClure's arrival in China, and McClure wrote an "Appreciation" in which he reveals the warmth of their relationship as well as the hardships of companionable plant collecting in that space and time.

"We always shared all things as they came to us—work, food, accommodations, extra burdens, extra sweets. Traveling, as we always did, with minimum baggage and personal effects, we were never able to make ourselves very comfortable. As for beds, or available planks or door-boards for beds, he always set aside the poorest for himself; with food, he was always frugal, never wasteful, in his purchases; nor did he ever take unfair advantage of my desire to provide well for those who worked for me. Spreading the table for a meal, he would always take for himself the broken bowl or the pair of chopsticks that were not mates. When using borrowed bowls and chopsticks, he always remembered to wash and scald them on my account.

"He always took the least attractive food, picked up the most inconvenient odds and ends to carry, and in every imaginable way strove to make my work as easy and as pleasant as possible. Needless to say, I often disputed these attentions, which my superior strength and endurance made unnecessary. But he often won out by sheer insistence."[1]

Gradually, over half a century, McClure became a botanical bridge between East and West, between unschooled farmers and the scientific elite, between business and government interests in dozens of countries. He was the ambassador of Bamboo to the human race, locally employed by the Department of Agriculture or other groups for a time, but always actually working simply for bamboo with the plodding patience of the plowboy. His roving eye and trilingual tongue eventually noticed and told more about bamboo in more places than anybody on the planet ever had before. He was noted by friends for his fanatic quest for the precise word and expression, for correcting rather than repeating the errors of earlier research, for getting down to the fundamentals of a subject by taking nothing for granted, and for a lifelong passion for plants and hard work.

Collecting plants and establishing living groves of bamboo for prolonged research and propagation were central aspects of his work, the foundation of everything else. He had six hundred species of bamboo in the Canton groves at Lignan University. They still are flourishing some sixty years after he

began them in the early 1920s. From this collection, the USDA accessioned 250 numbers of living plants between 1924–1940. They were carefully coddled by McClure from China to the USDA groves just south of Savannah, Georgia, from where they were distributed, for fifty years, around the world. This service has been unnecessarily interrupted for nearly twenty years owing to a supposed lack of funds; but a proper design for distribution could easily pay for itself. With the arrival of World War II, McClure's work moved West—with periodic returns to various oriental countries:

> During 1943–1944 I made a survey of useful bamboos in the United States, Mexico, Honduras, Colombia, Venezuela, Brazil, and Puerto Rico. . . . The post of field service consultant on bamboo with the USDA (1944–1954) gave me the opportunity to study and collect bamboos in six countries of Central and South America, as well as in India, East Pakistan, and the islands of Java and Luzon. Ultimately I was able to establish living collections of elite economic species in Guatemala, El Salvador, Nicaragua, Costa Rica, Ecuador, and Peru. As consultant on bamboo to Champion Papers, Inc., I made field studies in Jamaica and Trinidad, designed and supervised the establishment of an experimental bamboo plantation in Guatemala, and participated in elaborate studies based on it.[2]

The Bamboos: A Fresh Perspective (1966) was an attempt to cram the research of a lifetime into three hundred pages and sketch relevant lines of investigation it would take many other skillful lifetimes to complete. Much of the book is intelligi-

Bamboo slaves.

The cheapness of bamboo artifacts imported into the United States has been one of their most attractive features. Bamboo's abundant yield has contributed to their low cost, but the dark shadow of the bargain is the ancient and on-going exploitation of the lowest rung on the economic ladder everywhere. The people dealing closest with the groves, as usual profit from them least. For those who harvest and transport bamboo, the rural poor and unskilled urban workers, bamboo is not a cultural icon, but a life-long cage—with beautiful green bars 10,000 meters tall, when you look at their shadows in an economic light. Basket makers and other artisans are not much better off. Socially just forms of bamboo development will be an important aspect of future designs for exploiting the plant. McClure's description of Kang Peng's bamboo job in 1890s China can serve as a sample of what we don't want to perpetuate.

"At thirteen he became apprenticed to a maker of wooden farm implements. Growing restless before his three-year apprenticeship was up, however, he ran away, taking a job as laborer in a bamboo shop at Wong Lin Hui, Shun Tak, at $2.00 per year and board. The work was utter drudgery, consisting in part of carrying tremendous loads of bamboo poles from the river where they had been soaking, up through the village to the shop. As they were dripping wet and slimey with mud, he wore no garments except a shoulder pad and an apron of rush matting. People would come to their doorways and watch, amazed at his endurance, pulling down the sleeves of their garments in cold winters to protect their hands from the frostiness of the metal water pipes which they smoked. He was promoted the second year to $3.00 per year, and during his third and fourth years he received 50 cents per month. From this mere pittance he managed to save always a share to give to his parents."

ble to the common reader, but Beware of Latin Roots if you are not a botanist. McClure himself was an "economic botanist," a student of plants from the perspective of human need. His work was polycultural and interspecific. He was the brief custodian of a vigorously scientific intelligence motivated in choice of subject by a heart friendly and eager to be of true use to his species, which he could see multiplying more rapidly than its ability to feed and shelter the new arrivals.

He was planetary before his culture began to poke its nose out of narrow nationalism, a farmer who knew more than the folks in town about their own interests. He was in love with an order more complex and steady than prosperity—and in 1954 when the USDA cut off support for research that had been ripening for thirty years, his wife, Ruth Drury McClure, already in her sixties, went to work so that he could continue to woo the mysteries of bamboo, a second wife who in fifty years of all three living together, never roused the jealousy of the first.

Among those who worked and lived with him, such as Cleofe Calderon and Thomas Soderstrom, heirs of his work at the Smithsonian in Washington, D.C., he is remembered with a special affection and respect: for his brimming measure of the milk of human kindness; for a warm sense of humor and all-inclusive curiosity; for a modesty as constant as his friendliness; and for a delicate, generous spirit always ready to give and forgive.

FRIENDS OF BAMBOO. From his wife, Ruth, now eighty-five or so, we learned that McClure was, in fact, a "Friend," as the Quakers call themselves. She shares her husband's friendliness—and his love of labor. "I need—I *require* physical activity. If it snows at night, I get up early for fear some neighbor will shovel the sidewalk before me." She lives alone but is much visited in a small house on Quaker Lane in a Friends' "retirement neighborhood" in rural Maryland, woods out her window, a community of people she likes out her door. "Our church is called a Meeting—but the building isn't important. A Meeting can be anywhere, under a tree. The number doesn't matter, either. A Meeting can be two people. A Meeting is measured not by numbers but by openness.

"Quakers are seekers. We're open to anybody. Modest and unself-conscious. Those are the virtues we admire. Mickey was open. He didn't care if it was the janitor or the president. He could sit there and talk to the King of England with no trouble at all . . . We were refugees once, living in Hong Kong. My husband stayed on the campus and was made chief of police. They were bombing Canton everyday. There were nine thousand people on the campus as refugees; 500 acres is a lot to guard. It was the Japanese we were involved with. We had to bow to them every time we went out the gate. They brought big horses with them that ate up all the grass and then got sick. So this Japanese fellow wanted to study the veterinary books at the college.

My husband made friends with him and came home to ask the cook—who was Chinese—if he could bring a Japanese man home for dinner. He turned out to be a good fellow, a schoolteacher. Mickey made friends with everybody, even the enemy."

Surrounded by bamboo artifacts not yet donated to McClure's collection presently housed at the National Arboretum in Washington, D.C., Ruth's memories range from first meeting her husband at Ohio State around 1915 to his last years at the Smithsonian.

> Mickey wanted me to help him get his office organized. Everything interested him, so it wasn't easy. And one thing my husband wasn't good at was raising money. When I got down to the Smithsonian, I was ready to cry to see the number of projects he had and no one to help him. And his field work was done under difficulties you can't imagine. He finally used an aluminum-covered notebook. In China he made galvanized sheeting boxes with lids so the stuff could stay dry. You have all these things to carry, and suddenly your carriers run away. From carrying too much himself, he came home once with a bad back that lasted six months.
>
> At the Smithsonian, we'd go in at 5 in the morning to avoid traffic, and come home around 3:30. He always brought work with him, but he never did it. He always worked in the garden. We drove home one evening, around cherry blossom time. Mickey said, "Let's go out the George Washington Highway, along the Potomac, and come home the other side." We drove along. I was saying, "In a few days, these buds will be a soft, delicate pink . . . " He said, "I like it just as it is . . . " Later, I said, "Would you like some music on the radio, Mickey?" "No," he said, "I like it just as it is." We got home, it was the first day you could dig in the ground without getting it muddy, so he went out to the garden and came back once to ask how long till supper. About ten minutes later, I went to the window to call him. He was lying down across the sidewalk that goes down to the back gate, with his head pillowed on the grass. I called him, but he didn't move. I knew immediately he was dead. He had his nitroglycerin pills in his hand. A few days later, I went down and saw he had been digging out a rhizome for a child who wanted a bamboo plant. He was digging it out from all four sides of the culm, and the spade was stuck in the fourth side.

FILIAL OBJECTIVES. Three objectives of the compilers of this book are intended as acts of filial piety towards the bamboo student whose scattered notes and articles never vanish for long from these pages. We intend, with the help of all readers who can relate to these tasks:

1. To make McClure's published works more broadly available in a graphic popular form in a series of publications in both English and Spanish addressed primarily to people who will grow and use bamboo.
2. To care for, propagate, and *use* (in local crafts, popular architecture, and ecological gardening) the groves he helped establish from Bethesda to Brazil, many of which are in a state of neglect; that is, to realize their dual purpose as garden-schools and gene banks.
3. To extend the cultivation of *Arundinaria amabilis* throughout the temperate regions of the United States where it will survive as a green memory of what the creative industry of one lifetime worker can contribute to the beauty and handiness of a planet more handsome for his brief residence upon it.

"I don't know if you believe in reincarnation—probably you do," she said, eyeing my hair. "But if that's true, I think perhaps he's off somewhere, maybe on another planet, on business I know nothing about. I don't want to be a drag on him, a burden. I want him to be free and fulfill his destiny, in whatever form. But still, I would like to be able

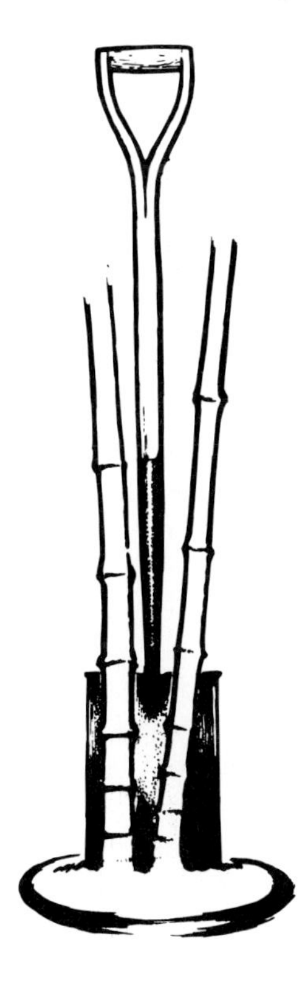

to communicate with him, somehow . . . I talk to him. 'But does he *answer?'* our daughter asks. He doesn't. And I only dream of him four or five times a year. I try to be accepting . . . to *ask* for nothing . . . but sometimes, I say to him, 'You know, Mickey, I would like it if you would come and be with me in a dream tonight—if you could . . .' "

Mickey, busy green ghost, by all means feel free to be with your wife in the evenings, but we have a daylight dream as well we'd like you to share, *si Dios quiere*—a reincarnation of your own.

Bamboo classics.

The German botanist, Charles Kunth, was the first to distinguish the bamboos as a separate category of grasses in a paper published in Paris (1815). In 1827 the curator of the Berlin botanical gardens published lists of plants under his care and included descriptions of *B. glaucescens* and *B. arundinacea.* In 1829, Christian Nees von Esenbeck, "a close friend of Goethe and a renowned botanist interested in all branches of natural history, philosophy and social ethics . . . one of the greatest agrostologists of all times,"[3] published a masterly treatise on the grasses of Brazil, in which bamboos were included. In 1835, he published the first taxonomic monograph in the world on bamboos, focused exclusively on Brazilian species, dividing the woody bamboos into two main groups, *Bambuseae* and *Arundinariae.* A third group, *Streptochaeteae,* covered the bambusoid grasses, diminutive bamboos found mainly as an understory plant in shady tropical forests between Mexico and Argentina, without economic importance and generally ignored by botanists since Nees.

The Russian Franz Ruprecht, in 1839, published the first monograph on the bamboos, treating them as a subfamily of grasses, dividing the sixty-seven worldwide species into two groups, following Nees, describing only woody bamboos. A distinguished English soldier and botanist, W. Munro, published a monograph in 1868 on world bamboos that divided temperate species into *Arundinariae, Arthrostylideae,* and *Chusqueae,* the last two covering New World species. He introduced a new group, the *Bacciferae,* containing eight Asiatic genera, and retained the *Bambuseae* of Nees as well.

S. Kurz, in two lengthy articles published in *The Indian Forester* on "Bamboo and Its Use" (1876), provided the first extended treatment on bamboo in relation to human need in a European language. In 1878 (Paris) the Rivieres, father and son, published a tome of 364 pages, *Les Bambous,* scattered with many beautiful illustrations unsurpassed in any bamboo volume till now. The text is rich in practical gardening experience with many species in their long years of work in Europe and North Africa encompassing both hardy and tropical bamboos.

James Gamble, director of the Imperial Forest School at Dehra Dun, India, published in 1896 an important monograph on Indian bamboos, frequently referred to in subsequent literature. In the same year *The Bamboo Garden* of Freeman-Mitford appeared, the classic Victorian description of hardy ornamental species that could survive in English weather. Ernest Satow's *The Cultivation of Bamboos in Japan,* mainly a translation of a Japanese horticultural work, appeared in 1899, laced with much bamboo lore and graced with a number of lovely watercolor plates. *Les Bambusees* (1913) of Edmund-Gustave Camus was largely a compilation of literature to that time. His daughter, Aimee Camus, in her 1935 *Classification des Bambusees,* published a number of new species and genera from Madagascar and other French colonies.[3a] F. A. McClure (*The Bamboos: a Fresh Perspective,* 1966) sums up much of the relevant past of bamboo literature in conjunction with his own global experience, and charts the necessary future for bamboo research and development.

CURRENT ASIAN RESEARCH: MODEL FOR WESTERN DEVELOPMENT

A three-day international bamboo meeting in Singapore (May 1980) provides a regional model for the sort of collaboration needed among Western nations with significant bamboo resources. A Canadian report on the gathering is abridged here for its own interest and as sample agenda for Western research.

At least a third of the human race is using bamboo extensively, so "it is remarkable that there has never been in the past an international meeting of research scientists interested in bamboo."[4] To review existing knowledge, consider problems preventing greater use, and identify regional research needs and priorities, the Singapore workshop was held, partly sponsored by the International Development Research Center (Ottawa, Canada) and the International Union of Forestry Research Organizations (Vienna). Twenty-two participants from the East (Bangladesh, India, Japan, China, Indonesia, Philippines, Sri Lanka, Thailand, and Malaysia) and the West (Canada, England, West Germany, and the Netherlands) surveyed research, distribution, and use country by country and established future research needs and priorities, such as an international seed exchange to transform flowering of important species from a local economic disaster into an international impetus for more extensive and efficient propagation.[5]

Some conclusions: Taxonomic work in classification and identification is basic to breeding and use. Increasing existing groves is the most effective method for gene conservation, and its importance should be stressed in national parks and other forested areas. Management of sympodial species in India and Bangladesh and the silviculture methods of China and Japan with temperate monopodial bamboos provide models to adopt or adapt elsewhere. Proper grove management includes: *clear* understanding of bamboo's growth; taking advantage of flowering for natural regeneration of groves; increasing yield, fertilization, and pest control studies; encouraging greater bamboo use in erosion control and in rural village economy; developing appropriate technologies in harvest, storage, and transport of culms.

Culm anatomy and the physical, mechanical, and structural properties of bamboos should be more studied, as should relationships between strength and anatomical characteristics. Chemical composition analyses will suggest new potential uses and pharmaceutical possibilities. Fiber studies will improve paper products. Attention should be given to research in natural durability of different species, traditional preservative methods, and design details in construction to maximize bamboo life expectancy in objects and dwellings. Protection by nonchemical methods needs further testing, as do studies on chemical preservatives that are widely available, economically feasible, and environmentally sound. Cheap fire retardants also deserve more study. Improved utilization of bamboo in manufactured products, pulp and paper, and architectural systems is important, and further exploration is needed in plybamboos, hardboards, and particle boards. Use of leftover bamboo stumps and biomass for novelty items deserves investigation, as does energy generation from tremendous harvest wastes: firewood, charcoal and charcoal briquettes, activated carbon, and fertilizer are some possibilities. Studies on bamboo use for human food should be increased and extended to more species. Long-term reinforced concrete research is needed; new processing techniques and machines to increase industrial use are important. Traditional bamboo musical instruments should be publicized and improved to preserve the heritage of the people and promote commercial development.

The present high cost of plastics may make possible a strong bamboo comeback if research discovers economic mass production, packaging, and innovative designs cheaper than plastic competitors.

Bangladesh.
The Forest Research Institute (FRI) at Chittagong began intensive bamboo research in 1971, establishing twenty-four species at the FRI Arboretum and four field stations, two in the northern deciduous forests where only cultivated groves are found and two in semievergreen regions in the southeast where spontaneous natural groves occur. Studies of flowering, seed longevity and storage, propagation, and pulping qualities of different species are among works in progress. One research focus is tissue culture propagation. Study has found that only buds are suitable for tissue culture. Bamboo culm buds are all multiprimordial, each primordium consisting of a rhizomelike structure. Entire primordial structures are used in tissue culture trials. Species, distribution, total acreage of groves, annual yield, and value of bamboo products are not reported.[6]

China.
With 2.9 million ha in twenty-two provinces from sea level to 3,000 m, China has nearly twenty-five

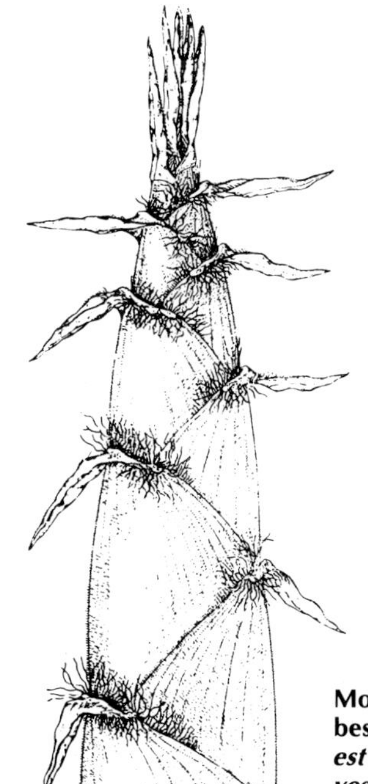

Moso (Phyllostachys pubescens) ***provides the largest part of the oriental harvest of edible shoots.***

times the bamboo reserves of Japan—and one third of India's (1 hectare = 2.47 acres). Two million ha are devoted to *Phyllostachys pubescens,* scattered in twenty-three production bases and introduced in 230 counties. Of 300 species in twenty-six genera in China, 200 bamboos from twenty-two genera are of commercial importance. Bamboo plantations form part of China's reforestation plans, the most effective in the world. In the Yellow River valley, for example, groves of *P. glauca, P. bambusoides,* and other bamboos now cover 30,000 ha—four times their former extent. High-yield tactics such as selective cutting to maintain grove nourishment, fertilizing, loosening soil, weeding, and rational harvests have also increased size and weight of individual culms.

Bamboo research departments have been established in the Chinese Academy of Forestry Services and related forestry colleges, as well as in provinces intensely cultivating *P. pubescens.* Bamboo hybrids have been realized at the Hunan Botanical Garden and the Guangdong Provincial Research Institute of Forestry. Recent anatomy and physical and chemical property studies now permit identification of bamboo fragments found in archeological excavations of sites seven thousand years old. Biological pest control is another research priority: There are 100 anti-bamboo bugs in China, and large-scale plantations invite large-scale problems. In the early 1960s, stag-head disease damaged 700,000 ha of *P. pubescens,* and in 1975 an equal area in the same province (Zhejiang) was infested by *Algedonia coclesalis.* Some nine million culms were lost in the two plagues.[7] (See also "Green China," pp. 288–290.)

India.

Twenty genera, 113 species, cover 9.57 million ha, 12.8 percent of India's total forest area, from sea level to 3,700 m in the Himalayas. Only 10 species are commercially exploited on a significant scale, with present estimated annual revenue from bamboo at approximately $8.53 million (U.S.), less than $1 per ha, which gives some notion of the underexploitation of bamboo resources. The extensive research of the Forest Research Institute and Colleges, Dehra Dun, has been treated in scattered portions of this book. Research with *D. strictus* over the years provides a valuable model for similar depth studies of other bamboos elsewhere.[8]

RESEARCH PRIORITIES. "For genetic upgrading of bamboos, a virgin field, the Forest Research Insti-

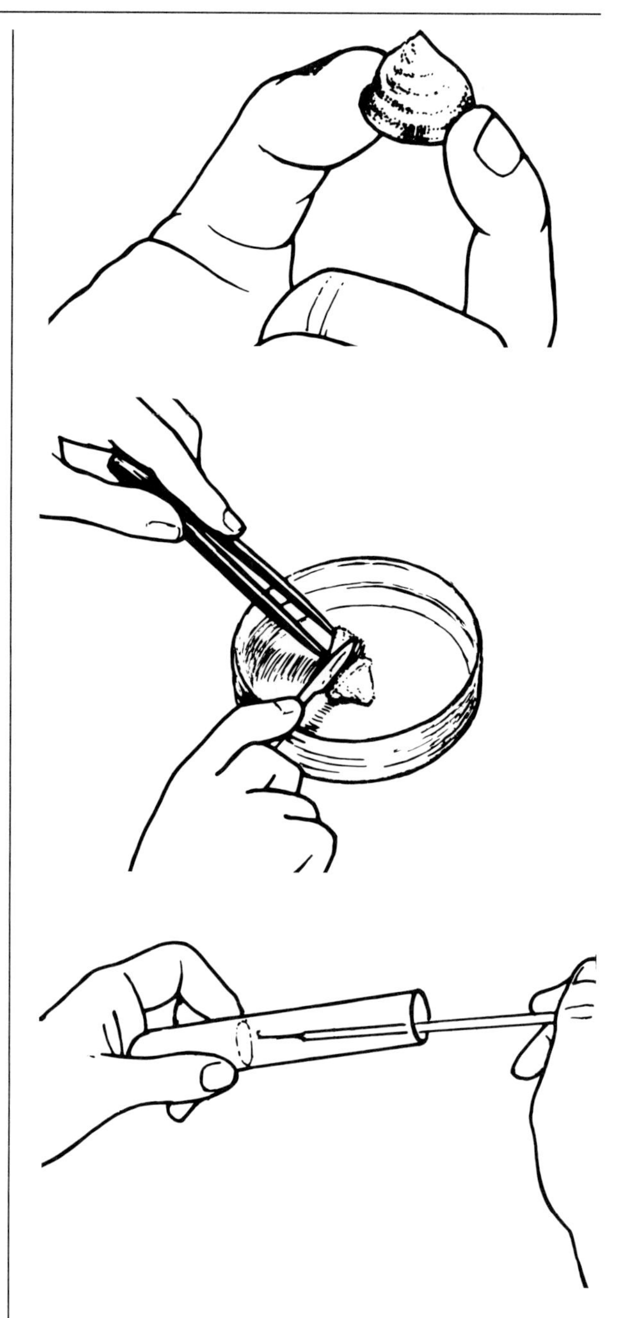

Tissue culture. (A) The tip of a bamboo shoot provides meristematic tissue—cells capable of division—that can produce, in a nurturing medium, thousands of times more propagules than conventional means. (B) A minute piece is severed from the tip—a microchip containing all the data and chemical codes needed to make this *species, complete with its special stripes on leaves or unique air canals in rhizomes. (C) A grove that can arrest erosion on a hillside while providing the raw material for a village industry begins with a speck. Tissue culture, realized in other plants, is still in its infancy with bamboo, but appears to be the relevant route to global sufficiency of elite bamboos. Miniaturization, seen by Fuller and others as the thrust of all technology, finds its present expression in bamboo cultivation in tissue culture of the plant.*

tute proposes three field stations in East, North, and South India for temperate, subtropical, and tropical species, with headquarters at Dehra Dun, to collect bamboo germ plasm from roughly one hundred wild Indian species, for growth in gene banks for evaluation and research. Genetic improvement of some ten top commercial species will be given high priority. Selected stands in certain areas will be conserved.

"Work will include surveying major bamboo areas, studying for genetic diversity, selecting suitable clones for conservation; collecting material for taxonomic, cytologic and palynologic studies; adding to banks variants arising as bud-sports; establishing one-parent progeny trials from seeds of species that may occasionally come into flower, evaluating progeny, selecting desirable seedlings and cloning for use in clonal trials which will also throw light on the breeding system of the species; exploratory hybridization using commercially important species of *Bambusa* with *B. atra* (a constant-flowering bamboo), species of *Arundinaria* with *Indocalamus wightianus,* an annual-flowering species, and so forth, studying the interspecific hybrids that may be realized.

"Nonavailability of seeds and transport of bulky propagules suggest the technique of tissue culture and/or shoot culture in bamboos to make the exchange and establishment of germ plasm much easier. This research should be given high priority, as well as the flowering mechanism in bamboos—one of the greatest botanical mysteries. Control over flowering would provide material to taxonomists to determine generic affinities of various species, to breeders for controlled hybridization and progeny testing. This feat could serve the cause of both applied and fundamental science."[9]

Indonesia.
Bamboo is in daily use by the people, and employment has "extended remarkably in the last few years." Industrial demand is also large, but researchers ignore bamboo, and forestry personnel turn groves into pine plantations. There is no full-fledged Indonesian expert on the biology or technology of bamboos. Many bamboo objects have been made obsolete by modern technology so the Lembaga Biologi Nasional (National Biologic Institute) in Bogor initiated emergency ethnobotany bamboo studies in 1975, focusing on musical instruments, fishing equipment, and wickerwork—particularly carrying baskets. A research priority is the role of bamboo in traditional Javanese customs. The increased use of bamboo in landscaping is under evaluation by students of the School of Garden Architecture at the University of Trissakti, Jakarta.

Use of bamboo for lightweight harvesting poles has resulted in plans for large-scale plantings of *M. baccifera* in oil palm areas of Sumatra. Construction remains the most extensive popular use: in Central Celebes split bamboo is used for roof tiles. Scaffolding, bridges, smokehouses for tobacco, and furniture are common bamboo uses, and tribes weave mats, baskets, and so on in distinctive patterns. Toys, vases, carvings, and a thousand bric-a-brac are produced, and *Tangerang* hats have become world famous. *G. apus* is much used, especially in West Java, for string. Waste branches provide valued firewood in some areas.

During the war for independence, a sharpened bamboo spear became a well-known weapon with "a respectable place in modern Indonesian history." There is an ample traditional arsenal of bamboo weapons, and bamboo arrows are still used in West New Guinea. For centuries of Indonesians, bamboo was the first piece of earth to touch them on entry: A freshly split bamboo knife severed the umbilical cord, and males were circumcised with the same sharp instant instrument.

West Java musical instruments depend on *G. atter (bambu hitam),* and only plants of this species grown in West Java are of the right quality and dimensions. Recently developed furniture and village handcraft industries, with enormous capital compared to instrument makers, are using this species and thus edging out the home-industry–scale instrument artisans. People have no extensive traditions of growing bamboo, so regeneration is left to nature—adequate only to earlier population densities and use. Quantities consumed in furniture are also a factor. A table and four chairs will use as much bamboo as an entire orchestra of instruments.[10]

Japan.
Makino, Tsuboi, and Muroi are among the early twentieth century bamboo students who laid the basis for the complex taxonomy of some 669 species in 13 genera which are native to Japan. Only a few are important in the national economy. Madake *(Phyllostachys bambusoides)* and moso *(P. pubescens)* make up 84 percent of Japan's 123,000 hectares of bamboo groves, and *Arundinaria simonii* provides some 7–8 percent of the harvest. Studies on pest control in living groves and preservation of harvested culms has intensified

since the 1960s. Harvest season greatly influences susceptibility to attack; late fall is best, April-June the worst time for cutting in the Japanese climate. Anatomical studies have disclosed characteristic culms for each species in terms of length, wall thickness, and in length, diameter, and number of internodes.

A paper mill was shut down in the 1960s, but research continues. An ancient technique using wood ash at ambient temperatures to defibrate young culms has been studied. A hand operated paper maker used by villagers combines bamboo pulp with other fibers successfully, and the institute of papermaking in Kochi Prefecture is researching small-scale paper production from waste of basketry and other uses. Japanese research is extensive and active. Lessard (1980:47–56) provides a scan of it which is suggestive for any country wishing to expand its bamboo research and development.

Malaysia.

Forestry services regard bamboos as weeds interfering with timber. Bamboo use is widespread but primarily small scale and cut from natural stands; cultivation is seldom practiced. Low durability is a problem. Incense sticks for Chinese religious rites are an important small industry, with 1978 exports valued at $654,080 (U.S.). The industrial sector is interested in canning shoots and producing lightweight bamboo harvesting poles for oil palms. A recently formed committee to coordinate research has priorities of performing taxonomic and ethnobotanic research; establishing a living collection for research, teaching, and gene conservation; and preparing an identification guide to forty-four regional species in seven genera.[11]

Philippines.

Of fifty-four species in ten genera, eight are commercially important: *B. blumeana, B. vulgaris; D. latiflorus, D. merrillianus; S. lima, S. lumampao; G. aspera,* and *G. levis.* From 200,000 ha mentioned by Gamble in 1910, bamboo areas dropped to 7,924 ha in 1978, or .03 percent of total land. It is not said whether private groves figure in this total. A research gap from 1930 to 1957 followed some earlier work, but revived interest since the late 1950s still leaves much unknown: distribution, yield, current and future needs of bamboo industries—all require more study. No Filipino taxonomist is involved with studies of Philippine bamboos. Harvests are not cut in a technically qualified manner. Rice, corn, sugar cane, and coconuts have edged out extensive nineteenth century bamboo plantations. Spontaneous forests are found in all parts of the Philippines but are not large and are diminishing from fires, grazing, and overcutting. There is probably more cultivated than spontaneous growth at present. Yield figures indicate the following culms per hectare: *B. blumeana,* 7,000–8,000; *B. vulgaris,* 9,000–10,000; *S. lumampao,* 9,000–10,000. Sawale, an important woven wall matting, is made from *S. lumampao.* One introduced species of *Phyllostachys—nigra,* var. *henonis*—grows well in the Philippines. Study center: Forest Research Institute, College, Laguna.[12]

Sri Lanka.

Fourteen species have been listed for the country, five widely used: *B. orientalis, B. vulgaris, B. vulgaris,* var. *vitata, D. giganteus,* and *O. stridula.* Bamboo has never been cultivated systematically, and there has hardly been any research on propagation or utilization. "In view of the importance of bamboo in the local economy, bamboo silviculture should be given more future importance . . . demand will soon outstrip supply. . . . The Forest Department is now contemplating the planting of bamboos." Report from Research Division, Forest Department, Colombo.[13]

Taiwan.

Export of bamboo shoots and some twenty-five bamboo products has been increasing steadily since 1960–61. With 1963 as a base year, the 1972 index on the export value of bamboo shoots stood at 503, poles and processed products at 1,218. Shoots are canned, preserved dry, boiled, salted, dehydrated, and shipped fresh. Development of new products include bamboo veneer, in 5-meter lengths (16.6 feet), thickness of 0.5 mm (.02 inch), used for plybamboo, trays, paneling and so forth. New processes include a dark brown coloring treatment for bamboo splits or veneers placed in 160°C kilns at 4 kg/cm^2 for 30–60 minutes. Hog pens, mushroom culture houses, oyster stakes, and banana supports are some of the main agricultural uses of bamboo, domestically.

Computerized yield studies estimate 21.06 metric tons of green, 15.35 metric tons of air-dried poles per hectare for a *makino* bamboo grove *(P. makioni)* with 9 m (3.0 feet) average height, an average density of three poles per square meter.

Among fifty-four reported species, the chief used are *P. makioni, P. edulis, D. latiflorus, B. oldhami, B. stenostachys,* and *B. dolichoclada.* Studies on genetic variability of Taiwan's 80,000 hectares of *D. latiflorus* and a selective breeding program for locating high-yield bamboo shoot clones have been initiated, using techniques of zymography. Analysis of peroxidase isozymes and microscopic study of leaf bristles and stomata reveal highly concentrated genetic variability in the west central part of Taiwan for this important species cultivated for both shoots and poles.[14]

Thailand.

About 2.5 million acres produce an annual harvest of some 600 million culms. Of the forty-one species in twelve genera reported, *T. siamensis,* 1.1 million acres (450,000 ha), and *B. arundinacea,* 625,000 acres (250,000 ha), are the most common, the first yielding 2,500–3,000 culms, 22–37 tons per acre (8.8–14.9 ton per hectare) annually; the second, around 62 tons per acre (24.7 tons per hectare). Groves are found in teak forests in the north, in evergreen or mixed deciduous forests in the south. Other important species include: *Bambusa blumeana, nana, polymorpha, tulda, vulgaris; Dendrocalamus asper, brandisi, hamiltonii, longispathus, membranaceus, strictus; Cephalostachyum pergracile, virgatum; Gigantochloa albociliata, nigrociliata, hasskarliana, macrostachys; Thyrsostachys oliveri.* Thai people eat a lot of bamboo, so much so that shoots taken from natural forests are a main source of depletion. Study center: Bamboo Research Station, Kanchanaburi Province, southwest Thailand.[15]

Fishing rods of *Thyrsostachys siamensis* are a large export item, and export of *D. asper* shoots have a "great future."[16] For recent work with bamboo cement, see "water storage" (pp. 66–67).

RESEARCH CENTERS AND GROVES: A SCANT SAMPLING

Note: Ministries of agriculture, forestry, and housing; botany and architecture departments of universities; arboretums and botanical gardens are all good places to start looking for active students of bamboo in any country. Other centers can be found, by country, in the reports above.

Centers.

FAR EAST.

Forest Research Institute
P.O. New Forest
Dehra Dun, India

Forest Research Institute
Chittagong, Bangladesh

Forestry Division
Joint Commission on Rural Reconstruction
37 Nan Hai Road
Taipei, Taiwan

Department of Forestry
Kyoto University
Kyoto, Japan

Department of Agriculture and Natural Resources
Manila, Philippines

National Forestry Research Bureau
Nanking, China

WEST EUROPE.

Tropical Products Institute
56–62 Gray's Inn Road
London, WC 1, England

LATIN AMERICA.

Instituto Nacional de Investigadores Forestales
Mexico, D.F.

Centro de Investigacion del Bambú (CIBAM)
Facultad de Artes, Universidad Nacional
Ciudad Universitaria, Apartado Aereo 54118
Bogotá, Colombia *(Oscar Hidalgo)*

UNITED STATES.

Division of Plant Exploration and Introduction
Bureau of Plant Industry
Washington, D.C.

Department of Botany
Smithsonian Institution
Washington, D.C. 20560
Main bamboo library in the West.

Groves.

UNITED STATES.

USDA Plant Introduction Station
Savannah, GA 31405
(Nine miles south of Savannah on Ogeechee Road.)

George Darrow Groves
5900 Bell Station Road
Glenn Dale, MD 20769

Jungle Gardens
Avery Island, Louisiana
(McIlhenny's Groves, begun in 1910. In mid 1940s, sixty-four species covered 80 acres.)

Arboretum
Golden Gate Park
San Francisco, California

Quail Botanical Gardens
230 Quail Gardens Drive
Encinitas, CA 92024
(American Bamboo Society Collection of some fifty-eight species, not all at the ¾ acre site at Quail.) (Cf. Journal ABS I.1:16–8.)

Green Gulch Zen Center
Rt. 1
Mill Valley, CA 94965

CHINA.
Lingnan University, Canton
(600 species; begun by McClure c. 1924.)

JAPAN.
Fuji Bamboo Garden
Nagahara, Gotemba
Shizuoka Prefecture, Japan

ENGLAND.
Pitt White Bamboo Gardens
Uplyme, East Devon, England *(southwest England)*

LATIN AMERICA.
Federal Agricultural Experiment Station
Mayaguez, Puerto Rico
(USDA)

Suchetepequez, Guatemala

Santa Ana, El Salvador

Lancetilla, Honduras

El Recreo, Nicaragua

Turrialba, Costa Rica

Summit Gardens, Canal Zone, Panama

Pichilingüe, Ecuador

Escuela Experimental de la Molina, Lima, Peru

VOLUME 1 NUMBER 2
THE
JOURNAL
OF THE
AMERICAN BAMBOO
SOCIETY
MAY 1980
ISSN 0197-3789

The American Bamboo Society.
The American Bamboo Society was formed in 1979, in California, "to fill a vacuum in bamboo culture and research that has existed since the USDA stopped large-scale bamboo research in 1965. Following the examples of the Lingnan University bamboo collection in Canton [begun by McClure in the 1920s] and the Fuji Bamboo Garden in Japan . . . the primary objectives are (1) To provide information on identification, propagation, and utilization of bamboos through a journal, library, and herbarium. (2) To promote use of elite species through distribution of plants to botanic gardens and eventually the general public. (3) To preserve and increase the number of bamboo species in the United States, establishing a bamboo quarantine greenhouse in the San Diego area to import selected species from foreign sources. (4) To maintain a bamboo garden to display beauty of mature plants and provide a means for research in the culture of as large a number of species as possible."[17]

Quarantine.
Early twentieth century bamboo plants from the Orient carried bamboo pests with them (see pp. 225–226, pests, smuts, and rusts). All bamboo plant introduction has, in consequence, been regulated strictly by the USDA since 1918: "The importation for any purpose of any variety of bamboo seed, plants, or cuttings thereof capable of propagation . . . is prohibited, except for experimental or scientific purposes by the Department of Agriculture."[18] Permission to import bamboo can be obtained, however, by groups responsibly following government quarantine regulations, such as the American Bamboo Society.

The *Journal* of the American Bamboo Society

(1.1:2–11) provides a useful survey of USDA introductions from 1898–1975. Some 189 identified species and variants are included, together with a large number of introductions in which only the genus is named.

Experimental bamboo farm.
Bamboo gardens are important for the spread of bamboo plants and information, but something in the very size of bamboo clamors for a bamboo farm. Apart from large urban parks, there are few areas in cities with enough space to establish large-scale plantings together with the curing sheds, preservative rooms, workshops, and toolsheds that are required in growing, processing, and using bamboo.

Granted the desire to extend the awareness of bamboo technology, a bamboo farm should include, as well, facilities for prolonged visits from those who would like to learn its large-scale management. The lack of such training facilities open to the public has resulted in the neglect of most ambitious plantings of bamboo in the West. Training bamboo cultivators, builders, inventors, artisans, salespeople, manufacturers, and messengers must be a priority of any bamboo enterprise that hopes to survive the enthusiasm of its initiators. All these roles are required for large-scale commercial production of bamboo plants, lumber, and artifacts. Generations of bamboo proponents have come and gone in the United States and the West, and many of the groves they established have been destroyed or have fallen into complete neglect or underproduction for lack of young hands and minds to follow the old man's affection for the plant. Where they still exist, they are the natural location in their region for a farm-training bamboo village.

A bamboo farm, by the nature of the plant, calls for a design quite different from a farm focused on wheat, corn, soybeans, or some other staple crop already well established in the agricultural economy of the West. The broad uses of the plant, still so scantily known among us, require a location where its versatility can be demonstrated. Bamboo's retarded cultural role implies a farm more experimental in nature than farming almost any other crop imaginable, with far more conscious public relations and public play than, say, a hog farm in the corn belt.

WHOO'S WHOO: BAMBOO GREEN PAGES

The World Directory of Bamboo Researchers lists 203 students of bamboo from twenty-four countries: Japan (54), the Philippines (31), India (24), Taiwan (22), Korea (15), Thailand (13), the United States (9), England (6), Bangladesh (5), China (4), West Germany (3), Australia, Malaysia, Mexico,

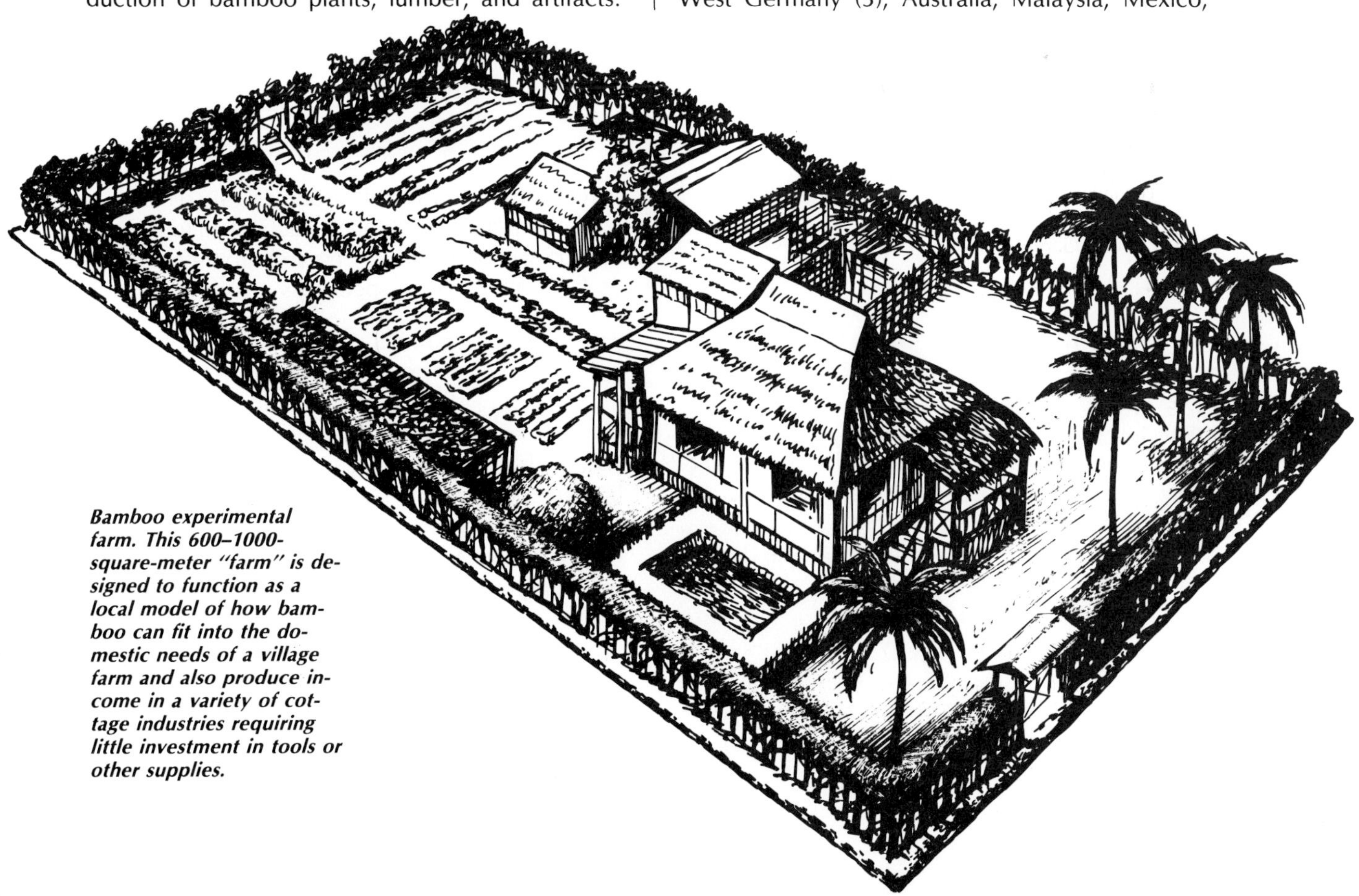

Bamboo experimental farm. This 600–1000-square-meter "farm" is designed to function as a local model of how bamboo can fit into the domestic needs of a village farm and also produce income in a variety of cottage industries requiring little investment in tools or other supplies.

Nepal (2 each), Brazil, Burma, Canada, Chile, Colombia, France, New Zealand, South Africa, and West Somoa (1 each). The number from Japan reflects a pardonable regional bias, natural for a Japanese editor for an event in Japan. The few names from mainland China indicate an unfortunate information gap. But the twenty-nine pages of bamboo students is a welcome start in the necessary direction of networking the planetary students of the plant.[19]

Obviously not every bamboo student in the world is in this directory. Still their numbers in relation to the population at large are few. How many doctors, lawyers, etcetera, whatever have you met? How many students of bamboo have you stumbled across? Even in countries where bamboo is economically important, bamboo researchers are the eye of the needle in the haystack. Lessard (1980) lists some twenty who would be valuable leads in many countries East and West. The *Journal* of the American Bamboo Society provides contact with its members. Bamboo flute people's addresses are available through Shepard and Levenson. (See also: *Také-no-Michi:* A Newsletter of Shakuhachi and Related Arts, published by Barry Weiss, 419 4th St., Brooklyn, NY 11215.)

The *World Directory* lists researchers by area of concern, such as distribution, ecological studies, cultivation (propagation, plantation methods, optimum felling cycles), harvest and transportation systems, anatomy, physical and chemical property studies, biochemical and biophysical aspects of bamboo formation, seasoning and preservation, pulp making, mechanization for processing in local industries for food, handicrafts, furniture, building materials, and bamboo-based panels.

From the listed categories, from published books and articles, and from experience of the Western bamboo community from the U.S. to Peru, it is apparent that bamboo is a circus with many rings but no barker. There are no bamboo communications specialists evolving tactics for increasing awareness of the plant on the part of governments and peoples. Oscar Hidalgo, in Colombia, is a rare exception to prove the rule. An architect, a teacher at the national university in Bogotá, he has managed time for three books (1974, 1978, 1981), and a film on a shoestring budget (1981). Parts of the filmscript are found in Chapter 5, pp. 122–123. He has conducted prolonged bamboo agitation among a broad circle of friends from many nations and supervised or inspired a number of projects that have made him *Papa Bambú* to many in Latin America. He works with an infectious warmth and enthusiasm, animated by a constant human interest in bamboo use among the poorest people. "Oscar is 200 percent crazy," an agronomist told us in Colombia. "Too bad there aren't many more with his madness."

The need to network: *Bamboo Tube.*
Apart from Hidalgo and his welcome ilk, elsewhere by whatever grove, many more scientists are fretting full time over their piece of the bamboo puzzle than are coordinating the vast masses of bamboo data, extracting a relevant vision of development design and communicating it in a way that can be heard by the masses whom all this research is ultimately meant to serve. A great gulf is fixed between the expert in the lab and the *campesino* in the field that somehow must be bridged for bamboo to assume its deserved prominence in the cultures and economies of the world's peoples.

Bamboo messengers are needed, as is a place and method to train them, and a publication to share relevant experience with the larger world. Those wishing to participate in the creation of such a center or centers in the West can communicate. Minimally, as an update some two years after publication, we intend mailing an irregular newsletter, whose first objective will be to make available, by name, address, and phone if desired, a list of those concerned with bamboo development in the West and caring to share their concern with others. Excerpts of bamboo information provided by correspondents will hopefully find room for inclusion. Send to *Bamboo,* Box 666, Bolinas, CA 94924.

MESSAGE DESIGN

This is not a book.
This artifact is composed of equal parts sweat and affection, collaged almost completely of recycled, even ancient, thoughts perched for the moment in English but ready as a resting wren to reenter the common air that bathes the globe. It has no chemical additives, is not toilet-trained, tamed, shaved, or built for sale. That those who need it most will sell hours of a brief life to buy it and read it is a chief defect of its design. It will knock down trees for pulp, and the nests in them, consume limited fossil fuels in its distribution. Some copies will be given to seduce the wrong lover or bought and given for Christmas to the wrong aunt, to lie silent seventeen years on a shelf and then burn to ash in a fire. Others will pass the hours of commuters, insomniacs, and other compulsive readers. One copy—

it could be the one in your hands—will meet the proper person at the proper time, infect a life with a life errand, repay the labor of compiling it, and compensate for the waste of ten thousand copies that do not reach their intended objective of a destined eye.

At a 1965 exhibition of Magritte in Chicago, a woman stopped with her husband next to a painting, squinted at the title, and read in midwestern French, "*'Ceci n'est pas une pipe. . .'* But it *is* a pipe!" And furthermore, Madame, this is not a book. A normal book is happy in a bookcase, but the very ink of these words aches to run over the margins into the world.

A quilt of quotes.
This work is in large measure a quilt of quotes. We have attempted an organic anthology drawn from bamboo literature that is rarely available even in the best libraries. A dialogue of many voices is always more useful than a monologue if the voices chosen are articulate. But this method implies much editing to make the pieces fit, filling gaps and pruning overlaps. We have made a virtue of the editing that necessity imposed. In addition to eliminating repetition from selected quotes, we condensed them to their core line of argument or information, rigorously pruning, always remembering Jack Sprat, who ate no fat, and Nietzsche, who hoped to say in a sentence what others said in a book.

We have inserted striking phrases or facts from the quoted author's own total work if it served to heighten the intensity, relevance, or eloquence of the cited observation. We have striven to preserve style as well as sense. For readability, we have eliminated the dots, brackets for minor supplied words, and other conventions of typography and scholarship. Our intent has not been—ever—to betray the original but to intensify the information per inch throughout a treatise whose amplitude demands economy of expression.

In the case of material from general literary sources and a range of contemporary culture, these nonbamboo books are provided as samplings of the directions bamboo development studies must take. The more subtle the grafting of any new crop or cultural innovation, the more chances for successful new life. These quotations—on shelter, education, and the whole spectrum of human values—are intended as a chorus of elders of the Global Village, which looks over our shoulders bent above this prolonged task. "The World wrote them, Goethe signed them," one author said of his work. Our intention has been to present the human reality of our time, in its roots and in its urgency, through the most complete voices of those who cared most deeply about the design choices confronting their species in its cultures and agricultures.

For lack of space, many remain unquoted, or barely mentioned, who are in fact the invisible underpinning of this work and who describe its desired future trajectory—they are usually noted and starred in the bibliography. Alexander, Illich, Kern, Montessori, Mollison, Prest—these are among them. Others, like Darrow, are sufficiently visible in a scan of footnotes to indicate our debt.

Fracture survival.
Survival in fragmented times is always precarious because survival is wholistic. But the judo of nature makes use of fracture in interesting ways, transforming breakdown into a technique of survival. Grab a lizard's tail, and you find he is equipped to leave it squirming in your hand as he returns to his family. Pieces of an artwork could imitate this design wisdom and, if possible, even improve it, leaving a tiny wriggly lizard living in the palm of your mind. Many pieces of this book are intended to form small wholes whose sense survives surgery from the surrounding animal. They can be broken off easily and used in many contexts with no direct relation to bamboo. The Eastern respect and feeling that surround bamboo can find Western expression in many forms with no physical presence of the plant. Bamboo is not the only ally on the path towards bamboo attitudes. (See, for example, *Earth Basketry,* Tod 1972.)

Radio rapidea.
Books should be designed to fit easily into radio. For thousands of millennia before written or printed words, there were spoken words. Books come long before radio in the chronology of invention, but what they deliver is more "modern"—the written word. Radio brings the ancient human voice to the ancient human ear. Present-day books should acknowledge radio as an ally by presenting information in a way that feels fluid on the tongue, with all the vigor and verve of speech. Information bits should be brief and form complete wholes in themselves, as well as portions of a larger whole.

A series of five-minute radio talks on bamboo is available in English and Spanish from the author on request. Twenty segments on a single cassette are designed for a four-week series but can be combined to form longer portions of a shorter run. Bamboo music, flute and percussion, provides a recurrent motif. Radio provides more broad and rapid idea diffusion than print and is particularly suited for village areas where bamboo is most relevant and books least used. The terrains for bamboo development imply both Spanish and English versions. Unfortunately, no such series is available elsewhere for the 800 million or so planet residents who listen to these two tongues, culturally and geographically densely related, which dominate the West: *Inglespañol.*

Explanations are bad Zen.

He was the Village Explainer,
excellent if you were a village;
if not, not.

—Gertrude Stein, of Ezra Pound

Until now, we have more or less assumed the semi-tedious role of Village Explainer, spelling it out as plain as possible: long lists of Latin names and so on, a superbore if you don't first feel the total bamboo conception. As this book moves to a close, we have resorted more to hint, hymn, fairy tale therapy, fragments of old masters, the shaggy, uncombed eyebrows of oracles to say it. Explanations are bad zen; they will die on the road home to our reknowable energy. Overtell yields underlisten. Lakes rest in the shape of land under them. Design the message by the ear.

Bamboo Ring

Once upon a tale, there was a morning
that round a wandering shoulder wrapped a road.
A bend of it revealed, requesting kisses,
a double ugly but seductive toad.

Which done, she hopped before him to a village,
and taught en route a simple song to sing
that turned the toad into a shining maiden,
and in her hand, a tiny bamboo ring.

All who tried it on went mildly crazy,
and felt a funny hum until they died;
the busy took the time to aid the lazy,
the friendly kept the lonely by their side.

Elites of greed began to feed the others.
Soldiers dropped the gun and hugged the foe.
Strangers treated strangers like their brothers—
but that was long away and far ago.

MESSENGER DESIGN: TACTICAL ATTITUDES

The point isn't merely to understand
history, but to enter and change it.

—Karl Marx

If you fall into a pit
and three uninvited guests
land later on top of you,
welcome them cheerfully.

—*I Ching*

Acupuncture is the art of effecting maximal positive reaction through minimal pressure, perfectly placed. The cultural equivalent of this economy of means is the study of all those who are poor or few in relation to the dimension of a prevailing error, but who, nevertheless, refuse to remain its victim.

Traditional oriental hints on the art of cultural acupuncture are densely embedded in a long literature on the Art of War, classically defined by Sun Tzu (c. 400 B.C.) and crowned by Miyamoto Musashi (1584–1645); and in a long scrutiny of tactical attitudes for students of personal or cultural change, in such summations of Eastern philosophy as the *I Ching.* Somewhat in the spirit of "flying white" brush strokes of *sumi-e* painting, we offer the barest suggestion of this rich heritage.

Learning to meet: the *I Ching.*

The bird, a nest.
The spider, a web.
The people, friendship.

—William Blake

Life winds a lot.
Sometimes it's easy,
sometimes it's not.
Sometimes you talk about it,
sometimes you can't.
But when two friends share one inside,
their hearts are just there,
like weather or air,
and their words smell like flowers.

—Confucius

The Chinese are the most numerous people on earth with the longest surviving cultural attic. They have the world's most extensive archives on the art of meeting and the most ample ongoing laboratory to explore human relationships: one billion people means many meetings. Since their culture has proved so durable, their earliest conclusions on the methods for durable and meaningful meeting deserve pondering. "Friendship" is our word for com-

The invention of the eight trigrams of the* I Ching *is attributed to the legendary Emperor Fu Hsi (2900 B.C.). The straight (yang) and open (yin) lines which form the trigrams are stylized genitals. The unbroken and broken lines also suggest the erect stalk and emptiness, or node and internode, of bamboo, which serve to sum up the basic rhythm of life.

plete meeting. We are friends in direct proportion to our capacity to meet the entirety of the other with the entirety of ourselves. Remaining friends is an art even more rare than becoming them, because friendships also, like all mortal things, are buffeted by Change, the main and most mutant divinity in China's ample pantheon. Change being the essence of life, learning to meet Change, becoming optimal mutants, the friend of Change rather than the victim, became the essence of learning for the Chinese.

> Hard to start? Invite helpers—but don't doze on their shoulders. Keep changing in the midst of danger: that's how complete people draw order from confusion, like thunder wringing rain from clouds.

They especially studied mood changes, rapid shifts in emotions, the inward weather of human nature, observable, with training, at first feel in ourselves and at second hand in the behavior of others.

> Tempers pop in the family, friends look like muddy pigs or a carful of hoods—but ugly's in the mind of the beholder. Friendship heals feuds, links families. Tight spots mean more room for friends; more friends mean more forgiveness.

> Forgive is lighter than grudge. First crying, then laughing. After fighting like crazy, we get sane and meet.

Friendship, the perfect meshing of weathers, was in some ways their ultimate art form, a process of mutual transformation. Friendship is an alchemy in which both lead and laboratory are another alchemist. It is the most subtle instrument or condition for one's personal growth available, an effective device of our emotions, rapid and sly beyond any art or science to arouse our slumbering faculties to their most flourishing development.

The more psychiatrists and "mental health personnel" multiply among us, the more we announce the erosion of friendship in our culture. The need for such doctoring is rare if the heart and ear of an alert friend are available. The most mentally healthy culture, in fact, will be that which most fosters friendship in home, school, market—the whole fabric of a culture is based on the experience of friendship, because without this intimate and personal school of union it is doubtful that more impersonal or abstract cultural unities like towns and countries can have much human vitality. Friendship is even more basic than the life, liberty, and quest for happiness declared by Jefferson as fundamental rights of a democratic society, because even death or any lesser oppression can be tolerable in the embrace of friendship, but a life or liberty or happiness without it is something not only "less than human," but less than animal or plant: friendship is inherent to all natural forms of life. It is distinguished from many human art forms in that, like true ritual or a sense of "ceremony," it can exist in the leanest times, with minimal economic base.

> The lotus blooms in muddy water. In danger, deep holes, thorns, three years of wall and knotted rope, live simple and forego decor. Broke is beautiful: a bowl of rice, a jug of wine—two small flowers are enough for the ceremony.
>
> —*I Ching.*

When the people of China called bamboo the Friend, it was no mean complement. Out of a bewildering abundance of species that botanists have declared the broadest repository of flora in the world, they considered three plants Friend: the pine, the plum, and the bamboo. If you act on this hint and actually regard bamboos not only as persons, but as friends energetically devoted to your interests and responding to your attentive affection, a very different dimension of gardening can bloom for you even on your windowsill; and you can begin to experience that interspecies affection is the essence not only of gardening but of being alive on the planet at all.

Each role or relationship has its place and labor —farmers to the field, lovers to bed or riverbank . . . And friends? *Shared work* proves the most durable bottle for this wine. Friendships without a shared labor of some kind usually grow thin and die, or remain feebly latent, a possibility glimpsed but not explored or embodied. "Coming to meet" must be followed by the "holding together" of active friendship.

> Hold together with clear motive and consistent energy that won't quit. Then the timid get sure. Make friends of neighbors. Offer truth, like a full earthen bowl. Hold on tight to the best in others. Holding together is not just hanging out.

FORGIVENESS: THE GENTLE CENTER OF FRIENDSHIP. Getting to know bamboo, you quickly get to want to know it better. The only thing that grows as quickly as bamboo is interest in it. This interest propels you to work with bamboo in some way, and working with this Friend you find that work of any significant scale implies working with human friends also—which can be a path to estrangement unless friends are skillful at fostering a durable relationship. Forgiveness, an ability to make a fresh start, is absolutely basic to this skill, and there is no more accurate measure of our capacity for friendship than our capacity to forgive.

Governments and large corporations have, in general, shown only sporadic interest in bamboo, which yields less rapid return on investment than armaments or automobiles. So immediate bamboo development in the West will depend much on an initial widespread popular concern, groups of people coordinating their creative energies. They must do this not for a week, month, or year, but in a durable manner, tuned to the ample rhythm and nature of bamboo, not to the spastic rhythms of whim or momentary fad. They will often find it necessary to forgive one another a certain maddening resistance to creative union. This resistance to significant union is a nuisance with a thousand faces that is built into us at deep levels of our competitive cultural imprint. It often compels groups to ravel at evening what they weave by day, like Penelope stalling the suitors, but without the steady sanity of her hidden cause.

For the modern West, the social unit is the individual, the supposedly separate, supposedly sacred "I." More traditional cultures generally see our human reality more as a web or relationship, a co-creation with others rather than a piece of Private Property surrounded, psychologically, by a cyclone fence. The *I Ching* is one ancient vision of a fruitful and harmonious inter-survival of the person, friendships, the family, the community, and the state. The sexist and hierarchic "superior man" of Wilhelm's familiar translation we would amend to the "complete person."

SMALL CHANGE.

> Complete people fix scenes in their act. Small change adds up. Come home again to the real little world around you. Keep your steps open and watch your eyes.
>
> Complete people are approachable, always ready to share knowledge, without limits of tolerance and care for the people. Yielding, but firm in movement; serious about lightening the inner life; treating the crude gently, crossing the river firmly; neither lost in the distant, nor enmeshed in the near; dwelling in essentials with devotion.
>
> Finish before starting again. Complete people make complete acts. Complete acts make incomplete people complete. But if things grind to a halt don't fight it. Time to breathe, watch. Complete people fall back on their insides to

escape difficulties. Mouth shut, hands active, stomach friendly, mind clear. Sky and earth united, everything flows free. Whoever hears the heart gets home again. Keep up the good mind.

Sun Tzu: The Art of War.

Hard to be soft, but it's the only armor worth wearing for the warrior.
—Buddhist Proverb

The Art of War, said to be composed by Sun Tzu around 400–320 B.C., is the earliest in a long lineage of Eastern classics on the martial arts. It has remained one of the most consulted to the present day, a bedside book of Mao that has deeply influenced both Japanese and Russian military theorists as well. For Sun Tzu, the best war is the one that successful diplomacy makes unnecessary: "To fight a thousand times and win a thousand times is not the blessing of blessings; it's to beat the other fellow without ever getting into the fight." This military version of the gentle art of *wu wei,* "not doing," reflects the same respect for emptiness and receptivity we have seen in oriental philosophy and art. The marrow of war for Sun Tzu, the fluid bones of the battle, is shapelessness—the same flexible emptiness described as the gate to the "wonder realm" in Zen. "Don't repeat a winning tactic. Rather, mold your actions to an infinite variety.

The good general: headquarters with heart.

Samurai such as Miyamoto Musashi staked their lives on their awareness. The buried treasure in his "map of tactics" is a spirit that refuses defeat.

Water takes the shape of ground below it. Design your battle by your enemy. War and water have no constant shape."

THE GOOD GENERAL: HEADQUARTERS WITH HEART. A crucial character in the art of war is "the good general." He is Sun Tzu's complete person, headquarters with heart.

Sun Tzu finds that the heart of leadership resides in physical solidarity with the led. This opinion departs vigorously from contemporary style, in which all countries, regardless of their political creed, believe firmly in soft couches in their embassies and an impressive limousine to lead the parade. The Minister of Housing is not homeless. The Minister of Agriculture is not hungry. The Minister of Health is not sick without doctor. The Minister of Transport is not hitching or riding a bus. The Minister of Labor is not unemployed. Hard to design a cultural pattern in which the leaders truly experience and remain in touch with the pain and possible exits from pain for the very poor, but here is what the oldest consulted classic of a flourishing contemporary art has to say about opening the third eye of enlightened management:

> The good general doesn't think of fame while advancing or of shame in retreat. He only thinks to protect the people. He is anxious as a grandmother for the army, so they will go into the deepest valleys. Thin coats feel warmer at his word. He eats and wears what the troops eat and wear. He doesn't ride if others are walking. He carries his own rations and sleeps on the ground like everyone else.
>
> He doesn't carry an umbrella in the summer, or get more blankets in the snow. When the whole army has water, he drinks. When the food is ready for the whole army, he eats. When all the walls in the camp are built, he thinks of his own roof then.

A map of tactics.

Hard to keep track of the true Way. Small and big, shallow and deep are all part of it. Here's a map like a straight road in front of your feet on the ground.

Words miss the way. You have to feel it in your body, from your heart. Spirit: determined and calm—in crisis or everyday life. Not tense, not reckless. Neither excited nor down in the mouth. To a friend—clear water in the wilderness; and to a foe—invisible.

Big people should know how it feels to be little; little people, how it feels to be big. But don't be fooled by roles or bodies: the open, unleashed spirit, the original energy, sees from above the body, can't be lied to, and stays steady even under stress.

Want to learn the way of strategy? Think about this: the teacher, the needle; the student, the thread. Practice day and night.

Big's easy to spot. Hard to see Small. Big groups move slow, so they're predictable, while one alone can change mind fast. Think about that.

Think honestly. Train in the Way. Know every art and tools of all trades. Judge spontaneously between profit and loss. See the unseen. Notice trifles. Do nothing useless. Scan on a large scale. The secret of strategy is a spirit which refuses defeat.

Time is in everything, even the void. All things rise and fall. Knowing the rhythm of things, you can strike naturally. Knowing their timing and using surprise timing, aware timing, a void timing—battles are won. Knowing the Way broadly, you see it everywhere: don't die with your weapon idle in your hand.

Masters of strategy know and deploy the people. They know abilities and imitations. Aware of limits, encouraging if necessary, asking nothing unreasonable, passing constantly among the people.

All ways have byways. Small shifts in directions get big. Straying from the Way, your spirit loses it. Look at the world: you'll find People for Sale. Eager to make it, anything for a buck. Somebody once said, "Unripe strategy is a guide to grief." Sure is true.

—Musahi

Miyamoto Musashi.
Miyamoto Musashi, the seventeenth century Japanese sword saint (1584–1645), distilled a life's experience in a brief work composed just before his death, *The Book of the Five Rings.* Musashi is still carefully consulted by executives of Japanese multinational corporations to plot global tactics, and has become a guru to Wall Street strategists as well.

THE EDEN OF EFFORT AND THE EDEN OF EASE. To those disturbed by the basic military metaphor in the Sun Tzu and Musashi material, we would suggest that even for the most nonviolent of all possible pacifists it serves to Know Thy Enemy—if these indeed be such—and salvage from their methods whatever tactical principles can be applied, in an inventive manner, to bloodless, cultural revolution on the planet.

William James, in an essay called "The Moral Equivalent of War," finds struggle so central an aspect of our nature that he believes we can gradually eliminate the gross horror of war from human cultures only by providing creative cultural expressions for its energy that are of equal intensity and magnitude. He suggests an Eden of creative Effort as a more interesting and engaging alternative to the Eden of Ease, the dream of effortless and expanding convenience infecting our late twentieth century Western imagination even more than it did the late nineteenth century psychology that James possessed and addressed.

Many shrewd people, perhaps correctly, feel it's sentimental to imagine that the Revolution needn't be red. But revolutionary style is itself evolving. A serious global effort could achieve a revolution in revolution design, perhaps, and blueprint a green, ecological revolution in such detail that it occurred. The accurate culture map becomes the territory.

But peace isn't lazy. We must fling ourselves into cultural revolution on the planet with the passion and energy that in the past have been reserved for war. Anything less than the everything of our unleashed capacities is perhaps, at this eleventh

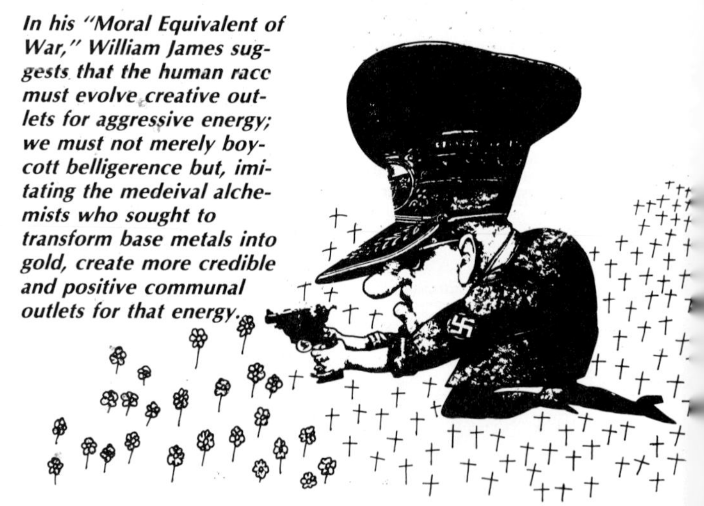

In his "Moral Equivalent of War," William James suggests that the human race must evolve creative outlets for aggressive energy; we must not merely boycott belligerence but, imitating the medeival alchemists who sought to transform base metals into gold, create more credible and positive communal outlets for that energy.

hour, not enough to make a revolution kind to the planet, and mending without blood.

As you read the words of this single sentence, four or five people, mostly children, have died of hunger, and forty or fifty more have been born. There is no place here anymore for people who don't want to share. How can we let them know that, quickly, and without a gun?

REKNOWABLE ENERGY

Cultural meristem.

Kids are kings in the
kingdom of Change.

The meristem, or meristematic tissue, of a plant refers to that part where the cells are still capable of division (word root is from Gr. *meristos,* divisable). All tissue can be divided, in our understanding, into three parts: dead, or merely living, or both living and productive of new life. This book is about innovation in culture (Barnett 1953), and so is addressed mainly to the cultural meristem, those still budding, still able to divide clearly in themselves the new from the old, scrutinize their attitudes and acts to see if they proceed from a frozen past or valid and fluid future, from parts of them dying, or from parts being born.

The inward meristem, where we are still breeding, have the innocence to enjoy ourselves and the courage to imagine that people can indeed create their world, is what Carlos Castaneda's holy Yaqui or Yankee hoax, Don Juan, calls "a path with heart," which feeds you every step of the way. At every step of it diverge the paths of phantoms, promising food and the way home, only to mislead you. Whether bamboo be heart or phantom lies in the hand behind the use, in you. But the bamboo path, unlike many, at least has the distinct possibility of heart, health, hearth, wholeness, a durable sanity, a way home.

We are less likely to lose our way on the way by following the true cultural meristem of our children, the budding point of social change. "Give me a lever long enough and place to stand," says Archimedes, "and I'll move the world." The world is moving on its own, but its trajectory is more deflectable than our directors dream. It is up to the people, not to "take charge" in imitation of our "leaders," but to learn the subtle art of nudging their lives and their history in more healing directions than are presently the general issue of the surrounding cultural design.

The long lever we have chosen to nudge the world is bamboo; the place to stand, child mind, the kingdom of kids.

KINDEREDEN: THE OPTIMAL ORPHANAGE. The cry of the global orphan is becoming a huge sound, not to be silenced, on the human doorstep. It seems more promising to regard this as an opportunity than as a problem.

Children of war, abandoned as garbage not worth looting, are in fact the neglected treasure of it, more valuable per acre than any land or resource the enemies could contest. Bucky Fuller computed in the '60s that it would require some 28 city blocks of computers to roughly approximate—never exhaust—the capacities of the human intellect, grandaddy computer stored in each of us, complete with that wondrous capacity to clone—with the help of an opposite equal to our biology.

So orphans can be seen as the chief booty of war, whose value is completely overlooked by the civil idiots who incite, supply, and wage wars to begin with. None or few of them are women, who know what it costs to bear and raise children only to say goodbye forever when the full possibility of their adult being has just begun, only to watch them leave to die for causes requiring ever more cosmetics to appear remotely glorious.

Nations—like the U.S.—that leave a sea of orphans in the wake of their fleets, could do better than erect marbled memorials back home to their wasted sons. A fitter monument to war would be an optimal orphanage, designed to foster the creative vitality of war's youngest victims so effectively that they become bright agents of peace, architects of the higher harmonies that our race must learn soon to erect in its fisty midst if we are to survive at all.

Bamboo for centuries in China was regarded as an optimal godfather. Placing a child under the protective adoption of a grove brought the best of all possible luck—so choosing bamboo as the patron plant of an orphanage seems in complete harmony with its ancient oriental role. We invite residents of the United States especially to participate in this task, because building a bamboo orphanage in Central America in 1984—famous as the date of Orwell's hell—seems an appropriate and durable expression of solidarity with the people of that region, whose death rate and orphan index is considerably heightened at present through United States aid. The most terrible apparent disadvantage is no lid to the original energy of new life, if fostered with alert love and earth awareness. Montessori in a slum of Rome, George Washington Carver with the chil-

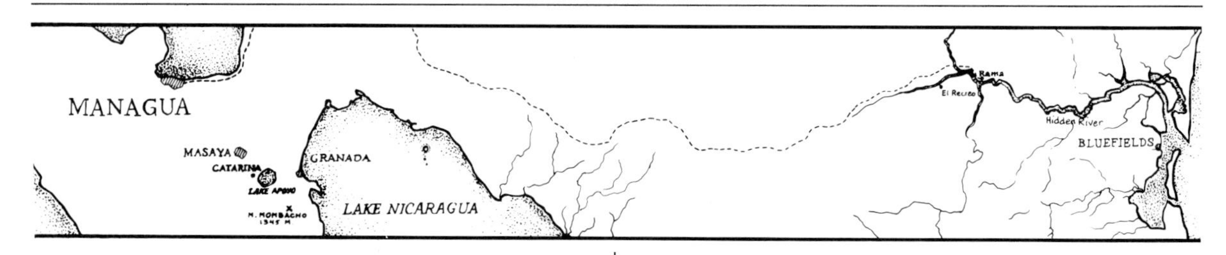

In Nicaragua, groves of elite bamboos exist in El Recreo, Zelaya, an easy float down the Hidden River to Rama, an inland port five days from Miami and New Orleans. Primary bamboo consumption is in Catarina, a basketmaking village in the low mountains south of Managua overlooking Lake Nicaragua. Introducing the best species of bamboo to the craftsmen most dependent on the plant for their livelihood is a priority in many Latin American countries. El Recreo and Catarina have been chosen as the poles of an experimental project exploring formats for effective bamboo development on a national level in the West.

dren of poor black farmers and exhausted soil of Alabama, have demonstrated that when ordinary human nature even in its most socially deprived forms is tended at the start in its *learning to learn* (what mainly makes us human) with affectionate and creative use of the methods of science—the miraculous can become the norm.

"The gate of the Great Learning lies in watching, with affection, how the people grow," say the Confucians. The greatest learning treats of the patterns of learning. These are best learned by focusing our lens on our own mirror: human nature in the form of new life, the reknowable energy of child-mind in full flourish of development. "When going into the woods to cut an ax handle, the model is not far off," say the Chinese. Human nature is at our elbow always, but most vibrantly so when we are surrounded and encouraged by new versions of our own creative energy. We have spoken often in these pages of appropriate technology, village technology, or technology "with a human face." Looking more deeply at the aims that direct cultures and define their stature, we could perhaps more accurately divide our human doing into the technology of service and the technology of profits—and pain.

> *Peoples' options on the earth are plain:*
> *wander a thousand valleys and you'll find*
> *a million villages, two states of mind—*
> *a state of service and a state of pain.*
>
> —Sufi Proverb

BAMBOO BOX SEAT ON A TINY WAR. Much of this book was composed in a country—Nicaragua—presently under heavy United States economic and military pressure. We were North American *tecnicos*—agricultural technical advisors—working from 1981 to 1983 with the Sandinista Ministries of Agriculture and Education. Our work included harvesting bamboo in groves in Zelayá 30 miles away from where our Cuban counterparts were being killed among their crops by *contras* (counter-revolutionaries) financed by the United States. We were frequently mistaken for Cubans in the small villages; we, too, could be killed by American aid.

One night, bridges suddenly blew up at several locations in the country, just before a scheduled field trip with our local school in Catarina near Masaya in western Nicaragua, to harvest at the government's collection of elite oriental bamboos in El Recreo, in the east. Preparations for the trip had been going on for three months, budgets submitted to the Ministry of Education, permissions formally secured from the Ministry of Agriculture, the cash, by some miracle, actually in our hands. We woke up, a day before departure, to a national state of emergency; militia or soldiers at all bridges in the country on major roads; the history of U.S. intervention summarized once more in the press; and parents unwilling to have their *muchachos* travel 150 dangerous miles from home to harvest bamboo, in *Zelaya,* of all places. We stopped trying to get them to distinguish between northern Zelaya, where most of the fighting was, and central Zelaya, where we were going, when the afternoon paper published accounts of the death of Cuban agronomists near the central Zelaya groves. Our work force dwindled from twenty to two.

In El Salvador, we had already found our research surrounded by war: a fresh corpse near the bamboo groves the day we arrived at the experimental station in Santa Ana, some 30 miles northwest of San Salvador, and another shot down on the street as we rode home on the bus. I reached for my camera, and a friend shoved it back in my lap. "Just shoot bamboo. And don't go out after dark. It's still supposedly legal for groups of three people or less, but you may meet a soldier who can't count" (July 1980). But we did go out, sleepless, around 11, and again around 3 A.M., and found all at peace in San Salvador, in our sector, only the lights of the funeral home burning between the Hotel of Saint Anthony (patron of pigs) and the bus station. The undertakers in Central America have their hands full of money and blood. . . .

We saw enough that it has become a struggle to maintain the orderly voice assumed in these pages. As Tolstoy said a long time ago in the face of equally intolerable outrage, "I cannot keep silent."

I have a scream. But better to sing it, and shape it into durable resistance of botanical activists. Better to keep a small fire burning in prevailing darkness, ready for the chilly children of the coming dawn.

Come without a book or banner.

In New York, there crooned a concrete siren
tunes that itched in ears across a sea:
"Send me your tears, your foiled and barren,
your thronging masses, longing to be free . . ."

Freeways arched above the startled farmer.
Fallout lurked beneath the leaves of grass.
The torch of Freedom in the harbor
was snuffed and shoved, still smoking, up Her past—

So we must act, but in a tranquil manner;
rest, but wakeful to the dream;
come without a book or banner:
how we move is what we mean.

Raise the child and weed the garden—
Nature's play is never done.
Praise the wild and free the warden;
cram a carrot up his gun.

Pile your bullets taller than a mountain;
pack the bloody oceans in a pail;
but life is not a bucket, but a fountain.
And the truly free are not for sale.

CHAPTER 10

0. The Fool; Tarot pack adapted from Waite.
1. McClure 1926:1.
2. McClure 1966:vii-viii.
3. Calderon 1980:5.
3a. Ibid.: 4–9.
4. Lessard 1980:5.
5. Ibid.: 205–6.
6. Ibid.: 15–8.
7. Ibid.: 57–62.
8. Ibid.: 19–46. Lessard basically summarizes Varmah's (1980) report on bamboo in India, available on request from Dehra Dun.
9. Varmah 1980:22.
10. Lessard 1980:63–8.
11. Ibid.: 91–5.
12. Ibid.: 69–80.
13. Ibid.: 81–4.
14. Kiang 1973.
15. Lessard 1980:85–90.
16. Ibid.: 109.
17. Condensed from the *Journal* of the American Bamboo Society 1980: I.1:1, "Statement of organization and purpose."
18. USDA Notice of Quarantine No. 34, Oct. 20, 1917. Effective Oct. 1, 1918.
19. Compiled by Higuchi of Kyoto University for the International Union of Forestry Research Organizations in preparation for their 1981 XVIII World Congress in Japan.

GLOSSARY

A

adventitious. occurring in an unusual position, "adventuring." "In the bamboos, and in other grasses as well, the principal complement of roots is adventitious, arising at the nodes of culms and rhizomes, and not from the primordial root." (McClure 1966:296)
aerial. above ground, as in "aerial roots."
albino. white, pale; without or with little chlorophyll.
albo-striatus. white lined.
angustifolius. narrow leaf.
argenteus. silvered.
arundo. reed.
aureus. golden.
auricle. earlike growth; in culm sheaths, a small growth on each side of the ligule (q.v.).
auricomus. golden haired.
axis. the central column of the inflorescence (q.v.). The mainstream and root. (*Oxford English Dictionary* 1971:600)

bloom. waxy deposit on new culms, white or blue white, excreted as waterproofing to protect the soft wood of young bamboos.
branch complement. group of branches at culm nodes.
branch sheath. protective covering borne at each node of branches, similar in form and function to culm and rhizome sheaths.

C

caespitose. (L. *caespitosus,* from *caespes,* sod) growing in tufts, or close clumps, as in bamboos with sympodial rhizomes.
calamus. reed.
Chimonobambusa. (Gr. *cheimōn,* winter) a hardy bamboo genus named for its rare habit of shooting in late fall or winter.
chino. of China.
chrysanthus. golden yellow.
circumaxial. around an axis; as a sheath or ring of adventitious roots that completely circles a culm node.
clone. "group of cultivated plants produced vegetatively from one original seedling or stock." (Lawson 1968: 178)
culm. (L. *culmus,* stalk) the stem of a bamboo plant that emerges from the underground rhizome and bears branches.
culm sheath. husklike protective covering attached at each node of culm whose distinctive coloring and shape serves perhaps more than any other organ to distinguish species within a genus of bamboos.
cultivar. "seedling sports from a species which have multiplied from a single clonal source." (Lawson 1968:40) A sport is a plant abnormally departing, especially in form or color, from the parent stock; a spontaneous mutation.

deciduous. falling off; shedding leaves annually; also applied to culm sheaths or other organs; the opposite of persistent.
dentate. "toothed"; with toothlike notches.
diaphragm. (Gr. *diaphragma,* a partition wall) the internal division between internodes at each node.
diffuse. (L. *difusus,* spread out) grove formation when growing with some distance between culms, typical of most temperate monopodial bamboos and those sympodial species with long-necked rhizomes such as *Guadua* species or *Melocanna baccifera.*
distichus. arranged in two rows.

E

edulis. edible.
efflorescence. powdery crust that covers a surface.
erectus. upright, erect.

F

falcatus. sickle-shaped.
fastuosus. bountiful, stately.
fistulose. (L. *fistula,* a pipe) hollow, like culm internodes and branches of most bamboos.
flexuosus. bending, zigzag.

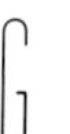

geniculate. (L. *geniculatus,* with bended knee) bent at an abrupt angle; zigzag.
giganteus. gigantic; larger than the type.
glabrous. free from hairs; smooth.
glaucous. a dull bluish green, usually with a matt surface.
gramineus. grasslike.
gregarious. (L. *gregarius,* of a flock or herd) simultaneous flowering of a given generation of bamboo plants. (Cf. sporadic.)

humilis. of low growth.

I

inflorescence. collective flower of a plant.
intermediary veins. fine veins between secondary veins of leaves, parallel to them.
internode. segment of culm, branch, or rhizome between nodes.

J

japonicus. of Japan.

L

lanceolate. lance-shaped.
leptomorph. (Gr. *leptos,* thin; *morphe,* form) McClure's term (1966:25) to describe the monopodial, temperate, running bamboo rhizome. (See pachymorph.)
ligule. (L. *ligula,* a little tongue) the projecting rim at the tip of a culm sheath between the auricles on the inner side where the sheath blade is attached.
lumen. (L., opening; pl. lumena) the hollow central cavity of culm, branch, or rhizome.

M

macrospermus. bearing long seeds.
marmoreus. marblelike.
meristem. (Gr. *meristos,* divisible) tissue in which cell division and growth are active or potential.
mitis. gentle, defenceless, without thorns or spines.
monopodial. (Gr. *monos,* one; *podos,* foot) a term coined by McClure (1925) to describe the long rhizome of running, temperate-zone bamboos.
morphology. (Gr. *morphe,* form; *logos,* word or account) study of plant or animal forms.
multiplex. manifold.

N

nanus. dwarf.
niger. black.
nipponicus. of Japan.
nitidus. shining, smooth, or clear.
nobilis. noble or stately.
node. (L. *nodus,* knot) the joint between hollow segments of a culm, branch, or rhizome; the point at which a rigid membrane of vascular bundles lends strength to an axis of bamboo by crossing it from wall to wall.

O

oral setae. the crinkly hairs on auricles of culm sheaths, sometimes called the shoulder bristles when auricles are absent.

P

pachymorph. (Gr. *pachys,* thick; *morphe,* form) term coined by McClure (1966:24) to designate the squat, short rhizome typical of tropical bamboos. Also called sympodial.
palmatus. lobed; divided into five lobes like a hand.
paniculatus. (L., tufted) having tufts or panicles (pyramidal clusters) of flowers.
parenchyma. (Gr. *parenchein,* to pour in beside) fundamental or ground tissue, such as pith, composed of thin-walled, undifferentiated cells.
persistent. (L. *persistens,* remaining in place) not deciduous; said of organs such as culm sheaths that remain in place after fulfilling their function.
Phyllostachys. (Gr. *phyllon,* leaf; *stachys,* spike) commercially important genus of bamboos native to China and Japan.
P.I. "plant introduction." abbreviation used by USDA, prefixed to a permanent identifying number assigned to each lot of living plant material (seeds, plants, or cuttings) accessioned in its record of plant introductions.
pith. spongy cellular tissue in culms and branches.
Poaceae. the grass family.
primordium. (L., the beginning; pl. primordia) an outgrowth—such as a branch or root—in an early dormant state or in its earliest recognizable condition.
proliferation. (L. *proles,* offspring; *fero,* I bear) "Bearing progeny as offshoots" (Jackson 1949). "The rapid multiplication of members of a branch complement by the prompt awakening of buds at the proximal nodes of the component members. The proliferation of the culm itself by the same process (without the intercalation of a rhizome) is called stooling or tillering" (McClure 1966:311).
prophyllum. (L., first leaf; pl. prophylla) a sheath at the first node of a branch; lacks blade, auricles, and ligule.
proximal. (L. *proximus,* nearest) located at or near the base of an axis or organ; basal. The opposite of distal.
puberulus. somewhat downy.
pubescens. hairy or downy.
pumilis. low, little, or dwarfed.
punctatus. marked with dots or spots.
pygmaeus. dwarfed; quite small.

R

ramosus. having many branches.
reticulatus. netted, resembling a network.
rhizome. (Gr. *rhizoma,* a mass of roots) a food-storing branch of the underground system of growth in bamboos from buds of which culms emerge above ground. Popularly known as *rootstock,* rhizomes are basically of two forms: sympodial (tropical, clumping, pachymorph) and monopodial (temperate, running, leptomorph).
rhizome sheath. husklike protective organ attached basally to each rhizome node.

S

secondary veins. veins on a leaf running from base to tip on both sides of midrib; considerably more prominent than intermediary veins.
septum. (pl. septa) partition or dividing wall of tissue.
sheath. an enveloping organ attached at its base to the nodes of rhizomes, culms, and branches in bamboos, designed to protect the tender tissue during growth. Mid-culm sheaths are valuable identification means for distinguishing species.
shoot. (AS. *sceotan,* to move rapidly) as a noun, signifies a young growing culm, branch, or twig; as a verb, designates the emergence of culms from the ground or branches from the culm.
silica bodies. pieces of silica secreted and remaining within cells of plant tissue, particularly the epidermis or outermost layer of cells. Sometimes small, without

defined form, and solitary or several in a cell; sometimes large enough to almost fill a cell completely ("silica cells") and having a characteristic shape. Some twenty different forms of silica bodies, useful for diagnostic and taxonomic purposes, have been described in the anatomy of grass leaves.

siliceous. containing silica.

Sino-. of China.

spathiflorus. bearing flowers in spathes (that is, one or several flowers within a bract).

sporadic. describes bamboos of a common generation of seeds that flower at different times or irregular intervals: as opposed to gregarious (q.v.).

stoma. (Gr., mouth, opening; pl. stomata) minute mouthlike openings in the epidermis—outer cell layer—that covers photosynthetic tissue. Two "guard cells" shrink or swell by loss or gain in fluid within, thus opening or closing a slit (the stoma) between them.

striatus. marked with fine lines or grooves.

sulcate (L. *sulcatus,* furrowed) grooved; said of culm internodes in which a depression occurs in the surface on alternate sides above the point of attachment of branches—a characteristic of *Phyllostachys* bamboos. Also occurs on branches.

sulcus. (pl., sulci) a groove or depression in internodes of culms or branches. See sulcate.

sulphureus. yellow or sulphur colored.

sympodial. (Gr. *sym,* together; *pous, podos,* foot) tropical, clumping bamboo rhizome form. Term coined by McClure (1925) and generally adopted by later botanists to describe the underground organ giving rise to culms in warmer climates. Later called pachymorph (McClure 1966:24). Distinguished from the monopodial rhizome of temperate bamboos.

T

tabaschir. (tabasheer) a porous, translucent, milky secretion of silica found in the stems of some tropical bamboos, anciently attributed medicinal and magical properties; "an opal of plant origin" (Jones 1966). Long considered an aphrodisiac in the Orient. Hidalgo (1974:xv) mentions the belief that it cures asthma and coughs.

taxon. (Gr. *taxis,* arrangement) a group of plants (or animals) with common characteristics indicating genetic relationship.

tectus. rooflike.

teratic. (Gr. *teras, teratos,* monster, wonder) abnormal, unexpected, deviating from the ordinary form; teratological. Tortoise-shell bamboo, a freak form of *Phyllostachys pubescens,* is one of the better known cases of oriental interest in and even reverence for bamboo's occasional departures from botanic norms.

tessellate. (L. *tessera,* a square piece of marble, glass, and so on used in pavements or walls). *Tessella,* its diminutive form, meant "little cube"—a bone, ivory, or wooden die used in gambling. Tessellate means to form mosaic, to lay checker work.

tessellation. "mosaic of small unequal squares or oblongs formed by crossveins of a leaf." (Lawson 1968: 182)

tillering. (AS. *telger,* a young branch; *tilian* to aim or strive, to work hard for; Dutch, *telen,* to propagate; Gothic, *til,* opportunity; *tils,* opportune, suitable; Old Irish, *dil,* agreeable, pleasant). (Partridge 1958:720) The word is an interesting example of the tillering of language, and the constellation of meanings suggests a more positive attitude towards effort and sweat than is our present norm.)

"Throwing out stems from the base of a stem." (Jackson 1949) "Proliferation of a culm from its basal (subterranean) buds, without the intercalation of a rhizome proper." (McClure 1966:317) (Tilth, a related word: act of tilling, cultivating; cultivated land. Tillage.)

vagans. roving or wandering.

variegation. diversity of color.

venation. arrangement of veins in a leaf.

verticillate. arranged in a whorl, as round a stem.

viridi-glaucescens. bluish green.

viridistriatus. marked or lined with green.

viridus. green.

vittatus. striped longitudinally.

witterings. small, often weak shoots growing at the base of a tree or shrub.

The philosophic glossary.

The complete and philosophic glossary which we have imagined but not compiled would be stuffed as a Christmas turkey with clear graphics and include generous examinations of the roots, where relevant. *Origins,* a short etymological dictionary of modern English, by Eric Partridge (1958), is a fascinating tool for the task. Returning to the roots of experience is a constant theme of the wise of all ages, and returning to the roots of language can give words the freshness of a spring morning for us again. The roots of *tillering,* above, suggest that the Old Irish and others found it "agreeable" to strive mightily and aim far, with themselves and their fields. It suggests the spirit in which we might till our own capacities. The related Gaelic word, *dil,* meant zealous. (See also the German *Ziel,* aim or goal: what we strive for with zeal.) On asking the wages for his effort as messenger, one of the Old Testament prophets was told by God, "Your wage is your struggle: watch the effort of your soul, and you'll be satisfied."

BIBLIOGRAPHY

A

Aero, Rita. *Things Chinese.* Garden City, N.Y.: Dolphin Books, Doubleday. 256 pp. illus. 1980.

Ahmed, K. J. Methods of increasing growth and obtaining natural regeneration of bamboo types in Asia. *Proceedings, IV World Forestry Congress* 1:100–7. 1954.

Alexander, Christopher, et al. *The timeless way of building.* 552 pp. illus. New York: Oxford Univ. Press, vol. I of three-part series from Center for Environmental Structure, Berkeley, Calif., 1979.

——. II. *A pattern language.* 1169 pp. illus. III. *The Oregon experiment.* 1977.

American Bamboo Society. *Journal.* 1101 San Leon Court, Solano Beach, CA 92075. 1980–Present.

Anonymous. *Annotated bibliography on bamboos compiled by Silviculture Branch, F.R.I., Dehra Dun.* 121 pp. (208 references.) 1960.

——. *Proceedings of the all India bamboo study tour and symposium, Dec. 1963– Jan. 1964.* Dehra Dun, India: Forest Research Institute. 1964.

——. *Recommendations of the all India bamboo study tour and symposium, Dec. 1963–Jan. 1964. Indian Forester* 91.9: 607–23. 1965.

Aprovecho. *Lorena owner-built stoves.* Appropriate Technology Project, Volunteers in Asia, Stanford, Calif.: Stanford University. 1979.

——. *Helping people in poor countries develop fuel-saving cookstoves.* From GATE (German Appropriate Technology Exchange): *Deutsche Gesellschaft fur Technische Zusammenarbeit* (GTZ) GmbH, Postfach 5180, D-6236 Eschborn 1, Federal Republic of Germany. Free to serious groups in Third World. 148 pp. illus. 1981.

Austin, R., and Koichiro Ueda. *Bamboo.* 215 pp. illus. Tokyo: Weatherhill. 1970.

Awaka, Yasuichi. *Zen painting.* New York: Harper & Row. 1970.

B

Bahadur, K. N. Bamboos in the service of man. *Biol. Contemp.* 1.2:69–72. 1974.

Ball, J. D. *Things Chinese.* (Bamboo: pp. 59–63, 311.) Shanghai: Kelly and Walsh. 1925.

Ball, Katherine M. *Bamboo: Its cult and culture.* Berkeley, Calif.: Gillick Press. 1945.

Barnett, H. G. *Innovation: The basis of cultural change.* 462 pp. New York: McGraw-Hill. 1953.

Beard, D. C. *The American boy's handy book.* Boston: David R. Godine. 441 pp. illus. 1983.

Bene, J. G., et al. *Trees, food and people: Land management in the tropics.* 52 pp. International Development Research Center, Box 8500, Ottawa, Canada k1G 3H9. 1977.

Bhargava, M. P. Substitutes for wood pulp and paper. *Proceedings of the Fifth Pacific Science Congress:* 3991–6. 1933.

Bhat, A. S. Need for an industrial outlook on management of bamboo. IPPTA Seminar, Dehra Dun, India. 1972.

Bowie, Henry P. *On the laws of Japanese painting.* New York: Dover. 1911.

Brand, Stewart, ed. *The next whole earth catalog.* 608 pp. illus. New York: Random House. 1980.

Brett, Lionel. *Architecture in a crowded world: Vision and reality in planning.* 181 pp. New York: Schocken Books. 1971.

Brown, Lester. *Food or fuel: New competition for the world's cropland.* 43 pp. Worldwatch Paper 35. Worldwatch Institute, 1776 Massachusettes Ave. N.W., Washington, DC 20036. 1980.

Buck, Peter H. (Te Rangi Hiroa) *Arts and crafts of Hawaii,* Sec. II, Houses. 39 pp. illus. Bishop Museum Press, Box 6037, Honolulu, Hawaii 96818. Request catalogue. 1964.

C

Cahill, James. *Hills beyond a river: Chinese painting of the Yuan dynasty (1279–1368).* 198 pp. illus. New York: Weatherhill. 1976.

Calderon, Cleofe E., and Thomas R. Soderstrom. *The genera of bambusoideae of the American continent: Keys and comments.* 27 pp. Contribution to Botany #44. Washington, D.C.: Smithsonian Institution Press. 1980.

Cameron, Alison Stilwell. *Chinese painting techniques.* 226 pp. illus. (Bamboo painting methods, 59–77.) (7th printing, 1980.) Rutland, Vt.: Charles E. Tuttle. 1968.

Camus, E. G. *Les bambusées.* Paris: Paul Lechevalier. 1913.

Castro, Dicken. *La Guadua.* 110 pp. illus. Bogotá, Colombia: Talleres Gráficos del Banco de la Republica. 1966.

Cave, Ray. Piercing the bamboo curtain. *The Baltimore Sun,* 24 April 1955.

Charney, Len. *Build a yurt. The low-cost Mongolian round house.* New York: Macmillan. 134 pp. illus. 1974.

Christian Relief and Development Association. *Bamboo piping: An experiment in Mezzan-Teferi, Ethiopia.* Appropriate Technology Unit Report #5. 16 pp. CRDA, Box 5674, Addis Ababa, Ethiopia. (Review, Darrow 1981: 213.)

Chuang Tzu. *Chuang Tzu; basic writings.* Trans. Burton Watson. 148 pp. illus. New York: Columbia Univ. Press. 1964.

Club du Sahel, CILSS. *Energy in the development strategy of the Sahel: Situation-perspectives-recommendations.* Paris, France. 1978.

Cocannouer, Joseph A. Bamboo and rattan furniture for primary and intermediate grades. *Philippine Craftsman* II.1 (July). 1913.

Crawford, Wallace. The salt industry of Tzeliutsing. *China Journal of Science and Arts* IV.4 (Apr.). 1926.

D

Darrow, Ken. *Alternative technology sourcebook,* vol. I and II. 816 pp. illus. Volunteers in Asia, Box 4543, Stanford, CA 94305. 1981.

Davidson, A. K. *The art of Zen gardens.* 151 pp. illus. Boston: Houghton Mifflin. 1983.

Davis, Ian. *Shelter after disaster.* 127 pp. Oxford Polytechnic Press, Headington, Oxford, United Kingdom OX3 OBP. 1978.

Day, C. L. *The art of knotting and splicing.* Annapolis, Md.: U.S. Naval Academy.

Deaver, Tom. *Ichiyo shakuhachi manual,* vol. I and II. 46-2 Shimizu Aza Oka Shita, Inagawa Cho, Kawabe Gun, Hyogo Ken, Japan 666-03. 1976.

Denyer, Susan. *African traditional architecture: An historical and geographical perspective.* 210 pp. illus. New York: Africana Publishing Co. 1978.

Deogun, P. N. The silviculture and management of bamboo *(Dendrocalamus strictus). Indian Forest Rec.* (n.s. Silvic.) 2.4:75–173. 1937.

Dickason, F. G. The bamboos of Burma. Undated typescript. 13 pp. Grass Library File # 00080300, Smithsonian Institution. Washington, D.C. 194?

Douglas, J. S., and Robert Hart. *Forest farming.* 197 pp. META Publications, Box 128, Marblemount, WA 98267. 1976.

Drew, W. B. *The ecological role of bamboos in relation to the military use of herbicides on forests in South Vietnam.* 14 pp. Washington, D.C.: National Academy of Sciences, National Research Council. 1974.

Duly, Colin. *The houses of mankind.* 96 pp. illus. London: Thames & Hudson. 1979.

E

Ecke, Tseng Yu-ho. *Chinese folk art* II. 159 pp. illus. Hawaii: The University Press of Hawaii. 1977.

Eckholm, Erik. *Losing ground.* 223 pp. New York: Norton. 1976.

——. *Planting for the future: Forestry for human needs.* 64 pp. Worldwatch Institute, 1776 Massachusettes Ave. N.W., Washington, DC 20036. 1979.

Eliot, John. Japan's warriors of the wind. *National Geographic* 151.4:550–61 (Apr.). 1977.

Engel, David. *Japanese gardens for today.* 270 pp. 274 pl. Rutland, Vt.: Charles E. Tuttle. 1959.

Engel, Heinrich. *The Japanese house: A tradition for contemporary architecture.* Rutland, Vt.: Charles E. Tuttle. 1964.

F

Fairchild, David. *Japanese bamboos and their introduction into America.* Washington, D.C.: USDA Bureau of Plant Industry Bulletin 43. 1903.

——. The Barbour Lathrop bamboo grove. *Journal of Heredity,* X.6 (June).1919.

——. *Exploring for plants.* New York: Macmillan. 1931.

FAO (United Nations Food and Agricultural Organization). *Progress report on the pulp and paper situation.* 1952.

——. *Raw materials for more paper.* 1953a.

——. *Summary of the world's pulp and paper resources and prospects.* 1953b.

——. *Annotated bibliography on bamboos.* (377 titles with abstracts.) Dehra Dun, India: Forest Research Institute. 1954.

——. *Handling and storage of food grains in tropical and subtropical areas.* 350 pp. (By D. Hall.). 1970.

——. *Aquacultural planning in Asia.* 154 pp. 1976.

——. *China: Forestry support for agriculture.* FAO Forestry paper No. 12. 103 pp. UNIDUB. (Review, Darrow 1981:498.) 1978.

——. *Manual on bamboos of Asia Pacific region.* (By Y. M. L. Sharma.). 1979.

Fathy, Hassan. *Architecture for the poor: An experiment in rural Egypt.* 232 pp. 132 illus. Chicago: Univ. of Chicago Press. 1973.

Federacion Nacional de Cafeteros de Colombia. *Evite la erosion con las barreras vivas.* 20 pp. illus. Bogotá, Colombia, Boletin de Extension No. 51. 1980.

Ferrar-Delgado, Roberto. The effect of bamboo on succeeding crops. *Tropical Agriculture* XXVIII (1–6). 1951.

Fisher, Sally. *The tale of the shining princess,* with illustrations from a late 18th century edition of *Taketori,* "The Bamboo Cutter," in the N.Y. Metropolitan Museum of Art. 71 pp. New York: Viking Press. 1980.

Fonblanque, E. B. de. *Two years in Japan and northern China. Niphon and Pe-che-li.* London: Saunders Otley and Co. 1863.

Frazer, Sir James George. *The golden bough: A study in magic and religion.* 1 vol. abridged edition. 864 pp. (Reprint 1963. 7th ed., 1974.) New York: Macmillan. 1922.

Freeman-Mitford, A. B. *The bamboo garden.* 224 pp. illus. London: Macmillan. 1896.

——. On the economic uses of bamboo. *J. of the Royal Horticultural Soc.* 22: 268–86. 1899.

Freire, Paolo. *Pedagogy of the oppressed.* 186 pp. (Reprint 1981.). New York: Continuum Pub. Co. 1968.

Fuji Bamboo Society. *Guidebook of the Fuji Bamboo Garden.* 76 pp. illus. Nagahara, Gotemba, Shizuoka Pref., Japan. 1963.

Fukuoka. *One straw revolution.* 181 pp. illus. Emmaus, Pa.: Rodale Press. 1978.

Fuller, Buckminster. *Education automation.* 85 pp. Garden City, N.Y.: Doubleday Anchor. 1971.

G

Galloway, B. T. *Bamboos: Their culture and uses in the United States.* Washington, D.C.: USDA Bureau of Plant Industry Bulletin 1329. 1925.

Gamble, J. S. Bambuseae of British India. *Ann. Roy. Botan. Garden* (Calcutta) 7: 1–133. Atlas, plates 1–119. 1896.

GATE. *Introduction to bamboo as a building material.* German Appropriate Technology Exchange. (Address: see Aprovecho 1981). 1979.

Giono, Jean. *The man who planted trees and grew happiness.* Friends of Nature. c/o D. Smith, Brooksville, ME 04617. (Reprinted in Brand 1980:78–9.)

Glenn, H. E., et al. *Bamboo reinforcement in Portland cement concrete.* 171 pp. illus. Bull. 4. Clemson Agricultural College, Eng. Exp. Sta., South Carolina. 1950.

——. *Seasoning, preservative treatment, and physical property studies of bamboos.* 47 pp. Bull. 7. Clemson Agricultural College, Eng. Exp. Sta., South Carolina. 1954.

——. *Seasoning, preservative, and water-repellant treatment and physical property studies of bamboo.* Bull. 8. Clemson Agricultural College, Eng. Exp. Sta., South Carolina. 1956.

Grunfeld, Frederic, et al. *Games of the world. How to make and play.* 280 pp. illus. New York: Holt, Rinehart & Winston. 1975.

Guha, S. R. D. Bamboo as a raw material for paper and board making. *Indian Buyer* 1. 2:142. 1961.

Guha, S. R. D., et al. Production of bleached pulps from mixture of bamboo and mixed hardwoods. *Indian Pulp and Paper* 20. 10:617. 1966a.

——. Wrapping, writing, and printing papers from bamboo dust. *Indian Pulp and Paper* 21. 3:187. 1966b.

H

Hammitzsch, Horst. *Zen in the art of the tea ceremony.* 104 pp. New York: St. Martin's Press. 1980.

Haun, J. R., et al. *Fiber and papermaking characteristics of bamboo.* Washington, D.C.: USDA Tech. Bulletin 1361. 19 pp. 1966.

Heronemus, W. *A survey of the possible use of windpower in Thailand and the Philippines.* 74 pp. Request from Agency for International Development, Washington, DC 20523. 1974.

Herrera, Eulogio O. *Feasibility study on the use of bamboo in pressurized water systems.* 23 pp. School of Engineering, Univ. of Massachusetts, Amherst, MA 01002. 1974.

Herrigel, Eugen. *Zen in the art of archery.* 109 pp. New York: Random House. 1971.

Herrigel, Gustie L. *Zen in the art of flower arrangement.* 124 pp. illus. Newton, Mass.: Charles T. Branford Co. 1958.

Herty Foundation. *Bamboo and its potential for papermaking.* Savannah, Ga.: Herty Foundation. 1956.

Heyne, K. *De Nuttige Planten van Indonesie,* 3d. ed. v.1: 285–93. Holland: 's-Gravenhage, van Hoeve. 1950.

Hidalgo, Oscar. *Bambu: su cultivo y aplicaciones en: fabricacion de papel, construccion, arquitectura, ingenieria, artesania.* 318 pp. illus. Estudios Tecnicos Colombianos (ETC), Apartado Aereo 50085, Bogotá, Colombia. 1974.

——. *Nuevas tecnicas de construccion con bambu.* 137 pp. illus. ETC (address above, Hidalgo 1974). 1978.

——. *Guadua: a bamboo on the edge of extinction.* Filmscript. Film available from O. Hidalgo, Facultad de Arquitectura, Univ. Nacional, Bogotá, Colombia. 1980.

——. *Manual de construccion con bambu. Construccion rural*-I. 71 pp. illus. ETC (address above, Hidalgo 1974). 1981.

Higuchi, Takayoshi. *World directory of bamboo researchers.* 37 pp. XVIII World Congress of International Union of Forestry Research Organizations. 1978.

Hodge, W. H., and David Bisset. Running bamboo for hedges. *National Horticultural Magazine* 34.3: 123–7 (Jul.). 1955.

——. Bamboos: Their economic future in America. *The Garden Journal* 7.4. (Jul.–Aug.). 1957.

Holttum, R. E. A Malayan blow-pipe bamboo. *Kew Bulletin* 493–6. 1953.

——. The bamboos of the Malay peninsula. *Bull., Botanic Gardens* (Singapore) 16:1–135. 1958.

——. The bamboos of New Guinea. *Kew Bulletin* 21. 2: 263–92. 1967.

Hommel, R. *China at work.* (Reissue 1969.) Cambridge, Mass.: MIT Press. (Review, Darrow 1981:71.) 1937.

Hooker, J. D. *Himalayan journals.* London: John Murray. 1854.

Hooper, David. Bamboo manna. *Indian Forester* 1900:363–4. 1900.

Hopkinson, Anthony. *Papermaking at home from recycled waste.* 94 pp. illus. Wellingborough, Northhamptonshire (England): Thorsons Publishers Ltd. 1978.

Horn, Claud. New bamboo revetment construction. *The Military Engineer.* (Jun.). 1943.

Houzeau de Lehaie, Jean. *Le bambou.* (journal) Mons, Belgium. 1906–1908.

——. *Memoir.* Grass Library Files # 00146850, Smithsonian Institution. Washington, D.C. 1907.

Howard, Sir Albert. *An agricultural testament.* (Reissue 1973.). 272 pp. Emmaus, Pa.: Rodale Books. 1940.

Hoyer, Fritz. Can bamboo avert the threatened shortage of papermaking fibre? *Tropenpflanzer* 34:403–33. 1931.

Huberman, M. A. Bamboo silviculture. *Unasylva* 13. 1:36–43. 1959.

Hughes, Ralph. Observations of cane *(Arundinaria)* flowers, seed, and seedlings in the North Carolina coastal plain. *Bull. Torrey Botanical Club* 78.2:113–21 (Mar.). 1951.

Hunt, Leslie L. *Twenty-five kites that fly.* 110 pp. illus. (Reprint 1971.) New York: Dover. 1929.

Hunter, Ilene, and Marilyn Judson. *Simple folk instruments to make and play.* 159 pp. illus. New York: Simon & Schuster. 1977.

Hutt, J. K. *Equipment for physically handicapped children in rattan and bamboo.* Disabilities Study Unit, Wildhanger, Amberley, Arundel, W. Sussex BN8 9NR England.

I

Illich, Ivan. *Deschooling society.* 167 pp. New York: Harper & Row. 1972.

Imle, E. P. What about bamboo? *Forest Farmer* (Jan.). 1958.

Innez, Jocasta. *Paint magic.* 240 pp. illus. (188–191, "bambooing" wooden furniture.) New York: Van Nostrand Reinhold. 1981.

Isaachsen, J. Some bamboos and their uses. *New Zealand Gardener:* 80–83 (Oct.). 1946.

J

Jackson, A. B. The best season for cutting bamboos. *Indian Forester:* 245–47. 1900.

Jackson, B. D. *A glossary of botanical terms.* 4th ed. London: Duckworth. 1949.

Jones, James L. Bamboo for northern gardens. *Horticulture* 57.7 (Jul.). 1979.

Jones, L. H. P. Tabashir: An opal of plant origin. *Science* 151. 3709:464–6 (Jan.). 1966.

Jue, David F. *Chinese kites: How to make and fly them.* 51 pp. illus. Rutland, Va.: Charles E. Tuttle. 1967.

Junka, Fujino. Bamboo. *The East* 8.9: 19–27. 1972.

K

Kawamura, S. On the periodical flowering of the bamboo. *Japanese Journal of Botany* 3:335–49. 1927.

Kern, Barbara and Ken; Mullan, Jane and Otis. *The earth sheltered owner-built home.* 268 pp. illus. Owner-Builder Publications, Box 817, North Fork, CA 93643. 1982.

Kern, Ken. *The owner-built home.* 374 pp. illus. New York: Scribner's. 1972.

Keswick, Maggie. *The Chinese garden.* 216 pp. illus. New York: Rizzoli International. 1978.

Kiang, Tao. *Bamboo production and research in Taiwan.* JCRR Special Technical Reports. (pp. 16–24.) (Jul.). 1973.

King, F. H. *Farmers of forty centuries.* (1973 reissue.) 456 pp. Emmaus, Pa.: Rodale Books. 1911.

Kirkpatrick, T. W., and N. W. Simmonds. Bamboo borers and the moon. *Trop. Ag.* 35. 4:299–301. 1958.

Kreider, Claude M. Bamboo rods—how to fix 'em. *Field & Stream,* Dec., 88–90. illus. 1944.

Krishnaswamy, V. S. Bamboos—their silviculture and management. *Indian Forester* 82. 6: 308–13. 1956.

Kudo, Kazuyoshi, and Kiyomi Suganuma. *Japanese bamboo baskets.* 88 pp. illus. New York: Kodansha International. 1980.

Kuehn, Dan. *Mongolian cloud houses: How to make a yurt and live comfortably.* 135 pp. illus. Cloud Houses, Box 2315, Taos, NM 87571. 1980.

Kurz, Sulpiz. Bamboo and its uses. *Indian Forester* 1.3:219–69; 335–62. 1876.

L

Laubin, Reginald and Gladys. *The Indian tipi; its history, construction, and use.* 350 pp. illus. Norman, Okla.: Univ. of Oklahoma Press. 1977.

Laufer, Berthold. *Chinese baskets.* Chicago: Field Museum of Natural History. 1925.

Lawson, A. H. *Bamboos: A gardener's guide to their cultivation in temperate climates.* 192 pp. illus. New York: Taplinger Publishing Co. 1968.

Lee, A. The utilization of bamboo in the construction of equipment for farm houses and dwellings. *Revista de Agricultura de Puerto Rico* 28.4:640–6. 1937.

——. Bamboo. *Agr. in the Americas* 4.7:127–9, 137. 1944.

——. Bamboos in the New World. In *New Crops for the New World,* ed. by G. M. Wilson. New York: Macmillan. 1945.

____. Equatorial currents—asset to middle American agriculture. *Agr. in the Americas* 7.1:3–6. 1947.

Lee, Sherman E. *Chinese landscape painting.* 159 pp. illus. New York: Harper & Row. 1960.

____. *A history of far eastern art.* 527 pp. 656 fig. 60 color pl. Englewood Cliffs, N.J.: Prentice-Hall. 1962.

Leon, Antonio J. de. Studies on the use of interwoven thin bamboo strips as stress-skin covering for aircraft. *Philippine J. of Science* 85:3. 1956.

Leong, Wing K. *How to paint bamboo.* 45 pp. illus. Chinese Art Studio, 332 S.W. 3rd Ave., Portland, OR 97204. 1979.

Lessard, Giles, and Amy Chouinard. *Bamboo research in Asia.* Proceedings of a workshop held in Singapore 28–30 May 1980. 228 pp. illus. International Development Research Centre, Box 8500, Ottawa, Canada. K1G 3H9. (Microfiche edition available.) 1980.

Leung, Woot-Tsuen Wu, et al. *Composition of Foods used in far eastern countries.* Washington, D.C.: Bureau of Human Nutrition, Agricultural Handbook 34, USDA. 1952.

Levenson, Monty. *The Japanese shakuhachi flute.* A. *Notes on the craft and construction.* 22 pp. illus. B. *A guide to the traditional music and notation.* 38 pp. illus. P.O. Box 294, Willits, CA 95490. 1974.

Limaye, V. D. Bamboo nails, their manufacture and holding power. *Indian Forest Records* (Utilization New Series) 3.3. 1943.

____. Strength of bamboo *(Dendrocalamus strictus). Indian Forester* 78: 558–75. illus. (Appendix on bamboo-reinforced concrete.) 1952.

Lin, Wei-Chih. *La culture du bambou à Madagascar. Publiée avec le concours du Centre de Formation pour l'Artisanat du Bambou Sino-Malagasy.* 1970.

____. *Répartition et utilisation des éspèces importantes du bambou dans le monde.* 24 pp. CFABS. 1972.

Lippert, Stanley. *Bamboo pipes for pressurized water systems, a feasibility study.* Amherst, Mass.: School of Engineering, Univ. of Mass. 1976.

Londoño, Francisco, and M. A. Montes. *La guadua: Su aplicacion en la construccion.* 187 pp. illus. Medellin, Colombia: Editorial Bedout. 1970.

Lutz, Walter. Bamboo brushpots. *Arts of Asia,* Sept./Oct., 25–33. 1975.

M

McClure, F. A. Notes on the island of Hainan. *Lingnan Ag. Rev.* 1:66–79. 1922.

____. *To Kang Peng, plant collector. 1877–1926. An appreciation.* Canton Christian College. 1926.

____. The native paper industry in Kwangtung. *Lingnan Science Journal (LSJ)* 5:255–68. 1927.

____. Some Chinese papers made on the ancient "wove" type of screen. *LSJ* 9:115–28. 1930.

____. Studies of Chinese bamboos: A new species of *Arundinaria (A. amabilis)* from southern China. Pt. I Diagnosis. *LSJ* 10:5–10. Pl. 1–8. Pt. II. Notes on culture, preparation for the market, and uses. *LSJ* 10:295–305. Pl. 34–39. 1931.

____. Methods and materials of Chinese table plant culture. *LSJ* 12 (Supp.): 119–49. Pl. 4–13. 1933.

____. Bamboo—a taxonomic problem and an economic opportunity. *Sci. Monthly* 51:193–204. illus. 1935.

____. Two new species of *Bambusa* from southeastern China. *(B. cerosissima* and *B. chungii.) LSJ* 15:637–45. illus. 1936.

____. *Bambusa ventricosa,* a new species with a teratological bent. *LSJ* 17:57–62. illus. 1938a.

____. Diary of a small experimental bamboo planting. *LSJ* 17:473–6. 1938b.

____. Notes on bamboo culture, with special reference to southern China. *Hong Kong Nat.* 9:1–18. 1938c.

____. *Western Hemisphere bamboos as substitutes for Oriental bamboos for the manufacture of ski-pole shafts.* Final Report, Project QMD-24, U.S. Natl. Res. Council, Committee on Quartermaster Problems. 92 pp. illus. 1944.

____. Suggestions on how to collect bamboos. (Abridged.) 2 pp. illus. Washington, D.C.: USDA Off. Foreign Agr. Relat. 1945a.

____. Bamboo culture in the Americas. *Agr. in the Americas* 5:3–7, 15–16. *Chemurg. Digest* 4:190–4. illus. 1945b.

____. Bamboo in Ecuador's lowlands. *Agr. in the Americas* 5:4, 190–2. 1945c.

____. Bambu para las Americas. *Hacienda* (Mar.), 119–21. (Spanish version of 1945b.) 1945d.

____. Bamboo in Ecuador's highlands. *Agr. in the Americas* 6:164–7 (Oct.). 1946a.

____. *The occurrence and exploitation possibilities of bamboo in Nicaragua; A confidential report based principally on field work carried out during the period March 20–30, 1946.* 26 pp. 3 pp. map. 1946b.

____. Bamboos for farm and home. *Grasses* (USDA yearbook): 735–40. 1948a.

____. Suggestions concerning the handling of bamboo plants shipped from a distance with soil removed from roots. 1 p. Washington, D.C.: USDA Off. Foreign Agr. Relat. 1948b.

____. Bamboo notes for El Salvador: For an exhibit of 17 species. 44 pp. Unpublished manuscript, available in Grass Library, Smithsonian Institution. Washington, D.C. 1948c.

____. Bamboo may prove new source of paper pulp for Guatemala. *Foreign Agr.* 13.6:143. 1949.

____. Bamboo propagation studies. Fertilizing bamboo. Puerto Rico Fed. Exp. Sta. *Annual Report,* pp. 27–8. 1950a.

____. Notes on guadua in the department of Caldas, Colombia. Unpublished manuscript. Grass Library, Smithsonian Institution. Washington, D.C. 1950b.

——. Bamboo propagation studies. Puerto Rico Fed. Exp. Sta. *Annual Report:* 34. 1951.

——. Bamboo in Latin America. *Turrialba* 2.3: 110–3. illus. 1952a.

——. Bamboo as a field for research. Talk, Oct. 16, for Office of Foreign Agricultural Relations. Typescript, Grass Library File # 00222750, Smithsonian Institution. Washington, D.C. 1952b.

——. *Bamboo as a building material.* Washington, D.C.: USDA Foreign Agr. Service. 52 pp. illus. 1953.

——. *Empalme perfecto para bambues. Hacienda* 49:54 (Jan.). illus. 1954a.

——. Simple new device for bamboo joints. Mimeograph. Grass Library File # 00228250, Smithsonian Institution. Washington, D.C. 1954b.

——. Observations on native and introduced bamboos in Florida and southeastern Texas. 25 pp. illus. (A private report.) 1954c.

——. Prospects and conditions for growing bamboo for paper pulp in the United States. Pt. I. Background. 41 pp. illus. (A private report.) 1954d.

——. Bamboos, in *Grasses of Guatemala,* by Swallen, J.R. in *Flora of Guatemala,* Pt.2. by Steyermark and Standley,. *Fieldiana (Botany):* (McClure's 24(2):2–4, 38–41, 41–4, 52–60, 86–93, 101–3, 139–44, 146–57, 204–6, 206–9, 308, 310–6, 328–31. 1955.

——. *El bambu como material de construccion. Serie:Traducciones, Adaptaciones y Reimpresiones* No. 6. Bogotá, Colombia: Centro Interamericano de Vivienda. *(Servicio de Intercambio Cientifico.)* 49 pp. illus. 1956.

——. Bamboo utilization in eastern Pakistan. (Some problems relating to long-term procurment of bamboo for the Karnaphuli Paper Mill.) *Pakistan J. Forestry* 6:182–6. 1956b.

——. Bamboo culture in the South Pacific. *South Pacific Comm. Quart. Bull.* 6:38–40. illus. 1956c.

——. *Bamboos of the genus* Phyllostachys *under cultivation in the United States.* Washington, D.C.: USDA Agr. Handbook 114. (June 1957.) 69 pp. illus. 1957a.

——. The taxonomic conquest of the bamboos with notes on their silvicultural status in the Americas. *FAO Tropical Silviculture* 2:304–8. 1957b.

——. Bamboo as a source of forage. *8th Pacific Science Congress Proceedings.* ivB:609–44. illus.1958a.

——. Bamboos for the Pacific Islands. *South Pacific Comm. Quart. Bull.* 8.2:20–2, 56–7 (April); 8.3:40–2 (July); 8.4:53–5 (Oct.). 1958b.

——. Bamboo in the economy of oriental peoples. *Smithsonian Institution Annual Report 1957,* pp. 391–412. 1958c.

——. Need for bamboos tolerant of salinity. *SPC Quart. Bull.,* July, 66. 1959.

——. Bamboo. *Encyclopaedia Britannica* 3:17–8. 1961a.

——. *Bamboo in the United States: Description, culture, and utilization.* (With Robert Young and J. Haun.) 74 pp. illus. Washington, D.C.: USDA Agr. Handbook 193. 1961b.

——. Bamboo plants: suggestions for selection, preparation, and packing for shipment. 3 pp. illus. Unpublished photo copy. (Available from the USDA.) 1963a.

——. A new feature in bamboo rhizome anatomy. *Rhodora* 65. 762: 134–6. 1963b.

——. *The bamboos: A fresh perspective.* 347 pp. illus. Cambridge, Mass.: Harvard Univ. Press. 1966.

——. Genera of bamboos native to the New World *(Gramineae: Bambusoideae). Smithsonian Contr. Bot.* 9. 148 pp. 48 fig. 1973.

MacFarlan, Allan A. *Living like Indians. A treasury of American Indian crafts, games, and activities.* 305 pp. illus. New York: Bonanza Books. 1961.

McHarg, Ian. *Design with nature.* 198 pp. illus. Garden City, N.Y.: Doubleday. 1971.

McIlhenny, E. A. Bamboo growing for the south. *Nat. Hort. Mag.,* Jan., 1–6. illus. 1945a.

——. Bamboo—A must for the south. *Nat. Hort. Mag.,* Apr., 120–5. 1945b.

McLuhan, Marshall. *Understanding media.* 318 pp. New York: New American Library. 1964.

Malm, William. *Music cultures of the Pacific, the Near East, and Asia.* 236 pp. Englewood Cliffs, N.J.: Prentice-Hall. 1977.

Mannix, Daniel. Quick, quiet, and deadly. *True,* Nov. 1956.

Marden, Luis. Bamboo, the giant grass. *Nat. Geographic* 158.4:502–28 (Oct.). 1980.

Marx, Leo. *The machine in the garden: Technology and the pastoral ideal in America.* 392 pp. illus. London: Oxford Univ. Press. 1964.

Mathur, G. C., et al. *Bamboo for house construction.* New Delhi, India: National Buildings Organization and Regional Housing Centre.

Matsuno, Masako. *Taro and the bamboo shoot.* illus. (Japanese fairy tale.) New York: Pantheon. 1964.

May, Elizabeth. *Musics of many cultures.* 434 pp. illus. (With three 7-inch 33⅓ rpm records.) Berkeley, Calif.: Univ. of California Press. 1983.

Medway, Lord. Bamboo control. Research at Univ. of Malaya Field Studies Center, Ulu Gombak. *Malaysian Forester 33* .1:70–9. 1970.

Meggers, B. *Amazonia: Man and culture in a counterfeit paradise.* 182 pp. Chicago: Aldine-Atherton. 1971.

Miller, C. A. *A market study of the current and potential demand for bamboo products in the United States.* Bellvue, CO 80512. Sponsored by: Forestry Division. Joint Commission on Rural Reconstruction. Taipei. Republic of China. 1974.

Miller, Hugo. Some commercial notes on baskets. *Philippine Craftsman* II.7 (Jan.). 1914.

Mohammed, S. *Bamboo pill or ointment boxes.* Indian Forest Leaflet 39. Dehra Dun, India: Forest Research Institute. 1943.

Moll, Lane de, and Gigi Coe. *Stepping stones; appropriate technology and beyond.* 204 pp. illus. New York: Schocken Books. 1978.

Mollison, Bill, and David Holmgren. *Permaculture One. A perennial agriculture for human settlements.* 127 pp. illus. International Tree Crops Institute, U.S.A. P.O. Box 888, Winters, CA 95694. 1978.

Mollison, Bill. *Permaculture Two.* 150 pp. illus. Same address. 1979.

Montaigne, Michel de. *Essays.* Trans. J. M. Cohen. 406 pp. Baltimore: Penguin.

Montessori, Maria. *The Montessori method.* 448 pp. Cambridge, Mass.: Robert Bentley.

____. *Spontaneous activity in education.* (7th printing 1972.) 355 pp. New York: Schocken Books. 1917.

____. *From childhood to adolescence.* 141 pp. New York: Schocken Books. 1948.

____. *The discovery of the child.* 365 pp. illus. (Orig. pub. 1948.) Notre Dame, Ind.: Fides Publishers. (See also Standing 1962.) 1967.

Montgomery, F. A. Oriental plant in the Occident: Bamboo as a farm crop to supply two million dollar demand. *Scientific American* 192:310–1 (Jun.). 1935.

Morse, Edward S. *Japanese homes and their surroundings.* 372 pp. illus. New York: Dover. 1886.

Mullan, Jane and Otis. *A low-cost earthquake resistant house.* Mullein Press, Box 533, Colfax, CA 95713. 1984.

Munro, W. A monograph of the *Bambusaceae,* including descriptions of all the species. *Transactions of the Linnean Society of London* 26. 157 pp. 1868.

Musashi, Miyamoto. *A book of five rings.* 96 pp. illus. Woodstock, N.Y.: The Overlook Press. 1982.

Myers, Gertrude. *Insects and mites recorded from bamboo in the United States.* 10 pp. Special Supplement No. 2. (Feb. 26.) Washington, D.C.: USDA, Bureau of Entomology and Plant Quarantine. 1947.

N

Narayanamurti, D. Building boards from bamboos. *Composite Wood* 3.1:1–13 (Jan.). 1956.

____. *Bamboo and reeds in building construction.* Centre for Housing, Building, and Planning, U.N. Dept. of Economic and Social Affairs. 95 pp. illus. (Sales Sect. Room A-3315, United Nations, New York, N.Y. 10017. Ask for sales no. E.72.IV.3.) 1972.

National Academy of Sciences. *Leucaena: Promising forage and tree crop for the tropics.* 121 pp. Free from Commission on International Relations JH 215, Nat. Acad. Sci., 2101 Constitution Ave. N.W., Washington, DC 20418. 1977.

Needham, Joseph. *Science and civilization in China.* Vols. I–V. New York: Cambridge Univ. Press. 1954–1980.

Nehrbass, N. Investigation of bamboo in a system of structure. 24 pp. illus. (Lafayette, Louisiana.) Manuscript at the Grass Library, Smithsonian Institute. Washington, D.C. c.1942.

Nehru, S.S. *Further experiments in electrofarming.* Chap. 10: The response of bamboo seedlings to electrical treatment. United Provinces, India: Dept. of Agriculture Bulletin. 1932.

Neihardt, John G. *Black Elk speaks: The life story of a holy man of the Oglala Sioux.* 280 pp. illus. Lincoln: Univ. of Nebraska Press. 1932.

Neutra, Richard. *Survival through design.* 384 pp. illus. (1969 ed.) New York: Oxford Univ. Press. 1954.

New York Times. A Philippine organ fashioned in 1821 is made of bamboo. 5 Feb. 1969.

Newman, Lee Scott. *Kite craft.* 214 pp. illus. New York: Crown Publishers. 1974.

Numata, Makoto, ed. *Ecology of grasslands and bamboolands in the world.* Jena, E. Germany: Gustav Fischer Verlag. 1979.

O

Odum, Howard T. and Elisabeth. *Energy basis for man and nature.* 296 pp. illus. New York: McGraw-Hill. 1976.

Oi, Motoi. *Brush strokes in sume-i painting.* Tokyo: Japan House of Art. 1963.

Oka, Hideyuki. *How to wrap five eggs: Traditional Japanese packaging.* New York: Weatherhill. 1975.

Okakura, Kakuzo. *The book of tea.* 76 pp. (Reprint 1964.) New York: Dover. 1906.

Orjala, Jim. *Bamboo in the Pacific Northwest.* Aprovecho, 359 Polk Street, Eugene, OR 97402. 1980.

Oshima, Jinsaburo. *Moso bamboo culture.* 100 pp. Washington, D.C.: USDA. [Reprinted, *Journal* of the American Bamboo Society 3.1:2–28 (Feb. 1982), 3.2:33–45 (May 1982).] 1931.

P

Papanek, Victor. *Design for the real world.* 294 pp. illus. New York: Pantheon, Random House. (Also Bantam paperback 1973.) 1971.

Parker, Jerry. What you can do with bamboo. *Popular Science,* May, 177–80. illus. 1955.

Parker, Luther. Primitive Philippine basketry. *Philippine Craftsman (PC)* II.2:71–82 (Aug.). illus. 1913.

____. The 1914 basketry exhibit. *PC* II.9:645–56 (Mar.). illus. 1914a.

____. Some common baskets of the Philippines. *PC* III.1:1–25 (Jul.). illus. 1914b.

Partridge, Eric. *Origins; a short etymological dictionary of modern English.* 970 pp. New York: Macmillan. 1958.

Peterson, John. *High flying kites.* 80 pp. illus. New York: Scholastic Book Services. 1969.

Piatti, Luigi. Manufacture of bamboo cellulose in Indochina. *Textil Rundschau* 2: 292–8, 330–40. English abstracts in *Chem. Abst.* 42:1736 (1948); *Bull. Pap. Inst.* 18:145 (1947). 1947a.

——. Liquid fuel from bamboo. *Schweizer Archiv fur Angewandte Wissenschaft und Technik* 13: 370–6 (1947). Abstract: *Chem. Abst.* 42:3153 (1948). (See also Sineath 1953:55.) 1947b.

Plank, Howard K. DDT for powder post beetle control in bamboo. *Science* 106.2733:317 (Oct.). 1947.

——. *Biology of the powder post beetle in Puerto Rico.* Mayaguez, Puerto Rico: USDA Fed. Exp. Sta. Bulletin 44. 29 pp. illus. 1948.

——. Control of the bamboo powder-post beetle in Puerto Rico. *Tropical Agriculture* 26. 1–6:64–7. 1949.

——. Permanence of DDT in powder-post beetle control in bamboo. *Journal of Economic Entomology* 42.6:963–5. 1950.

Porterfield, William M. *Bamboo and its uses in China.* 77 pp. illus. Chinese Govt. Bureau of Economic Information. Booklet Series 2. 1927.

——. Bamboo, the universal provider. *Scientific Monthly* 36:176–83 (Feb.). 1933.

Pound, Ezra. *Confucius.* New York: New Directions. 1969.

Prest, John. *The garden of Eden: The botanic garden and the re-creation of paradise.* 122 pp. illus. New Haven, Conn.: Yale Univ. Press. 1981.

R

Raitt, William. Investigation of bamboo as material for production of of paper pulp. *Indian Forest Records* 3:3. 1912.

——. Summary of investigations on bamboos and grasses for paper pulp. *Indian Forest Records* 11.9:271–81. 1925.

Raizada. World distribution of bamboos with special reference to the Indian species and their more important uses. *Indian Forest Leaflet* No. 151:1–4. Dehra Dun, India: Forest Resesrch Institute. 1956.

Ramaswamy, N. S. *The management of animal energy resources and the modernization of the bullock cart system.* 137 pp. illus. (Photos of old and new designs.) Available on request to serious groups in the Third World from author: Director, Indian Institute of Management, 33 Langford Road, Bangalore 560027, India. (Review, Darrow 1981:566.) 1979.

Rao, R. S. Investigation on the manufacture of paper from bamboos. Master's thesis, Univ. of Maine. 1917.

Raval, Gil. A year's course in elementary bamboo weaving. *Philippine Craftsman* III.4:247–62. illus. 1914.

Reader's Digest. *The world's last mysteries.* 320 pp. illus. Pleasantville, N.Y.: Reader's Digest Press. 1978.

Reck, David. *Music of the whole earth.* 545 pp. illus. New York: Scribner's. 1977.

Rehman, M. A., and S. M. Ishaq. Seasoning and shrinkage of bamboo. *Indian Forest Records* 4.2. 1947.

Riviere, A., and C. Riviere. *Les bambous.* 363 pp. illus. Paris. 1879.

Roi, J. Bibliography of bamboo. *Peking Natural History Bulletin* 16. 1:1–16. 1941.

Rothschild, Emma. *Paradise lost: The decline of the auto-industrial age.* 264 pp. New York: Random House. 1973.

Routledge, T. *Bamboo considered as a paper-making material.* 40 pp. London: E. & F. N. Spon. 1875.

Rudofsky, Bernard. *Architecture without architects.* 157 pp. illus. Garden City, N.Y.: Doubleday. 1964.

——. *The prodigious builders: Notes towards a natural history of architecture.* 383 pp. illus. New York: Harcourt Brace Jovanovich. 1977.

Ruprecht, F. J. *Bambuseas monographice exponit.* 75 pp. 18 plates. St. Petersburg: Typis Academiae Caesareae Scientiarum. 1839.

Rykwert, Joseph. *On Adam's house in paradise: The idea of the primitive hut in architectural history.* 250 pp. illus. (1981 ed.) Cambridge, Mass.: MIT Press. 1972.

S

Saint Barbe Baker, Richard. *My life my trees.* 167 pp. illus. Scotland: Findhorn Publications, Findhorn, Moray. 1970.

Samaka Service Center. *The Samaka guide to homesite farming.* 168 pp. illus. Samaka Service Center, Box 2310, Manila, Philippines. 1962.

Sanger, Clyde. *Trees for the people.* 52 pp. IDRC-094e. Ottawa, Canada: International Development Research Center. (For address, see Bene 1977.) 1977.

Sarjeant, Peter T. *Papermake: How to make paper in your kitchen sink.* 72 pp. Covington, VA 24426. 1976.

Satow, Ernest. *The cultivation of bamboos in Japan.* 127 pp. The Asiatic Society of Japan. Vol. 27, pt. 3. 1899.

Sawyer, Leroy. Bamboo and rattan furniture. (Projects for schools.) *Philippine Craftsman* I.4:247–75. 1912.

Schaarschmidt-Richter, Irmtraud. *Japanese gardens.* 325 pp. illus. New York: William Morrow. (The tea garden as an environment for art, 76–88, 198–204. The garden as a manifestation of paradise, 171–8.) 1979.

Schuman, Jo Miles. *Art from many hands: Multicultural art projects for home and school.* 256 pp. illus. Englewood Cliffs, N.J.: Prentice-Hall. 1981.

Schurmacher, Emile. Blow guns. *Parade,* 7 Aug. 1949.

Seike, Kiyoshi, et al. *A Japanese touch for your garden.* 80 pp. illus. Tokyo & New York: Kodansha International. 1980.

Semple, A. T. Improving the world's grasslands. *FAO Agr. Stud.* 16. 147 pp. illus. 1952.

Sen, Soshitsu. *Chado: The Japanese way of tea.* 186 pp. illus. New York: Weatherhill, Tankosha. 1979.

Seth, S. K. Natural regeneration and treatment of bamboos. *FAO Forest Prod. Stud.* 13 *(Tropical Silviculture)* 2:298–303. 1957.

Sharma, Y. M. L. *Manual on bamboos of Asia Pacific region.* Rome: FAO. 1979.

Shaw, George Russell. *Knots useful and ornamental.* 194 pp. illus. New York: Bonanza Books. 1933.

Shelter Publications. *Shelter.* 176 pp. illus. 1973.

____. *Shelter II.* 224 pp. illus. Shelter Publications, Box 279, Bolinas, CA 94924. 1978.

Shepard, Mark. *Flutecraft: An artisan's guide to flute acoustics and bamboo flutemaking.* San Francisco: Simple Press. 1978.

____. *How to love your flute: A guide to modern and folk flutes and flute playing.* San Francisco: Panjundrum Press. (Copies of these booklets available from Monty Levenson, Box 294, Willits, CA 95490.) 1979.

Shepherd, W. D., and E. U. Dillard. Best grazing rates for beef production on cane range. *North Carolina Agr. Expt. Sta. Bull.* 384. 1953.

Shigemori, Kanto. *The Japanese courtyard garden: Landscapes for small spaces.* 224 pp. illus. New York: Weatherhill. 1981.

Simmonds, N. W. The virtues of the bamboos. *New Scientist* 352:334–6. 1963.

Sindall, R. W. *Bamboo for papermaking.* London: Marchant Singer.

Sineath, H. H., et al. *Industrial raw materials of plant origin. V.: A survey of the bamboos.* 230 pp. Atlanta, Ga.: Engineering Exp. Sta. of the Georgia Institute of Technology. Vol. XV.No. 18. 1953.

____. Bamboo—plant with a future. *Chemurgic Digest* XIII.2. 1954.

Singh, Man Mohan, et al. Some investigation on pressed boards from bamboo. *Indian Pulp and Paper,* June, 651–5. 1969.

Singh, Sagar. Notions about bamboos *(Dendrocalamus strictus)* vis-à-vis the paper industry. *Indian Pulp and Paper,* Feb., 497–500. 1969.

Sirén, Osvald. *Gardens of China.* 141 pp. 207 pl. New York: Ronald Press. 1949.

____. *Chinese painting: Leading masters and principles.* 7 vols. London. 1956–58.

Sloane, Eric. *A reverence for wood.* 111 pp. illus. New York: Ballantine Books. 1979.

Smith, Henry Nash. *Virgin land: The American west as symbol and myth.* 305 pp. (1957 ed.) New York: Vintage Books. 1950.

Smith, Russell. *Tree crops: A permanent agriculture.* 408 pp. (Reprinted 1978.) New York: Harper & Row. 1950.

Soderstrom, T. R., and Cleofe Calderon. *Report on the national madake bamboo survey undertaken by the Smithsonian International Environmental Alert Network.* 8 pp. Center for Short-Lived Phenomena. Cambridge, MA 02138. 1974.

____. A commentary on the bamboos. (Poaceae: Bambusoideae.) *Biotropica* 11.3:161–72. 1979a.

____. The bamboozling *Thamnocalamus. Garden Magazine,* Jul./Aug., 22–7. 1979b.

Spörry, Hans, and C. Schröter. *Die Verwendung des Bambus in Japan und Katalog der Spörry'schen Bambus-Sammlung.* 198 pp. 62 fig. 8 pls. Zürich: Zürcher und Furrer. 1903.

Squibb, R. L. *Progress report on investigation of bamboo leaves as source of vitamin A in chick rations.* Guatemala: Instituto Agropecuario Nacional. Monthly report, July. 1953.

Squibb, R. L., et al. Utilization of the carotenoids of bamboo leaves, teosinte and ixbut by New Hampshire chicks. *Poultry Science* 36 .6: 1241–4 (Nov.). 1957.

Stalberg, Robert Helmer, and Ruth Nesi. *China's crafts; The story of how they're made and what they mean.* 200 pp. illus. New York: Eurasia Press. 1980.

Standing, E. M. *Maria Montessori. Her life and work.* 382 pp. illus. New York: Mentor Books, New American Library. 1962.

Stebbing, E. P. *Note on the preservation of bamboos from the attacks of the bamboo beetle.* Calcutta, India: Govt. of India, Forest Zoology Series #2. Forest Pamphlet 15 (2d ed.). 1910.

Stein, Richard G. *Architecture and energy.* 322 pp. illus. Garden City, N.Y.: Doubleday. (See especially Chaps. 2. A history of comfort with low technology; 9. Learning from the schools; 16. The needs of the United States versus the needs of the world.) 1978.

Stevens, R. H. Bamboo or wood? *Southern Pulp and Paper Manufacturer,* 10 March, 93–4, 130. 1958a.

____. Bamboo: The facts and the problem. *Chemurgic Digest* 17:8–12 (Jan.). 1958b.

Sturkie, D. G., et al. *Bamboo growing in Alabama.* Alabama Agric. Exp. Sta. Bulletin No. 387. 30 pp. illus. 1968.

Sullivan, Michael. *Symbols of eternity; The art of landscape painting in China.* 205 pp. illus. Stanford, Calif.: Stanford Univ. Press. 1979.

Sun Tzu. *The art of war.* Trans. Samuel Griffith. 197 pp. illus. New York: Oxford Univ. Press. 1971.

Suzuki, D. T. *Zen and Japanese culture.* 477 pp. illus. Princeton, N.J.: Bollingen Series #64, Princeton Univ. Press. 1959.

____. *Sengai the Zen master.* New York: New York Graphic Society. 1971.

Suzuki, Sadao. *Index to Japanese Bambusaceae.* 384 pp. illus. Tokyo: Gakken Co. Ltd. 1978.

Sze, Mai-mai. The book of bamboo, in *The mustard seed garden manual of painting.* 624 pp. illus. Princeton, N.J.: Bollingen, Princeton Univ. Press. 1956.

____. *The Tao of painting.* New York: Viking. 1977.

T

Takama, Shinji. *The world of bamboo.* 236 pp. illus. S. San Francisco: Heian International. 1983.

Taloumis, George. Maybe there's a bamboo for you. *Flower and Garden,* Jul., 38–40, 46. 1974.

——. The inscrutable and handsome bamboo: Not just in Asia. *Boston Sunday Globe,* 9 Feb. 1975.

Tankrush, Sangchai. *The work in promoting the construction of bamboo-cement water tank as a solution to the problem of water shortage in the rural area of Thailand.* 22 pp. illus. Thailand: Adaptive Technology Group. 1981.

Thoria, Lavji. Selection of site for a rayon factory. *Indian Textile Journal,* LVII:536–52 (Mar.). 1947.

Tillman, Gus. The end of the line for 2,4,5-T. *Garden Magazine,* Jul./Aug., 11–4. 1979.

Tod, Osma Gallinger. *Earth basketry: A handbook of concise directions with clear designs for the beginner or experienced weaver.* 169 pp. illus. New York: Bonanza Books. 1972.

Tompkins, Peter, and Christopher Bird. *The secret life of plants.* 416 pp. New York: Avon Books. 1974.

TRANET. *Tranet.* A newsletter-directory of, by, and for individuals and groups worldwide actively developing appropriate/alternative technologies. No. 18. Transnational Network for Appropriate Technology, P.O. Box 567, Rangeley, ME 04970. 1981.

Trotter, H., and C. F. C. Benson. Liability of solid bamboo staves to attack by borers. *Indian Forester* 59:709–12. 1933.

Tseng-Tseng Yu, Leslie. *Chinese painting in four seasons: A manual of aesthetics and techniques.* 195 pp. illus. (See How to paint bamboo in the oriental manner, pp. 38–96.) Englewood Cliffs, N.J.: Prentice-Hall. 1981.

Tsuboi, Isuke. *Experimental method of establishing bamboo groves.* Gifu Prefecture, Japan: Forestry Association of Gifu Prefecture. (Trans. Saburo Katsura and Robert Young. Washington, D.C.: Division of Plant Exploration and Introduction, USDA. 1936.) 1913.

Tze-chiang, Chao. *A Chinese garden of serenity. Epigrams from the Ming dynasty "Discourses on vegetable roots."* 60 pp. illus. Mt. Vernon, N.Y.: Peter Pauper Press. 1959.

U

U Pe Kin. Durable bamboo lacquerware in Burma. *Indian Forester* 59: 635–38 (Oct.). 1933.

Ueda, Koichiro. *Studies on the physiology of bamboo; with reference to practical application.* 167 pp. illus. Kyoto, Japan: Kyoto Univ. Forests Bulletin 30. 1960.

——. *Bamboo resources for pulp and papermaking in Thailand.* 1966.

——. *Culture of bamboo as industrial raw material.* 47 pp. illus. Association for Overseas Technical Scholarship. No. 2-A. 1968.

——. *Bamboo.* (See Austin.) 1970.

Ukiah Daily Journal. Acres of bamboo will rise in Talmage. 12 April. (Ukiah, Calif.) 1981.

USDA: booklets and bulletins.
See Fairchild. 1903.
See Galloway. 1925.
Bamboo as a building material. See McClure. 1953.
Bamboos of the genus Phyllostachys under cultivation in the United States. See McClure. 1957.
Bamboo in the United States. See Young. 1961.
Growing ornamental bamboo. USDA Handbook 76. 1963.
Bamboo production research at Savannah, Georgia, 956–77. 1978.

USDA: leaflets.
American nurseries listing bamboos. 1979.
Artificial culture of running bamboos for hedge use.
Bamboo culms should be at least three years old before harvesting.
Bamboo—general information.
Bamboo quarantine. 1918.
Composition of foods used in Far Eastern countries.
Description of bamboos for testing at Plant Introduction Garden, Savannah, Georgia. 1947.
Directions for planting and care of hardy bamboos.
Edible bamboo sprouts.
Method of large-scale propagation from rhizome cuttings of the hardy running bamboos.
New method of joining bamboo in furniture and other construction.
Phyllostachys bambusoides.
Phyllostachys aurea.
Plant introductions: Annual descriptive list. 1942–43.
Preparation of bamboo for fishing poles and certain other purposes.

V

Van Briessen, Fritz. *The way of the brush: Painting techniques of China and Japan.* Rutland, Vt.: Charles E. Tuttle. 1975.

Van Dine, Alan. *Unconventional builders.* 184 pp. illus. Chicago: J. G. Ferguson. 1977.

Varmah, J. C., and K. N. Bahadur. *Country report and status of research on bamboos in India.* 28 pp. Dehra Dun, India: Forest Research Institute. Indian Forest Records (New Series) 6.1. 1980.

Vequaud, Yves. *Women painters of Mithila.* London: Thames & Hudson. 1977.

VITA (Volunteers in Technical Assistance). *Village technology handbook.* (Reprinted 1981.) 387 pp. illus. VITA, 3706 Rhode Island Ave., Mt. Rainer, MD 20822. 1963.

——. *Playground manual.* 29 pp. illus. VM6-69. (By Don L. Bartlett.) Mt. Rainer, Md: VITA. 1968.

W

Washington Post. Riders on research boondoggles: Bamboo is out. (Drew Pearson). 12 Feb. 1965.

Weber, Fred. *Reforestation in arid lands.* 258 pp. Mt. Rainer, Md.: VITA. (See VITA, above for address.) 1977.

West, E. M. Canebrakes of the southeastern U.S. *Abstracts of Doctors' Dissertations* 16: 253–65. Ohio State Univ. Press. 1935.

White, David G., et al. Bamboo for controlling soil erosion. *J. of the American Society of Agronomy* 37.10:839–47. 1945.

———. The relation between curing and durability of *Bambusa tuldoides. Caribbean Forester* 7.3:253–73 (Jul.). 1946a.

———. *El bambu para finca y hogar. Revista de Agricultura,* Jan.–Mar., 18–22. illus. 1946b.

———. Longevity of bamboo seed under different storage conditions. *Tropical Agriculture* 24:51–3. 1947.

———. *Bamboo culture and utilization in Puerto Rico.* Mayaguez, Puerto Rico: USDA Fed. Exp. Sta. Circular 29. 34 pp. illus. (April). 1948.

Wilckens, Leonie von. *Mansions in miniature: Four centuries of dolls' houses.* 252 pp. 324 pl. New York: Viking Press. 1980.

Wilhelm, Richard. *The i ching, or book of changes.* 3d ed. 740 pp. Princeton, N.J.: Bollingen, Princeton Univ. Press. (1st ed. 1950. 3rd ed., 10th printing, 1973.) 1973.

Williams, C. A. S. *Outlines of Chinese symbolism and art motives.* 472 pp. illus. (Reprint 1976.) New York: Dover. 1931.

Wilson, E. H. *A naturalist in western China.* 1913.

Worcester, G. R. G. *Sail and sweep in China.* London: Her Majesty's Stationery Office. 1966.

Wright, Frank Lloyd. *The living city.* 255 pp. illus. (1970 ed.) New York: New American Library. 1958.

Wright, H. L. Preparing bamboos for the market in India. *Scientific American,* 6 Aug. 1921.

Yagi, Koji. *A Japanese touch for your home.* 84 pp. illus. Tokyo, New York, San Francisco: Kodansha International. 1982.

Yoshida, Tetsuro. *Gardens of Japan.* New York: Praeger. 1957.

Young, R. A. Bamboos for American horticulture. *National Horticultural Magazine* (NHM). Five articles. 1945–46.

———. *Arundinaria* and *Sasa.* NHM 24:171–96. 1945a.

———. *Phyllostachys.* NHM 24:274–91. 1945b.

———. *Phyllostachys.* NHM 25:40–64. 1946a.

———. *Bambusa.* NHM 25:257–83. 1946b.

———. *Cephalostachyum, Dendrocalamus, Gigantochloa, Guadua,* and *Sinocalamus.* NHM 25:352–65. 1946c.

———. Bamboos for northern gardens. *Arnoldia: Bulletin of Popular Information* 6.7–9:29–43. (Arnold Arboretum. Harvard Univ.) 1946b.

———. Table of bamboos useful in American horticulture. Prepared for Montague Free, Staff Horticulturist for *The Home Garden* (Sept.). 1947.

———. *Bamboo in the United States: Description, culture, and utilization.* (With Joseph Haun and F. A. McClure.) 74 pp. illus. Washington, D.C.: USDA Agricultural Handbook 193. 1961.

Books by Subject.

1. INTRODUCTORY ARTICLES, MISCELLANEOUS TOPICS, AND SPECIALIZED AREA STUDIES. American Bamboo Society *Journal,* Anonymous, Austin, Bahadur, Cave, Dickason, Fairchild,* FAO 1979, Galloway, Higuchi, Hodge 1957, Holttum, Hooper, Houzeau de Lehaie, Huberman,* Hughes, Imle, Isaachsen, Junka, Kawamura, Kiang, Kirkpatrick, Kreider, Kurz,* Lee, A., Lessard,* Limaye, Lin, McClure 1931, 1935, 1957a, 1958c,* 1966,* 1973; McIlhenny, Marden, Miller, C. A., Myers, Nehru, Plank, Porterfield,* Raizada, Rehman, Roi, Satow, Sharma, Simmonds,* Sineath,* Stebbing, Stevens, Trotter, Tsuboi, USDA 1903, 1925, 1961,* 1963, 1978, Ueda,* Varmah,* White,* Wright, H. L.

2. BOTANY. Calderon, Camus, Gamble, McClure* 1957, 1966, 1973, Munro, Numata, Ruprecht, Soderstrom, Suzuki, S., Ueda 1960.*

3. CULTIVATION. Ahmed, Bhat, Deogun, Hodge, Jackson, A. B., Krishnaswamy, Oshima,* Seth, Sturkie.*

4. ARTIFACTS. Aero, Ball, J. D., Ecke, Eliot, Mannix, Mohammed, Needham,* Raval, Schurmacher, Spörry. *Baskets:* Kudo,* Laufer, Miller, H., Parker, L., Tod.*

5. ARTS. (Painting, carving, tea, house and ceremony.) Awaka, Ball, K. M., Bowie, Cahill, Cameron, Hammitzsche, Lee, S. E., Leong, Lutz, Oi, Oka, Okakura, Sen, Sirén, Sullivan, Suzuki, D. T., Sze,* Tseng-Tseng, U Pe Kin, Van Briessen, Vequaud, Williams.

6. MUSIC. Deaver,* Hunter, Levenson, Malm, May, Reck, Shepard.

7. CONSTRUCTION METHODS. Castro,* GATE, Glen, Hidalgo,* Limaye, Londoño,* McClure 1953,* Mathur, Narayanamurti,* Nehrbass, Orjala, Singh, M. M.

**Recommended; especially useful.*

8. PAPER. Bhargava, FAO 1952, 1953, Guha, Haun, Herty, Hoyer, McClure 1927, 1930, 1949, 1954d, Piatti, Raitt, Rao, Routledge, Sarjeant, Sindall, Sineath, Singh, S., Stevens, Ueda 1966.

9. SCHOOL USE. Beard, Cocannouer, Day, Grunfeld, Hunt, Jue, Laubin, MacFarlan, Matsuno, Newman, Parker, J., Sawyer, Schuman, Shaw.

10. GARDENS. Davidson, Engel, D., Freeman-Mitford,* Fuji Bamboo Society, Jones, J. L., Keswick, Lawson,* McClure 1933, 1957a,* Prest,* Riviere,* Schaarschmidt-Richter, Shigemori,* Sirén,* Takama,* Taloumis, Yoshida, Young 1961.*

Contexts

11. ALTERNATIVE TECHNOLOGY. Aprovecho,* Brand, Buck, Christian Relief, Club du Sahel, Crawford, Darrow,* Davis, Heronemus, Herrera, Hommel, Horn, Hutt, Kern, Leon, Lippert, McHarg, Moll, Needham, Odum, Piatti, Ramaswamy, Stein, Tankrush, Thoria, Tillman, TRANET, U Pe Kin, VITA, Wilson, Worcester.*

12. ARCHITECTURE. Alexander,* Brett, Buck, Charney, Denyer,* Duly, Engel, H.,* Fathy,* Kern,* Kuehn, Morse,* Neutra, Rudofsky,* Rykwert,* Shelter,* Wilckens, Wright, F. L.

13. AGRICULTURE. Brown, FAO 1970, 1976, 1978, Ferrar-Delgado, Fukuoka,* Howard,* King,* McClure 1958a, Mollison,* Samaka Service Center, Semple, Shepherd, Squibb, Tillman, Tompkins.

12. FORESTRY. Bene, Brand, Douglas,* Drew, Eckholm, Frazer, Giono,* Medway, Saint Barbe Baker,* Sanger, Sloane, Smith, R.,* Tillman, Weber.

13. CULTURAL DESIGN. Chuang Tzu,* Freire, Illich,* McLuhan, Marx, Meggers, Montessori,* Neihardt, Papanek,* Pound, Rothschild, Smith, H. N., Standing,* Sun Tzu,* Tompkins, Wilhelm,* Worcester.*

ILLUSTRATION SOURCES

Many illustrations have been drawn, retouched or redrawn by Stuart Chapman (SC), and a number provided by Julia Foster (JF). Two or more illustrations on a page are credited in their order within the text. Dates are given with authors only when two or more bibliographic entries by a single author require a date to clarify the source.

CHAPTER 1

*2:*SC. *4:*Terry Bell. Hidalgo 1974. *5:* JF. *7,8:* Cameron. *10:* Hidalgo 1974. *11:*Aero.

CHAPTER 2

*12:*Hidalgo 1974. *14:*Needham (SC). *15:*Needham. *17:* Hidalgo 1974. *19:*Ecke 1977. *20:*Ecke. SC. Darrow. *21:*Sze, Hidalgo 1974. *22:*Sze, Hidalgo 1978. *23:*Hidalgo 1974. Sze. JF. *24:*Worcester. SC. *25:*Sze. *26:*Hutt, JF. *29:*Austin (SC). *30:*Darrow. *32:*SC. *33:*Samaka Guide. *34:*Ibid., SC. *35:* Samaka Guide. *36:*Seike. *37:*Aero, SC. *38,39:*Aprovecho(SC). *40:*Worcester (SC). *41:*Sze, Samaka Guide. *42:*Worcester. *43:*Needham (JF, SC). *45:*Ecke (SC). *46:*Worcester. *47:* Worcester, Hidalgo 1974. *48:*Hidalgo 1974. Aero. *49:* Rudofsky 1977 (SC). *50:*SC. *51:*Reader's Digest 1978. *52:* Sze. *54, 55:*JF. *56:*Sze. *57:*Worcester (SC). *58:*Japanese print, Smithsonian Institution, Grass Library. *59:*Worcester (SC). *60:* Needham. *61:*Aero. *62:*Ecke (SC) *63:*Sze. Needham. *65, 66:* Sze, *68:*Needham(SC). Aero. *69:*Shelter 1973.

CHAPTER 3

*72:*Vequaud. *74:*Hidalgo 1974. *75:*Williams. *76:*Sze. Rider-Waite tarot pack. *77:*Suzuki, D.T. 1959. Sze. *78, 79:* San Francisco Asian Art Museum library. *80:*Cameron. Williams. *81:*Sze. *82:*Cameron. *83:*Sze, Vequaud. *84:* McClure 1953 (SC). *85:*McClure 1973(SC). *86:*SC. *87:*Sze (SC). *89:*San Francisco Asian Art Museum library. *90:*Sze, Suzuki 1959. *91:*Cameron. Morse. *92:*Suzuki 1959.

CHAPTER 4

*94:*McClure 1966. *96:*SC. *97:*McClure 1966. *98:*Van Dine. Hidalgo 1974. *99:*Londono. *100:*Orjala. *101:* Hidalgo 1974. *102:*McClure 1966. *105:*Hidalgo 1978. *106:*Hidalgo 1974. Nehrbass. *107, 108, 109:* Morse. *110:* SC. *111:*Morse.

CHAPTER 5

*112:*Van Dine (SC). *114:*Rudofsky 1964. A Van Dine. McClure 1953 (SC). *116–121:*Denyer. *123–125:*Hidalgo 1974. *126:*Londono. *127:*Londono. McClure 1953. *128:* CRAMSA (SC). *129:*Hidalgo 1978. *130:*Kevin Quigley. *131–133:*(SC).

CHAPTER 6

*136:*Japanese print in Smithsonian Institution, Grass library. *138, 139:*McClure 1966. *140:*McClure 1966. Ueda 1960. *141:*McClure 1966. Hidalgo 1978. *142:*McClure 1966. *144, 145:*McClure 1966. *146:*Kevin Quigley. *147, 148:*McClure 1966. *149:*McClure 1966 (JF) *150:*McClure 1966 (SC). *151:* Austin (SC). *152:*McClure 1966 (JF). *153:*McClure 1966.

CHAPTER 7

*156:*Riviere. *159:*McClure. *160:*West. *161:*McClure 1963b. *162:*Suzuki, D.T. 1959. *163:*Riviere. McClure 1966. *165:* Lawson 1968. McClure 1957a. *166:*Japanese print, 1829. Smithsonian Institution, Grass library. *167:*Fuji Bamboo Society. *168:*Lawson. Fuji Bamboo Society. *169, 170:*Fuji Bamboo Society. *171:*Lawson. *172:*Fuji Bamboo Society. *173:*Japanese print, Smithsonian Institution, Grass library. *174, 175:*McClure 1957a. *177:*Riviere. *178:*Fuji Bamboo Society. *179:*McClure 1966 (JF). *180:*McClure 1966. Fuji Bamboo Society. *181:*McClure 1966 (JF). *182:*Riviere. *183, 184:*McClure 1966. *186:*McClure 1938a. *187:*McClure 1966. Riviere. *189:*Lawson. McClure 1966. *190:*McClure 1966(SC). *191:*McClure 1966. *193:*Ibid. *194:*Ibid. *195:*Ibid. *197:*Lawson. *199:*McClure 1966. *201:*McClure 1945a.

CHAPTER 8

*206:*Riviere. *208:*Ibid. *209:*Ibid. *210:*McClure 1963a. *211:* Austin. *213:*Japanese print, Smithsonian Institution, Grass library. *214:*San Francisco Asian Art Museum library. *215:* Young 1961 (Kevin Quigley). *216:*McClure 1966. *217:* McClure 1966. *218:*Hidalgo 1974 (SC). *219:*SC. *220:* Londono. McClure 1957a. *221:*McClure 1931. *223:*Londono. *224:*Ibid. *225:*Londono (SC). *227:*Riviere. *228:*Ibid. *229:*Ibid. *230:*Fuji Bamboo Society. *231:*Fuji Bamboo Society.

CHAPTER 9

*234:*SC. *236:*SC. *238:*McClure 1953. Hidalgo 1974 (SC). *239:*Hidalgo 1978 (SC). Hidalgo 1974. *240:*Hidalgo 1974. *241:*Ibid. *242:*Ibid. *243:*Ibid. *244:*Grunfeld (SC). *245:* Grunfeld (SC). JF. *246:*SC. *247:*Denyer. *248:*Ibid. *249:* Rudofsky 1977 (SC). *251:*Rykwert. Shelter 1973. Rudofsky 1977. *252:*SC. JF. *253:*Aero. Stalberg. *254:*JF. *255:*SC. JF. *256:*SC. *257:*Deaver. *258:*Sze. Denyer. *259:*Morse. *260:* Morse. Rudofsky 1964 (SC). *261:*Morse. Shelter 1973. *262:* Denyer. Morse. Hidalgo 1974. *263:*SC. Hidalgo 1974. *264:* Morse. Samaka Guide. *266:*Sze. Ecke(SC). *269:*Tod. *270:* Tod. Rudofsky 1977. *271:*Rudofsky 1977. *272:*Hidalgo 1974 (Kevin Quigley). *273:*Hopkinson. *274:*Smithsonian Institution, Grass Library. *277:*Hopkinson. *278:*Cameron. Lawson. *279:* SC. JF. *280:*JF. *281:*Van Dine (SC). *282:*SC. *284:*Sze. *285:* SC. Suzuki 1971. *286:*Sze. *287:*Sze. *288:*Stalberg. *289:*Sze. Stalberg. *290:*Sze. *291:*Ibid. *292:*Ibid.

CHAPTER 10

*294:*Terry Bell. *296:*Hidalgo 1974. *297:*Ibid. *299:*SC. *301:* Fuji Bamboo Society. *302:*Hidalgo 1978 (SC). *306:*Journal, American Bamboo Society. *307:*Samaka Guide (SC). *311:* Aero. *313:*Suzuki 1957. Musashi. *314:*Naranjo, Rogelio: *Me vale madre.* Ediciones de Cultura Popular. Mexico City 1980. *316:*SC. *317:*Naranjo, *op. cit.*

DISCOGRAPHY

1. *A Bell Ringing in the Empty Sky* (Nonesuch H 72025), by Goro Yamaguchi, one of the foremost virtuoso players in Japan today.
2. *Japanese Masterpieces for the Shakuhachi* (Lyrichord LL-176), by master musicians of Nanzenji and Meianji in Kyoto.
3. *Music for Zen Meditation and Other Joys* (Verve V/V6-8634). Improvization for clarinet, shakuhachi, and koto, by Tony Scott, Shinichi Yuize and Hozan Yamamoto.
4. *Musique Traditionelle du Japan* (Disques Vogue CLVLX 326), by Kofu Kikusui, Nonko Nada, and Yayoi Nishimura. Kikusui plays a three-foot shakuhachi and his own style plastic shakuhachi.
5. *The Mysterious Sounds of the Japanese Bamboo Flute* (Everest 3289). Watazumido-Shuso plays oversize flutes in an unorthodox, highly personal style.
6. *Japanese Treasures with Shamisen and Shakuhachi* (Lyrichord LLST-7228). Performed by the master musicians of the Ikuta-Ryu.
7. *Koto Music of the One-String Ichigenkin* (Folkways FTS-31800), by Isshi Yamada and Fuzan Sata.
8. *Musique de Japon Imperial* (Editions de la Borte a Musique BAM LD 058-M), by Shinichi Yuize, Yasuko Nakashima, and Hozan Yamamoto.
9. *Japanese Poetry Chant* (Lyrichord LLST-7164), by Shufu Abe.
10. *Festival of Japanese Music in Hawaii* (Folkways FW 8885-86).
11. *Tozan-ryu Honkyoku* (Nippon Columbia CLS 72, 81, 88; CL 103–115).
12. *Shakuhachi no Ayumi to Sono Rekishi* (Teichiku Records SL-1046, 1047). "The Way of the Shakuhachi and its History," edited by Tanaka Inzan, contains music for the t'ung hsiao and hito-yogiri as well as samples from all major schools.
13. *Nihon Dento no Ongaku* (Nippon Victor JL 32–34). A broad historical survey of Japanese traditional music edited by Kishibe Shigeo.
14. *Hogaku Kessaku Shin-shu (Nippon Victor JV 96–97). Masterpieces of Traditional Music. An anthology of Tozan honkyoku.*
15. *Shakuhachi Gaku* (Crown SWS-3).
16. *Escale au Japon* (Pathe CCTZ 240, 763).
17. *Japanese Koto Music* (Lyrichord LLST-7131).
18. *Valiha—Madagascar* (Ocora, OCR 18). This recording of the tubular bamboo zither, which is the national instrument of Madagascar, includes an extensive introduction.

This list is lifted from Levenson (1974:31–32) with gratitude and small changes. It barely scratches the surface of bamboo recordings.

INDEX

Boldface numbers indicate illustrations. More complete reference to individual authors can be found by a scan of footnotes ending each chapter.

A

ABC bamboo catalogue, 15–69
A-frame construction, **105**
A Naturalist in Western China (Wilson), 22
A Pattern of Language (Alexander), 113, 248, 250
Accordians, 84
Acupuncture needles, 16
  and cultural acupuncture, 310
Adaptive Technology Group (ATG), 67
Adoption by bamboo, 74, 315
Agroforestry, 284–285. *See* Reforestation
Alabama yields, 232
Alcohol fuel, 28–29
Aerial roots, 140
African Traditional Architecture (Denyer), 116–121, 134–135, 248ff.
Airfoil kites, 49
Airplane skins, 16, **17**
Alexander, Christopher, 113, 248, 250
American Bamboo Society, **306**
Ancient Learning, 23
Animal cages, **264**–265
Animal pests, 225–226
Angklung, 16, 53
Aphrodisiac, 9, 17
Appleseed, Johnny, 284, **295**–296
Appropriate nostalgia, 110
  technology, 13–15
  toys, 29–30, 268
  travel, 76–77, 216, 265
Aprovecho, Lorena, 39, 44
Aquaculture, 40–43
Aqueducts, **264**. *See also* Pipe
Archimedes, 315
Architecture, 112–135
  African, 116, 248
  for beginners, 248–249
Architecture for the Poor (Fathy), 116–122
Architecture without Architects (Rudofsky), 115–116, 248
Aripana, **72**
Arnold, Matthew, 116
Arrow bamboo, 176
Arrows, 17
The Art of War (Sun Tzu), 310, 312–313
The Art of Zen Gardens (Davidson), 213
Arundinaria sp., 158
A. alpina, 159
A. amabilis, 30, **159**–160
  processing in China, 221
  ski poles, 66
A. anceps, 159
A. angustifolia, 159
A. auricoma, 159–160
A. callosa, 160
A. falcata, 160
A. fastuosa, **144,** 160
A. fortunei, 160
A. elegans, 160
A. giantea, 160–**161**
  yield, 231
A. graminea, 161–162
A. griffithiana, 162
A. hindsii, **162**
A. hookeriana, 20, 162
  "rice" from, 44
A. humilis, 162–163
A. intermedia, 163
A. khasiana, 163
A. prainii, 163
A. pusilla, **139(3)**
A. racemosa, 163
A. simonii, **144, 156, 163**–164
A. tecta, **139(8), 144**
  distribution, **160**
  Rhizome, **161**, 164
A. tesselata, 164
A. vagans, 164
A. variegata, **145, 148(5,6)**
A. viridi-striata, 164
A. wightiana, 164
Arundo donax, 198
Atomic survival (bamboo at Hiroshima), 27
Aulonemia queko, **85**
Austin, R., 17, 32, 79, 212
Autoarchitecture, 95, **247–251**, 265, 267, **270**

B

Bagasse, 277
Bamboo (Austin), 212
Bamboo ambush, 66
Bamboo and Reeds in Building Construction (Narayanamurti), 104, 106
Bamboo beetles. *See* Beetles
Bamboo-clay walls, 259
Bamboo cutter myth, 269
The Bamboo Garden (Freeman-Mitford), 16, 300
Bamboo smut, 225–226
Bamboo tamboo stick bands, 53
Bamboophilia, 9
Bamboos (Lawson), 150. *See also* Lawson
The Bamboos: A Fresh Perspective (McClure), 297, 300
Bamboo womb, **4**
Bambusa sp., **145**, 180–188
B. arundinacea, 5, 180–181 **182**
  as feed, 280
  Plybamboo, 97
  seasoning, 243
  seed longevity, 218–219
  tabasheer from, 64
B. balcooa, 181
B. beecheyana, **138** 181
B. blumeana, 27, 181
B. burmanica, 181
B. dissimulator, 181, 183
B. dolichomerithalla, 183
B. edulis, 183
B. glaucescens, **150(1), 183**
  for pot culture, 212
  seedling, **152**
  yield, 231
B. khasiana, 183–184
B. longispiculata, **146, 150(2)** 184
B. malingensis, 184
B. multiplex. See B. glaucescens.
B. Nutans, 184
B. oldhami, 184
B. pachinensis, **139(1), 147, 184**
B. pallia, 184
B. polymorpha, 37, 97, 184–185
B. pervariabilis, 184, **199**
B. sinospinosa, 61
B. spinosa, 16
B. stenostachya, 185
B. textilis, 103, 185, **199**
  site conditions, 215–216
B. tulda, 44, 185, **199**
  strength of, 143
B. tuldoides, 138, 139(2), 185–186, **199**
B. ventricosa, 186
  as feed, 280
  drought resistance, 102
  pot culture, 212
B. vulgaris, 16, 27–28, 186, **187** 188
  banana props, 55
  erosion control, 286
  as feed, 280
  growth rates, 103
  in Managua, 102
  pot culture, 212
  propagation, **216, 217**
  reforestation with, 40
B. wrayi, 21, 188
Banana props, 55
Bangladesh research, 301
Bare-root bamboo, 210–211
Baseball bats, 96–97
Baskets, **12**, 17, **18 19–20**, 246
  as elevators in playgrounds, **270**
  early baskets, 250
  uses of, 19–20
  in various countries, 19–20
  weaving patterns, **267**
Battens for sails, 59
Beehive house, **112**
Beehives, 20
Beer, **20**
Beethoven, Ludwig von, 83
Beetles, 125
  clump-cured culms and, 219
  and DDT, 222
  in India, 222–223
  powder post beetle, 221
  roundheaded beetle, 226
Bell, Alexander Graham, 49
Belluschi, Pietro, 115
Bending bamboo, 220, **221, 239**
Bene, J. G., 284
Berry, Wendell, 225
Bicycles, 20–21
Big node bamboo, 170
Birdhouse, **270**
Birthday plantings, 288
Bismarck's oak, 7
Black Bamboo, 5, **171**
Black Elk, 268
Blake, William, 310

Bleak cube, 268
Blimps, 29
Blocks, 268. *See also* Toys
Blooming cycle, 149
Blowguns, 21
Boards. *See Esterilla*
Boat cabins, 21
Boats, 21–22, **57**, 63
  booms, **21**
  cabins, **21**
  caulking material, 26
  design, **43**
  pictograms for, **42**
  sails, 59
  thatch boat, **59**
  see also Home on the wave, Junks, Rafts
Bodhidharma, **90**
Bonsai of bamboo, **211**–212
Book of Songs, 19
Book of Tea (Okakura), 110
The Book of the Five Rings (Musashi), 313
Boucherie process, 224
Bouffier, Elzeard, 132–133
Box kites, 49
Boy scout camp, 265–271
Branches, 144–145
  removal of, 219–**220**
  sheaths, 145
Branching of culms, 142
Breedlove, Dennis, 155
Bridges, **10, 22–23, 123, 266**
Brine buckets, 14
Brine conduits, 60
Broom bamboo, 170–171
Brooms, **22**, 23
  from moso, 227
Brown, Lester, 28
Brushes, **23**
  of Meng Tien, 79
  tempo of, **80–81**
Buckets, 24
Build a Yurt (Charney), 69
Bullock carts, 25

C
Cables, **22, 23** 24
  for salt well drilling, 60
Cage culture, 41
Calderon, Cleofe, 179, 285, 298
Camps, 265–266
Camus, Aimee, 300
Camus, Edmund-Gustave, 300
Canal embankments, 216
Candlesticks, **24**–25
Candlewicks, 36
Canebrakes, slave's hideout, 161
Canoes, 21
Canteens, **24, 124**
Carnival, 52–53
Carrying poles, **25**
Carts, 25–26
Carver, George Washington, 30, 315
Castillon, **169**
Categorical imperative, 14
Catenary suspension bridges, 22
"Cathedrals" of bamboo, **114**
Caulking, 26
  of junks, 46
Cayley, Sir George, 43
Cement, 66–67
Cephalostrachyum sp., 188
C. pergracile, 50, 188
C. virgatum, 188
Chado, The Japanese Way of Tea (Sen), 213
Chairs, 266
Chao Meng-fu, 78
Chapman, John, 284, 295–296
Charcoal, 26
Charney, Len, 69
Chasen, **246**
Chauchala, **101**
Chemical methods
  dangers of, 224
  for insect control and cracking, 243
Changtu Research Organization, 16
Chestnuts and bamboo, 154
Ch'i, 78
Chickens and bamboo, 32
Chime-mobiles, **252**
Chimonobambusa sp., 164–165
C. marmorea, 150(3) 164–165
C. quadrangularis, **140, 165**
  for pot culture, 212
The Chinese Garden (Keswick), 213
Chinese goddess bamboo, **183**
Chinese research, 301–302
Chisels, 26
Chuang Tzu, 83, 84, 89, 290
Chu-fa chuan, 57
Chusquea sp., 188–189;
  branches, **144**
  culm sheath, **147**
  as feed, 280
C. culeou, 153, 188–**189**
C. fendleri, **139(10), 189**
C. longipendula, 189
C. pilgeri, 189
C. ramosissima, 189
C. simpliciflora, 189
Cities, foliage in, 7–8, 103
Clambering bamboos, 142
Classification des Bambusees (Camus), 300
Classes, bamboo as a bridge between, 7, 74, 91
Clear cutting, 219
Clemson College concrete experiment, 99
Climber-bridge, **252**
Clump cured culms, 219
Coating culms, 224
Cock cage, 19
Collecting specimens, 199–200, **201**, 202
Color coding culms, 219
Communal architecture, 115
Community workshops, 235–**236**
Compost bin, **34**
Concrete, **98–100**
  for containing bamboo, 210
Confucius, 88, 272, 310
Containment of bamboo, 210
Cookbook, **279–280**
Cookstove design, 39–**40**
Cracking problem, 243
Creation myths, 74
Creosoted racks, 223
Cricket fights, 62
Crooked culms, 220–221
Crop storage, **30**–32
Crutches, **26**–27
Culm, 140–143
  anatomy of, 224–225
  curing, **220**–222, **223**
  dating of, 219
  height estimate, **141**
  of moso, 228
  of *Phyllostachys* sp., 175
  sheaths, 145, 147
  shooting culm diagram, 153
  stance, **142**
  storage, **225**
  strength of, 143
Culm-cup cure, **224**
Cultivation
  of moso, 226–231
  table for, 217
The Cultivation of Bamboos in Japan (Satow), 300
Cultural acupuncture, 310–314
Curare, 21
Curtain, **109**
Cutting new shoots, 209

D
Darrow, Ken, 41, 104
Darwish, Mahmud, 287
Dating culms, 219
Davidson, A. K., 213
Davis, Ian, 104, 107
DDT treatment, 222
Decline of the West (Spengler), 6
Defense. *See* War
Deforestration, 37–40, 207, 281–285
  See Reforestation
Dehra Dun, India, 214,
  concrete report, 99
  harvest rules, 219
Dendrocalamus sp., 189–192
D. asper, 20, 189
  seeds of, **190**
D. brandisii, 189–190
D. giganteus, 190
D. hamiltonii, 190, 243
D. hookerii, 190
D. latiflorus, **141**, 190
D. longispathys, 190
D. membranaceus, 190
D. merrillianus, 190
D. sikkimensis, 190
D. strictus, 40, 63–64, 190–**191**
  curing, 243
  in Managua, 102
  plybamboo, 97
  seed, **150(5)**
  structure of, 191
  wind and, 192
Denyer, Susan, **116–121**, 134–135, 248 **247–248**
Design for the Real World (Papanek), 6
Dewi Sri (Java rice goddess), 53
Diablo, **245**
Diagonal lashing, 241
Diesel fuel, 28–29
Dinochloa sp., 142
D. maclellandii, 192
Dinoderus minutus, 221, 222–223
Dirigibles, 29
Disabled persons, implements for, **26**–27
Disasters, 96
  erosion control for, 285
  in Managua, 100–103
  shelter designs, 104–107
Diseases, 226
Distribution of bamboo, 152–154
Dixie defined, 207
Dolls, 29–30
Domes. *See* Polyhedra
Door curtains, 109
Dowel pins, 30
Drainage, 229
Dramatic architecture, **251, 270, 271**
Drying crops, 30–32
Dumont, Alberto Santos, 29
Dwarf dwellings, 247–248

E
Earthquakes, 100–107
Earthskin, 281
Easel, **32**
Eckholm, Erik, 282
Ecoprotection, 281–290
Ecoscab, 285
Edison, Thomas, 5, 50–51, 64
Education Automation (Fuller), 8
Eggcups, 32
Egypt, mud housing, 117–118

Electricity
and pedal power, 20–21
wind energy, 68, 79
El Salvador, 316–317
Elephant grass, 195–196
Emptiness, 82, 86–89, 91
forgiveness, 312
shapelessness, 313
time in the void, 314
Engel, David, 213
English names for common species, 158
Environment for bamboos, 154
Erosion control, 285–286
Esenbeck, Christian Nees von, 300
"Essay Concerning Human Understanding" (Locke), 9
Esterilla, 128, 258–259
in kiosks, 47
Expanding groves, 211–214, 230–231
Experimental bamboo farm, 4, **307**
Extending length lashing, 241

F
Fairchild, David, 9, 154, 278
Fairyland bamboo, 167, 188
Famine relief, 149–150
Fans, 32–33
FAO report, 290
Farm buildings, 33–35
Farming, **33–35**
Fathy, Hassan, 115, 116, 117–118
Federal Experiment Station, Mayaguez, 187, 221–222, 226, 306
Fences, **35–36**, 261, **262**, 263
patterns, 36
Fenders, 36
Fernleaf Young cultivar, 183
Ferrar-Delgado, Roberto, 36
Fertilizer, 33, **34**, 36, 209–210, 215–216
moso, 230
Fiber grass, 142–143, 257–260
Fibers, 225
Field planting, 216–217
Fiestas, 36
Fighting kites, 49
Filaments for light bulbs, 5, 50–51
Firearms, 36–**37**
Firewood, **37–39**, 40
Firework towers, 36
Fisher, Sally, 269
Fishing, **40–43**
kites for, 49
myth, 42
net floats, 54, 227
Pseudosasa japonica and, 177
Flakeboards, 97
Flexibility, 106
Floating castle, 57
kitchen, **47**
village, 57, **271**
Floor construction, 258
Floret, 149
Flowering, 76–77, 148, **149**, 150, **151**
collection of branches, 200
Flutes, **83**–84
construction of, **86**–87, **255–257**
"flute bamboo," 162
Flying dragon toy, **43**
Flying white, 78
Foliage, 7–8, 103, 147–148
collection of specimens, 200
cross section, 148
of seedlings, 152
sheath, 145, 147
Fonblanque, Colonel Barrington de, 56
Food or Fuel: New Competition for the World's Croplands (Brown), 28
Food value of sprouts, 280
Forage, 280
Forgiveness, 312
Fossil bamboos, 155
Foundations, **127**, 258
Fountain bamboo, 196
Four treasures, 79–80
Framing, **101, 102, 123, 126, 127**
Franklin, Benjamin, 35, 49
Frazer, James George, 288
Freeman-Mitford, A. B., 16, 85, 157, 174, 196, 300
Friend (bamboo in Chinese folklore), 3, 74, 76
Friends of bamboo, 236, 298
Friendship, 298–299 310–312
Friendship ceremony, 91–92
Fröbel, Friedrich, 267
Fuke, 86
Fuller, Buckminster, 8, 46, 51, 99, 106, 315
Fungi infestation, 225
Furniture, **3**, 43–44, **45, 63**
construction of, **266**, 270
cracking of, 243
joinery for, 240
from plybamboo, 97

G
Gabions, **43**–44
Galvanized metal sheets, 210
Gamble, James, 300
Gambling, 44
Games, 44
Gandhi, Mahatma, 86
Gardens of China (Sirén), 213
Gate of Ganda, Uganda, 262
Gate rattles, **109**
Genera of bamboo, 202–203
Georgia yields, 232
Giant pandas, **61**
Giant thorny bamboo. *See Bambusa arundinacea*
Gide, André, 108
Gigantochloa sp., 192–193, **206, 209**
G. levis, 192
G. ligulata, 192
G. macrostachya, 192
G. maxima, 21
G. scortechinii, 192
G. verticillata, 192–193
Glenn, H. E., 99
The Golden Bough (Frazer), 288
Golden brilliant bamboo, 169
Grain, 44
mills, 14
Greenhouses, 44
Gregarious flowering, 148–149
Groves, 138–140
culm size and, 141
flowering and, 149;
form, **139**
Guadua sp., 115, 193–194
G. aculeata, **150(6)** 193
G. amplexifolia, 193
G. angustifolia, 47, 56, **121–131, 193**–194
"feet" of, **95**
fence, **262**
in St. Louis, Missouri, climatron, **130**
Manizales construction, **128**
G. capitata, 194
G. inermis, 194
G. latifolia, 194
G. paniculata, 194
G. superba, 194
G. tagoara, 194
G. tomentosa, 194
G. virgata, 194
G. werbertauneri, 194
Guerrilla economics, 244
Gunpowder, 37
Gutters, 260
Gynerium sagittatum, 198–199

H
Hah Shan, 162
Hacksaw, 238
Hammitzsch, Horst, 213
Hard energy paths, 15
Harvest methods, **219–221**
Hats, 44–45
Hawley, Walter, 155, 176
Hay, 280
Hayes, Denis, 15
Headrests, **45**
Heating crooked culms, 220–221
Hedge bamboo, 183
Hedges, **213**–214
Helmets, 45
Helping People in Poor Countries Develop Fuel-Saving Cookstoves (Aprovecho), 39
Hen house, 264–265
Henonis, 171–172
Herbicides, 283
Herrigel, Eugen, 17
Hexagonal plantings, 230
Hidalgo, Oscar, 23, 41, 48, 67, 99, 250, 308
Hillside moso culture, 228
Himalayan Journals (Hooker), 20
Hiroshima atomic explosion, 27
Hollow design, 87–90
Home on the road, 76, 251
Home on the wave, 46–47, 51–52, **271**
Hooker, J. D., 20, 44
Hoops, **245**
Hopi Indians
fiestas, 96
kachinas, 29
Horizontal joining, 240
Hornbostel, Eric von, 66, 74
Horseshoe harvest, **219**
Horseshoes, 252
Houckgeest, Van Braam, 14
Housing. *See* Shelter
How to Wrap 5 Eggs (Oka), 55
Hsiang-yen (Kyogen), 89–90
Huang chung (the Yellow Bell) 82, 86
Hui-neng, **92**
Huizinga, 115
Humidity requirements, 154
Hummer, **254**
Hutt, J. K., 26–27

I
I Ching, 52, 310, **311** 312
Iceless cooler, 45
Identification of specimens, 199–200
Ideograms, **23, 42**
Immortals, **75**
India
research in, 302–303;
yields, 232;
see also Dehra Dun, India
The Indian Forester, 276
Indonesian research, 303
Ink, 78
Inro, 54
Insects, 221–226
International seed exchange, 218–219
Internodes, 142
collection of, 200
Introduction to Tensegrity (Pugh), 107
Invasion of bamboo, 215
Irrigation, **33, 34**

J
Jacket, **45**
Jackstraws, **252**
Jalbert, Domina, 49

James, William, 314
Japan yields, 232
The Japanese Courtyard Garden, Landscapes for Small Spaces (Shigemori), 212
Japanese Gardens (Schaarschimidt-Richter), 213
Japanese Gardens for Today (Engel), 213
Joinery, **239**–240
Junks, 45–47
  description of, **46**
  sails, 59
Justinian, 63

K
Kafka, Franz, 8
Kant, Immanuel, 14
Kachinas, 29, 96
Keswick, Maggie, 212, 213
Kiln seasoning, 225
Kindereden, 315–316
Kindershelter, **251, 270**
Kiosk, **47–48**
Kitchen boats, **47**
Kitchen racks, **109**
Kites, **48–49,** 252–254
  whistle kite, 68
Knots, **241**
Kokushi, Daito, **77**
Kunth, Charles, 300
Kurz, Sulpiz, 23, 36, 50, 64, 300
Kutung units, 149
Kyogen, 89–**90**

L
Lacquerware, 49–50
Ladders, 50
Ladles, **50**
Lamination, **96**–98
Landslide control, 286
Lanterns, 74
Lao Tzu, 74, 82, 87, 88
Lashing, **21, 102, 241–243**
Lattice, 99–**100**
Laundry poles, 50
Lawson, A. H., 148, 150–151, 157, 196
Leaching, **223**–224
Leaf fall, 286
Leaf/people disequilibrium, 7–8, 103–104
Leaves. *See* Foliage
Leisure shelter, 110
Lembaga Biologi Nasional, 303
Les Bambusees (Camus), 300
Lessard, Giles, 215, 300–305, 308
Level conditions, 216
Levenson, Monty, 257
Light bulb filament, **5**, 50–51
Linnaeus, 181, 199
Locke, John, 9
Longevity of culms, 141–142
Louisiana
  polyhedra, 106–107
  yields, 231–232
Lovins, Amory, 15
Low-shrub communities, 283–284
Lutz, Walter, 243

M
Machete, 239
McClure, F. A., 9, 32, 36, 66, 99, 122, 150, 152, 155, 167, 169, 192, 193, 195, 280, 300, 306
  memorial to, 296–298
  taxonomy and, 199
McClure Ruth D., 210, 298–300
McLuhan, Marshall, obsolete technology as art, 109–110
McIlhenny, E. A., 9, 207, 231–232
Madake. *See Phyllostachys bambusoides*
Mahjong, 44
Makino bamboo, 52. *See Phyllostachy makioni*
Malaysian research, 304
Management of groves, 214–219
Managua. *See* Nicaragua
Mangle stilts, 6
Manizales construction, 6, 128
Manna from bamboo, 63–64
Marconi, 49
Marcotting, 218
Markers, 51
Market preparation, 220–221
Marx, Karl, 293(81) 310
Matcha, 91
Mattangs, **51**–52
Matting, 19
  farm produce drying, 31
Meat production, 282–283
Medicine, 52
  inro, 54
  tabasheer, 64–65
Melocalamus compactiflorus, 194
Melocanna baccifera, 44, 64, **97, 139(5), 150(7)**, 194–195
  seed of, 151–152
  seedling, **181**
Meristem, 152, **153**
  cultural meristem, 315
Message design, 5
Messenger, 4
  bamboo, 5
  design, 6
Metal consumption, 14–15
Meyer, Frank, 170, 172, 176
Minakami, Tsutomu (bamboo childhood), 244
Miniaturization, 51, 272
Mites, 226
Möbius strip, 108–109
Model houses, **247–251**
Moisture requirements, 154
Molybdenum blades, 238
Monasteries, 88
Mongolian Cloud Houses (Kuehn), 69
Monocrop mentality, 8
Monopodial bamboos, 137, 158–180
  rhizome of, 138
  propagation of, 208
Montessori, Maria, 245, 246, 315
"The Moral Equivalent of War" (James), 314
Morse, Edward S., 212
Moso. *See Phyllostachys pubescens*
Mountain-water paintings, 78
Mountain, symbolism in China, **52**
Mud housing, 117–118
Mud-stuffed bamboo walls, 259–260
Mulch, 230
Muli. *See Melocanna baccifera.*
Munro, W., 84, 164, 197, 300
Musashi, Miyamoto, 310, **313**–314
Mushroom culture, 52
Music of the Whole Earth, 54
Musical instruments, 52–54, 82–87, 254–257
  accordians, 84
  angklung, 16, 53
  bocina, **85**
  bulu perindu, 84
  bundle pipes, 85
  calung, 53
  castanets, 255
  celempung, 53–54
  ch'ih, 86
  chime mobiles, **252**
  didgeridoo, 85, 255
  fiddles, 66, **255**
  gambang, 53
  guitar, 85
  hatong, 53
  hsiao, 86
  hsuan, 86
  hu ch'in, **255**
  hummer, **254**
  jew's harp, 85
  khāen, **84**
  kohkohl, 53
  kolecer, 53
  organ, 84–85
  p'ai hsiao, 86
  pan pipes, 255
  pitch pipes, 82, 86
  quenas, 85
  raft pipes, 85–86
  rattles, 85, **245**
  rengkong, 53
  rhythm sticks, 254
  rondadores, 85
  sebi, 86
  shakuhachi, 52, **86**, 244, 256
  sheng, 84
  shepherd's pipes, 255
  sho, 84
  stick bands ("bamboo tamboo"), 52–53
  ti, 86
  tinakling poles, 44, 254–255
  trumpets, 85
  valiha, 65–66, 85
  violin, 66
  yueh, 86
  zithers, 85–86. *See also* Flutes
  homemade, 254
  in bamboo orchestra, 85–87
  of West Java, 53–54, 303
Musics of Many Cultures, 54
Mustard Seed Garden Manual of Painting, 78

N
Nailing bamboo, 239
Nails, 54
Narayanamurti, D., 97, 104
Nastus elongatus, 195
Needham, Joseph, 13, 41–42, 52
Needles, **54**
Nehrbass, Neil, 106–107
Neihardt, John G., 268
Nemagari-dake, 177–178
Net floats, 54
Netsuke, 54
Nicaragua
  bamboo development design, 6
  carts, 25
  earthquake, 102–104
  electricity in rural areas, 21
  groves in, 316–317
Night growth, 140–141
Nodes, 200
Nursery planting, 216

O
Oaks and bamboo, 154
Oblique axis windmill, 68
Ochlandra sp., 195–196
O. capitata, 195
O. travancorica, 150(8) **195**–196
Oil lamp stand, 266
Oka, Hideyuki, 55
Okakura, Kakuzo, 110
Orchestra, bamboo, 85–87
Orjala, Jim, 99–100
Oshima, Jinsaburo, 40, 226–231
Outrigger canoes, 21
Outward Bound, 266
Overthinning, 209
Oviedo, 123
Oxytenanthera sp., 196

O. abyssinica, 196
O. albociliata, 196
O. nigrociliata, 196
Oxygen and foliage, 147–148
Oyster cultivation, 41, 54

P
Packaging, 54–55
Paint brushes, 23
Painting, 77–78
  sumi-e, 80
Palembang lacquerware, 50
Pandas, **61**
Panpipes, 255
Papanek, Victor, 6
Paper, 3, 4, 40, 271, **272–274**, 275–276, **277**, 278
Papercuts, **61**, 74, **289**
Parafoil kites, 49
Parenchyma tissue, 225
Parquets, 96–97
Partridge, Eric, 320
Party supplies, 36
Path (teahouse), **91**, 110–111
Peach of eternal life, 75, 92
Pedal power, 20
Pens, **55**
Permaculture, 291–292
Permashelter, 115
Persimmon trees, 154
Philippine research, 304
Philosophic glossary, 320
Phoenix bamboo, 167
Photographs of bamboos, 202
Phyllostachys sp., 158, **165**–176
  culms, 175
  as feed, 280
P. angusta, 167
P. arcana, 167
P. aurea, 5, 140, **167–168**, 175(1)
  for pot culture, 212
P. areosulcata, 168
P. bambusoides, 86, **138, 139(9), 140, 145, 168–169, 170**
  for pot culture, 212
P. bissetii, 169
P. congesta, 169
P. decora, 170
P. dulcis, 170
P. edulis, 172–174
P. elegans, 170
P. flexuosa, 170, 175(2)
P. glauca, 170
P. makioni, 52, 170
P. meyeri, 170
P. nidularia, **149** 170–171, 175(3)
P. nigra, 5, **140 171**
  cultivar Bory, 171
  Henon, 171, **172**
P. nuda, 172
P. propinqua, 172
P. pubescens, 30, **172**–174
  culm sheath, **227**
  cultivation, **226–231**
  leaf blade, **148**
  leaf sheath, **148(7,8)**
  paper from, 275
  ritual objects from, **58**
  seed, **150(9)**
  shoot, 172, **227, 228**
  turtleshell mutant, **173, 230, 231**
P. purpurata, **174**–176
  culm sheath, **175(4)**
P. rubromarginata, **175(5)**, 176, 208
P. viridis, **175(7)**, 176
P. viridi-glaucescens, **175(6)**, 176, **208**
P. vivax, **175(8)**, 176
Picnic hut, **270**
Pig shed, **34**
Pilgrimages, 76–77
Pipe, **33**, 54, **60, 263–264**
Planetary norm, 4, 14–15, 282
Plank, Howard K., 185, 221
Plant therapy, 7–11, 103
Planting, 230
Plastics, 97
Plato, 6, 29
Play, 249–250
Playgrounds, **250–252**
Plum blossoms, 76
Plybamboo, 55, 97–98
Poison, 56
Polo balls, 55
Polo, Marco, 24, 26, 37, 49
Polyhedra
  after earthquake, 107
  greenhouses, 44
  in Louisiana, **106**–107
Polyhedra: A Visual Approach (Pugh), 107
Pond borders, 216
Porterfield, William, 64, 143, 246
Pot culture, 211–212
Pots, 55
Powder post beetle, 221
Prefabricated walls, 6, 33, **84, 97**
Pre-school building project, 248
Primary root and culm, 152
Prodigious Builders (Rudofsky), 57, **249**, 268, 250 **251, 270, 271**
Propagation, **208–210**, 217
Props, 55–56
Pseudosasa japonica, 17, 139–140, 147, 150(10) 176–**177**
  for pot culture, 212
  as windbreak, 213–214
Pseudostachyum polymorphum, 74, 196
Pugh, Anthony, 107
Punishment, 56
Punting poles, **56**, 185
Puppetry, 268
Pygmy bamboos, 153–154

Q
Quakers, 298
Quarantine of plants, 306–307
Quindio, 123–124
Quoits, 252

R
Rackets, 96–97
Racks, 56
Radio, 309–310
Rafts **40**, 56–**57**
  from *Guadua,* **125**
Raitt, William, 275, 292
Rakes, **34**
Rama, 108
Ramaswamy, N. S., 25
Ratadas, 44
Rats, 225–226
Rayon, 58
Rays of Hope: Transition to a Post-Petroleum World (Hayes), 15
Record needles, 58
Reforestration, 39, 281, 284, 290
  See Deforestation, Revegetation
Reinforced concrete. *See* Concrete
Reps, Paul, 91
The Republic (Plato), 6
Research centers, 305–306; research in Asia, 300–305
Revegetation, 7–8, 102–104 284–285. *See* also Reforestation
Revolution design, 314
Rhizome, 137–**138**
  of *Arundinaria gigantea,* **161**
  collection of, 200
  of moso, 227–228
  propagation and, 208–209
  of seedlings, 152
  sheaths, 145
  systems, **139**
  uprooting of, 229–230
Rhythm sticks, 254
"Rice" from bamboo, 44
Right angle unions, 240–241
Rikyu, 90, 110–111, 212, 236
Rilke, T., 249–250
Ritual objects, 58
River Min bridge, 22–23, 257
Riverbank cultivation, 216
Roof, **58, 101, 102** 260
  sections, 127, 129
  thatch, 34, 58, 260–261
Roots, 140
Rosewarne Experimental Horticultural Station, 214
Roshi, Suzuki, 8
Rossetti, C. G., 76
Routledge, T., 275
Rudofsky, Bernard, 57, 113, 115, 248, 251. *See also* Prodigious Builders.
Ruprecht, Franz, 300
Rusts, 225–226

S
Sachs, Curt, 74
Sahara desert, 38
Sail and Sweep in China (Worcester), 26, 46
Sail tying, 242
Sailbarrow, **14**
Sails, 59
Saint Barbe Baker, Richard, 207, 281, 283, 289
Sake, 59, 69
Salt well drilling, 59–**60**
Samaka Guides, 33, 41
Sand curing, 224
Sandals, 61
Sasa bamboo thatch, 58
Sasa sp., 158, 176–180
  for bonsai, 211
  branches of, 144
S. disticha, 177
S. kurilensis, **145**, 153 177–**178**
S. palmata, **144** 178
S. paniculata, 66
  as feed, 280
  wax, 66
S. pumila, 178
S. pygmaea, 178–179
S. tessellata, 179
S. variegata, 179
S. veitchii, **179**
Satow, Ernest, 69, 229, 300
Sawale, 6, 16–17, 33
Scaffolding, 61–62
  lashing for, **242**
Scales, 62
Scarecrows, 62
Schaarschimidt-Richter, Irmtraud, 212–213
Schiller, F., 249–250
Schools, 3–4, 8–9, 244–257, 265–270
  paper and, 277–278
Science and Civilization in China (Needham), 13, 41–42, 43
Scissors lashing, 241
Schizostachyum sp., 98, 196
S. gracile, **153**
S. hainanense, 196
S. lima, 196
S. lumampao, 196
Screen, **62**
Sea water leaching, 223
Secret Life of Plants (Tompkins), 30, 83
Sedan chairs, **63**

Seedbox, **34**
Seedlings, 151–**152, 153**
collection of, 200
of *Dendrocalamus* strictus, **191**
Seeds, 44, 150
of *Dendrocalamus asper,* **190**
storage of, 218–219
Selective cutting, 219
Selective introduction, 215
Sen, Soshitsu, 213
Sharma, Y. M. L., 243
Sheaths, 145, 147, 200
Shed curing, 222
Shelter, 94–111
architecture for, 112–135
early shelters, 250
flexible shelter, 126–128
Guadua angustifolia, 122–126
psychology of, 107–111
taxonomy of housing, 134–135
temporary, **21**, 65
Shelter After Disaster (Davis), 104, 107
Shepard, Mark, 257
Shibataea kumasaca, **139(7), 148(11, 12)**, 180
Shigemori, Kanto, 212
Shingles for roofs, 58
Ship design, 63
Shipping bamboo, **210**–211
culm bundles, 221
moso plants, 230
Shoji screen, 109
Shooting culm, **136**
diagram, **153**
Shoots, 278–280
Silence, 88
Silk, 63
Silverstripe Fernleaf Young cultivar, 183
Silverstripe Young cultivar, 183
Silviculture summary, 214–215
Simons, Polly McIlhenny, 232
Sinarundinaria nitida, 139(4) 196
Singapore meeting, May, 1980, 300
Sirén, Osvald, 213
Site conditions, **215**–216
Ski poles, 66
Skin of bamboo, 143
Skyscraper scaffolding, 61–62
Slash and burn cultivation, 282
Slope conditions, 216
Smith, Russell, 284
Smoke curing, 223
Smuts, 225–226
Soderstrom, Thomas, 151, 179, 285, 298
Soft Energy Paths (Lovins), 15
Soil conditions, 209, 215–216
erosion control, 285–286
for moso, 228
Somoza, Anastasio, 104
Specialization, 8–9
Spengler, Otto, 6
Sphagnum, 211
Splints, 63
Splitting culms, **238**, 239
Spörry, Hans, 13
Square bamboo, **218**
See also Chimonobambusa quadrangularis
Squibb, R. L., 32
Sri Lanka research, 304
Steel, bamboo replacing, 100
Stein, Gertrude, 310
Stilts, **244, 252**
Stockades, 123
Stone bamboo, 167
Storage,
of crops, 30–32
of culm bundles, 221
of seeds, 218
Stove design, 39–40
Straightening bamboo bends, 220–**221**
Stripestem Fernleaf cultivar, 183
Su Tung-p'o, 73, 79, **80**–81, 188
Sugar, 63–64
Sumi-e, 80
Sun curing, 223
Sun Tzu, 310, 312–313
Suspension bridges, 22
Swimming pools, 64
Swings, **242, 252**
Switch cane, 164
Sympodial bamboos, 137, 180–198
propagation of, 208
rhizome, 138
Sze, Mai-Mai, 77, 79

T

T joints, 240
Tabasheer, 64–65
Tactics maps, 310–314
Taiwanese research, 304–305
Taj Mahal, **101**
Takama, Shinji, 212
Tao Te Ching, 75, 88
Tarp tying, **21**
Taxonomy, 150, 199–202
of housing, **116–121** 134–135
Tea, **90–91**
ceremony, 91–93, 96
Teahouses, **91**, 96, **110–111, 236**
Teinostachyum dullooa, 197
Tesselation, **148**
Thailand research, 305
Thamnocalamus sp., 197–198
T. falconeri, 197
T. spathaceus, 151, **197**–198
T. spathiflorus, 198
Thatch, **260**
roof, 34, 58;
therapy, 110;
transport, **59**
"Theseus" (Gide), 108
Thoreau, Henry David, 107–108, 126
Thorns, 145
Thyrsosstachys oliverii, 198
T. siamensis, 198
Tile roofs, **58**
Tiles, parquet, 97
Tillering, 180, 320
Tillman, 283
Tipis, 270
Tissue culture, **302**
Tod, Osma Gallinger, 309
Tokonoma, 111, 227
Tompkins, Peter, 30
Tonkin cane, 158
See also Ai amabilis
Tools, 236–238
as toys, 29
Towel racks, **109**
Toys, **2, 29**–30, **43**
a toy history, 268
Tractors, 289
Transplanting bamboo, 229
Transporting bamboo, 210–211
Travel, **76, 77**, 216, 265
Trays, 56
Tree houses, **270**
Trees, 286–287
for paper, 276–277
preservation of, 283
worship of, **287–288**
Trestles, 65
Tripod lashing, **241**
Triumphal arches, 36
Trumpets, **85**
Tsuboi, Isuke, 58
Turtleshell bamboo, **58**, 230, 231
2,4,5-T, 283

U

Ueda, Koichiro, 212
Umbrellas, **65**
Universidad Nacional Autonoma de Nicaragua (UNAN), 103
U.S. Naval Academy, 8

V

Vahila, 65–66, 85
Valdez, Jose, 155
Varmah, J. C., 17
Violin, 66

W

Walden (Thoreau), 107–108, 126
Walkers, 26
Walls, **84, 258–259**
War, 27–28, 303
blowguns, 21
concrete for, 98
design fallout, 98–99
rangyoos, 66
Warmth requirements, 154
Warnings on bamboo, 215
Water
carriers, **24**
curing, 222
farming, **33, 40–43**
guns, 14
leaching, 223–224
level, **63**
living on, **46–47, 270–271**
and papermaking, 275
storage, 66–67
systems, **33**
Watersheds, 124
Waterwheels, 14, **15**, 67
Waxes, 67–68
Weather
kites, 49
and trees, 154, **287–288**
Weaving patterns, 267
Weeping bamboos, 84
Weightlessness of bamboo, 76
Wells, 14–**15**, **33**
Wheelbarrows, **14**
Whips, **66**
Whistle kite, 68
White, David G., 230
Whole-culm planting, 217–218
Wick action, 224
Willowy Young cultivar, 183
Wilson, E. H., 22, 197
Windbreaks, 213–214
Windmills, 14, **68**
Winds, 154, 192
Wine storage, 68–69
Worcester, G. R. G., 26, 46, 57
The World Directory of Bamboo Researchers, 307–308
The World of Bamboo (Takama), 212
Wright, Orville, 49
Wright, Wilbur, 49
Wu Wei (the art of minimal busyness), 82, 312

Y

Ya River raft, **57**
Yellow bell, 82, 86
Yields, 231–233
Young, R. A., 157, 186, 226
Yurts, **69**
Yushania niitakayamensis, 139(6)

Z

Zen in the Art of Archery (Herrigel), 17
Zen in the Art of the Tea Ceremony (Hammitzsch), 213
Zendo, as teahouse prototype, 111